Asymptotic Analysis and Perturbation Theory

Asymptotic Analysis and Perturbation Theory

William Paulsen

Arkansa State University, USA

CRC Press is an imprint of the
Taylor & Francis Group, an **informa** business
A CHAPMAN & HALL BOOK

CRC Press
Taylor & Francis Group
6000 Broken Sound Parkway NW, Suite 300
Boca Raton, FL 33487-2742

Printed on acid-free paper
Version Date: 20130524

International Standard Book Number-13: 978-1-4665-1511-6 (Hardback)

Visit the Taylor & Francis Web site at
http://www.taylorandfrancis.com

and the CRC Press Web site at
http://www.crcpress.com

Contents

List of Figures

List of Tables

Preface

The goal of this textbook is to present the topics of asymptotic analysis and perturbation theory to a level obtainable to students who have only completed the standard calculus sequence. Even though the most common application of asymptotics is in analyzing differential equations, students need not have prior knowledge of differential equations for this text. Rather, the book begins by immediately introducing the asymptotic notation, and applying this new tool to problems that the students will already be familiar with: limits, inverse functions, and integrals. In fact, only the simplest differential equations, such as first order linear or separable equations, will the students need to learn how to solve exactly. Hence, there is very little overlap between this text and a standard differential equation textbook.

The text follows the traditional organization, with plenty of exercises at the end of each section, and the answers to the odd numbered problems in the back. However, it also includes an abundance of computerized graphs and tables that will illustrate how well the asymptotic approximations approach the actual solutions. These graphs and charts enhance the student's learning of the material, giving them visual evidence that these approximation methods can be applied to the many types of problems that the student will encounter in his or her field.

This book will benefit instructors in that it will allow them to offer a course in Applied Mathematics that does not require a differential equations prerequisite. It will benefit students by bringing this difficult subject material to an easy to comprehend level. The book will benefit the mathematics department by making a course which is attractive to both majors and non-majors alike. The fields of engineering, physics, and even computer science utilize the study of asymptotic analysis and perturbation theory.

Although the emphasis of this book is problem-solving, there are some proofs scattered throughout the book. The purpose of these proofs is to give the students a justification for the methods that they will be using. Just as there are some proofs in a freshman level calculus book which are not as rigorous as the corresponding proofs in an advanced calculus text, these proofs are more informal, and often will refer the students to other sources for the details. These proofs enrich the students understanding of the material.

Another focus of this textbook is flexibility. Knowing that the readership will be extremely diverse, the aim was to include material that would be beneficial to both beginning students and researchers. Also, the book was designed to be completely self-contained, requiring only a calculus sequence background. There is a section giving the necessary background material for complex variables, since this knowledge tends to be lacking in the undergraduate curriculum. References to differential equations is deferred until chapter 4, where the small amount of background is covered, with minimal duplication of a standard differential equations course. Since the goal is to only *approximate* the solutions to such equations, it is not necessary for the students to know how to solve differential equations exactly, except for first order linear or separable equations. Hence, an undergraduate course can easily be designed using this text.

There is also more than enough material needed for a semester course. Professors may choose to skip chapter 3, (or even chapter 4, if differential equations is a prerequisite,) in order to reach the latter chapters. On the other hand, the first 6 chapters will make a good undergraduate course on asymptotics. There are a myriad of possibilities between these two extremes.

Finally, there are plenty of homework problems of various levels of difficulty. Most sections have between 20 to 30 problems, giving professors enough choices for assignments. Also, the answers to the odd numbered problems appear in the back of the book.

Acknowledgments

I want to begin by thanking my mentor, Carl Bender, for lighting the spark that got me excited about asymptotics and perturbation theory. He inspired me in the direction of my research, which eventually gave me the motivation for writing this book.

I also would like to express my appreciation to my wife Cynthia and my son Trevor for putting up with me during this past two years, since this project ended up taking much more of my time than I first realized. They have been very patient with me and are looking forward to me finally being done.

About the Author

William Paulsen is a Professor of Mathematics at Arkansas State University. He has taught asymptotics in a dual level undergraduate/graduate level course since 1994. He received his B.S. (summa cum laude), M.S., and Ph.D. degrees in Mathematics at Washington University in St. Louis. He was on the winning team for the 45th William Lowell Putnam Mathematical Competition.

Dr. Paulsen has authored over 15 papers in applied mathematics. Most of these papers make use of *Mathematica*®, including one which proves that Penrose tiles can be 3-colored, thus resolving a 30-year old open problem posed by John H. Conway. He has authored an abstract algebra textbook, "Abstract Algebra: an Interactive Approach," also published by CRC press.

Dr. Paulsen has also programmed several new games and puzzles in Javascript and C++. One of these puzzles, Duelling Dimensions, was syndicated through Knight Features. Other puzzles and games are available on the Internet.

Dr. Paulsen lives in Harrisburg, Arkansas with his wife Cynthia, his son Trevor, and two pugs and a dachshund.

Symbol Description

Symbol	Description	Page
$\sim$	behaves similar to	1
$\ll, \gg$	much less than, much greater than	3
$O(g(x))$	is of order $g(x)$	6
$o(g(x))$	is less than order $g(x)$	7
$S(x)$	Stieltjes integral function	16
$\sinh(x)$	hyperbolic sine function	18
$\sinh^{-1}(x)$	inverse hyperbolic sine	18
$\cosh(x)$	hyperbolic cosine function	18
$\operatorname{sech}(x)$	hyperbolic secant function	19
$\tanh(x)$	hyperbolic tangent function	20
$\coth x$	hyperbolic cotangent function	297
$\operatorname{csch}(x)$	hyperbolic cosecant function	313
$W(x)$	Lambert W function	26
$\mp$	minus or plus sign (vs. $\pm$)	32
$E_1(x)$	exponential integral function	41
γ	Euler-Mascheroni constant	43
$\operatorname{Si}(x)$	sine integral	43
$\operatorname{Ci}(x)$	cosine integral	51
$\operatorname{erf}(x)$	error function	48
$\operatorname{erfc}(x)$	complementary error function	482
$\Gamma(x)$	gamma function	59
$\prod_{n=1}^{\infty}$	infinite product	59
$\zeta(x)$	Riemann zeta function	77
$\tilde{z}$	point on a different sheet of Riemann surface	78
$\int_C f(z)\,dz$	complex contour integral	77
$J_v(x)$	Bessel function of the first kind	96
$\operatorname{Ai}(x)$	Airy function of the first kind	109
$S^m(A_n)$	iterated Shanks transformations	112
$S_m(A_n)$	generalized Shanks transformation	115
$R_m(A_n)$	generalized Richardson's extrapolation	120
$\delta(t)$	Dirac delta function	139

$\overset{\infty}{\underset{n=0}{\mathrm{K}}}$	infinite continued fraction	146
$\Big/\!\!\Big($	divide by the quantity	146
$P_M^N(x)$	N, M-Padé approximate	155
$W(y_1, y_2)$	Wronskian of two (or more) functions	168
$\sum_n$	sum over all integers n	191
$\mathrm{Bi}(x)$	Airy function of the second kind	210
$D_v(x)$	parabolic cylinder function	212
$Y_v(x)$	Bessel function of the second kind	224
δ_i	Kronecker delta function	228
Δa_n	difference operator on a sequence	254
$\Delta^k a_n$	iterated difference operator	254
$\Delta^{-1} a_n$	indefinite anti-difference operator	256
$\binom{n}{i}$	binomial coefficients	254
$\psi(x)$	digamma function	269
B_n	Bernoulli numbers	289
$B_n(x)$	Bernoulli polynomials	289
$\mathrm{Cthi}(x)$	hyperbolic cotangent integral function	298
$n^{[-k]}$	$\Gamma(n)/\Gamma(n+k)$	305
ϵ	small parameter	317
δ	small scaled parameter	321
$y_{\mathrm{out}}(x)$	outer solution	339
$y_{\mathrm{in}}(t)$	inner solution	339
$y_{\mathrm{comp}}(x)$	uniformly valid composite approximation	345
$y_{\mathrm{out},n}(x)$	outer solution expanded to order ϵ^n	356
$y_{\mathrm{in},n}(t)$	inner solution expanded to order ϵ^n	356
y_{match}	solution in the overlapping region	357
$y_{\mathrm{comp},n}$	n^{th} order composite approximation	357
y_{exact}	exact solution	357
$(f(x))_{\mathrm{out},n}$	convert to outer variable, expanded to order ϵ^n	369
$(f(x))_{\mathrm{in},n}$	convert to inner variable, expanded to order ϵ^n	369
$y_{\mathrm{comp},m,n}$	mixed order composite approximation	369
$y_{\mathrm{in,left}}$	left hand inner solution	382
$y_{\mathrm{in,right}}$	right hand inner solution	382
$y_{\mathrm{leftmatch}}$	left hand matching function	382
$y_{\mathrm{rightmatch}}$	right hand matching function	382
$y_{\mathrm{WKBJ},n}$	n^{th} order WKBJ approximation	394
$\max_x$	maximum over all possible x	409

y_{right}	right hand WKBJ approximation	419
y_{left}	left hand WKBJ approximation	419
$G_0(x)$	modified parabolic cylindrical equation	429
$\hbar$	Planck's constant	436
$y_{\text{strain},n}$	strained coordinate approximation of order n	448
$y_{\text{multi},n}$	multi-variable unstrained approximation of order n	464
$y_{\text{two-var},n}$	two variable expansion of order n	468
$\text{Ein}(x)$	entire exponential integral function	483
$\text{Cin}(x)$	entire cosine integral function	483
$\text{Shi}(x)$	hyperbolic sine integral	483
$\text{Chi}(x)$	hyperbolic cosine integral	483
$\text{Chin}(x)$	entire hyperbolic cosine integral	483
$\text{Thi}(x)$	hyperbolic tangent integral	483
$I_v(x)$	modified Bessel function of the first kind	488
$K_v(x)$	modified Bessel function of the second kind	488
$\text{He}_n(x)$	Hermite polynomial of order n	491

Chapter 1

Introduction to Asymptotics

Asymptotics has been called the "calculus of approximations." It provides a powerful tool for approximating the solutions to wide classes of problems, including limits, integrals, differential equations, and difference equations. Although the basic definitions are easy to understand, it requires skill to use asymptotics effectively and accurately. The goal of this chapter is to teach the necessary skills for a basic understanding of asymptotics. In later chapters we will apply the techniques to more difficult problems that cannot be solved any other way. In the process, we will learn the properties of some very important functions in applied mathematics.

1.1 Basic Definitions

Since the foundation of standard calculus is the concept of a limit, it is not surprising that the "calculus of approximations" will also hinge on limits. Usually we will consider a finite limit $\lim_{x\to a} f(x)$, but we can also have infinite limits, so a can be ∞ or $-\infty$.

1.1.1 Definition of $\sim$ and $\ll$

We begin with the two fundamental definitions of asymptotics.

DEFINITION 1.1 Given two functions, $f(x)$ and $g(x)$, we say that $f(x)$ is *similar* to $g(x)$ as x approaches a, written

$$f(x) \sim g(x) \quad \text{as} \quad x \to a,$$

if

$$\lim_{x\to a} \frac{f(x)}{g(x)} = 1.$$

For example, $\sin x \sim x$ as $x \to 0$, since $\lim_{x\to 0} \frac{\sin x}{x} = 1$. Also, $x + 1 \sim x$ as $x \to \infty$, since $\lim_{x\to\infty} \frac{x+1}{x} = \lim_{x\to\infty} 1 + \frac{1}{x} = 1$.

PROPOSITION 1.1

The relation $\sim$ as $x \to a$ is an *equivalence relation* on the set of all functions that are non-zero near a. That is, $\sim$ obeys the *reflexive property*:

$$f(x) \sim f(x) \text{ as } x \to a, \tag{1.1}$$

the *symmetric property*:

$$\text{if } f(x) \sim g(x) \text{ as } x \to a, \text{ then } g(x) \sim f(x) \text{ as } x \to a, \tag{1.2}$$

and the *transitive property*:

$$\text{if } f(x) \sim g(x) \text{ and } g(x) \sim h(x) \text{ as } x \to a, \text{ then } f(x) \sim h(x) \text{ as } x \to a. \tag{1.3}$$

PROOF: Since $f(x)$ is non-zero near a, we have

$$\lim_{x \to a} \frac{f(x)}{f(x)} = 1,$$

so $f(x) \sim f(x)$ as $x \to a$. To prove the symmetric property, note that

$$\lim_{x \to a} \frac{g(x)}{f(x)} = \frac{1}{\lim\limits_{x \to a} (f(x)/g(x))} = 1.$$

Finally, the transitive property follows from the fact that

$$\lim_{x \to a} \frac{f(x)}{h(x)} = \lim_{x \to a} \frac{f(x)}{g(x)} \cdot \frac{g(x)}{h(x)} = \lim_{x \to a} \frac{f(x)}{g(x)} \cdot \lim_{x \to a} \frac{g(x)}{h(x)} = 1.$$

□

Note that the reflective property only applies to functions that are non-zero near a. Unfortunately, we *cannot* say that $0 \sim 0$. We will stress this point by highlighting the following statement.

$$\boxed{\text{There is no function } f(x) \text{ such that } f(x) \sim 0 \text{ as } x \to a.} \tag{1.4}$$

Otherwise, we would have

$$\lim_{x \to a} \frac{0}{f(x)} = 1,$$

which is impossible. This marks a stark difference between asymptotics and standard limit notations. We can say that

$$\sin(x) \sim 1 \text{ as } x \to \pi/2,$$

but we *cannot* say that

$$\sin(x) \sim 0 \text{ as } x \to \pi.$$

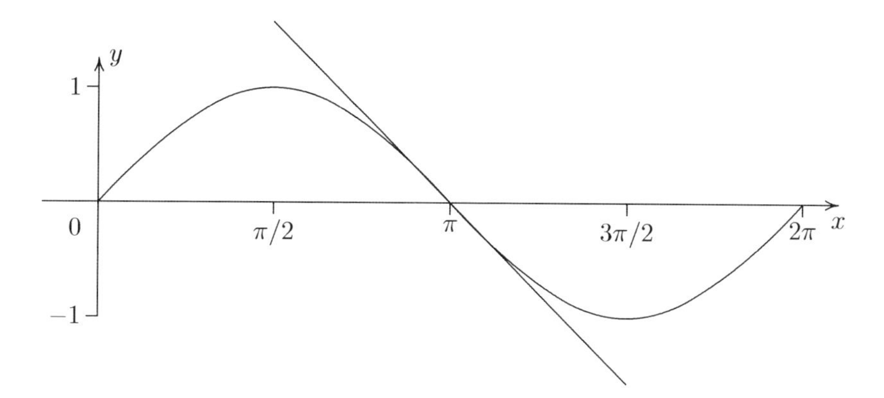

FIGURE 1.1: The two graphs reveal that $\sin(x) \sim \pi - x$ as $x \to \pi$. Note that the ratio of the two functions approaches 1 as x approaches π.

In fact, there is a linear function that is similar to $\sin(x)$ as $x \to \pi$, namely, the tangent to the curve at that point

$$\sin(x) \sim \pi - x \text{ as } x \to \pi.$$

Figure 1.1 gives a visualization of this asymptotic relationship.

The second main notation is also defined in terms of limits:

DEFINITION 1.2 We say that $f(x)$ is *much less than* $g(x)$ as x approaches a, written

$$f(x) \ll g(x) \qquad \text{as} \qquad x \to a,$$

if

$$\lim_{x \to a} \frac{f(x)}{g(x)} = 0.$$

We can think of this as "$f(x)$ is a drop in the bucket compared to $g(x)$, when x is close enough to a." Likewise, if a is ∞, we can say that "$f(x)$ is a drop in the bucket compared to $g(x)$, for sufficiently large x." We can similarly define $f(x) \gg g(x)$ as $x \to a$.

For example, $x^2 \ll x$ as $x \to 0$, since $\lim\limits_{x \to 0} x^2/x = 0$. However, $x^2 \gg x$ as $x \to \infty$, since $\lim\limits_{x \to \infty} x/x^2 = 0$.

The $\ll$ notation also has a special property:

PROPOSITION 1.2
If $f(x) \ll g(x)$ and $g(x) \ll h(x)$ as $x \to a$, then $f(x) \ll h(x)$ as $x \to a$. This property is called the *partial ordering property* of $\ll$ as $x \to a$.

PROOF: If $f(x) \ll g(x)$ and $g(x) \ll h(x)$ as $x \to a$, then

$$\lim_{x\to a} \frac{f(x)}{h(x)} = \lim_{x\to a} \frac{f(x)}{g(x)} \cdot \frac{g(x)}{h(x)} = \lim_{x\to\infty} \frac{f(x)}{g(x)} \cdot \lim_{x\to a} \frac{g(x)}{h(x)} = 0 \cdot 0 = 0.$$

□

The two fundamental notations of asymptotics are in fact related.

PROPOSITION 1.3

If $f(x) \sim g(x)$ as $x \to a$, then the relative error between the functions is going to zero, that is,

$$f(x) - g(x) \ll g(x) \qquad \text{as} \qquad x \to a.$$

Likewise, if $h(x) \ll f(x)$ as $x \to a$, then adding (or subtracting) $h(x)$ to $f(x)$ will produce a function similar to $f(x)$. That is,

$$f(x) \pm h(x) \sim f(x) \qquad \text{as} \qquad x \to a.$$

PROOF: If $f(x) \sim g(x)$ as $x \to a$, then

$$\lim_{x\to a} \frac{f(x) - g(x)}{g(x)} = \lim_{x\to a} \frac{f(x)}{g(x)} - \frac{g(x)}{g(x)} = 1 - 1 = 0.$$

So $f(x) - g(x) \ll g(x)$ as $x \to a$. On the other hand, if $h(x) \ll f(x)$ as $x \to a$, then

$$\lim_{x\to a} \frac{f(x) \pm h(x)}{f(x)} = \lim_{x\to a} \frac{f(x)}{f(x)} \pm \frac{h(x)}{f(x)} = 1 \pm 0 = 1.$$

So $f(x) \pm h(x) \sim f(x)$ as $x \to a$. □

Comparing algebraic functions such as polynomials is particularly easy. If $a > b$ then as $x \to \infty$, $x^a \gg x^b$. However, if we consider the limit as $x \to 0$, this reverses the direction: $x^a \ll x^b$. Thus $x^3 + 3x^2 - 2x \sim x^3$ as $x \to \infty$ since $3x^2 - 2x \ll x^3$. In fact, any polynomial is similar to its highest order term as $x \to \infty$. However, as $x \to 0$, $x^3 + 3x^2 - 2x \sim -2x$, the lowest order term.

1.1.2 Hierarchy of Functions

It will be important to understand how fast different functions grow, particularly as $x \to \infty$. Given two functions, we could ask whether one function grows more rapidly than another. This gives us a type of hierarchy to the different functions. We have already seen that given two polynomials of different degrees, the one with the larger degree will be much greater than the other as $x \to \infty$. However, functions which increase exponentially will grow faster than any polynomial.

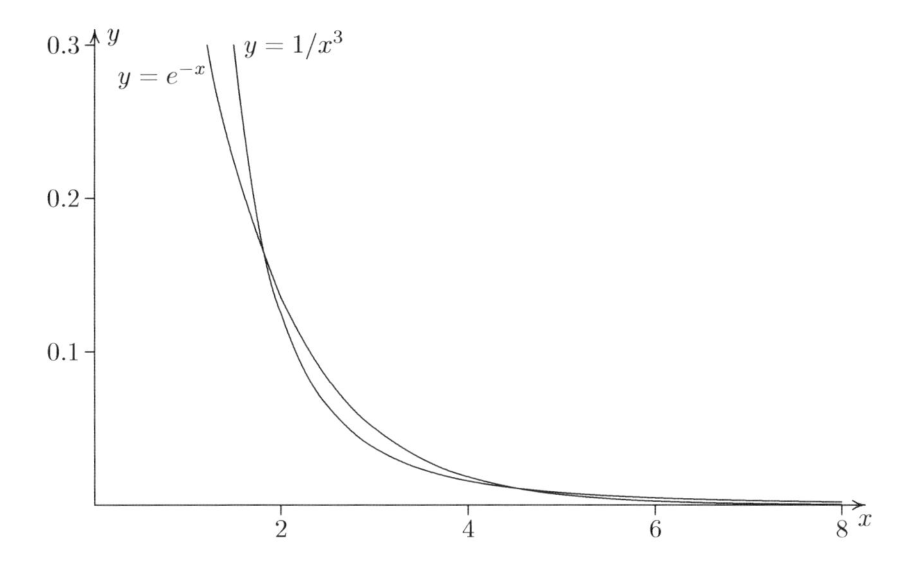

FIGURE 1.2: The graphs of $y = e^{-x}$ and $y = x^{-3}$. Although both converge to 0 as $x \to \infty$, the e^{-x} approaches zero faster as x increases, showing that $e^{-x} \ll x^{-3}$ as $x \to \infty$.

PROPOSITION 1.4
If $a > 0$ and b is a real constant, then $e^{ax} \gg x^b$ as $x \to \infty$. Also, $e^{-ax} \ll x^b$ as $x \to \infty$.

PROOF: Pick a positive integer n bigger than b. We can apply L'Hôpital's rule n times to the limit

$$\lim_{x\to\infty} \frac{x^n}{e^{ax}} = \lim_{x\to\infty} \frac{nx^{n-1}}{ae^{ax}} = \lim_{x\to\infty} \frac{n(n-1)x^{n-2}}{a^2e^{ax}} = \cdots = \lim_{x\to\infty} \frac{n!}{a^n e^{ax}} = 0.$$

Thus, $e^{ax} \gg x^n$ as $x \to \infty$, and since $x^n \gg x^b$, $e^{ax} \gg x^b$ as $x \to \infty$. Also note that

$$\lim_{x\to\infty} \frac{e^{-ax}}{x^b} = \lim_{x\to\infty} \frac{x^{-b}}{e^{ax}} = 0,$$

so $e^{-ax} \ll x^b$ as $x \to \infty$. □

Figure 1.2 is a graphical illustration that $e^{-x} \ll x^{-3}$ as $x \to \infty$. Note that this does *not* mean that $e^{-x} < x^{-3}$ for all x. In fact, the curves cross each other in two places. Only what happens for large values of x counts towards deciding which function is much smaller as $x \to \infty$.

Finally, we can compare two exponential functions by determining which has the larger exponent.

$$\boxed{\text{If } a > b, \text{ then } e^{ax} \gg e^{bx} \text{ as } x \to \infty.} \tag{1.5}$$

The proof is left as an exercise. See problem 31.

To compare two exponential functions with different bases, we can convert all of the bases to e. Thus, to compare e^{2x} and 8^x we observe that $8^x = (e^{\ln 8})^x = e^{(\ln 8)x}$. Since $\ln 8 \approx 2.079$, we have $e^{2x} \ll 8^x$ as $x \to \infty$.

We can determine how the logarithm function fits into the ranking by using L'Hôpital's Rule. It is easy to see that $\ln(x) \gg 1$ as $x \to \infty$, and we can observe that, for any $a > 0$, $\lim\limits_{x\to\infty} \frac{\ln(x)}{x^a} = 0$. (See problem 32.) Therefore,

$$\boxed{\text{If } a > 0, \text{ then } 1 \ll \ln x \ll x^a \text{ as } x \to \infty.} \tag{1.6}$$

Thus, as $x \to \infty$, $x^{1/1000} \gg \ln(x) \gg 1 = x^0$. The logarithm function squeezes in between the tight gap between x^0 and x^ϵ as $x \to \infty$, where ϵ is an extremely small positive number. This fact is helpful for computing limits involving logarithms. For example, $\lim_{x\to\infty} \frac{\ln x}{\sqrt{x}} = 0$, since $\ln x \ll \sqrt{x}$ as $x \to \infty$.

When $x \to 0$, $\ln(x)$ approaches $-\infty$, and we can use the property that $\ln(x) = -\ln(1/x)$ to derive the following result.

$$\boxed{\text{If } a > 0, \text{ then } 1 \ll \ln x \ll x^{-a} \text{ as } x \to 0.} \tag{1.7}$$

See problem 33. For example, $\lim_{x\to 0} \sqrt[3]{x}\ln(x) = 0$, since $\ln x \ll x^{-1/3}$ as $x \to 0$.

1.1.3 Big O and Little o Notation

Two more useful notations that are sometimes used are referred to as the "big O" and "little o" notation.

DEFINITION 1.3 We say that a function $f(x)$ is *of order* $g(x)$ as x approaches a, denoted by $f(x) = O(g(x))$ as $x \to a$, if the ratio $f(x)/g(x)$ is *bounded* for x near a. If a is finite, we can say this by saying that there are M and ϵ such that

$$|f(x)| \le M|g(x)| \text{ whenever } 0 < |x - a| < \epsilon. \tag{1.8}$$

Likewise, we say that $f(x) = O(g(x))$ as $x \to \infty$ if there are M and N such that

$$|f(x)| \le M|g(x)| \text{ whenever } x > N. \tag{1.9}$$

It is clear that if $f(x) \sim g(x)$ or $f(x) \ll g(x)$ as $x \to a$, then $f(x) = O(g(x))$ as $x \to a$. In fact, if $f(x) \sim kg(x)$ for some constant k, then $f(x) = O(g(x))$ as $x \to a$. However, the big O notation is useful when there is no clear asymptotic behavior of $f(x)$.

For example, $\sin x = O(1)$ as $x \to \infty$, since we can pick $M = 1$ and $N = 0$. Then of course, $|\sin x| \leq 1$ for all $x > 0$. Note in this example that there is *no* function $f(x)$ for which $f(x) \sim \sin(x)$ as $x \to \infty$, because $\sin x$ is not non-zero near ∞. The ratio $f(x)/\sin(x)$ would be undefined whenever x is a multiple of π, so technically, the limit as $x \to \infty$ does not exist. We will later see in subsection 2.4.4 how we can asymptotically analyze periodic and near-periodic functions.

The big O notation is often used with series to show the order of the first term left out. For example, the familiar Maclaurin series for $\cos(x)$ can be written

$$\cos x = 1 - \frac{x^2}{2} + \frac{x^4}{24} + O(x^6) \qquad \text{as } x \to 0.$$

The $O(x^6)$ in this equation replaces some function that is of order x^6 as $x \to 0$. In other words, the function

$$\cos x - 1 + \frac{x^2}{2} - \frac{x^4}{24}$$

must be of order x^6. In fact,

$$\cos x - 1 + \frac{x^2}{2} - \frac{x^4}{24} \sim \frac{x^6}{720} \qquad \text{as } x \to 0,$$

as indicated by the next term in the Maclaurin series.

The little o notation is similar, except that the function must be strictly smaller than the function inside the o.

DEFINITION 1.4 We say that a function $f(x)$ is *less than order* $g(x)$ as x approaches a, denoted by $f(x) = o(g(x))$ as $x \to a$, if the ratio $f(x)/g(x)$ approaches 0 as x approaches a.

To say that $f(x) = o(g(x))$ as $x \to a$ is equivalent to saying that $f(x) \ll g(x)$, but the little o notation can also be used for series to indicate the accuracy of the series. For example, one can write

$$\cos x = 1 - \frac{x^2}{2} + \frac{x^4}{24} + o(x^5) \qquad \text{as } x \to 0,$$

which emphasizes that there is no x^5 term.

Problems for §1.1

For problems **1** through **12**: State whether the following statements are true or false.

1 $x^2 - 2 \sim 2$ as $x \to 2$
2 $x^2 - 4 \sim 0$ as $x \to 2$
3 $x^2 \sim x$ as $x \to 0$
4 $x^2 + x \sim x$ as $x \to 0$
5 $x^2 + x \sim x^2$ as $x \to \infty$
6 $x^2 + x \sim 2x$ as $x \to 1$
7 $x^2 \ll x$ as $x \to 0$
8 $\frac{x}{1000} \ll x$ as $x \to \infty$
9 $x \ll -2$ as $x \to 0$
10 $\sqrt{x+1} \sim \sqrt{x}$ as $x \to \infty$
11 $e^{x+1} \sim e^x$ as $x \to \infty$
12 $\sin(x+1) \sim \sin(x)$ as $x \to \infty$

For problems **13** through **20**: Find the polynomial of lowest degree that is similar to the following functions as $x \to a$.

13 x^2 $a = 2$
14 $\sin x$ $a = \pi/2$
15 $\cos x$ $a = \pi/2$
16 $\sin x$ $a = 0$
17 $2x^3 - 3x^2 + 1$ $a = 1$
18 $\sin x - 1$ $a = \pi/2$
19 $1 + \cos x$ $a = \pi$
20 0 $a = 0$

For problems **21** through **26**: Find the polynomial $p(x)$ of lowest degree so that the equation is a true statement.

21 $e^x = p(x) + O(x^3)$ as $x \to 0$
22 $e^x = p(x) + o(x^3)$ as $x \to 0$
23 $\sin(x) = p(x) + o(x^5)$ as $x \to 0$
24 $\sin(x) = p(x) + O(x^5)$ as $x \to 0$
25 $\cos(x) = p(x) + o(x^5)$ as $x \to 0$
26 $\sqrt{x} = p(x) + O((x-1)^3)$ as $x \to 1$

27 Is there a function $f(x)$ for which no other function can be much greater than $f(x)$ as $x \to \infty$? Why or why not?

28 Is there a function $f(x)$ for which no other function can be much smaller than $f(x)$ as $x \to \infty$? Why or why not?

29 Is it possible for a function to be $O(0)$ as $x \to \infty$? Why or why not?

30 Is it possible for a function to be $o(0)$ as $x \to \infty$? Why or why not?

31 Prove equation 1.5. That is, show that if $a > b$, then $e^{ax} \gg e^{bx}$ as $x \to \infty$.

32 Prove equation 1.6. That is, use L'Hôpital's rule to show that $\ln x \ll x^a$ as $x \to \infty$, where a is a positive constant.

33 Prove equation 1.7. That is, use L'Hôpital's rule to show that $\ln x \ll x^{-a}$ as $x \to 0$, where a is a positive constant.

1.2 Limits via Asymptotics

One of the basic applications of asymptotics is as an alternative to L'Hôpital's rule for finding limits. The basic principal is to replace parts of a limit, such as the numerator or denominator, with another function that is similar as $x \to a$. This usually will not affect the limit, and we will cover the exceptions as we encounter them.

$$\boxed{\text{If } f(x) \sim g(x) \text{ and } h(x) \sim k(x) \text{ as } x \to a, \text{ then } \lim_{x\to a} \frac{f(x)}{h(x)} = \lim_{x\to a} \frac{g(x)}{k(x)}} \tag{1.10}$$

provided either of these limits exist. The reasoning is simple:

$$\lim_{x\to a} \frac{f(x)}{h(x)} = \lim_{x\to\infty} \frac{f(x)}{g(x)} \cdot \frac{g(x)}{k(x)} \cdot \frac{k(x)}{h(x)} = \lim_{x\to\infty} \frac{f(x)}{g(x)} \cdot \frac{k(x)}{h(x)} \cdot \lim_{x\to\infty} \frac{g(x)}{k(x)} = \lim_{x\to\infty} \frac{g(x)}{k(x)}.$$

Example 1.1
Find

$$\lim_{x\to\infty} \frac{x^3 + 3x^2 - 2x + 1}{3x^3 + 2x^2 - 5x}.$$

SOLUTION: Since $x^3 + 3x^2 - 2x + 1 \sim x^3$ and $3x^3 + 2x^2 - 5x \sim 3x^3$ we have

$$\lim_{x\to\infty} \frac{x^3 + 3x^2 - 2x + 1}{3x^3 + 2x^2 - 5x} = \lim_{x\to\infty} \frac{x^3}{3x^3} = 1/3.$$

Note that L'Hôpital's Rule would have to be applied 3 times to solve this problem. □

We can also replace a function inside a square root or other radical with a similar function as $x \to a$, without affecting the limit.

$$\boxed{\text{If } f(x) \sim g(x) \text{ as } x \to a, \text{ and } c \text{ is real, then } [f(x)]^c \sim [g(x)]^c \text{ as } x \to a.} \tag{1.11}$$

This is easy to verify with limits.

$$\lim_{x\to\infty} \frac{[f(x)]^c}{[g(x)]^c} = \lim_{x\to\infty} \left(\frac{f(x)}{g(x)}\right)^c = \left(\lim_{x\to\infty} \frac{f(x)}{g(x)}\right)^c = 1^c = 1.$$

Example 1.2
Find

$$\lim_{x\to\infty} \frac{\sqrt{4x^2 + 3x - 2}}{3x + 1}.$$

SOLUTION: The plan is to replace the complicated function $4x^2 + 3x - 2$ inside the square root with the simpler function $4x^2$. Since $4x^2 + 3x - 2 \sim 4x^2$ as $x \to \infty$, this substitution will not affect the limit. So

$$\lim_{x\to\infty} \frac{\sqrt{4x^2 + 3x - 2}}{3x + 1} = \lim_{x\to\infty} \frac{\sqrt{4x^2}}{3x} = \lim_{x\to\infty} \frac{2x}{3x} = 2/3.$$

□

We can also substitute one factor of the numerator or denominator with a similar function, without affecting the limit. See problem 25.

Example 1.3
Find

$$\lim_{x\to\infty} \frac{(2x+3)\sqrt{3e^x + \cos x}}{(e^{x/2} + x^{100})\sqrt{x^2+4}}.$$

SOLUTION: Note that $|\cos x| < 1$ for all x, so $\cos x \ll e^x$ as $x \to \infty$. Also, $x^{100} \ll e^{x/2}$ as $x \to \infty$, so we have

$$\lim_{x\to\infty} \frac{(2x+3)\sqrt{3e^x + \cos x}}{(e^{x/2} + x^{100})\sqrt{x^2+4}} = \lim_{x\to\infty} \frac{2x\sqrt{3e^x}}{e^{x/2}\sqrt{x^2}} = 2\sqrt{3}.$$

□

Can we substitute a similar function inside of a logarithm? That is,

$$\text{if } g(x) \ll f(x) \text{ as } x \to a, \text{ is } \ln(f(x) + g(x)) \sim \ln(f(x))?$$

The answer is usually yes, but not always. As long as $f(x)$ approaches 0 or ∞ as $x \to a$, then the substitution will produce a similar function. However, if $f(x)$ approaches 1, there is a complication. To see this, consider the limit

$$\lim_{x\to 0} \frac{\ln(1+x)}{x}.$$

It is certainly true that $x \ll 1$ as $x \to 0$, but we cannot say that $\ln(1+x) \sim \ln(1)$, because $\ln(1) = 0$, and we already established that *no* function is similar to 0.

So what do we do in this situation? By looking at the Maclaurin series for $\ln(1+x)$,

$$\ln(1+x) = x - \frac{x^2}{2} + \frac{x^3}{3} - \frac{x^4}{4} + \frac{x^5}{5} + O(x^6), \tag{1.12}$$

we see that $\ln(1+x) \sim x$ as $x \to 0$. From this result, we can establish the following:

$$\boxed{\text{If } g(x) \ll 1 \text{ as } x \to a, \text{ then } \ln(1+g(x)) \sim g(x) \text{ as } x \to a.} \tag{1.13}$$

Note that this is the exceptional case, not the rule, for dealing with logarithms. If $g(x) \ll f(x)$ as $x \to a$, and $f(x)$ is approaching 0 or ∞, or even some constant other than 1, then $\ln(f(x) + g(x)) \sim \ln(f(x))$ as $x \to a$. See problem 27.

Example 1.4
Find the limit

$$\lim_{x\to\infty} x \ln\left(\frac{x+3}{x+1}\right).$$

SOLUTION: Since $x + 3 \sim x$ and $x + 1 \sim x$ as $x \to \infty$, we see that the expression within the logarithm function is approaching 1. So we must first decompose the improper rational function, and rewrite the limit as

$$\lim_{x\to\infty} x \ln\left(1 + \frac{2}{x+1}\right).$$

Now, $2/(x+1)$ is approaching 0 as $x \to \infty$, so we can use equation 1.13:

$$x \ln\left(1 + \frac{2}{x+1}\right) \sim x\frac{2}{x+1} \sim \frac{2x}{x} = 2.$$

□

Example 1.5
Find the limit

$$\lim_{x\to\infty} \frac{\ln(3x^5 + 2x^2 + 10)}{\ln(5x^7 + 3x^4 + 10)}.$$

SOLUTION: Since the argument of the logarithms are going to ∞, we can substitute $3x^5 + 2x^2 + 10 \sim 3x^5$ and $5x^7 + 3x^4 + 10 \sim 5x^7$ as $x \to \infty$ without affecting the limit. Thus, as $x \to \infty$,

$$\frac{\ln(3x^5 + 2x^2 + 10)}{\ln(5x^7 + 3x^4 + 10)} \sim \frac{\ln(3x^5)}{\ln(5x^7)} = \frac{\ln 3 + 5\ln x}{\ln 5 + 7\ln x}.$$

Since $\ln x \gg 1$ as $x \to \infty$, this simplifies further:

$$\frac{\ln 3 + 5\ln x}{\ln 5 + 7\ln x} \sim \frac{5\ln x}{7\ln x} = \frac{5}{7}.$$

□

Is it possible to substitute a similar function in an exponent without affecting the limit? Note that even though $x + 1 \sim x$,

$$\lim_{x\to\infty} \frac{e^{x+1}}{e^x} = \lim_{x\to\infty} \frac{e^x \cdot e}{e^x} = e \neq 1.$$

Hence, the answer is a resounding *no*.

Just because $f(x) \sim g(x)$ as $x \to a$ does *not* mean that $e^{f(x)} \sim e^{g(x)}$.

(1.14)

Even though we cannot make simplifications withing the exponent, we can instead use the properties of exponentials to simplify a limit involving complicated powers.

Example 1.6
Find the limit

$$\lim_{x\to\infty} \frac{(4x)^{x-2}x^3}{x^{x+1}2^{2x+3}}.$$

SOLUTION: Expanding the exponentials, we obtain

$$\frac{(4x)^{x-2}x^3}{x^{x+1}2^{2x+3}} = \frac{(4x)^x(4x)^{-2}x^3}{x^x x^1 2^{2x}2^3} = \frac{4^x x^x 4^{-2}x^{-2}x^3}{x^x x(2^2)^x 8} = \frac{4^{-2}}{8} = \frac{1}{128}.$$

□

There is also an issue as to whether we can substitute the *base* of a complicated exponential function with a similar function without affecting the limit. For example, in the limit

$$\lim_{x\to 0}(1+x)^{1/x}, \tag{1.15}$$

can we replace the $1+x$ with 1, since these are similar as $x \to 0$? Calculus students should recognize this as the indeterminate 1^∞ form, so the answer is no.

$$\boxed{f(x) \sim g(x) \text{ does } \textit{not} \text{ imply that } f(x)^{h(x)} \sim g(x)^{h(x)} \text{ if } h(x) \to \infty.} \tag{1.16}$$

The situation where the exponent is going to ∞ is best handled by taking the logarithm of both sides of an equation, so that the properties of logarithms can be used. In the limit of equation 1.15, we set $y = \lim_{x\to 0}(1+x)^{1/x}$, so that $\ln y = \lim_{x\to 0} \ln(1+x)/x$. Using equation 1.13 quickly produces $\ln y = 1$, so $y = e$ is the original limit.

Example 1.7
Find the limit

$$\lim_{x\to\infty} \frac{(x+3)^{x+1}x^{x-1}}{(x+1)^{2x}}.$$

SOLUTION: We can first simplify this expression:

$$\frac{(x+3)^{x+1}x^{x-1}}{(x+1)^{2x}} = \frac{(x+3)^x(x+3)x^x x^{-1}}{((x+1)^2)^x} = \frac{x+3}{x}\left(\frac{x^2+3x}{x^2+2x+1}\right)^x.$$

The limit of the first factor is clearly one, so we will set y equal to the second factor. Then

$$\ln y \sim x \ln\left(\frac{x^2+3x}{x^2+2x+1}\right) = x \ln\left(1 + \frac{x-1}{x^2+2x+1}\right) \sim x\frac{x-1}{x^2+2x+1} \sim 1.$$

Since $\ln y = 1$, the original limit is $e^1 = e$. □

Problems for §1.2

For problems **1** through **12**: Find the following limits, using asymptotics. Note that most of these cannot be done with L'Hôpital's rule alone.

1 $\lim_{x\to\infty} \frac{3x^3-5x^2+4x-6}{2x^3+7x^2-4x+5}$

2 $\lim_{x\to\infty} \frac{4x^5+3x^2+5}{5x^5-8x^4+3x}$

3 $\lim_{x\to\infty} \frac{e^x+x^{100}}{e^x+\cos x}$

4 $\lim_{x\to\infty} \frac{2^{x-1}+3^{x-1}+x^3}{x^5-3^{x+1}}$

5 $\lim_{x\to\infty} \frac{\sqrt{3x^2-3x+4}}{2x-1}$

6 $\lim_{x\to\infty} \frac{3x^2-4}{\sqrt{x^4+3x-1}}$

7 $\lim_{x\to\infty} \frac{\sqrt{e^{2x}+x^4}}{e^x+2^x}$

8 $\lim_{x\to\infty} \frac{2^x-x^5}{\sqrt{4^x+e^x+x^2}}$

9 $\lim_{x\to\infty} \frac{(e^x+\cos x)\sqrt{4x^2-5x+3}}{(3x+2)\sqrt{e^{2x}+7^x}}$

10 $\lim_{x\to\infty} \frac{(x^2+\sin x)\sqrt{e^{4x}+54^x}}{(e^{2x}+x^{10})\sqrt{5x^4+3x-3}}$

11 $\lim_{x\to\infty} \frac{(3x+3)^{20}(x-2)^{10}}{(2x+5)^{30}}$

12 $\lim_{x\to\infty} \frac{(5x^2+4x-3)^6(4x^2+3x-2)^8}{(3x+5)^{18}(x-5)^{10}}$

For problems **13** through **24**: Find the following limits involving logarithms. Some of these require using equation 1.13.

13 $\lim_{x\to\infty} \frac{\ln(x^3-3x^2+4)}{\ln(x^2-4x+6)}$

14 $\lim_{x\to\infty} \frac{\ln(5x^2-4x+3)}{\ln(3x^5+2x^3+4x-1)}$

15 $\lim_{x\to\infty} \frac{\ln(e^{2x}+2^x+x^9)}{\ln(e^{3x}+(\ln x)^3+x)}$

16 $\lim_{x\to\infty} \frac{\ln(4x^2e^{3x}+x^3e^{2x})}{\ln(5x^5e^{2x}+3xe^{4x})}$

17 $\lim_{x\to\infty} \frac{\ln(8x^3e^{3x}+4x^7 20^x)}{\ln(7x^5e^{2x}+2x8^x)}$

18 $\lim_{x\to\infty} e^x \ln(1+e^{-x})$

19 $\lim_{x\to 0} \frac{\ln(1+x)}{\ln(1+2x)}$

20 $\lim_{x\to 2} \frac{\ln(x^2+2x-7)}{\ln(x^2-x-1)}$

21 $\lim_{x\to\infty} \frac{(x+1)^{2x}}{(x^2+3x)^x}$

22 $\lim_{x\to\infty} \frac{(x^2-5x)^x}{(x+3)^{2x}}$

23 $\lim_{x\to\infty} \frac{(x^2+3x+2)^x}{(x^2+5x-6)^x}$

24 $\lim_{x\to\infty} \frac{(x^3+4x^2+x\ln x)^{2x}}{(x^2+3x+5)^{3x}}$

25 Show that if $f(x) \sim g(x)$ as $x \to a$, and $h(x)$ is a non-zero function, then $f(x)h(x) \sim g(x)h(x)$ as $x \to a$.

26 If $f(x) \sim g(x)$ as $x \to a$, can we always say that $f(x)+h(x) \sim g(x)+h(x)$ as $x \to a$? Why or why not?

27 Show that if $f(x) \sim g(x)$ with $f(x)$ either approaches 0 or ∞ as $x \to a$, then $\ln(f(x)) \sim \ln(g(x))$ as $x \to a$. You can assume that both $f(x)$ and $g(x)$ have a derivative.

Hint: In these two cases, $\frac{\ln(f(x))}{\ln(g(x))}$ is of the form $\frac{\infty}{\infty}$, so we can use L'Hôpital's rule.

1.3 Asymptotic Series

Knowing the asymptotic behavior of a function gives us an idea of what is happening with the function as $x \to a$. However, we can get a much sharper picture of the behavior with an *asymptotic series*.

DEFINITION 1.5 We say that

$$f(x) \sim g_1(x) + g_2(x) + g_3(x) + \cdots \text{ as } x \to a, \tag{1.17}$$

or

$$f(x) \sim \sum_{n=1}^{\infty} g_n(x) \text{ as } x \to a, \tag{1.18}$$

provided each term of the series is asymptotic as $x \to a$ to the function created by subtracting the previous terms of the series from $f(x)$. That is,

$$\lim_{x\to a} \frac{f(x)}{g_1(x)} = 1, \qquad \lim_{x\to a} \frac{f(x) - g_1(x)}{g_2(x)} = 1, \qquad \lim_{x\to a} \frac{f(x) - g_1(x) - g_2(x)}{g_3(x)} = 1,$$

$$\lim_{x\to a} \frac{f(x) - g_1(x) - g_2(x) - g_3(x)}{g_4(x)} = 1, \text{ etc.} \tag{1.19}$$

So an asymptotic series gives us an infinite number of asymptotic relations, each giving a sharper picture to the behavior of the function $f(x)$ as $x \to a$.

Example 1.8
Find an asymptotic series for $\cos(x)$ as $x \to 0$.
SOLUTION: Since $\cos(0) = 1$, the first order approximation is $\cos(x) \sim 1$ as $x \to 0$. If we peel away this approximation, what is the behavior of $\cos(x) - 1$ near 0? From the Taylor series,

$$\cos(x) = 1 - \frac{x^2}{2} + \frac{x^4}{24} - \frac{x^6}{720} + \cdots,$$

we see that $\cos(x) - 1 \sim -x^2/2$, so this is the second term in the asymptotic series. In a sense, the asymptotic series keeps peeling away approximations from the function like an onion, except that onions don't have an infinite number of layers. It is clear that each time we subtract a term of the Taylor series, the next term of the series will describe the behavior. Thus,

$$\cos(x) \sim \sum_{n=0}^{\infty} \frac{(-1)^n x^{2n}}{(2n)!} \text{ as } x \to 0. \tag{1.20}$$

□

In this case, the Taylor series is the same as the asymptotic series. In fact, if a non-truncating Taylor series centered at a converges to a function $f(x)$, then the asymptotic series of $f(x)$ as $x \to a$ will be the same as the Taylor series. But there are some important differences between Taylor series and asymptotic series.

First of all, each term in a Taylor series must be a polynomial, in particular, one of the form $c_n(x-a)^n$. Asymptotic series, on the other hand, have no such

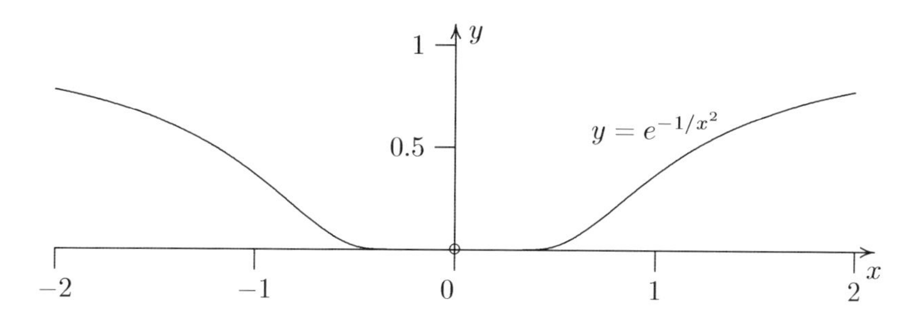

FIGURE 1.3: The graph of e^{-1/x^2}. This function approaches 0 as $x \to 0$ faster than any power of x, so this function will be *subdominant* to any Maclaurin series.

restriction. Often the terms involve fractional powers, exponential functions, or even logarithms. The only restriction is that $g_{n+1}(x) \ll g_n(x)$ as $x \to a$ for all terms in the series.

But more importantly, an asymptotic series is a *relative* property of a function, whereas a convergent Taylor series is an *absolute* property. In order to prove that a given series is the asymptotic series of $f(x)$, one must consider both $f(x)$ and the terms of the series. On the other hand, one can determine whether or not a Taylor series

$$\sum_{n=0}^{\infty} c_n(x-a)^n$$

converges or not without knowing the function that it converges to. Hence, the convergence is an absolute property intrinsic to the coefficients c_n.

Let us clarify this distinction. Suppose we are given a series of functions with $g_{n+1}(x) \ll g_n(x)$, and ask what function is asymptotic to that series. The answer is that there are infinitely many functions that have that series as its asymptotic series! For example, $\cos(x) + e^{-1/x^2}$ has the same asymptotic series as $\cos(x)$.

$$\cos(x) + e^{-1/x^2} \sim \sum_{n=0}^{\infty} \frac{(-1)^n x^{2n}}{(2n)!} \quad \text{as } x \to 0.$$

What is happening here? Note that because e^x grows faster than any polynomial as $x \to \infty$, $e^{-1/x^2} \ll x^n$ as $x \to 0$ for *all* n. Figure 1.3 shows the graph of this function. Hence e^{-1/x^2} will be smaller as $x \to 0$ than *all* of the terms of the cosine series, so no matter how many terms of the cosine series are subtracted from $\cos(x)$, the next largest factor will be the next term in the cosine series.

Any function which is smaller than all of the terms of an asymptotic series is said to be *subdominant* to the series. Because of subdominance, we cannot have a unique function associated with an asymptotic series.

Not every function has an asymptotic series as $x \to a$. Consider, for example, the hyperbolic function $\cosh(x)$ as $x \to \infty$. Since

$$\cosh(x) = \frac{e^x + e^{-x}}{2},$$

the first order approximation as $x \to \infty$ is $e^x/2$. If we subtract off this first term, we get $e^{-x}/2$, the second order term. But when this term is subtracted, we get 0, and by equation 1.4 this cannot be asymptotic to any function. Hence, $\cosh(x)$ does not have an asymptotic series as $x \to \infty$. So unlike Taylor series, an asymptotic series must contain an infinite number of non-zero terms.

Another important difference between Taylor series and asymptotic series is that Taylor series must *converge* if they are to be useful. For example, consider the power series

$$\sum_{n=0}^{\infty} (-1)^n n! x^n = 1 - x + 2x^2 - 6x^3 + 24x^4 - 120x^5 + \cdots.$$

We can use the ratio test to see if this converges, that is, if $\lim_{n\to\infty} |a_{n+1}/a_n| < 1$. But

$$\left| \frac{a_{n+1}}{a_n} \right| = \frac{(n+1)! x^{n+1}}{n! x^n} = (n+1)x.$$

Unless $x = 0$, this will go to ∞ as $n \to \infty$, so the series converges only for $x = 0$. As a Taylor series, this is not very useful.

However, there *is* a function for which this is the asymptotic series! In chapter 2, we will determine that the function

$$S(x) = \int_0^\infty \frac{e^{-t}}{1 + xt}\, dt, \tag{1.21}$$

which is called the *Stieltjes integral function* has the asymptotic series

$$\int_0^\infty \frac{e^{-t}}{1 + xt}\, dt \sim \sum_{n=0}^{\infty} (-1)^n n! x^n \text{ as } x \to 0^+.$$

See example 2.7. The reason for the one sided limit is that $S(x)$ is undefined for negative x. (The integrand is undefined at the point $t = -1/x$.) In spite of the fact that the series diverges, the asymptotic series precisely describes the behavior of $S(x)$ near $x = 0$, namely, the ratios

$$\frac{S(x)}{1}, \qquad \frac{S(x) - 1}{-x}, \qquad \frac{S(x) - 1 + x}{2x^2}, \qquad \frac{S(x) - 1 + x - 2x^2}{-6x^3}, \qquad \text{etc.,}$$

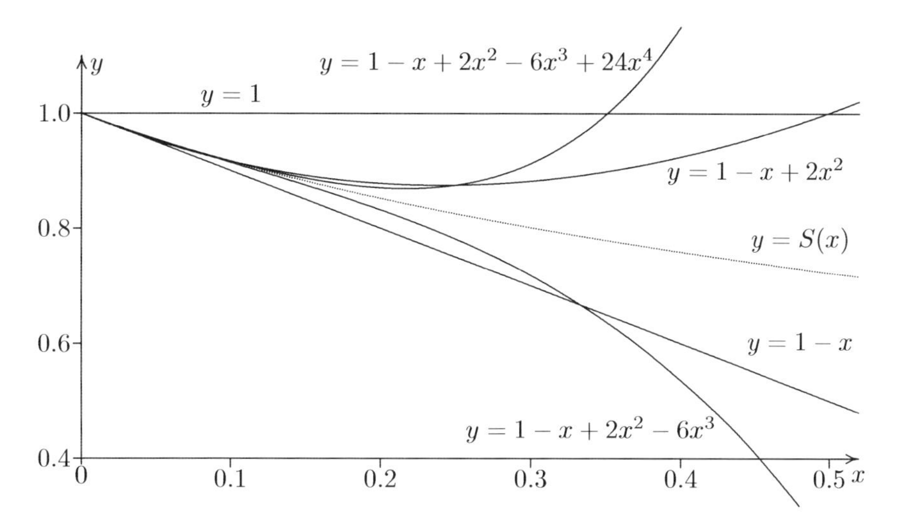

FIGURE 1.4: The graph shows the successive asymptotic approximations to the Stieltjes integral function $S(x)$. Note that the more terms of the series included, the better the approximation near $x = 0$, yet the approximation pulls away from $S(x)$ sooner.

all approach 1 as $x \to 0^+$. See figure 1.4 for a graphical illustration.

These limits illustrate the key difference between Taylor series and asymptotic series. For a Taylor series, if we pick a value x close to a, then the more terms we add gets us closer and closer to the function. For an asymptotic series, we first pick a number of terms, and we have an accurate approximation to the function, getting more accurate as x approaches a. Hence, asymptotic series can be used to compute complicated limits, regardless of whether the series converges or diverges.

For limits in which there is cancellation in the first order approximation, we can replace a function with not just a similar function, but with the first several terms of its asymptotic series.

Example 1.9

Find the limit

$$\lim_{x\to 0^+} \frac{S(x)}{x^2 - x^3} - \frac{1}{x^2}$$

where $S(x)$ is the Stieltjes integral function.

SOLUTION: If we consider just the first order approximation, we find that $S(x)/(x^2 - x^3) \sim 1/x^2$, causing cancellation to occur. The solution is to keep more terms of the asymptotic series. Although it is possible to find the asymptotic series for $S(x)/(x^2 - x^3)$ via long division, it is easier to rewrite the original limit in terms of a single fraction.

$$\frac{S(x)}{x^2 - x^3} - \frac{1}{x^2} = \frac{x^2 S(x) - x^2 + x^3}{x^4 - x^5}.$$

There is still cancellation in the numerator as $x \to 0$, but if we keep three terms of the asymptotic series for $S(x)$,

$$x^2 \left(1 - x + 2x^2 + O(x^3)\right) - x^2 + x^3 = 2x^4 + O(x^5).$$

Thus,

$$\frac{S(x)}{x^2 - x^3} - \frac{1}{x^2} = \frac{2x^4 + O(x^5)}{x^4 + O(x^5)} = 2 + O(x).$$

Thus, the limit as $x \to 0^+$ is 2. ▯

Unfortunately, it is impossible to know ahead of time how many terms of the asymptotic series must be used to compute the limit. The only advise is to try a reasonable number of terms, and if all of these cancel out, try again with more terms.

Example 1.10
Find the limit

$$\lim_{x \to 0} \frac{\sin(x) \sin^{-1}(x) - \sinh x \sinh^{-1}(x)}{x^2(\cos(x) - \cosh(x) + \sec(x) - \mathrm{sech}(x))}.$$

SOLUTION: We will need to use equation 1.20, along with the following:

$$\sin(x) \sim \sum_{n=0}^{\infty} \frac{(-1)^n x^{2n+1}}{(2n+1)!} \sim x - \frac{x^3}{6} + \frac{x^5}{120} - \frac{x^7}{5040} + \cdots \text{ as } x \to 0. \tag{1.22}$$

$$\sinh(x) \sim \sum_{n=0}^{\infty} \frac{x^{2n+1}}{(2n+1)!} \sim x + \frac{x^3}{6} + \frac{x^5}{120} + \frac{x^7}{5040} + \cdots \text{ as } x \to 0. \tag{1.23}$$

$$\sin^{-1}(x) \sim \sum_{n=0}^{\infty} \frac{(2n)! x^{2n+1}}{2^{2n}(n!)^2(2n+1)} \sim x + \frac{x^3}{6} + \frac{3x^5}{40} + \frac{5x^7}{112} + \cdots \text{ as } x \to 0. \tag{1.24}$$

$$\sinh^{-1}(x) \sim \sum_{n=0}^{\infty} \frac{(-1)^n (2n)! x^{2n+1}}{2^{2n}(n!)^2(2n+1)} \sim x - \frac{x^3}{6} + \frac{3x^5}{40} - \frac{5x^7}{112} + \cdots \text{ as } x \to 0. \tag{1.25}$$

$$\cosh(x) \sim \sum_{n=0}^{\infty} \frac{x^{2n}}{(2n)!} \sim 1 + \frac{x^2}{2} + \frac{x^4}{24} + \frac{x^6}{720} + \cdots \text{ as } x \to 0. \tag{1.26}$$

$$\sec(x) \sim 1 + \frac{x^2}{2} + \frac{5x^4}{24} + \frac{61x^6}{720} + \frac{1385x^8}{8!} + \frac{50521x^{10}}{10!} + \cdots \text{ as } x \to 0. \tag{1.27}$$

$$\operatorname{sech}(x) \sim 1 - \frac{x^2}{2} + \frac{5x^4}{24} - \frac{61x^6}{720} + \frac{1385x^8}{8!} - \frac{50521x^{10}}{10!} + \cdots \text{ as } x \to 0. \quad (1.28)$$

Since the denominator does not involve any products, let us begin there, keeping three terms in each of the series.

$$x^2\left[\left(1 - \frac{x^2}{2} + \frac{x^4}{24} + O(x^6)\right) - \left(1 + \frac{x^2}{2} + \frac{x^4}{24} + O(x^6)\right)\right.$$
$$\left. + \left(1 + \frac{x^2}{2} + \frac{5x^4}{24} + O(x^6)\right) - \left(1 - \frac{x^2}{2} + \frac{5x^4}{24} + O(x^6)\right)\right]$$
$$= O(x^8).$$

We discover that all terms canceled, so we will need to keep more terms.

$$x^2\left[\left(1 - \frac{x^2}{2} + \frac{x^4}{24} - \frac{x^6}{720} + O(x^8)\right) - \left(1 + \frac{x^2}{2} + \frac{x^4}{24} + \frac{x^6}{720} + O(x^8)\right)\right.$$
$$\left. + \left(1 + \frac{x^2}{2} + \frac{5x^4}{24} + \frac{61x^6}{720} + O(x^8)\right) - \left(1 - \frac{x^2}{2} + \frac{5x^4}{24} - \frac{61x^6}{720} + O(x^8)\right)\right]$$
$$= \frac{x^8}{6} + O(x^{10}).$$

Now we can proceed with the numerator, using the distributive law to perform the product of two series. By doing the denominator first, we learn that we must keep terms of order x^8. The numerator of the limit can now be computed to be

$$\left(x - \frac{x^3}{6} + \frac{x^5}{120} - \frac{x^7}{5040} + O(x^9)\right)\left(x + \frac{x^3}{6} + \frac{3x^5}{40} + \frac{5x^7}{112} + O(x^9)\right)$$
$$- \left(x + \frac{x^3}{6} + \frac{x^5}{120} + \frac{x^7}{5040} + O(x^9)\right)\left(x - \frac{x^3}{6} + \frac{3x^5}{40} - \frac{5x^7}{112} + +O(x^9)\right)$$
$$= \left(x^2 + \frac{x^6}{18} + \frac{x^8}{30} + O(x^{10})\right) - \left(x^2 + \frac{x^6}{18} - \frac{x^8}{30} + O(x^{10})\right)$$
$$= \frac{x^8}{15} + O(x^{10}).$$

Putting the pieces together, we get

$$\frac{\sin(x)\sin^{-1}(x) - \sinh x \sinh^{-1}(x)}{x^2(\cos(x) - \cosh(x) + \sec(x) - \operatorname{sech}(x))} \sim \frac{x^8/15 + O(x^{10})}{x^8/6 + O(x^{10})} = \frac{2}{5} + O(x^2). \quad \square$$

This example reviewed many of the classical Maclaurin series, but there are a few more that may be used in future exercises, so let us cover these here. All of these series are for $x \to 0$.

Exponential series, converges for all x:

$$e^x \sim \sum_{n=0}^{\infty} \frac{x^n}{n!} = 1 + x + \frac{x^2}{2!} + \frac{x^3}{3!} + \frac{x^4}{4!} + \cdots \quad (1.29)$$

Logarithmic series, converges for $|x| < 1$:

$$\ln(1+x) \sim \sum_{n=1}^{\infty} \frac{(-1)^{n+1}x^n}{n} = x - \frac{x^2}{2} + \frac{x^3}{3} - \frac{x^4}{4} + \cdots \tag{1.30}$$

Binomial series, converges for $|x| < 1$:

$$(1+x)^k \sim 1 + kx + \frac{k(k-1)}{2!}x^2 + \frac{k(k-1)(k-2)}{3!}x^3 + \cdots \\ + \frac{k(k-1)\cdots(k-n+1)}{n!}x^n + \cdots \tag{1.31}$$

$$\frac{1}{\sqrt{1-4x}} \sim \sum_{n=0}^{\infty} \frac{(2n)!x^n}{(n!)^2} = 1 + 2x + 6x^2 + 20x^3 + 70x^4 + \cdots \tag{1.32}$$

Inverse tangent series:

$$\tan^{-1} x \sim \sum_{n=0}^{\infty} \frac{(-1)^n x^{2n+1}}{2n+1} = x - \frac{x^3}{3} + \frac{x^5}{5} - \frac{x^7}{7} + \cdots \tag{1.33}$$

Inverse hyperbolic tangent series:

$$\tanh^{-1} x \sim \sum_{n=0}^{\infty} \frac{x^{2n+1}}{2n+1} = x + \frac{x^3}{3} + \frac{x^5}{5} + \frac{x^7}{7} + \cdots \tag{1.34}$$

Tangent series:

$$\tan(x) \sim x + \frac{x^3}{3} + \frac{2x^5}{15} + \frac{17x^7}{315} + \frac{62x^9}{2835} + \frac{1382x^{11}}{155925} + \cdots \tag{1.35}$$

Hyperbolic tangent series:

$$\tanh(x) \sim x - \frac{x^3}{3} + \frac{2x^5}{15} - \frac{17x^7}{315} + \frac{62x^9}{2835} - \frac{1382x^{11}}{155925} + \cdots \tag{1.36}$$

Some limits require plugging one asymptotic series into another asymptotic series.

Example 1.11
Find the limit

$$\lim_{x\to\infty} \sqrt{x^4 + 4x^3 + 7x^2} - x^2 - 2x.$$

SOLUTION: Again, a first order approximation yields $\sqrt{x^4} - x^2 = 0$, which cannot be the asymptotic approximation of this function. So we must use an asymptotic series.

The asymptotic series for the square root function is found by plugging $k = 1/2$ into the binomial series.

$$\sqrt{1+\epsilon} \sim 1 + \frac{1}{2}\epsilon - \frac{1}{8}\epsilon^2 + \frac{1}{16}\epsilon^3 - \frac{5}{128}\epsilon^4 + \cdots \quad \text{as } \epsilon \to 0. \tag{1.37}$$

To utilize this series, we rewrite $\sqrt{x^4 + 4x^3 + 7x^2}$ as $x^2\sqrt{1 + 4/x + 7/x^2}$, and as $x \to \infty$, $4/x + 7/x^2 \to 0$. So we replace ϵ with $4/x + 7/x^2$ in the series.

$$\sqrt{1 + \frac{4}{x} + \frac{7}{x^2}} = 1 + \frac{1}{2}\left(\frac{4}{x} + \frac{7}{x^2}\right) - \frac{1}{8}\left(\frac{4}{x} + \frac{7}{x^2}\right)^2 + \frac{1}{16}\left(\frac{4}{x} + \frac{7}{x^2}\right)^3 + O(1/x^4).$$

In expanding, we only have to keep terms up to order $1/x^3$. Thus,

$$\sqrt{1 + \frac{4}{x} + \frac{7}{x^2}} = 1 + \frac{2}{x} + \frac{3}{2x^2} - \frac{3}{x^3} + O(1/x^4).$$

Thus,

$$x^2\sqrt{1 + \frac{4}{x} + \frac{7}{x^2}} - x^2 - 2x = \frac{3}{2} - \frac{3}{x} + O(1/x^2).$$

So the limit is $3/2$. $\square$

Problems for §1.3

For problems **1** through **14**: By replacing functions with a few terms of their asymptotic series, find the following limits.

1 $\displaystyle\lim_{x\to 0} \frac{e^x - \sqrt{2x+1}}{x^2}$

2 $\displaystyle\lim_{x\to 0} \frac{\cos(x) - \sqrt{1-x^2}}{x^4}$

3 $\displaystyle\lim_{x\to\infty} \sqrt{x^2+x} - x$

4 $\displaystyle\lim_{x\to\infty} \sqrt{2x^2+3x} - \sqrt{2x^2+x}$

5 $\displaystyle\lim_{x\to 0} \frac{e^x}{x^2+x^3} - \frac{1}{x^2}$

6 $\displaystyle\lim_{x\to 0} \frac{\cosh(x)}{2x^4+x^6} - \frac{1}{2x^4}$

7 $\displaystyle\lim_{x\to 0} \frac{e^x \ln(x+1) + \ln(1-x)}{\cos(x)\cosh(x) - 1}$

8 $\displaystyle\lim_{x\to 0} \frac{\tan(x) - \sin(x)\cosh(x)}{x^5}$

9 $\displaystyle\lim_{x\to 0} \frac{\sin(x)\sin^{-1}(x) - x^2}{\tan(x)\tan^{-1}(x) - x^2}$

10 $\displaystyle\lim_{x\to\infty} \sqrt{x^4+2x^3} - x^2 - x$

11 $\displaystyle\lim_{x\to\infty} \sqrt[3]{x^3+2x^2} - x$

12 $\displaystyle\lim_{x\to 0} \frac{1}{\sin^2 x} - \frac{1}{\sinh^2 x}$

13 $\displaystyle\lim_{x\to 0} \frac{1}{\tanh^2 x} - \frac{1}{\tan^2 x}$

14 $\displaystyle\lim_{x\to 0} \frac{e^{-x^2}\cosh(x) - \cos(x)}{\sin(x)\sinh(x) - x^2}$

15 Find the limit

$$\lim_{x\to 0} \frac{S(x)e^x - 1}{x^2},$$

where $S(x)$ is the Stieltjes integral function, defined by equation 1.21.

16 Note that in figure 1.4, the curves $y = 1$ and $y = 1 - x + 2x^2$ cross at $x = 1/2$, the curves $y = 1-x$ and $y = 1-x+2x^2-6x^3$ cross at $x = 1/3$, and the curves $y = 1 - x + 2x^2$ and $1 - x + 2x^2 - 6x^3 + 24x^4$ cross at $x = 1/4$. Show that the pattern continues. That is, show that the nth degree polynomial approximation and the $(n-2)^{\text{nd}}$degree polynomial approximation cross at $x = 1/n$.

For problems **17** through **28**: Find the first three (non-zero) terms of the asymptotic series as $x \to a$ for the following functions.

17 $\ln(x^2 + x^3)$ as $x \to \infty$
18 $\ln(e^x + 1)$ as $x \to \infty$
19 $\sqrt{x^4 + 2x^3 + 4x^2}$ as $x \to \infty$
20 $\sqrt{x^4 + 2x^3 + 4x^2}$ as $x \to 0$
21 $\sin(x)\sin^{-1}(x)$ as $x \to 0$
22 $\sin(x + x^3)$ as $x \to 0$
23 $e^{(x+x^2)}$ as $x \to 0$
24 $\ln(\sin(x))$ as $x \to 0$
25 $\sqrt[3]{x^3 + 4x^2 + 3x}$ as $x \to \infty$
26 $1/\ln(1 + x)$ as $x \to 0$
27 $\csc(x)$ as $x \to 0$
28 $\sinh(x)/\sin(x)$ as $x \to 0$

1.4 Inverse Functions

Although we have used asymptotics to calculate limits, we still have not applied this tool for what it is mainly designed to do: approximate functions that cannot be calculated any other way. One application of asymptotics comes from finding inverses of tricky functions. Recall that an inverse of a function $f(x)$ is the function $f^{-1}(x)$ such that $f(f^{-1}(x)) = f^{-1}(f(x)) = x$, at least for part of the domain. For a function like $f(x) = x^3 + x$, it is difficult to find a formula for $f^{-1}(x)$. Yet we can find the asymptotic series for $f^{-1}(x)$ as $x \to \infty$.

Example 1.12
Find the first three terms for the inverse of the function $f(x) = x^3 + x$ as $x \to \infty$.
SOLUTION: Since $x^3 + x \sim x^3$ as $x \to \infty$, it is natural to assume that the inverse function will be similar to $\sqrt[3]{x}$ as $x \to \infty$. But what will be the next term in the series? The plan is to peel off this first term, writing $f^{-1}(x) = \sqrt[3]{x} + g(x)$, and find the asymptotic approximation for $g(x)$. Since we know that $f(f^{-1}(x)) = x$, we have

$$(\sqrt[3]{x} + g(x))^3 + \sqrt[3]{x} + g(x) = x.$$

We can expand this out asymptotically, utilizing the fact that $g(x) \ll \sqrt[3]{x}$. As a general rule, we do not need to keep terms that involve the square (or higher power) of the unknown function.

$$x + 3g(x)\sqrt[3]{x^2} + O\left(g(x)^2\sqrt[3]{x}\right) + \sqrt[3]{x} + g(x) = x.$$

The x's cancel out, and $g(x)$ is small compared to $g(x)\sqrt[3]{x^2}$ as $x \to \infty$. By throwing out terms that are known to be smaller than a *non-canceling* term as $x \to \infty$, we get

$$3g(x)\sqrt[3]{x^2} \sim -\sqrt[3]{x},$$

which tells us that $g(x) \sim -x^{-1/3}/3$. So we now have two terms of the asymptotic series:

$$f^{-1}(x) \sim \sqrt[3]{x} - \frac{1}{3\sqrt[3]{x}} \quad \text{as } x \to \infty.$$

To find the next term in the series, we repeat the process, assuming that $f^{-1}(x) = \sqrt[3]{x} - x^{-1/3}/3 + h(x)$. Since $f(f^{-1}(x)) = x$,

$$\left(\sqrt[3]{x} - \frac{1}{3\sqrt[3]{x}} + h(x)\right)^3 + \sqrt[3]{x} - \frac{1}{3\sqrt[3]{x}} + h(x) = x.$$

Cubing a trinomial is a bit tricky, but any term involving $h(x)^2$ will almost certainly be small. So we can first rewrite this as

$$\left(\sqrt[3]{x} - \frac{1}{3\sqrt[3]{x}}\right)^3 + 3\left(\sqrt[3]{x} - \frac{1}{3\sqrt[3]{x}}\right)^2 h(x) + O\left(h(x)^2\sqrt[3]{x}\right) + \sqrt[3]{x} - \frac{1}{3\sqrt[3]{x}} + h(x) = x.$$

It is clear that the largest term involving $h(x)$ is $3\sqrt[3]{x^2}h(x)$, which does not cancel with any other terms. But we will have to expand the cube, to give us

$$\left(x - \sqrt[3]{x} + \frac{1}{3\sqrt[3]{x}} - \frac{1}{27x}\right) + 3\sqrt[3]{x^2}h(x) + \sqrt[3]{x} - \frac{1}{3\sqrt[3]{x}} \sim x.$$

Since the x's canceled before, we expect them to cancel again. But this time, the $\sqrt[3]{x}$ and $x^{-1/3}/3$ also cancel, giving us

$$3\sqrt[3]{x^2}h(x) \sim \frac{1}{27x}.$$

So $h(x) \sim x^{-5/3}/81$. Thus, we have

$$f^{-1}(x) \sim x^{1/3} - \frac{1}{3\sqrt[3]{x}} + \frac{1}{81\sqrt[3]{x^5}} \quad \text{as } x \to \infty.$$

Figure 1.5 shows how each successive term gives a better approximation to the inverse function. Although the approximations are designed to be excellent for large values of x, even at $x = 1$ there is only a 0.5% error. □

Let us recap the steps that were used in this last example, since the same steps will be used for a variety of different types of problems.

1) Determine the first term of the asymptotic series. Many times, this can be done via simple approximations, but for more complicated problems

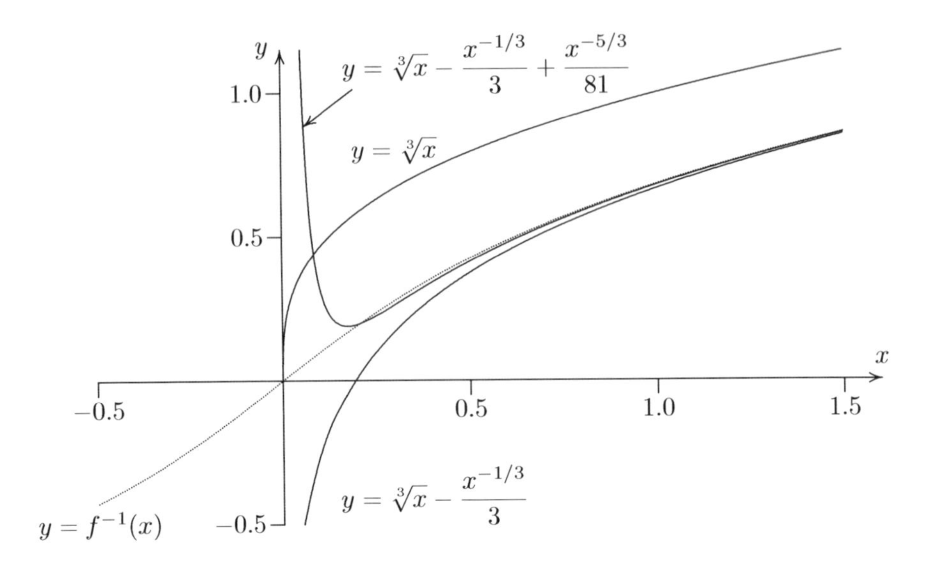

FIGURE 1.5: Comparing the inverse function of $x+x^3$ with the first three asymptotic approximations as $x \to \infty$. Note that the final approximation is indistinguishable from the inverse function for $x > 1$.

it may require more sophisticated methods such as a trial and error method called dominant balance. This first term is called the *leading behavior* of the solution.

2) Add an unknown function to the series so far. We can assume that this function is smaller than the previous term, which will help in later steps.

3) Plug this series into the equation that the function must satisfy.

4) Carefully expand this equation asymptotically, canceling terms whenever possible. Note that usually terms involving the unknown function squared need not be considered.

5) After the terms have canceled, we want to throw out any terms that are asymptotically smaller than a term that did not cancel.

6) The remaining terms should give an equation for the unknown function that is now easy to solve. This gives us the next term in the series.

7) Repeat steps 2-6 to get more terms in the series.

One can see that each term in the series is progressively harder to obtain. Nonetheless, it usually only takes a few terms of an asymptotic series to achieve incredible accuracy in the approximation.

Example 1.13
Analyze the inverse of the function $f(x) = xe^x$ as $x \to \infty$.
SOLUTION: Although there is only one term in $f(x)$, it is clear that the

dominant factor that controls the behavior is the exponential function, so it is natural to assume that the inverse function will behave like a logarithm. So let us try $f^{-1}(x) \sim \ln(x)$. If we successfully find the next order term, this will verify that our guess is correct.

If we set $f^{-1}(x) = \ln(x) + g(x)$, then since $f^{-1}(f(x)) = x$, we have

$$\ln(xe^x) + g(xe^x) = x.$$

Using the properties of logarithms, we can simplify this to the exact equation $\ln(x)+g(xe^x) = 0$. Again, since the dominant factor of xe^x is the exponential, we get $g(e^x) \sim -\ln(x)$, so $g(x) \sim -\ln(\ln(x))$. Indeed, this is smaller than $\ln(x)$, so we have confirmed that the first term was correct.

Since we have an exact equation for $g(x)$, we can use this as a shortcut for finding the next term. We can let $g(x) = -\ln(\ln(x)) + h(x)$, to produce the equation

$$\ln(x) - \ln(\ln(xe^x)) + h(xe^x) = 0.$$

Now

$$\ln(\ln(xe^x)) = \ln(x + \ln(x)) = \ln\left(x\left(1 + \frac{\ln(x)}{x}\right)\right) = \ln(x) + \ln\left(1 + \frac{\ln(x)}{x}\right),$$

so the $\ln(x)$ cancels to produce

$$h(xe^x) = \ln\left(1 + \frac{\ln(x)}{x}\right). \tag{1.38}$$

Since $\ln(x)/x \ll 1$ and $\ln(1+\epsilon) \sim \epsilon$, we see that $h(xe^x) \sim \ln(x)/x$, or $h(x) \sim \ln(\ln(x))/\ln(x)$.

This is proceeding well enough to brave yet another term. Substituting $h(x) = \ln(\ln(x))/\ln(x) + k(x)$ into equation 1.38 and expanding the inner logarithms produces

$$\frac{\ln(x + \ln(x))}{x + \ln(x)} + k(xe^x) = \ln\left(1 + \frac{\ln(x)}{x}\right). \tag{1.39}$$

Expanding the right hand side asymptotically is easy using equation 1.12, but the first term has to first be rewritten as

$$\frac{\ln(x + \ln(x))}{x + \ln(x)} = \frac{1}{x}\left(\ln(x) + \ln\left(1 + \frac{\ln(x)}{x}\right)\right)\left(1 + \frac{\ln(x)}{x}\right)^{-1}.$$

Then equation 1.39 can be approximated asymptotically to give us

$$\frac{1}{x}\left(\ln(x) + \frac{\ln(x)}{x} + O\left(\frac{\ln(x)^2}{x^2}\right)\right)\left(1 - \frac{\ln(x)}{x} + O\left(\frac{\ln(x)^2}{x^2}\right)\right) + k(xe^x)$$
$$= \frac{\ln(x)}{x} - \frac{\ln(x)^2}{2x^2} + O\left(\frac{\ln(x)^3}{x^3}\right).$$

Expanding the product, we find that the $\ln(x)/x$ terms will cancel, and we have

$$k(xe^x) = \frac{\ln(x)^2}{2x^2} - \frac{\ln(x)}{x^2} + O\left(\frac{\ln(x)^3}{x^3}\right).$$

Thus, we get the first two terms for $k(x)$,

$$\ln(\ln(x))^2/2(\ln(x))^2 - \ln(\ln(x))/(\ln(x))^2.$$

Putting all of the terms together, we get

$$f^{-1}(x) \sim \ln(x) - \ln(\ln(x)) + \frac{\ln(\ln(x))}{\ln(x)} + \frac{\ln(\ln(x))^2}{2\ln(x)^2} - \frac{\ln(\ln(x))}{\ln(x)^2}. \quad (1.40)$$

□

The inverse function of $y = xe^x$ is called the Lambert W function. Its asymptotic series as $x \to \infty$ is more complicated than the ones we have seen before, since the logarithm terms complicate matters a bit. In fact, later terms will all have a power of $\ln(\ln(x))$ in the numerator, and a power of $\ln(x)$ in the denominator. Yet the simple function $1/x$ goes to zero faster than all of these terms. So how well does the asymptotic series approximate $W(x)$? Figure 1.6 shows that in spite of the fact that the individual terms to go 0 very slowly, the first few terms give a fairly accurate approximation to the function. There is less than 0.4% error for $x > 10$.

1.4.1 Reversion of Series

For a function in which $f(x) \sim cx$ as $x \to 0$, it is fairly clear that the inverse function will have $f^{-1}(x) \sim x/c$ as $x \to 0$. In fact, if we have a power series expansion for such a function, we can compute the power series expansion for its inverse.

Often the powers of x will appear in an arithmetic progression. For example, an odd function will have only odd powers of x in its Maclaurin series. In general, suppose that

$$f(x) \sim a_0 x + a_1 x^{b+1} + a_2 x^{2b+1} + \cdots = \sum_{n=0}^{\infty} a_n x^{bn+1}.$$

Then the inverse of $y = f(x)$ is given by

$$f^{-1}(y) = A_0 y + A_1 y^{b+1} + A_2 y^{2b+1} + \cdots = \sum_{n=0}^{\infty} A_n y^{bn+1},$$

where

$$A_0 = \frac{1}{a_0}, \quad A_1 = \frac{-a_1}{a_0^{b+2}}, \quad A_2 = \frac{1}{a_0^{2b+3}}((b+1)a_1^2 - a_0 a_2),$$

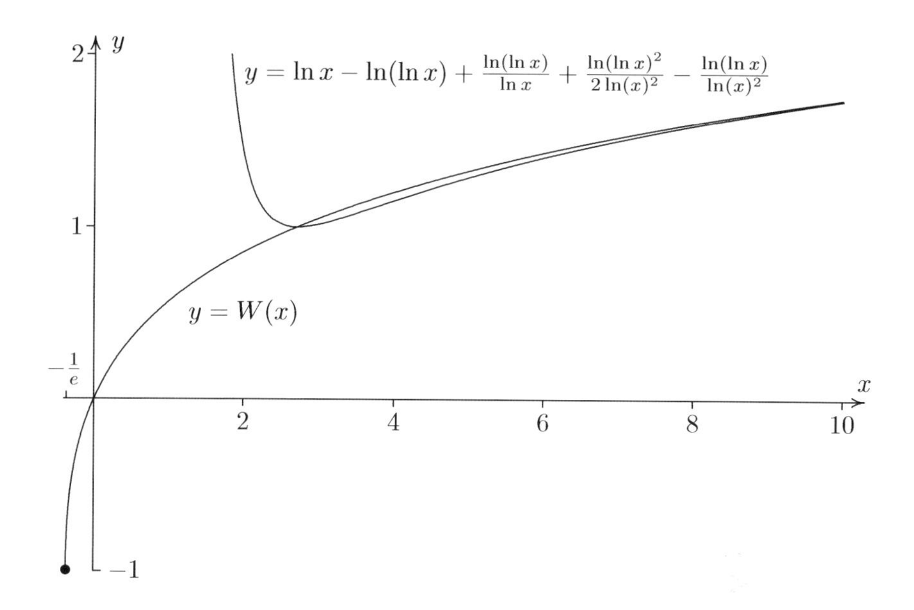

FIGURE 1.6: Comparing the Lambert W function (the inverse of $y = xe^x$) with the asymptotic approximation as $x \to \infty$. $W(x)$ is defined for $x \geq -1/e$, but the approximation is fairly good for $x \geq e$.

$$A_3 = \frac{1}{a_0^{3b+4}}\left(-\frac{(3b+3)(3b+2)}{6}a_1^3 + (3b+2)a_0a_1a_2 - a_0^2a_3\right), \tag{1.41}$$

$$A_4 = \frac{1}{a_0^{4b+5}}\left(\frac{(4b+4)(4b+3)(4b+2)}{24}a_1^4 - \frac{(4b+3)(4b+2)}{2}a_0a_1^2a_2 \right.$$
$$\left. + \frac{(4b+2)}{2}a_0^2a_2^2 + (4b+2)a_0^2a_1a_3 - a_0^3a_4\right),$$

$$A_5 = \frac{1}{a_0^{5b+6}}\left(-\frac{(5b+5)(5b+4)(5b+3)(5b+2)}{120}a_1^5\right.$$
$$+ \frac{(5b+4)(5b+3)(5b+2)}{6}a_0a_1^3a_2 - \frac{(5b+3)(5b+2)}{2}(a_0^2a_1a_2^2 + a_0^2a_1^2a_3)$$
$$\left. + (5b+2)(a_0^3a_2a_3 + a_0^3a_1a_4) - a_0^4a_5\right).$$

For most functions, we use $b = 1$, but occasionally we may want to take the inverse of an odd function, in which case we can use $b = 2$. The pattern for these coefficients can be given by

$$\boxed{A_n = \frac{1}{a_0^{bn+n+1}}\sum(-1)^{n+p_0}\frac{(bn+n-p_0)!}{(bn+1)!p_1!p_2!\cdots p_n!}a_0^{p_0}a_1^{p_1}a_2^{p_2}\cdots a_n^{p_n},} \tag{1.42}$$

where the sum is taken over all combinations of non-negative integers p_0, p_1, $\ldots p_n$ such that both $p_0 + p_1 + p_2 + \cdots p_n = n$, and $p_1 + 2p_2 + \cdots np_n = n$. See problem 19. If the original series has a non-zero radius of convergence, then the series for the inverse function will also have a non-zero radius of convergence, but it may be difficult to determine exactly what that radius is.

Example 1.14

We have already seen Lambert's W function, which is the inverse of $y = xe^x$. Although we have seen the asymptotic series for when $x \to \infty$, what is the behavior as $x \to 0$?

SOLUTION: We have

$$xe^x \sim x + x^2 + \frac{x^3}{2} + \frac{x^4}{6} + \frac{x^5}{24} + \cdots \text{ as } x \to 0.$$

Thus, we can plug in $b = 1$ and $a_n = 1/n!$ into equations 1.41 to get $A_0 = 1$, $A_1 = -1$, $A_2 = 3/2$, $A_3 = -8/3$, $A_4 = 125/24$, and $A_5 = -54/5$. Thus,

$$W(x) \sim x - x^2 + \frac{3x^3}{2} - \frac{8x^4}{3} + \frac{125x^5}{24} - \frac{54x^6}{5} + \cdots \text{ as } x \to 0.$$

In fact, there turns out to be a nice pattern:

$$W(x) \sim \sum_{n=1}^{\infty} \frac{(-1)^{n-1} n^{n-1} x^n}{n!} \text{ as } x \to 0.$$

□

Example 1.15

Find the Maclaurin series for $\tan(x)$.

SOLUTION: We can use the fact that the series for its inverse has a simple pattern.

$$\tan^{-1}(x) = \sum_{n=0}^{\infty} \frac{(-1)^n x^{2n+1}}{2n+1} = x - \frac{x^3}{3} + \frac{x^5}{5} - \frac{x^7}{7} + \cdots.$$

Because this is an odd function, we can use $b = 2$, and replace

$$a_n = \frac{(-1)^n}{(2n+1)}$$

into equations 1.41 to get $A_0 = 1$, $A_1 = 1/3$, $A_2 = 2/15$, $A_3 = 17/315$, $A_4 = 62/2835$, and $A_5 = 1382/155925$. This gives use the first 6 terms of the tangent series:

$$\tan(x) \sim x + \frac{x^3}{3} + \frac{2x^5}{15} + \frac{17x^7}{315} + \frac{62x^9}{2835} + \frac{1382x^{11}}{155925} + \cdots \text{ as } x \to 0.$$

□

Problems for §1.4

For problems **1** through **12**: Find the first three (non-zero) terms of the asymptotic series as $x \to \infty$ for the inverse of the following functions.

1 $x^2 + 5x + 3$
2 $x^3 + 2x$
3 $x^3 + x + 1$
4 $x^3 + x^2$
5 $x^4 + x$
6 $x^4 + x^3$
7 $x^2 e^x$
8 $x + \ln(x)$
9 $e^x + \ln(x)$
10 $x \ln(x)$
11 $e^x + x$
12 $e^x \ln(x)$

For problems **13** through **18**: Many inverse functions can be expressed in terms of the Lambert W function. Express the inverse of the following functions in terms of $W(x)$.

13 $x + \ln(x)$
14 $x \ln(x)$
15 $x + e^x$
16 $\sqrt{x} e^x$
17 $x^2 + \ln(x)$
18 $x^2 \ln(x)$

19 Use equation 1.42 to find the formula for A_6 when $b = 1$

Hint: A *partition of n* is a set of positive integers that add up to n. For example, $1+1+1+3$ is a partition of 6. Rearrangement of the integers are not considered as different partitions. For example, there are seven partitions of 5: $1+1+1+1+1$, $1+1+1+2$, $1+2+2$, $1+1+3$, $2+3$, $1+4$, and 5. Each partition of n can be padded with 0's to produce a set of n integers adding to n. For example, $1+1+1+3$ becomes $0+0+1+1+1+3$. Every partition of n becomes one term in equation 1.42. For each partition padded with 0's, we let p_i be the number of i's in the partition. For example, the partition $0+0+1+1+1+3$ produces $p_0 = 2$, $p_1 = 3$, $p_3 = 1$, and $p_2 = p_4 = p_5 = p_6 = 0$.

20 Find all of the partitions of 7. See problem 19 for an explanation of the partitions of n.

For problems **21** through **26**: Use equations 1.41 to find the first four non-zero terms in the Maclaurin series for the inverse of the function.

21 $x \cos x$
22 $x \cosh x$
23 $x + \ln(x+1)$
24 $1/\sqrt{1-4x} - 1 - x$
25 $x + x^4 + x^7$
26 $x e^{x^3}$

27 Use equation 1.42 to find a formula for the nth term in the Maclaurin series of the inverse of $f(x) = x + x^3$.

Hint: Using $b = 2$, the only partition of n that produces a non-zero term is the one where $p_1 = n$. See problem 19 for an explanation of the partitions of n.

28 The equations 1.41 can be used for a divergent series as well, producing an asymptotic series for the inverse function. Find the first six (non-zero) terms for the inverse of the function $S(x) - 1$, where $S(x)$ is the Stieltjes integral function defined by equation 1.21. Show that this series is divergent for all positive x.

1.5 Dominant Balance

In finding the asymptotic series for the solution of an equation, we must first determine its leading behavior. Sometimes this is easy, but usually this is a non-trivial problem. In these situations, a strategy that is very effective is the method of dominant balance.

The principle behind the dominant balance is quite simple. If there are three or more terms in an equation, usually two of the terms will be asymptotically larger than the others, that is, they will dominate the other terms. Also, these terms will balance each other, so we can form an asymptotic equation with only two terms. Such equations are usually very easy to solve.

The problem, of course, is determining *which* two terms are the ones that are dominant. This can only be determined by trial and error. In each case, we can test to see if the other terms are indeed small compared to the ones that we assumed were dominant.

Example 1.16

Find the behavior of the function defined implicitly by $x^2 + xy - y^3 = 0$ as $x \to \infty$.

SOLUTION: Since there are three non-zero terms, there are three choices for a pair of terms. If we assume y^3 is small as $x \to \infty$, then we have $x^2 \sim -xy$. This quickly leads to $y \sim -x$ as $x \to \infty$. But then $y^3 \sim -x^3$, which is *not* small compared to a term that we kept, x^2, as $x \to \infty$.

Suppose instead that we assume $x^2 \ll xy$. Then $xy \sim y^3$, producing $y \sim \pm\sqrt{x}$. But then $xy \sim \pm x^{3/2}$, so x^2 is larger as $x \to \infty$. So this possibility is ruled out.

The final case to try is to assume that xy is the smallest term. Then $x^2 \sim y^3$, which tells us that $y \sim x^{2/3}$. To check to see if this is consistent, we need to check that $xy \ll x^2$. Indeed, $xy \sim x^{5/3}$ which is smaller than x^2 as $x \to \infty$.

At this point, we have shown that $y \sim x^{2/3}$ is *consistent*, but this alone is not proof that the leading behavior is indeed $x^{2/3}$. In order to show that this is the correct leading behavior, we must find the next term in the series, and show that it is smaller than $x^{2/3}$.

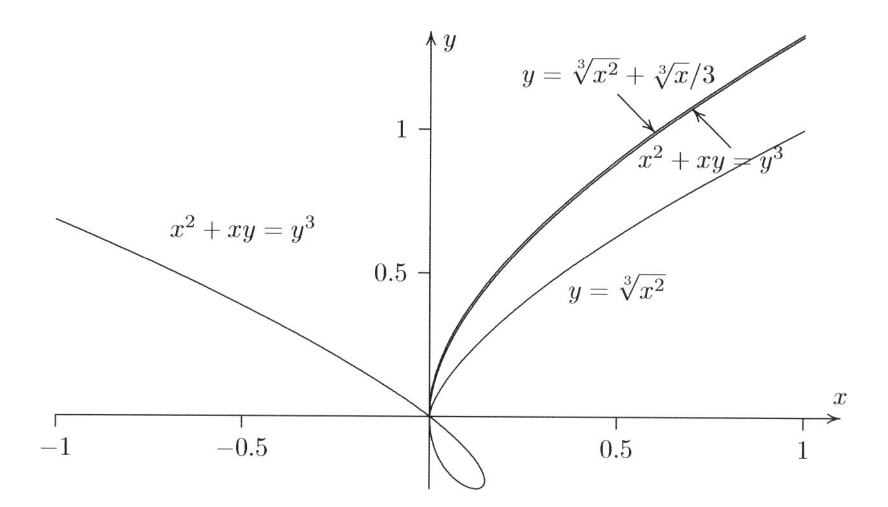

FIGURE 1.7: Comparing the function defined implicitly by $x^2 + xy = y^3$ with its second order approximation $y = \sqrt[3]{x^2} + \sqrt[3]{x}/3$. The two graphs are almost indistinguishable. The error at $x = 1$ is only 0.6%.

Letting $y = x^{2/3} + g(x)$ gives us

$$x^2 + x(x^{2/3} + g(x)) = (x^{2/3} + g(x))^3,$$

or

$$x^2 + x^{5/3} + xg(x) \sim x^2 + 3g(x)x^{4/3} + O(x^{2/3}g(x)^2).$$

The x^2's cancel, and keeping the higher order terms, we get $x^{5/3} \sim 3g(x)x^{4/3}$, which simplifies to $g(x) \sim x^{1/3}/3$. This is indeed smaller than $x^{2/3}$ as $x \to \infty$, so we have confirmed the leading behavior, along with the next term $y \sim x^{2/3} + x^{1/3}/3$. Figure 1.7 compares the two term approximation with the implicitly defined function. ▯

Here is a summary of the steps that are used in the method of dominant balance.

1) Make a guess as to which terms may be negligible in the limit.

2) Drop these terms to form a simpler equation. Solve this equation exactly.

3) Check to make sure that the solution is consistent with step 1. If not, try dropping different terms in step 1.

4) Find the next term to verify that the leading behavior is indeed correct.

If this seems like circular reasoning, it's because it is! This is why step 4 is so important. Just because we find an approximation that is consistent with the equation does not mean that there is a true solution to the equation which is asymptotic to our approximation.

Even after we find a combination of two terms which seem to be dominant, we still have to check all other possibilities, for there may be more than one solution!

Example 1.17

Find the possible behaviors as $x \to \infty$ of the function defined implicitly by $y^3 - xy^2 + x^2 - 2xy + 1 = 0$.

SOLUTION: Even though there are five non-zero terms, it is clear that $1 \ll x^2$ as $x \to \infty$. Thus, we have to choose two terms out of four to be the dominant terms, giving us six choices.

Case 1) If we assume that y^3 and xy^2 are small compared to x^2 and $2xy$, then $x^2 \sim 2xy$ as $x \to \infty$. This implies that $y \sim x/2$, but then $y^3 \sim x^3/8$ is larger than x^2.

Case 2) If we assume that y^3 and x^2 are small compared to xy^2 and $2xy$, then $xy^2 \sim -2xy$. Since it is impossible for $y \sim 0$, this would indicate that $y \sim -2$ as $x \to \infty$, but this contradicts that x^2 is smaller than xy^2.

Case 3) If we assume that y^3 and $2xy$ are small compared to xy^2 and x^2, then $x^2 \sim xy^2$ as $x \to \infty$. This is readily solved to give us $y \sim \pm\sqrt{x}$. Under these assumptions, both y^3 and $2xy$ are $O(x^{3/2})$, so indeed this choice is consistent.

To determine whether these are possible leading terms, we need to show that the next term in the series is asymptotically smaller. If we let $y = \pm\sqrt{x} + g(x)$, and plug this into the original equation, we get

$$(\pm\sqrt{x} + g(x))^3 - x(\pm\sqrt{x} + g(x))^2 + x^2 - 2x(\pm\sqrt{x} + g(x)) + 1 = 0.$$

Expanding this, we get

$$\pm\sqrt{x}^3 + 3xg(x) + O(xg(x)^2) - x^2 \mp 2\sqrt{x}^3 g(x) + x^2 \mp 2\sqrt{x}^3 - 2xg(x) + 1 = 0. \tag{1.43}$$

This notation needs a bit of explanation. Whenever both $\pm$ and $\mp$ appear in an equation (or more than one $\pm$ sign), either the top symbol is always used, or the bottom symbol is always used. In this way, two very similar equations can be combined into a single equation. For example, the $\pm\sqrt{x}^3$ can combine with the $\mp 2\sqrt{x}^3$ to produce $\mp\sqrt{x}^3$.

Canceling out the x^2's, and throwing out terms that are known to be small produces $\mp 2\sqrt{x}^3 g(x) \sim \pm\sqrt{x}^3$, giving $g(x) \sim -1/2$ with either choice of signs. Since this is smaller than $\pm\sqrt{x}$ as $x \to \infty$, we have confirmed that two possible behaviors of y are $\pm\sqrt{x} - 1/2$.

However, we are not done, since the other three cases may produce a consistent leading behavior.

Case 4) If we assume that xy^2 and x^2 are small compared to y^3 and $2xy$, then $y^3 \sim 2xy$, which simplifies to $y \sim \pm\sqrt{2x}$. But then $y^3 = O(x^{3/2})$, so x^2 would not be small as $x \to \infty$.

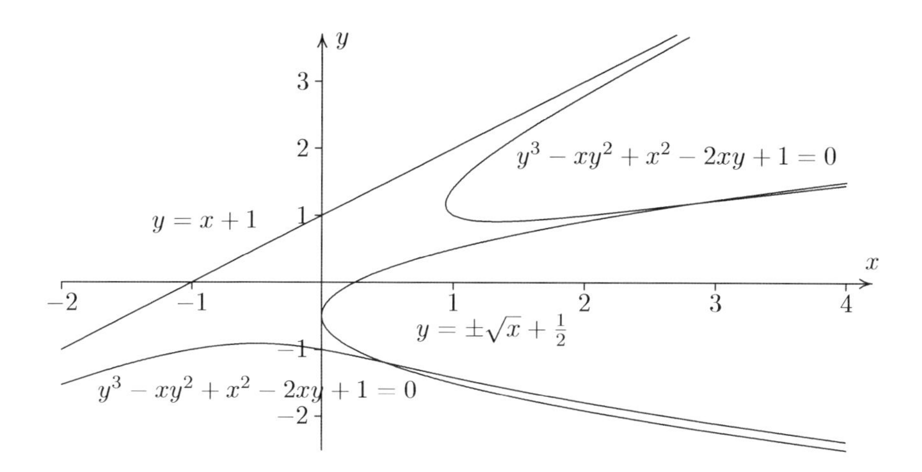

FIGURE 1.8: The curve $y^3 - xy^2 + x^2 - 2xy + 1 = 0$ has two parts, each of which has both the line $y = x + 1$ and the parabola $y = \pm\sqrt{x} + 1/2$ as asymptotes.

Case 5) If we assume that xy^2 and $2xy$ are small compared to y^3 and x^2, then $y^3 \sim -x^2$, or $y \sim -x^{2/3}$. But then $xy^2 \sim x^{7/3}$ would be larger than x^2 as $x \to \infty$.

Case 6) If we assume that x^2 and $2xy$ are small compared to y^3 and xy^2, then $y^3 \sim xy^2$, or $y \sim x$. Then x^2 and $2xy$ would indeed be smaller than $y^3 \sim x^3$ as $x \to \infty$. To confirm that this is a possible behavior, we work to find the next term. If $y = x + g(x)$, then

$$(x + g(x))^3 - x(x + g(x))^2 + x^2 - 2x(x + g(x)) + 1 = 0.$$

Expanding this, we get

$$x^3 + 3x^2 g(x) + O(xg(x)^2) - x^3 - 2x^2 g(x) + x^2 - 2x^2 - 2xg(x) + 1 = 0.$$

Canceling out the x^3's, and noting that $xg(x) \ll x^2 g(x)$, this simplifies to $x^2 g(x) \sim x^2$, so $g(x) \sim 1$. Indeed, this is smaller than x as $x \to \infty$, so we have a third possible behavior, $y \sim x + 1$.

Figure 1.8 shows that the exact solution curve indeed approaches all three of the asymptotic behaviors as $x \to \infty$. □

Usually with dominant balance, there will be two terms that are dominant over the other terms. But occasionally there may be three terms that are all the same size, that is, the ratios between two of the three terms approach a non-zero constant. Here is an example where this happens.

Example 1.18
Find the possible behaviors as $x \to \infty$ of the function defined implicitly by $y^4 - y^3 - 2xy^2 + x^2 = 1$.
SOLUTION: Since $1 \ll x^2$, there are only four terms which might be dominant, giving us six cases. But all six cases yield a contradiction:

Case 1) y^4 and y^3 are small compared to xy^2 and x^2. Then $2xy^2 \sim x^2$, making $y \sim \sqrt{x/2}$. But then $y^4 \sim x^2/4$, which is *not* $\ll x^2$ as $x \to \infty$.

Case 2) y^4 and xy^2 are small compared to y^3 and x^2. Then $y^3 \sim x^2$, so $y \sim x^{2/3}$, which makes $y^4 \gg x^2$ as $x \to \infty$.

Case 3) y^4 and x^2 are small compared to y^3 and xy^2. Then $y^3 \sim -2xy^2$, so $y \sim -2x$, which makes $y^4 \gg y^3$ as $x \to \infty$.

Case 4) y^3 and xy^2 are small compared to y^4 and x^2. Then $y^4 \sim -x^2$, which yields no real solutions.

Case 5) y^3 and x^2 are small compared to y^4 and xy^2. Then $y^4 \sim 2xy^2$, making $y \sim \sqrt{2x}$. But then $y^4 \sim 4x^2$, which is *not* $\gg x^2$ as $x \to \infty$.

Case 6) xy^2 and x^2 are small compared to y^4 and y^3. Then $y^4 \sim y^3$, so $y \sim 1$, which makes $x^2 \gg y^4$ as $x \to \infty$.

Even though all six cases failed, we notice that in cases 1 and 5, the problem was that a term that we thought was small turned out to be *the same order* as the terms we thought were large. In both of these cases, $y \sim c\sqrt{x}$ for some constant c. If this were the case, then $y^3 \ll y^4$, so we have to solve $y^4 + x^2 \sim 2xy^2$. If we plug in $y \sim c\sqrt{x}$, we get $c^4x^2 + x^2 \sim 2c^2x^2$, making $c^4 - 2c^2 + 1 = 0$. This factors to $(c^2 - 1)^2 = 0$, so $c = \pm 1$. Thus, there are two possible leading behaviors, $y \sim \pm\sqrt{x}$.

But are both of these behaviors actually realized by the curve? To find out, we must find the next order term. If we plug $y = \pm\sqrt{x} + g(x)$ into $y^4 + x^2 = y^3 + 2xy^2 + 1$ and expand, we get

$$
\begin{aligned}
&(x^2 \pm 4\sqrt{x}^3 g(x) + 6xg(x)^2 \pm 4\sqrt{x}g(x)^3 + g(x)^4) + x^2 = \\
&\qquad (\pm\sqrt{x}^3 + 3xg(x) \pm 3\sqrt{x}g(x)^2 + g(x)^3) + 2x(x \pm 2\sqrt{x}g(x) + g(x)^2) + 1.
\end{aligned}
$$

The x^2 terms cancel as expected, but surprisingly, so do the $\sqrt{x}^3 g(x)$ terms. This means that the $xg(x)^2$ terms may in fact be important. To explain why this is happening, look back at how we solved for our first order term, solving $(c^2 - 1)^2 = 0$. Both roots were in fact *double roots,* so it is not surprising if there are two possible second order terms for each of the leading terms.

Once we cancel and combine like terms, we get

$$
4xg(x)^2 \pm 4\sqrt{x}g(x)^3 + g(x)^4 = \pm\sqrt{x}^3 + 3xg(x) \pm \sqrt{x}g(x)^2 + g(x)^3 + 1. \quad (1.44)
$$

This looks like a formidable dominant balance problem, but we have the advantage of being able to assume that $g(x) \ll \sqrt{x}$. Thus, $\sqrt{x}g(x)^3 \ll xg(x)^2$, $g(x)^4 \ll xg(x)^2$, $xg(x) \ll \sqrt{x}^3$, $\sqrt{x}g(x)^2 \ll \sqrt{x}^3$, $g(x)^3 \ll \sqrt{x}^3$, and of course $1 \ll \sqrt{x}^3$. So we have $4xg(x)^2 \sim \pm\sqrt{x}^3$, making $g(x)^2 \sim \pm\sqrt{x}/4$. If $y \sim -\sqrt{x}$, then there will be no real second order term. Hence, the only leading order

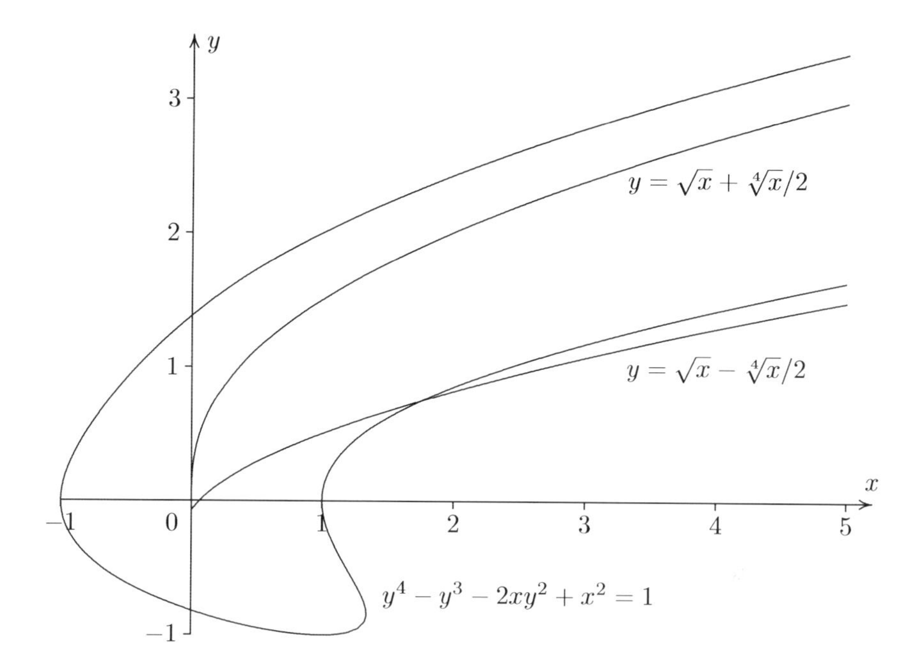

FIGURE 1.9: The curve $y^4 - y^3 - 2xy^2 + x^2 = 1$, along with the second order approximations as $x \to \infty$. Since the third order correction is a constant, the approximation curves become parallel to the curve, instead of approaching it. See problem 15 for determining this constant.

to the curve is $y \sim \sqrt{x}$, and the two possible second order terms are $\pm\sqrt[4]{x}/2$. Figure 1.9 shows the two curves approaching the implicitly defined function. □

Problems for §1.5

For problems **1** through **14**: Find all possible leading behaviors of the following curves as $x \to \infty$, along with the second order corrections.

1 $y^3 - x^3 + xy = 0$
2 $y^3 + y^2 - x^2 = 0$
3 $y^3 - x^2y + x^2 = 0$
4 $xy^3 - x^3 + y^2 = 1$
5 $xy^3 + y - x^2 = 1$
6 $xy^3 + y - x^2 = x$
7 $xy^3 - yx^2 - x = 0$
8 $xy^3 + y^2 - x^2y - 2x^2 = 1$
9 $y^3 - 2xy^2 + x^2y - x = 0$
10 $y^4 + y^2 - xy - x^2 = 1$
11 $y^4 + xy^2 - x^2y - x^3 = 1$
12 $y^4 - y^3 - 5xy^2 + 4x^2 = -1$
13 $y^4 - y^3 - 4xy^2 + 4x^2 = -1$
14 $y^4 - y^3 - 2xy^3 + x^2y^2 + x^2 = 1$

15 Show that the next term in the asymptotic series from example 1.18 is a constant, and find this constant. Note that one can start with equation 1.44 with all $\pm$ signs converted to $+$.

For problems **16** through **25**: If we consider only those terms in the asymptotic series which are not going to 0 as $x \to \infty$ or $x \to -\infty$, we obtain an *asymptote* to the curve. For example, the asymptotes of the classical hyperbola $y^2 - x^2 = 1$ are $y = x$ and $y = -x$, because these are the terms not going to 0 in the asymptotic series

$$y \sim \pm x \pm \frac{1}{2x} \mp \frac{1}{8x^3} + \cdots.$$

Find all of the non-vertical asymptotes for the following curves.

16 $y^2 - x^2 y - x^3 = 0$
17 $y^3 - x^2 y - x^2 = 0$
18 $y^3 - 3xy - x^2 - y^2 = 0$
19 $y^3 + 2xy^2 - x^3 y = 1$
20 $y^3 - 4xy^2 + 4x^2 y + xy = 1$
21 $y^4 - 2x^2 y^2 + x^4 + xy = 1$
22 $y^4 - 3xy^2 - 4x^2 - y^3 = 1$
23 $y^4 - 8xy^2 + 16x^2 - y^3 = 1$
24 $y^4 - 2x^2 y^2 + x^4 - y^3 = 1$
25 $y^4 - xy^2 - x^3 - y^3 = 1$

Chapter 2

Asymptotics of Integrals

In this chapter, we will apply the techniques of asymptotics to integration problems. Many integrals do not have an exact anti-derivative that can be expressed in terms of the standard functions, such as exponential, logarithmic, and trig functions. Thus, an approximation method must be used to compute such integrals. Although there are standard integration approximation methods, such as Simpson's rule, it is usually more efficient to find the asymptotic series of the integral.

Furthermore, many new functions can be defined in terms of integrals. Although most integrals involve only one variable, we can integrate a function with two variables,

$$\int_a^b f(x,t)\,dt$$

to produce a function involving only x. There are several techniques for finding the asymptotic series for this integral.

2.1 Integrating Taylor Series

The easiest way of forming an asymptotic series of an integral is by first expanding the integrand into an asymptotic series, and then integrating term by term. To show that this is a valid method, we will utilize a result from advanced calculus.

THEOREM 2.1
Suppose that each the functions $u_n(x)$ is integrable on the interval $a \le x \le b$, and the series $\sum_{n=1}^{\infty} u_n(x)$ converges uniformly to a function $f(x)$ on this interval. Then $f(x)$ is integrable on the interval $a \le x \le b$, and

$$\int_a^b f(x)\,dx = \sum_{n=1}^{\infty}\left(\int_a^b u_n(x)\,dx\right).$$

PROOF: For a given $\epsilon > 0$, because the series is uniformly convergent, we can find an M such that for $m \geq M$

$$\left| f(x) - \sum_{n=1}^{m} u_n(x) \right| < \frac{\epsilon}{b-a}.$$

Hence,

$$\int_a^b \left| f(x) - \sum_{n=1}^{m} u_n(x) \right| dx < \epsilon.$$

If we bring the integral inside of the absolute value sign, it can only cause the result to get smaller. Thus,

$$\left| \int_a^b f(x)\, dx - \sum_{n=1}^{m} \left(\int u_n(x)\, dx \right) \right| < \epsilon.$$

Since we have established this for all ϵ, we see that the series

$$\sum_{n=1}^{\infty} \left(\int u_n(x)\, dx \right)$$

converges to $\int_a^b f(x)\, dx$. Notice that in the process, we have proven that $f(x)$ is integrable. □

It should be remarked that even though the condition of uniform convergence was sufficient to prove the theorem, we can actually get by with a weaker condition that the partial sums be *bounded*. That is, there is a K for which

$$\left| \sum_{n=1}^{m} u_n(x) \right| \leq K \qquad \text{for all } m, \text{ and for all } a \leq x \leq b.$$

This result is known as the Lebesgue's dominated convergence theorem, and the proof becomes much more difficult [22, p.92]. Theorem 2.1 will be sufficient for our needs.

Example 2.1

Find the asymptotic series as $x \to 0$ for the function defined by the definite integral

$$f(x) = \int_x^1 \frac{\cos t}{t}\, dt.$$

SOLUTION: The first step is to find the asymptotic series for $\cos t/t$. Using equation 1.20 we see that

$$\frac{\cos t}{t} \sim \frac{1}{t} - \frac{t}{2} + \frac{t^3}{24} - \frac{t^5}{6!} + \frac{t^7}{8!} + O(t^9).$$

For a fixed $x > 0$, this is uniformly convergent on the interval $x \le t \le 1$, so we can integrate term by term, to get

$$\int \frac{\cos t}{t}\,dt \sim \ln(t) - \frac{t^2}{4} + \frac{t^4}{4\cdot 4!} - \frac{t^6}{6\cdot 6!} + \frac{t^8}{8\cdot 8!} + O(t^{10}).$$

Finally, we evaluate this at the endpoints. When $t = 1$, we get a constant, which we find by summing the convergent series

$$-\frac{1}{2\cdot 2!} + \frac{1}{4\cdot 4!} - \frac{1}{6\cdot 6!} + \frac{1}{8\cdot 8!} + \cdots \approx -0.239811742.$$

Of course, the lower endpoint x produces the same series only with x instead of t. Thus,

$$\int_x^1 \frac{\cos t}{t}\,dt \sim -\ln(x) - C + \frac{x^2}{4} - \frac{x^4}{4\cdot 4!} + \frac{x^6}{6\cdot 6!} - \frac{x^8}{8\cdot 8!} + \cdots,$$

where $C \approx 0.239811742$. □

It should be noted that while we can *integrate* an asymptotic series term by term to produce the asymptotic series of the integral, we cannot always *differentiate* an asymptotic series term by term. The problem has to do with subdominance. For example, let $f(x)$ be any function with a Maclaurin series, and let

$$g(x) = f(x) + e^{-1/x^2}\sin(e^{1/x^2}).$$

We have already established that e^{-1/x^2} is subdominant to all of the terms in the Maclaurin series, and since the sine function always evaluates to a number between -1 and 1, $e^{-1/x^2}\sin(e^{1/x^2})$ will also be subdominant. Thus, $f(x)$ and $g(x)$ will have the same asymptotic series as $x \to 0$. However, the derivative is another matter. We find that

$$g'(x) = f'(x) + \frac{2e^{-1/x^2}\sin(e^{1/x^2})}{x^3} - \frac{2\cos(e^{1/x^2})}{x^3}.$$

The last term is actually *larger* than all of the terms of the Maclaurin series as $x \to 0$. Hence, differentiating an asymptotic series term by term does not necessarily give us the asymptotic series for the derivative. There are, however, circumstances in which differentiating an asymptotic series term by term is justified. If we can rule out the presence of small, highly oscillating terms, such as a convergent power series expansion, then differentiating term by term is allowed.

When the integrand involves two variables, we keep in mind which variable will be involved in the final series. Any other variables will be temporarily treated as a constant when forming the series of the integrand.

Example 2.2
Find the Maclaurin series for the function defined by

$$f(x) = \int_0^1 \frac{e^{xt} - 1}{t}\,dt.$$

SOLUTION: Since we want the behavior as $x \to 0$, we form a series of $(e^{xt} - 1)/t$ in powers of x. We start by replacing x with xt in the series for e^x

$$e^{xt} \sim 1 + xt + \frac{x^2t^2}{2} + \frac{x^3t^3}{6} + \frac{x^4t^4}{24} + O(x^5).$$

Subtracting 1, and dividing by t gives us

$$\frac{e^{xt} - 1}{t} \sim x + \frac{x^2t}{2} + \frac{x^3t^2}{6} + \frac{x^4t^3}{24} + O(x^5) \text{ as } x \to 0.$$

Although the series involves two variables, only x is approaching 0 in the integral, whereas t will range from 0 to 1. We can now integrate the series with respect to t. There is no need to add an arbitrary constant, since this is a definite integral.

$$\int \frac{e^{xt} - 1}{t}\,dt \sim xt + \frac{x^2t^2}{4} + \frac{x^3t^3}{3 \cdot 6} + \frac{x^4t^4}{4 \cdot 24} + O(x^5t^5).$$

Currently this series is valid as $xt \to 0$, but we now evaluate the series at $t = 1$, and subtract the series evaluated at $t = 0$.

$$\int_0^1 \frac{e^{xt} - 1}{t}\,dt \sim \sum_{n=1}^{\infty} \frac{x^n}{n \cdot n!} = x + \frac{x^2}{4} + \frac{x^3}{18} + \frac{x^4}{96} + \frac{x^5}{600} + \cdots. \qquad \square$$

For an integral of the form $\int_a^b f(x,t)\,dt$, expanding $f(x,t)$ in a series of x can yield to a power series

$$f(x,t) \sim f_0(t) + f_1(t)x + f_2(t)x^2 + f_3(t)x^3 + \cdots,$$

where each $f_n(t)$ can be any function of t. This can make integration term by term more difficult than when $f_i(t)$ was always a polynomial.

Example 2.3
Find the asymptotic series as $x \to 0$ for the integral

$$\int_0^1 \frac{e^{xt}}{1 + t^2}\,dt.$$

SOLUTION: We find the series for the integrand using powers of x,

$$\frac{e^{xt}}{1+t^2} \sim \frac{1}{1+t^2} + \frac{xt}{1+t^2} + \frac{x^2t^2}{2(1+t^2)} + \frac{x^3t^3}{6(1+t^2)} + \cdots.$$

Now we integrate each term with respect to t from 0 to 1.

$$\int_0^1 \frac{dt}{1+t^2} = \tan^{-1} t\Big|_0^1 = \frac{\pi}{4}, \qquad \int_0^1 \frac{t\,dt}{1+t^2} = \frac{1}{2}\ln(1+t^2)\Big|_0^1 = \frac{\ln 2}{2},$$

$$\int_0^1 \frac{t^2\,dt}{2(1+t^2)} = \frac{4-\pi}{8}, \qquad \int_0^1 \frac{t^3\,dt}{6(1+t^2)} = \frac{1-\ln 2}{12},$$

$$\int_0^1 \frac{t^4\,dt}{24(1+t^2)} = \frac{3\pi-8}{288}, \ldots.$$

Note that many of these integrals required long division to integrate. Hence, we have as $x \to 0$,

$$\int_0^1 \frac{e^{xt}}{1+t^2}\,dt \sim \frac{\pi}{4} + \frac{\ln 2}{2}x + \frac{4-\pi}{8}x^2 + \frac{1-\ln 2}{12}x^3 + \frac{3\pi-8}{288}x^4 + \cdots. \qquad \square$$

As simplistic as this method is, it does have its limitations. For example, if one of the endpoints is $\pm\infty$, one has to worry about whether all of the integrals will converge for each term. For an example of this, see problem 16. The best way to deal with an infinite endpoint is to split the integral up into two pieces, where one piece has finite endpoints, and the other is known to converge to a constant.

Example 2.4

Find the behavior of the *exponential integral*

$$E_1(x) = \int_x^\infty \frac{e^{-t}}{t}\,dt$$

as $x \to 0$.

SOLUTION: The strategy is to rewrite the integral as two pieces

$$\int_x^1 \frac{e^{-t}}{t}\,dt + \int_1^\infty \frac{e^{-t}}{t}\,dt.$$

The latter integral converges by the comparison test, since it is clearly smaller than $\int_1^\infty e^{-t}\,dt = 1/e$. So we can denote the latter integral by a constant C_1.

The series for the first integral can now be found by integrating the asymptotic series term by term. Since

$$\frac{e^{-t}}{t} \sim \frac{1}{t} - 1 + \frac{t}{2!} - \frac{t^2}{3!} + \frac{t^3}{4!} - \cdots,$$

we can integrate term by term to obtain

$$\int \frac{e^{-t}}{t}\,dt \sim \ln(t) - t + \frac{t^2}{2\cdot 2!} - \frac{t^3}{3\cdot 3!} + \frac{t^4}{4\cdot 4!} + \cdots.$$

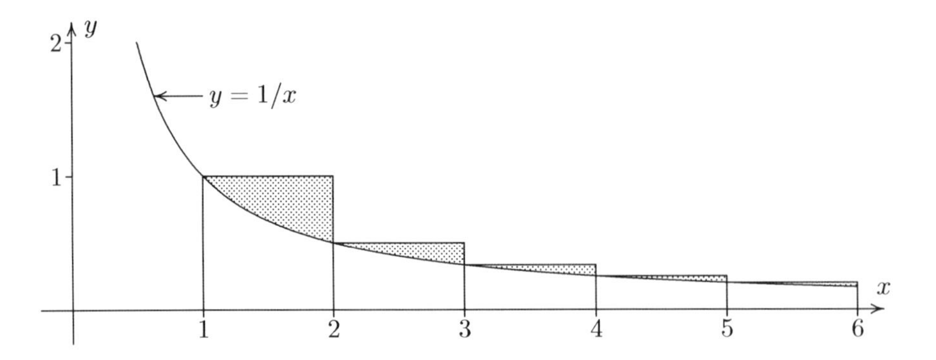

FIGURE 2.1: The area inside the rectangles represents $\sum_{i=1}^{n} 1/i$, whereas the area under the curve is $\ln(n)$. Hence, the shaded region has the area of γ as $n \to \infty$.

When we evaluate this series at $t = 1$, we get a convergent series that will sum to another constant C_2. For the lower endpoint, we replace the t with an x. Thus,

$$\int_x^\infty \frac{e^{-t}}{t}\,dt \sim -\ln(x) + C_3 + x - \frac{x^2}{2\cdot 2!} - \frac{x^3}{3\cdot 3!} + \frac{x^4}{4\cdot 4!} - \cdots,$$

where $C_3 = C_1 + C_2$.

Finding the approximation for the constant C_3 is a bit trickier, since to find C_1 we must approximate an improper integral. Simpson's rule cannot be used directly, but if we make the substitution $u = 1/t$, then the integral for C_1 becomes

$$C_1 = \int_1^\infty \frac{e^{-t}}{t}\,dt = \int_0^1 \frac{e^{-1/u}}{u}\,du.$$

The last integral appears to have a problem at the endpoint $u = 0$, but in fact $\lim_{u\to 0^+} e^{-1/u}/u = 0$, so this becomes a proper integral. Most graphing calculators can approximate this using Simpson's rule to produce $C_1 \approx 0.2193839344$.

The constant C_2 is easier to approximate, since it is the sum of the alternating series

$$C_2 = -1 + \frac{1}{2\cdot 2!} - \frac{1}{3\cdot 3!} + \frac{1}{4\cdot 4!} + \cdots \approx -0.7965995993.$$

Thus, $C_3 = C_1 + C_2 \approx -0.5772156649$. □

The absolute value of the constant C_3 in the last example turns out to be a very important mathematical constant that we will see again many times throughout this book. It is known as the *Euler-Mascheroni constant*, and is denoted by the Greek letter γ. Its formal definition is given by

$$\boxed{\gamma = \lim_{n\to\infty}\left(1 + \frac{1}{2} + \frac{1}{3} + \cdots + \frac{1}{n} - \ln(n)\right).} \tag{2.1}$$

Since we can consider $\ln(n)$ as the area under the curve $y = 1/x$ from $x = 1$ to $x = n$, γ can be represented graphically as the area between this curve and the sequence of rectangles representing the sum of $1/x$ up to $x = n$, as shown in figure 2.1.

This constant crops up in some very unexpected places, and is considered the third most important irrational constant in mathematics after π and e. Actually, this assumes that γ is indeed irrational, since this has never been formally proven. The first 50 digits are given by.

$$\gamma \approx 0.57721566490153286060651209008240243104215933593992.$$

To show that C_3 from the last example is indeed $-\gamma$, see problem 25 from section 2.3.

Problems for §2.1

1 Express the constant found in example 2.1 in terms of a definite integral. Hint: This constant can be expressed by the limit

$$\lim_{x\to 0}\left(\ln x + \int_x^1 \frac{\cos t}{t}\,dt\right).$$

Convert $\ln x$ into another integral with the same endpoints.

2 The *sine integral*, denoted by $\mathrm{Si}(x)$, is defined by the integral

$$\mathrm{Si}(x) = \int_0^x \frac{\sin t}{t}\,dt.$$

Find the asymptotic series as $x \to 0$ for this function.

For problems **3** through **14**: Find the first four terms of the asymptotic series as $x \to 0^+$ for the following integrals.

3 $\displaystyle\int_0^x e^{-t^2}\,dt$

4 $\displaystyle\int_0^x \sin(t^2)\,dt$

5 $\displaystyle\int_0^x \cosh(t^2)\,dt$

6 $\displaystyle\int_0^1 \frac{\sin xt^2}{t}\,dt$

7 $\displaystyle\int_0^1 \sqrt{1+xt^3}\,dt$

8 $\displaystyle\int_0^1 e^{xt^2}\,dt$

9 $\displaystyle\int_x^1 \frac{e^{xt}}{t}\,dt$

10 $\displaystyle\int_x^1 \frac{\sqrt{1+xt}}{t}\,dt$

11 $\displaystyle\int_x^1 \frac{\cos(xt)}{t}\,dt$

12 $\displaystyle\int_0^1 \frac{\cos(xt)}{1+t}\,dt$

13 $\displaystyle\int_0^1 \frac{e^{xt}}{1+t}\,dt$

14 $\displaystyle\int_0^1 \frac{e^t}{1+xt}\,dt$

15 Find the first three terms as $x \to 0$ for the integral

$$\int_0^1 \frac{\cos(xt)}{1+x^2 t}\,dt.$$

Hint: Multiply two series in x together.

16 Show that if we try to find the asymptotic series for the integral

$$f(x) = \int_0^\infty \frac{e^{-xt}}{(1+t^2)^2}\,dt$$

via expanding the integrand as a series in x, and integrating term by term, only the first three terms can be found. Find the first three terms. You may use the integral formula

$$\int \frac{dt}{(1+t^2)^2} = \frac{\tan^{-1} t}{2} + \frac{t}{2(1+t^2)} + C.$$

For problems **17** through **22**: Verify the the following integrals converge for $x > 0$, and find the asymptotic series as $x \to 0^+$ up to the x^5 term, using the technique of example 2.4. You do not have to solve for the constant term in the series.

17 $\displaystyle\int_x^\infty \frac{e^{-t}}{t^2}\,dt$

18 $\displaystyle\int_x^\infty \frac{e^{-t^2}}{t}\,dt$

19 $\displaystyle\int_x^\infty \frac{\sin(t)}{t^2}\,dt$

20 $\displaystyle\int_x^\infty \frac{\cos(t)}{t^2}\,dt$

21 $\displaystyle\int_x^\infty \frac{\tan^{-1}(t)}{t^2}\,dt$

22 $\displaystyle\int_x^\infty \frac{\sin(t)}{t^3}\,dt$

23 The Stieltjes integral function is defined by

$$S(x) = \int_0^\infty \frac{e^{-t}}{1+xt}.$$

Use the methods of this section to find the asymptotic series as $x \to 0^+$. Note that each term will involve an improper integral in t, but all of the integrals manage to converge. Does the resulting asymptotic series converge?

2.2 Repeated Integration by Parts

Standard integration by parts

$$\int u\,dv = uv - \int v\,du$$

is a powerful tool for finding the exact integral of a function. Repeated integration by parts is useful for integrating a product involving a polynomial. However, the same tool can be used to find the asymptotic series for integrals, even if the integrand does not involve a polynomial.

Example 2.5

Find the asymptotic series as $x \to \infty$ for the exponential integral

$$E_1(x) = \int_x^\infty \frac{e^{-t}}{t}\, dt.$$

SOLUTION: If we let $u = 1/t$, $dv = e^{-t}\,dt$, so $du = -1/t^2\,dt$ and $v = -e^{-t}$, we get

$$\int_x^\infty \frac{e^{-t}}{t}\,dt = \left.\frac{-e^{-t}}{t}\right|_x^\infty - \int_x^\infty \frac{e^{-t}}{t^2}\,dt = \frac{e^{-x}}{x} - \int_x^\infty \frac{e^{-t}}{t^2}\,dt.$$

Although this last integral is harder to solve than the original integral, it is progress because this integral is asymptotically smaller than the original integral as $x \to \infty$. So we have that

$$\int_x^\infty \frac{e^{-t}}{t}\,dt \sim \frac{e^{-x}}{x} \qquad \text{as} \quad x \to \infty.$$

This process can be repeated, always using the same $dv = e^{-t}\,dt$, and adjusting u accordingly.

$$\begin{aligned}\int_x^\infty \frac{e^{-t}}{t}\,dt &= \frac{e^{-x}}{x} - \left.\frac{-e^{-t}}{t^2}\right|_x^\infty + \int_x^\infty \frac{2e^{-t}}{t^3}\,dt \\ &= \frac{e^{-x}}{x} - \frac{e^{-x}}{x^2} + \left.\frac{-2e^{-t}}{t^3}\right|_x^\infty - \int_x^\infty \frac{6e^{-t}}{t^4}\,dt \\ &= \frac{e^{-x}}{x} - \frac{e^{-x}}{x^2} + \frac{2e^{-x}}{x^3} - \left.\frac{-6e^{-t}}{t^4}\right|_x^\infty + \int_x^\infty \frac{24e^{-t}}{t^5}\,dt...\end{aligned}$$

Notice that each term is smaller than the previous as $x \to \infty$. Thus, we have the asymptotic series of the integral as $x \to \infty$ to be

$$\int_x^\infty \frac{e^{-t}}{t}\,dt \sim e^{-x}\left(\frac{1}{x} - \frac{1}{x^2} + \frac{2}{x^3} - \frac{6}{x^4} + \frac{24}{x^5} - \cdots\right) \sim e^{-x}\sum_{n=0}^\infty \frac{(-1)^n n!}{x^{n+1}}. \quad \square$$

In general, if $u'(t)$ is asymptotically smaller than $u(t)$, then $u\,dv$ will usually be smaller than $v\,du$.

Example 2.6

Find the asymptotic series as $x \to \infty$ for the integral

$$\int_1^x \frac{e^t}{t}\,dt.$$

SOLUTION: The key difference here is the endpoints. As $x \to \infty$, the integral becomes unbounded, since the integrand itself goes to ∞ as $t \to \infty$. If we let $u = 1/t$, $dv = e^t\,dt$, then $du = -1/t^2\,dt$ and $v = e^t$. So

$$\int_1^x \frac{e^t}{t}\,dt = \left.\frac{e^t}{t}\right|_1^x - \int \frac{-e^t}{t^2}\,dt = \frac{e^x}{x} - e - \int \frac{-e^t}{t^2}\,dt$$

Once again, the new integral is asymptotically smaller than the original as $x \to \infty$. Since we are on the right track, we can repeat the process, each time using $dv = e^t\,dt$.

$$\begin{aligned}
\int_1^x \frac{e^t}{t}\,dt &= \frac{e^x}{x} - e + \left.\frac{e^t}{t^2}\right|_1^x - \int_x^\infty \frac{-2e^t}{t^3}\,dt \\
&= \frac{e^x}{x} - e + \frac{e^x}{x^2} - e + \left.\frac{2e^t}{t^3}\right|_1^x - \int_x^\infty \frac{-6e^t}{t^4}\,dt \\
&= \frac{e^x}{x} - e + \frac{e^x}{x^2} - e + \frac{2e^x}{x^3} - 2e + \left.\frac{6e^t}{t^4}\right|_1^x - \int_x^\infty \frac{-24e^t}{t^5}\,dt.
\end{aligned}$$

Notice that this time we have constants in the integration by parts. However, each constant would be asymptotically smaller than *all* of the exponential terms of the series. Hence, these constants are *subdominant* to the series, and can be ignored. Thus, we have

$$\int_1^x \frac{e^t}{t}\,dt \sim e^x\left(\frac{1}{x} + \frac{1}{x^2} + \frac{2}{x^3} + \frac{6}{x^4} + \frac{24}{x^5} + \cdots\right). \tag{2.2}$$

□

Example 2.7

Equation 1.21 introduced the Stieltjes integral function

$$S(x) = \int_0^\infty \frac{e^{-t}}{1+xt}\,dt.$$

Use integration by parts to derive the asymptotic series as $x \to 0^+$.

SOLUTION: If we let $u = 1/(1+xt)$, $dv = e^{-t}$, then $v = -e^{-t}$ and $du = -x/(1+xt)^2\,dt$.

$$\begin{aligned}
\int_0^\infty \frac{e^{-t}}{1+xt}\,dt &= \left.\frac{-e^{-t}}{1+xt}\right|_0^\infty - \int_0^\infty \frac{xe^{-t}}{(1+xt)^2}\,dt \\
&= 1 + \left.\frac{xe^{-t}}{(1+xt)^2}\right|_0^\infty + \int_0^\infty \frac{2x^2e^{-t}}{(1+xt)^3}\,dt \\
&= 1 - x - \left.\frac{2x^2e^{-t}}{(1+xt)^3}\right|_0^\infty - \int_0^\infty \frac{6x^3e^{-t}}{(1+xt)^4}\,dt \\
&= 1 - x + 2x^2 + \left.\frac{6x^3e^{-t}}{(1+xt)^4}\right|_0^\infty + \int_0^\infty \frac{24x^4e^{-t}}{(1+xt)^5}\,dt
\end{aligned}$$

$$= 1 - x + 2x^2 - 6x^3 - \left.\frac{24x^4 e^{-t}}{(1+xt)^5}\right|_0^\infty - \int_0^\infty \frac{120x^5 e^{-t}}{(1+xt)^6}\,dt.$$

The terms decrease in magnitude as $x \to 0$, so we have

$$S(x) \sim 1 - x + 2x^2 - 6x^3 + 24x^4 - 120x^5 \sim \sum_{n=0}^{\infty} (-1)^n n!\, x^n \text{ as } x \to 0^+.$$

□

Example 2.8
Find the behavior of the integral

$$\int_0^x e^{-t^2}\,dt \qquad \text{as } x \to \infty.$$

SOLUTION: It is clear that the integrand goes to 0 faster than e^{-t} as $t \to \infty$, so $\int_0^\infty e^{-t^2}\,dt$ converges to some constant C. Thus, as $x \to \infty$, the leading behavior will be the constant C. But to find the other terms in the asymptotic expansion, we can write

$$\int_0^x e^{-t^2}\,dt = \int_0^\infty e^{-t^2}\,dt - \int_x^\infty e^{-t^2}\,dt,$$

and apply integration by parts on the final integral. Even though we cannot integrate $e^{-t^2}\,dt$, we *can* integrate $dv = te^{-t^2}\,dt$, yielding $v = -e^{-t^2}/2$. So $u = 1/t$, and $du = -1/t^2\,dt$.

$$\begin{aligned}
\int_x^\infty e^{-t^2}\,dt &= \left.\frac{-e^{-t^2}}{2t}\right|_x^\infty - \int_x^\infty \frac{e^{-t^2}}{2t^2}\,dt \\
&= \frac{e^{-x^2}}{2x} - \left.\frac{-e^{-t^2}}{4t^3}\right|_x^\infty + \int_x^\infty \frac{3e^{-t^2}}{4t^4}\,dt \\
&= \frac{e^{-x^2}}{2x} - \frac{e^{-x^2}}{4x^3} + \left.\frac{-3e^{-t^2}}{8t^5}\right|_x^\infty - \int_x^\infty \frac{15e^{-t^2}}{8t^6}\,dt \\
&= \frac{e^{-x^2}}{2x} - \frac{e^{-x^2}}{4x^3} + \frac{3e^{-x^2}}{8x^5} - \left.\frac{-15e^{-t^2}}{16t^7}\right|_x^\infty + \int_x^\infty \frac{105e^{-t^2}}{16t^8}\,dt.
\end{aligned}$$

Again, the terms are decreasing in magnitude as $x \to \infty$, so we have

$$\int_0^x e^{-t^2}\,dt \sim C - e^{-x^2}\left(\frac{1}{2x} - \frac{1}{4x^3} + \frac{3}{8x^5} - \frac{15}{16x^7} + \frac{105}{32x^9} - \cdots\right).$$

It is actually possible to compute C by forming a double integral of its square and converting to polar coordinates.

$$C^2 = \left[\int_0^\infty e^{-x^2}\,dx\right]\left[\int_0^\infty e^{-y^2}\,dy\right]$$

$$= \int_0^\infty \int_0^\infty e^{-x^2-y^2}\, dx\, dy$$

$$= \int_0^{\pi/2} \int_0^\infty r e^{-r^2}\, dr\, d\theta$$

$$= \frac{-e^{r^2}}{2}\Big|_0^\infty \cdot \frac{\pi}{2} = \frac{\pi}{4}.$$

Since $C^2 = \pi/4$, we see that

$$\int_0^\infty e^{-x^2}\, dx = \frac{\sqrt{\pi}}{2}. \tag{2.3}$$

The normalized form of this integral, scaled by $2/\sqrt{\pi}$, is called the *error function*, or $\mathrm{erf}(x)$, and is very useful in statistics. Its asymptotic expansion is

$$\mathrm{erf}(x) = \frac{2}{\sqrt{\pi}} \int_0^x e^{-t^2}\, dt$$
$$\sim 1 - \frac{2e^{-x^2}}{\sqrt{\pi}} \left(\frac{1}{2x} - \frac{1}{4x^3} + \frac{3}{8x^5} - \frac{15}{16x^7} + \frac{105}{32x^9} - \cdots \right)$$

as $x \to \infty$. ▯

2.2.1 Optimal asymptotic approximation

The series produced by repeated integration by parts are often divergent for all x. At first glance, this appears to be a totally useless series, but there is a simple way to gain a good approximation from the series in spite of its divergent nature. Consider the asymptotic power series,

$$f(x) \sim a_0 + a_1 x + a_2 x^2 + a_3 x^3 + \cdots \qquad \text{as } x \to 0.$$

If we terminate the series after the x^n term to form the polynomial approximation

$$P_n(x) = a_0 + a_1 x + a_2 x^2 + a_3 x^3 + \cdots + a_n x^n,$$

then the error in this approximation is $f(x) - P_n(x) \sim a_{n+1} x^{n+1}$. Thus, the error can be approximated by the first term left out.

In the case of a divergent asymptotic series, typically the terms will begin to decrease, but then the terms will start increasing in magnitude and grow without bounds. For example, plugging in $x = 1/10$ into the series for the Stieltjes integral function $S(x) = \int_0^\infty e^{-t}/(1 + xt)\, dt$ produces

$$S(1/10) = 1 - \frac{1}{10} + \frac{2}{100} - \frac{6}{1000} + \frac{24}{10000} - \frac{120}{100000} + \cdots.$$

TABLE 2.1: This table shows the results of adding more and more terms of the divergent series for the Stieltjes integral function. The true value of $S(1/10)$ is 0.9156333393978808181876.

n	$P_n(0.1)$	error
0	1	0.08436666
1	0.9	0.01563334
2	0.92	0.00436666
3	0.914	0.00163334
4	0.9164	0.00076666
5	0.9152	0.00043334
6	0.91592	0.00028666
7	0.915416	0.00021734
8	0.9158192	0.00018586
9	0.91545632	0.00017702
10	0.9158192	0.00018586
11	0.915420032	0.00021331

Clearly the first terms are decreasing rapidly, so initially, each term we add will reduce the error between the true function $S(x)$ and its polynomial approximation. Table 2.1 shows the result of adding more and more terms, and the corresponding error. Eventually, the terms will start increasing, since the series is divergent. Yet the error is as low as 0.000177 by including up to the x^9 term.

This example suggests a simple strategy for getting the most accuracy out of a divergent series. Since the error is approximately the first term left out, we search the series for the term with the smallest magnitude. We then add all of the terms up to, *but not including*, this term. In case of a tie, as in this present example ($9!/10^9 = 10!/10^{10}$), we add one term but not the other. This strategy is called the *optimal asymptotic approximation* of the divergent series. Figure 2.2 shows the graph of the optimal asymptotic approximation for the asymptotic series of $S(x)$. For small values of x, the accuracy is impressive, even though the series diverges.

Example 2.9

In example 2.8, we defined the error function, and found that for $x \to \infty$,

$$\operatorname{erf}(x) \sim 1 - \frac{2e^{-x^2}}{\sqrt{\pi}}\left(\frac{1}{2x} - \frac{1}{4x^3} + \frac{3}{8x^5} - \frac{15}{16x^7} + \frac{105}{32x^9} - \cdots\right).$$

Use the optimal asymptotic approximation to find $\operatorname{erf}(2)$.

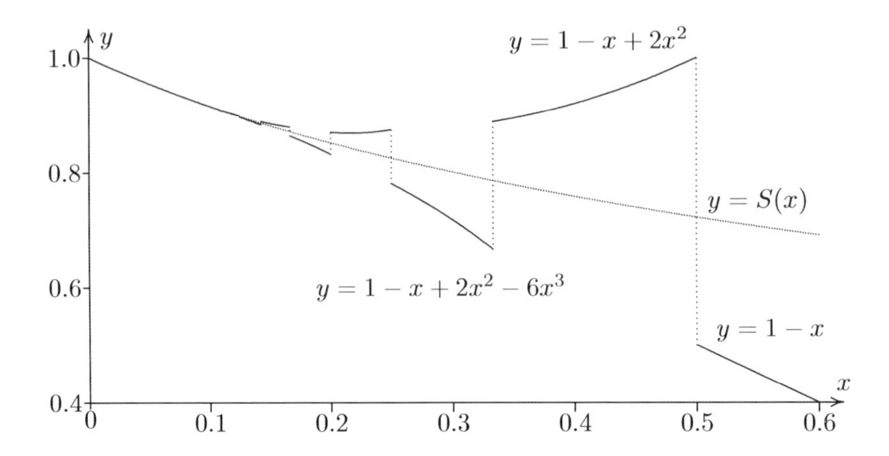

FIGURE 2.2: Plot of the optimal asymptotic approximation to the Stieltjes integral function $S(x)$, in comparison to $S(x)$. The graphs of the two appear to be indistinguishable for $x < 0.1$. Compare this to figure 1.4.

SOLUTION: Plugging in $x = 2$ into each term of the series inside the parenthesis, we find the terms are

$$\frac{1}{4} = 0.25, \qquad \frac{-1}{32} \approx -0.03125, \qquad \frac{3}{256} \approx 0.0117188,$$

$$\frac{-15}{2048} \approx -0.00732422, \qquad \frac{105}{16384} \approx 0.00640869, \qquad \frac{-945}{131072} \approx -0.00720978.$$

So the 5th term is the smallest in magnitude, hence we use the first four terms for the optimal asymptotic approximation. We find that

$$1 - \frac{2e^{-4}}{\sqrt{\pi}}\left(\frac{1}{4} - \frac{1}{32} + \frac{3}{256} - \frac{15}{2048}\right) \approx 0.99538827524.$$

The exact value of $\operatorname{erf}(2)$ is $0.99532226502\ldots$, so we have three places of accuracy. ∎

It should be noted that there is a Taylor series for $\operatorname{erf}(x)$, found by integrating the series for e^{-t^2} at $t = 0$.

$$\operatorname{erf}(x) = \frac{2}{\sqrt{\pi}}\left(x - \frac{x^3}{3} + \frac{x^5}{2! \cdot 5} - \frac{x^7}{3! \cdot 7} + \frac{x^9}{4! \cdot 9} - \frac{x^{11}}{5! \cdot 11} + \cdots\right).$$

This series converges for all x, so we can ask how well does this series do at $x = 2$? It takes no fewer than 14 terms (up to the x^{29} term) to achieve the same 3 places of accuracy. Hence, a divergent series can outperform a convergent series, as shown in figure 2.3.

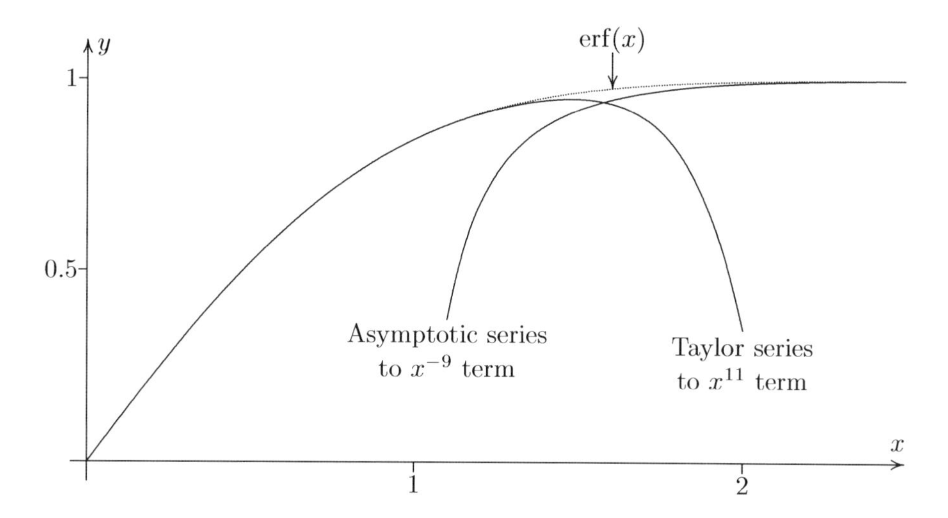

FIGURE 2.3: The function $\mathrm{erf}(x)$ has a convergent Taylor series about $x = 0$ and a divergent asymptotic series about $x = \infty$. For $x > 2$, the asymptotic series outperforms the Taylor series.

Problems for §2.2

For problems **1** through **11**:

Part a: Use repeated integration by parts to find the first three terms of the asymptotic series as $x \to \infty$.

Part b: Find the next two terms of the series.

1 $\mathrm{Ci}(x) = -\int_x^\infty \frac{\cos t}{t}\, dt.$

2 $\mathrm{Si}(x) = \int_0^x \frac{\sin t}{t}\, dt.$ Hint: Use the fact that $\mathrm{Si}(\infty) = \pi/2$.

3 $\int_x^\infty \frac{\cos t}{t^2}\, dt.$

4 $\int_x^\infty \frac{e^{-t}}{t^2}\, dt.$

5 $\int_1^x \frac{e^t}{t^2}\, dt.$

6 $\int_x^\infty \sqrt{t}e^{-t}\, dt.$

7 $\displaystyle\int_1^x e^{t^2}\,dt$. Hint: Always let $dv = 2te^{t^2}$.

8 $\displaystyle\int_0^x e^{t^2}\,dt$. Hint: Consider $\displaystyle\int_0^1 e^{t^2}\,dt + \int_1^x e^{t^2}\,dt$.

9 $\displaystyle\int_1^x e^{t^3}\,dt$.

10 $\displaystyle\int_0^\infty \frac{e^{-t}}{x+t}\,dt$.

11 $\displaystyle\int_0^\infty \frac{e^{-xt}}{1+t^2}\,dt$.

For problems **12** through **14**: Use repeated integration by parts to find the first five terms of the asymptotic series as $x \to 0$.

12 $\displaystyle\int_0^\infty e^{-t}\ln(1+xt)\,dt$.

13 $\displaystyle\int_0^\infty e^{-t}\sqrt{1+xt}\,dt$.

14 $\displaystyle\int_0^\infty \frac{e^{-t}}{\sqrt{1+xt}}\,dt$.

For problems **15** through **22**: Use the optimal asymptotic approximation to approximate the divergent asymptotic series as $x \to \infty$ at (part a) $x = 3$ and (part b) $x = 6$.

15 $\displaystyle\int_x^\infty \frac{e^{-t}}{t}\,dt \sim e^{-x}\sum_{n=0}^\infty \frac{(-1)^n n!}{x^{n+1}}$

16 $\displaystyle\int_1^x \frac{e^{t}}{t}\,dt \sim e^{x}\sum_{n=0}^\infty \frac{n!}{x^{n+1}}$

17 $\displaystyle\sum_{n=0}^\infty \frac{(-1)^n(2n)!}{n!(2x)^n}$

18 $\displaystyle\sum_{n=0}^\infty \frac{(-1)^n(2n)!}{(6x)^n}$

19 $\displaystyle\sum_{n=1}^\infty \frac{(-1)^n(n+1)^n}{x^n}$

20 $\displaystyle\sum_{n=1}^\infty \frac{(3n)!}{(2n)!(3x)^n}$

21 $\displaystyle\sum_{n=1}^\infty \frac{(-1)^n(2n)!}{(2n+1)^n x^n}$

22 $\displaystyle\sum_{n=1}^\infty \frac{(n!)^2}{(n+1)^n x^n}$

23 One may have noticed a similarity between the asymptotic series for $E_1(x)$ as $x \to \infty$ and the Stieltjes integral function $S(x)$ as $x \to 0$. In fact, the series for $e^{1/x}/xE_1(1/x)$ will be identical to the asymptotic series for $S(x)$. Prove that indeed,

$$S(x) = \int_0^\infty \frac{e^{-t}}{1+xt}\,dt = \frac{e^{1/x}}{x}E_1\left(\frac{1}{x}\right).$$

2.3 Laplace's Method

Many of the integrals from section 2.2 involved exponentials, and in fact this exponential function dominated all of the terms in the series. For example, all of the terms in the series for example 2.6,

$$\int_1^x \frac{e^t}{t}\,dt,$$

was dominated by e^x. In fact, the lower endpoint at 1 was irrelevant, since the constants were *subdominant* to the series. This illustrates an important principle:

For integrals involving an exponential function, the only significant part of the integral is the neighborhood where the exponent is at its maximum.

For example, the integral in example 2.6, the exponent t is largest over the interval from 1 to x when $t = x$. So in fact, the only significant part of the integral is

$$\int_{x-\epsilon}^x \frac{e^t}{t}\,dt,$$

where ϵ is any small positive value. To see this, we have

$$\int_1^x \frac{e^t}{t}\,dt = \int_1^{x-\epsilon} \frac{e^t}{t}\,dt + \int_{x-\epsilon}^x \frac{e^t}{t}\,dt.$$

but the integral from 1 to $x - \epsilon$ will have all terms of order $e^{x-\epsilon}$. This part is asymptotically small compared to *all* of the terms in equation 2.2.

Now that the integral is over a very small interval, t will always be close to x as we do the integration. The plan is to replace part of the integrand with its Taylor series. However, it is more convenient if we first do a simple linear substitution which sends the maximum of the exponent to the origin, and use a Maclaurin series. In this case, we would let $t = x - s$, so $dt = -ds$.

$$\int_1^x \frac{e^t}{t}\,dt \sim \int_0^\epsilon \frac{e^{x-s}}{x-s}\,ds = e^x \int_0^\epsilon \frac{e^{-s}}{x-s}\,ds.$$

We now replace $1/(x-s)$ with its Maclaurin series in powers of s:

$$\frac{1}{x-s} = \frac{1}{x} + \frac{s}{x^2} + \frac{s^2}{x^3} + \frac{s^3}{x^4} + \cdots,$$

giving

$$e^x\left(\int_0^\epsilon \frac{e^{-s}}{x}\,ds + \int_0^\epsilon \frac{se^{-s}}{x^2}\,ds + \int_0^\epsilon \frac{s^2e^{-s}}{x^3}\,ds + \int_0^\epsilon \frac{s^3e^{-s}}{x^4}\,ds + \cdots\right).$$

Now, for each of these integrals, the integrand is defined for all s. Since the only significant part of these integrals is when s is close to 0, we can change the upper endpoint without affecting the asymptotic series. Thus, we can in fact change the upper endpoints to ∞, which will only add a subdominant function.

$$e^x\left(\int_0^\infty \frac{e^{-s}}{x}\,ds + \int_0^\infty \frac{se^{-s}}{x^2}\,ds + \int_0^\infty \frac{s^2e^{-s}}{x^3}\,ds + \int_0^\infty \frac{s^3e^{-s}}{x^4}\,ds + \cdots\right).$$

Each of these integrals can now be evaluated via repeated integration by parts.

$$\int_0^\infty e^{-s}\,ds = -e^{-s}\Big|_0^\infty = 1, \qquad \int_0^\infty se^{-s}\,ds = -(s+1)e^{-s}\Big|_0^\infty = 1,$$

$$\int_0^\infty s^2e^{-s}\,ds = -(s^2+2s+2)e^{-s}\Big|_0^\infty = 2,$$

$$\int_0^\infty s^3e^{-s}\,ds = -(s^3+3s^2+6s+6)e^{-s}\Big|_0^\infty = 6.$$

We can prove by induction that the pattern continues (see problem 1). Thus, we have

$$\boxed{\int_0^\infty s^n e^{-s}\,ds = n!.} \tag{2.4}$$

By putting all of the pieces together, we have

$$\int_1^x \frac{e^t}{t}\,dt \sim e^x\left(\frac{1}{x} + \frac{1}{x^2} + \frac{2!}{x^3} + \frac{3!}{x^4} + \cdots\right).$$

This gives us the same series as example 2.6, but using a different technique, called *Laplace's method*. Let us recap the steps that were taken:

1) Determine the point in the interval of integration where the exponent of the exponential is at a maximum.

2) Change the range of the integral to cover just the vicinity of the maximum. This will not affect the asymptotic series, since the difference will be subdominant to the entire series.

3) (Optional) Make a linear substitution which sends the maximum point to 0. This allows a Maclaurin series to be used in the next step.

4) Use the fact that t must be close to a known value to convert the remaining part of the integrand into a Taylor or Maclaurin series. Then integrate term by term, so that each term involves an integral.

5) Change the range of the integrals again to a more convenient endpoint, such as $\pm\infty$. Again, this will only change the integral by a subdominant amount.

6) Use either equation 2.4 or one of the integrals

$$\boxed{\int_{-\infty}^{\infty} s^{2n} e^{-s^2}\, ds = \frac{(2n)!\sqrt{\pi}}{2^{2n} n!}} \tag{2.5}$$

or

$$\boxed{\int_{0}^{\infty} s^{2n+1} e^{-s^2}\, ds = \frac{n!}{2}} \tag{2.6}$$

to evaluate the integrals producing the asymptotic series. (See problems 2 and 3 for the derivation of these integrals.)

Usually, the maximum of the exponent occurs at one of the endpoints of integration, but occasionally this maximum occurs in the interior of the interval.

Example 2.10

Find the asymptotic series for

$$\int_{-1}^{2} \frac{e^{-xt^2}}{t+2}\, dt.$$

SOLUTION: The maximum of the exponential component clearly occurs at $t = 0$. So the asymptotic series will be unaltered if we adjust the endpoints of the integral to

$$\int_{-\epsilon}^{\epsilon} \frac{e^{-xt^2}}{t+2}\, dt.$$

Over this region, it is valid to replace $1/(t+2)$ with its Maclaurin series

$$\frac{1}{t+2} = \frac{1}{2} - \frac{t}{4} + \frac{t^2}{8} - \frac{t^3}{16} + \cdots.$$

Thus, we have

$$\int_{-\epsilon}^{\epsilon} \frac{e^{-xt^2}}{t+2} \sim \int_{-\epsilon}^{\epsilon} \frac{1}{2} e^{-xt^2}\, dt - \int_{-\epsilon}^{\epsilon} \frac{t}{4} e^{-xt^2}\, dt + \int_{-\epsilon}^{\epsilon} \frac{t^2}{8} e^{-xt^2}\, dt - \int_{-\epsilon}^{\epsilon} \frac{t^3}{16} e^{-xt^2}\, dt + \cdots.$$

At this point, we can make an important observation. For those integrals with an odd power of t, we are integrating an odd function over the interval $-\epsilon \le t \le \epsilon$. By symmetry, all of these integrals will evaluate to 0. For the other integrals, we can keep the symmetry by changing the endpoints from $-\infty$ to ∞. So we have

$$\int_{-\epsilon}^{\epsilon} \frac{e^{-xt^2}}{t+2} \sim \int_{-\infty}^{\infty} \frac{1}{2} e^{-xt^2}\, dt + \int_{-\infty}^{\infty} \frac{t^2}{8} e^{-xt^2}\, dt + \int_{-\infty}^{\infty} \frac{t^4}{32} e^{-xt^2}\, dt + \cdots.$$

Now we substitute $u = t\sqrt{x}$ so that the integrals will be in the form of equation 2.5. We have

$$\int_{-\epsilon}^{\epsilon} \frac{e^{-xt^2}}{t+2} \sim \int_{-\infty}^{\infty} \frac{1}{2\sqrt{x}} e^{-u^2}\, du + \int_{-\infty}^{\infty} \frac{u^2}{8x^{3/2}} e^{-u^2}\, dt + \int_{-\infty}^{\infty} \frac{u^4}{32x^{5/2}} e^{-u^2}\, du + \cdots.$$

Using equation 2.5, we can evaluate each of the integrals to be

$$\int_{-\epsilon}^{\epsilon} \frac{e^{-xt^2}}{t+2} \sim \frac{\sqrt{\pi}}{2\sqrt{x}} + \frac{\sqrt{\pi}}{16x^{3/2}} + \frac{3\sqrt{\pi}}{128x^{5/2}} + \cdots.$$

This gives us the complete asymptotic series for the original integral:

$$\int_{-1}^{2} \frac{e^{-xt^2}}{t+2}\, dt \sim \sum_{n=0}^{\infty} \frac{(2n)!\sqrt{\pi}}{2^{4n+1} n! x^{n+1/2}}.$$

□

Although the Laplace's method for finding the asymptotic series is straight forward, the most confusing part is validating the step where the endpoints are changed. As long as no new maximums are introduced or destroyed in the exponential component, then changing the endpoints will only affect the result by a *subdominant* function as $x \to \infty$.

Example 2.11

Find the first two terms of the asymptotic series as $x \to \infty$ for

$$f(x) = \int_0^{\pi} e^{x \sin t}\, dt.$$

SOLUTION: The maximum of the exponent occurs at $t = \pi/2$, but if we adjust the endpoints to $\pm\infty$ as this point, we would introduce an infinite number of maximums ($t = 5\pi/2, 9\pi/2$, etc.) So before we change the endpoints to $\pm\infty$, we must have only *one* maximum for the exponent.

We first change the interval of integration to the neighborhood of the maximum $t = \pi/2$.

$$f(x) \sim \int_{\pi/2-\epsilon}^{\pi/2+\epsilon} e^{x \sin t}\, dt.$$

Next, we substitute $u = t - \pi/2$ to shift the maximum to the origin.

$$f(x) \sim \int_{-\epsilon}^{\epsilon} e^{x \sin(u+\pi/2)}\, du = \int_{-\epsilon}^{\epsilon} e^{x \cos u}\, du.$$

Since the interval of integration is close to 0, we can replace $\cos u$ with its Maclaurin series.

$$f(x) \sim \int_{-\epsilon}^{\epsilon} e^{x - xu^2/2! + xu^4/4! - xu^6/6! + \cdots}\, du.$$

At this point, we can use the Maclaurin series for the exponential function to simplify this expression. However, we must be careful not to change the location and nature of the maximum point. Because of this, *we must keep the first non-constant term in the exponent.* For example, x is considered a constant term, since it does not depend on the integration variable u. The next term, $-xu^2/2$, is the first non-constant term, so it must stay in an exponent. The other terms can be exponentiated via the Maclaurin series. That is, we can replace $e^{xu^4/4!-xu^6/6!+\cdots}$ with

$$1+\left(\frac{xu^4}{4!}-\frac{xu^6}{6!}+\cdots\right)+\frac{1}{2!}\left(\frac{xu^4}{4!}-\frac{xu^6}{6!}+\cdots\right)^2 + \frac{1}{3!}\left(\frac{xu^4}{4!}-\frac{xu^6}{6!}+\cdots\right)^3+\cdots.$$

Luckily, we only need two terms of the asymptotic series for $f(x)$, so we will only keep the leading two terms of this series. Hence

$$f(x)\sim\int_{-\epsilon}^{\epsilon} e^x e^{-xu^2/2}\left(1+\frac{xu^4}{24}+\cdots\right)du.$$

Note that $u=0$ is still the maximum for the exponential function, but now it is the only maximum. Thus, we can safely adjust the endpoints to $\pm\infty$, which only adds a subdominant function. We can also move the e^x to the outside of the integral, since it does not depend on u.

$$f(x)\sim e^x\int_{-\infty}^{\infty} e^{-xu^2/2}\left(1+\frac{xu^4}{24}+\cdots\right)du.$$

At this point the substitution $u=s\sqrt{2/x}$ will convert the problem to one where equation 2.5 can apply to each term.

$$f(x)\sim e^x\int_{-\infty}^{\infty} e^{-s^2}\left(1+\frac{s^4}{6x}+\cdots\right)\sqrt{\frac{2}{x}}\,ds\sim e^x\sqrt{\frac{2}{x}}\left(\sqrt{\pi}+\frac{\sqrt{\pi}}{8x}+\cdots\right).$$ □

So far, we have only dealt with cases where the maximum of the exponent does not depend on x. The next example shows the situation for which the maximum depends on the value of x. This is called a *movable maximum.*

Example 2.12

Use the integral representation of the factorial function,

$$n! = \int_0^{\infty} t^n e^{-t}\,dt,$$

to find the leading behavior of $n!$ as $n\to\infty$.

SOLUTION: Since there are two power functions involved, both of which get large as $n \to \infty$, we must first combine them into one exponential function so we can find the maximum of the exponent.

$$t^n e^{-t} = e^{n \ln t} e^{-t} = e^{(n \ln t - t)}.$$

So the maximum occurs when $(n \ln t - t)' = n/t - 1 = 0$, or when $t = n$. Thus, the dominant contribution of the integral is

$$n! \sim \int_{n-\epsilon}^{n+\epsilon} e^{n \ln t - t}\, dt.$$

Note that the location of the maximum changes with n.

By making the linear substitution $u = t - n$, we move the maximum to the origin, eliminating the problem of the moving maximum.

$$n! \sim \int_{-\epsilon}^{\epsilon} e^{n \ln(n+u) - (n+u)}\, du.$$

Using a Maclaurin series, we see that when u is close to 0,

$$n \ln(n+u) - (n+u) \sim (n \ln n - n) - \frac{u^2}{2n} + \frac{u^3}{3n^2} - \frac{u^4}{4n^3} + \cdots .$$

The first non-constant term is $-u^2/2n$, so the terms beyond this point can be expanded using Maclaurin series for e^x. This gives us $n! \sim$

$$\int_{-\epsilon}^{\epsilon} e^{(n \ln n - n - u^2/(2n))} \left(1 + \left(\frac{u^3}{3n^2} - \frac{u^4}{4n^3} \right) + \frac{1}{2!} \left(\frac{u^3}{3n^2} - \frac{u^4}{4n^3} \right)^2 + \cdots \right) dt.$$

We now can replace the endpoints of the integral with $-\infty$ and ∞, introducing only subdominant terms. To get the leading order of the integral, we integrate

$$e^{n \ln n - n} \int_{-\infty}^{\infty} e^{-u^2/(2n)}\, dt.$$

Substituting $s = u/\sqrt{2n}$ gets the integral into a form which can utilize equation 2.3.

$$n! \sim n^n e^{-n} \sqrt{2n} \int_{-\infty}^{\infty} e^{-s^2}\, ds = n^n e^{-n} \sqrt{2\pi n}, \qquad \text{as } n \to \infty. \tag{2.7}$$

□

Equation 2.7 is known as *Stirling's formula*. Note that the integral in equation 2.4 can be used to take factorials of fractional n, as long as $n > -1$.

Because the singularity is at -1, it is convenient to create a new function shifted to the right one unit, called the *gamma function*,

$$\Gamma(z) = \int_0^\infty t^{z-1} e^{-t}\, dt. \tag{2.8}$$

This integral converges as long as the real part of z is positive. Because of the importance of the gamma function, we will devote a subsection to studying its properties.

2.3.1 Properties of $\Gamma(x)$

Since we defined the gamma function as a shift of the factorial function, we have that for positive integers, $\Gamma(n) = (n-1)!$. But the integral in equation 2.8 gives a way of computing the gamma function for fractions. For example, $\Gamma(1/2)$ can be computed to be

$$\int_0^\infty t^{-1/2} e^{-t}\, dt = \int_0^\infty 2e^{-s^2}\, ds = \sqrt{\pi}$$

using the substitution $t = s^2$.

Since $n! = n(n-1)!$, there is a similar recursion formula for the gamma function:

$$\Gamma(n+1) = n\Gamma(n). \tag{2.9}$$

Although the integral in equation 2.8 only converges for positive z, we can use this recursion formula to define the gamma function for negative values. For example, since $\Gamma(1/2) = (-1/2)\Gamma(-1/2)$, we have that $\Gamma(-1/2) = -2\sqrt{\pi}$. Figure 2.4 shows a graph of the gamma function. There are vertical asymptotes at the non-positive integers, but the gamma function is defined for all other complex numbers via equations 2.8 and 2.9.

There are other expressions for $\Gamma(z)$ that are valid everywhere except at the singularities $z = 0, -1, -2, \ldots$. One of these is due to Euler:

$$\Gamma(z) = \frac{1}{z} \prod_{n=1}^{\infty} \left[\left(1 + \frac{1}{n}\right)^z \left(\frac{n}{n+z}\right) \right]. \tag{2.10}$$

Here, using the giant Π symbol is similar to the Σ notation, indicating that we are to take the *product* of the expressions instead of the sum. In order for an infinite product to converge, the expressions must approach 1 instead of 0. In fact $\prod_{n=1}^\infty a_n$ will converge if, and only if, $\sum_{n=1}^\infty (a_n - 1)$ converges. If we plug in $z = 2$, we get that

$$\Gamma(2) = \frac{1}{2}\left[2^2 \frac{1}{3}\right] \cdot \left[\left(\frac{3}{2}\right)^2 \frac{2}{4}\right] \cdot \left[\left(\frac{4}{3}\right)^2 \frac{3}{5}\right] \cdot \left[\left(\frac{5}{4}\right)^2 \frac{4}{6}\right] \cdots.$$

It is not hard to see that this product telescopes, so the product after n terms is $(n+1)/(n+2)$. Hence, this product converges to 1.

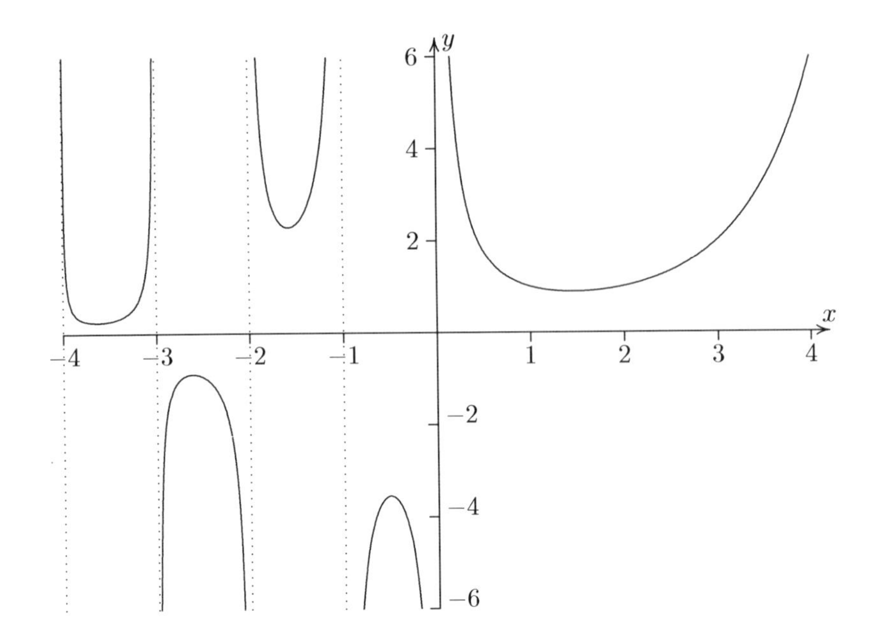

FIGURE 2.4: Graph of $\Gamma(x)$ for x between -4 and 4. Note that there are vertical asymptotes at $x = 0, -1, -2, \ldots$

There are various identities for the gamma function besides the recursion formula $\Gamma(n+1) = n\Gamma(n)$.

Example 2.13

Derive a "double angle" formula for the gamma function. That is, express $\Gamma(2n)$ in terms of other gamma functions.

SOLUTION: If n is a positive integer, then

$$\Gamma(2n) = (2n-1)! = 1 \cdot 2 \cdot 3 \cdots (2n-2) \cdot (2n-1).$$

Because of the $2n$'s that appear, let us divide each of the factors on the right hand side by 2. Since there are $2n - 1$ factors, both sides will be divided by 2^{2n-1}.

$$\frac{\Gamma(2n)}{2^{2n-1}} = \frac{1}{2} \cdot 1 \cdot \frac{3}{2} \cdots (n-1) \cdot \left(n - \frac{1}{2}\right).$$

We can group the integers together, and also multiply both sides by $\Gamma(1/2) = \sqrt{\pi}$.

$$\frac{\sqrt{\pi}\,\Gamma(2n)}{2^{2n-1}} = \Gamma\left(\frac{1}{2}\right) \cdot \frac{1}{2} \cdot \frac{3}{2} \cdots \left(n - \frac{1}{2}\right) \cdot 1 \cdot 2 \cdots (n-1).$$

An important technique associated with the gamma function is repeated use of the recursion formula, which has the effect of a "zipper." Since $\Gamma(1/2) \cdot (1/2) =$

$\Gamma(3/2)$, this sets up the simplification of the next factor: $\Gamma(3/2)\cdot(3/2) = \Gamma(5/2)$, and so on. All of the fractions on the right hand side then "zip up" to form

$$\frac{\sqrt{\pi}\,\Gamma(2n)}{2^{2n-1}} = \Gamma\left(n+\frac{1}{2}\right)\cdot(n-1)!.$$

Finally, replacing $(n-1)!$ with $\Gamma(n)$, and isolating the $\Gamma(2n)$ gives us

$$\Gamma(2n) = \frac{2^{2n-1}\Gamma(n)\Gamma(n+\frac{1}{2})}{\sqrt{\pi}}. \tag{2.11}$$

This formula is referred to as the *duplication formula* for the gamma function. Although we demonstrated this identity for positive integers, it is in fact valid for all complex numbers provided $2n$ is not $0, -1, -2, \ldots$. See problems 20 and 21 for the proof. □

There is another important property of the gamma function, called the *reflection formula*, valid for all non-integer z:

$$\Gamma(z)\Gamma(1-z) = \frac{\pi}{\sin(\pi z)}. \tag{2.12}$$

See problem 22 for a proof. The combination of the double angle and reflection formulas allows us to calculate some points of the gamma function in terms of other points.

Example 2.14
Given that $\Gamma(1/6) \approx 5.56631600178$, find an approximation for $\Gamma(1/3)$.
SOLUTION: The duplication formula shows that

$$\Gamma(1/3) = 2^{-2/3}\Gamma(1/6)\Gamma(2/3)/\sqrt{\pi}.$$

This almost gets $\Gamma(1/3)$ expressed in terms of $\Gamma(1/6)$, but the equation also involves $\Gamma(2/3)$. However, the reflection property shows that

$$\Gamma(1/3)\Gamma(2/3) = \frac{\pi}{\sin(\pi/3)} = \frac{2\pi}{\sqrt{3}}.$$

So there is a way of expressing $\Gamma(2/3)$ in terms of $\Gamma(1/3)$. With this substitution, we find that

$$\Gamma(1/3)^2 = \Gamma(1/6)2^{1/3}\sqrt{\pi/3}.$$

Since $\Gamma(1/3)$ is known to be positive, we have $\Gamma(1/3) \approx 2.67893853471$. □

2.3.2 Watson's Lemma

As powerful as Laplace's method is, it has yet to be put on a solid foundation. We are, after all, changing the order of an infinite sum and an integral, often producing a divergent series. Although Laplace's method applies to many different cases, most of these cases can be converted to a form for which a rigorous proof can easily be given. Often a substitution can convert the problem to finding the behavior as $x \to \infty$ of

$$\int_0^b e^{-xu} f(u)\, du, \tag{2.13}$$

for some function $f(u)$. For example, for the integral

$$\int_0^\pi e^{x \sin t}\, dt,$$

we can substitute $u = 1 - \sin t$, or $t = \sin^{-1}(1-u)$, so $dt = -du/\sqrt{1-(1-u)^2}$. As t ranges from 0 to π, u will go from 1 to 0 and back to 1, so the interval from 0 to 1 is covered *twice*. Hence

$$\int_0^\pi e^{x \sin t}\, dt = 2\int_0^1 \frac{e^{x(1-u)}}{\sqrt{1-(1-u)^2}} = 2e^x \int_0^1 \frac{e^{-xu}}{\sqrt{2u-u^2}}.$$

This last integral in the form of equation 2.13.

To find the full asymptotic expansion as $x \to \infty$ of equation 2.13, (assuming that this integral converges) we only need the asymptotic series for $f(u)$ as $u \to 0$. The simplest case is that $f(u)$ has a Maclaurin series

$$f(u) \sim \sum_{n=0}^\infty a_n u^n,$$

but there may be cases (like $f(u) = 1/\sqrt{2u-u^2}$), for which there may be fractional powers in the expansion. So we will consider the more general form

$$f(u) \sim \sum_{n=0}^\infty a_n u^{\alpha n + \beta} \qquad \text{as } u \to 0, \tag{2.14}$$

where α and β are constants.

Clearly, the maximum exponent of the exponential function in equation 2.13 occurs at $t = 0$, so by Laplace's method we only need to consider the integral from 0 to ϵ for small ϵ. For this range of t, we can replace $f(t)$ by its series.

$$\int_0^b e^{-ut} f(u)\, du \sim \int_0^\epsilon e^{-ut} f(u)\, dt \sim \int_0^\epsilon e^{-ut} \sum_{n=0}^\infty a_n u^{\alpha n + \beta}\, dt.$$

Now, if we interchange the sum and the integral, we get

$$\sum_{n=0}^\infty a_n \int_0^\epsilon e^{-ut} u^{\alpha n + \beta}\, du.$$

Since the upper endpoint of the integral doesn't matter in the asymptotic expansion as $x \to \infty$, we can replace the ϵ with ∞ and compute this integral in terms of the gamma function.

$$\int_0^\epsilon e^{-xu} u^{\alpha n+\beta}\,dt \sim \int_0^\infty e^{-xu} u^{\alpha n+\beta}\,dt = \int_0^\infty e^{-w}\left(\frac{w}{x}\right)^{\alpha n+\beta}\frac{dw}{x}$$
$$= \frac{\Gamma(\alpha n+\beta+1)}{x^{\alpha n+\beta+1}}.$$

Thus,

$$\sum_{n=0}^{\infty} a_n \int_0^\epsilon e^{-xu} u^{\alpha n+\beta}\,dt \sim \sum_{n=0}^{\infty} \frac{a_n\Gamma(\alpha n+\beta+1)}{x^{\alpha n+\beta+1}}.$$

Hence, we have a result known as *Watson's Lemma.*

LEMMA 2.1: Watson's Lemma

If the integral

$$\int_0^b e^{-ut} f(u)\,du$$

converges for sufficiently large x, where $f(u)$ has the asymptotic series given by

$$f(u) \sim \sum_{n=0}^{\infty} a_n u^{\alpha n+\beta} \qquad \text{as } u \to 0,$$

then the full asymptotic expansion as $x \to \infty$ of the integral is given by

$$\int_0^b e^{-ut} f(u)\,du \sim \sum_{n=0}^{\infty} \frac{a_n\Gamma(\alpha n+\beta+1)}{x^{\alpha n+\beta+1}}.$$

PROOF: The only problem with the above argument is that we interchanged the infinite sum and the integral. So let us do this step carefully.

In order to verify the first N terms of the asymptotic series, we pick ϵ small enough so that the first N terms of the series for $f(u)$ is a good approximation to $f(u)$, namely, the remainder term must be $\le K u^{\alpha(N+1)+\beta}$ for some constant K. That is,

$$\left| f(u) - \sum_{n=0}^{N} a_n u^{\alpha n+\beta} \right| \le K u^{\alpha(N+1)+\beta} \qquad \text{whenever } |u| < \epsilon.$$

So

$$\left| \int_0^\epsilon e^{-xu} f(u)\,du - \sum_{n=0}^{N} a_n \int_0^\epsilon e^{-xu} u^{\alpha N+\beta}\,du \right| \le K \int_0^\epsilon e^{-xu} u^{\alpha(N+1)+\beta}\,dt$$

$$\le K \int_0^\infty e^{-xu} u^{\alpha(N+1)+\beta}\, dt$$
$$= \frac{K\Gamma(\alpha(N+1)+\beta+1)}{x^{\alpha(N+1)+\beta+1}}.$$

Since we already have shown that

$$\int_0^\epsilon e^{-xu} u^{\alpha n+\beta}\, du \sim \frac{\Gamma(\alpha n+\beta+1)}{x^{\alpha n+\beta+1}}$$

with the error subdominant to this series, we have that the first N terms of the series are

$$g(x) \sim \sum_{n=0}^{N} \frac{a_n \Gamma(\alpha n+\beta+1)}{x^{\alpha n+\beta+1}} + O(x^{-(\alpha(N+1)+\beta+1)}).$$

□

Example 2.15
Let us find the full asymptotic series for the integral

$$\int_0^\infty \frac{e^{-xt}}{1+t} dt.$$

By expanding the function $f(x) = 1/(1+t)$ into a Maclaurin series, we have

$$\frac{1}{1+t} = \sum_{n=0}^{\infty} (-1)^n t^n.$$

So $a_n = (-1)^n$, $\alpha = 1$, and $\beta = 0$, so Watson's lemma gives us the full expansion

$$\int_0^\infty \frac{e^{-xt}}{1+t} dt \sim \sum_{n=0}^{\infty} \frac{(-1)^n n!}{x^{n+1}} \sim \frac{1}{x} - \frac{1}{x^2} + \frac{2}{x^3} - \frac{6}{x^4} + \cdots.$$

□

Watson's lemma can also be used if the series has fractional powers of x. The resulting series must be expressed in terms of the gamma function with fractional arguments.

Example 2.16
Use Watson's lemma to find the full asymptotic expansion as $x \to \infty$ of

$$\int_0^\pi e^{x \sin t}\, dt.$$

SOLUTION: The substitution $u = 1 - \sin t$ converts the integral to

$$2e^x \int_0^1 \frac{e^{-xu}}{\sqrt{2u-u^2}},$$

so we need the asymptotic series for the function

$$\frac{1}{\sqrt{2u-u^2}} = \frac{1}{\sqrt{2u}}\frac{1}{\sqrt{1-(u/2)}}$$

We have from equation 1.32 that

$$\frac{1}{\sqrt{1-4x}} \sim \sum_{n=0}^{\infty} \frac{(2n)!x^n}{(n!)^2} = 1 + 2x + 6x^2 + 20x^3 + 70x^4 + \cdots,$$

so replacing x with $u/8$ gives us part of this function:

$$\frac{1}{\sqrt{1-(u/2)}} \sim \sum_{n=0}^{\infty} \frac{(2n)!u^n}{8^n(n!)^2}.$$

Finally, we divide by $\sqrt{2u}$ to get

$$\frac{1}{\sqrt{2u-u^2}} \sim \sum_{n=0}^{\infty} \frac{(2n)!u^{n-1/2}}{8^n\sqrt{2}(n!)^2} = \frac{u^{-1/2}}{\sqrt{2}} + \frac{u^{1/2}}{4\sqrt{2}} + \frac{3u^{3/2}}{32\sqrt{2}} + \frac{5u^{5/2}}{128\sqrt{2}} + \cdots.$$

Applying Watson's lemma, we get

$$\begin{aligned}\int_0^{\pi} e^{x\sin t}\,dt &\sim 2e^x \sum_{n=0}^{\infty} \frac{(2n)!\Gamma(n+1/2)}{8^n\sqrt{2}(n!)^2 x^{n+1/2}} \\ &\sim e^x\sqrt{2\pi/x}\left(1 + \frac{1}{8x} + \frac{9}{128x^2} + \frac{75}{1024x^3} + \cdots\right).\end{aligned}$$

□

Example 2.17

The modified Bessel function of order 0, I_0, can be expressed by the integral

$$I_0(x) = \frac{e^x}{\pi}\int_0^2 \frac{e^{-xt}}{\sqrt{2t-t^2}}\,dt. \tag{2.15}$$

Thus, we need the asymptotic series for $f(x) = 1/\sqrt{2t-t^2}$, which we saw in example 2.16.

$$\frac{1}{\sqrt{2t-t^2}} \sim \sum_{n=0}^{\infty} \frac{(2n)!t^{n-1/2}}{8^n\sqrt{2}(n!)^2} = \frac{t^{-1/2}}{\sqrt{2}} + \frac{t^{1/2}}{4\sqrt{2}} + \frac{3t^{3/2}}{32\sqrt{2}} + \frac{5t^{5/2}}{128\sqrt{2}} + \cdots.$$

By applying Watson's lemma and using the fact that $\Gamma(1/2) = \sqrt{\pi}$, we get

$$\begin{aligned}I_0(x) &\sim \frac{e^x}{\pi}\sum_{n=0}^{\infty} \frac{(2n)!\Gamma(n+1/2)}{8^n\sqrt{2}(n!)^2 x^{n+1/2}} \\ &\sim \frac{e^x}{\sqrt{2\pi x}}\left(1 + \frac{1}{8x} + \frac{9}{128x^2} + \frac{75}{1024x^3} + \cdots\right).\end{aligned}$$

Notice that the original integral was improper at both of the endpoints, but as long as the original integral converges, Watson's lemma can be used.

Although the series for $I_0(x)$ differs from the series of example 2.16 by only a constant, the integral of example 2.16 does not evaluate to $2\pi I_0(x)$. Rather, it differs from $2\pi I_0(x)$ by a subdominant function. This gives another example of two different functions having the same asymptotic series. □

Problems for §2.3

1 Use induction to show that equation 2.4 is true for all integers $n \geq 0$.
Hint: If $f(n)$ is the integral, show that $f(0) = 1$, and using integration by parts, $f(n+1) = (n+1)f(n)$.

2 Use induction to show that equation 2.5 is true for all integers $n \geq 0$.
Hint: Use equation 2.3 to do the case $n = 0$, then use integration by parts with $dv = -2se^{-s^2}$, so $v = e^{-s^2}$.

3 Show that equation 2.6 is true for all integers $n \geq 0$.
Hint: A simple substitution converts it to equation 2.4.

For problems **4** through **11**: Use Laplace's method to find the first three terms of the asymptotic series as $x \to \infty$ for the following integrals:

4 $\displaystyle\int_0^\infty \frac{e^{-xt}}{t+1}\,dt$

5 $\displaystyle\int_0^1 \frac{e^{-xt}}{t^2+1}\,dt$

6 $\displaystyle\int_1^\pi \frac{e^{-xt}}{t^2}\,dt$

7 $\displaystyle\int_{-1/e}^e \frac{e^{-xt^2}}{t+1}\,dt$

8 $\displaystyle\int_0^2 \frac{e^{-xt^2}}{t+2}\,dt$

9 $\displaystyle\int_{-1/2}^1 e^{-2xt^2}\ln(t+1)\,dt$

10 $\displaystyle\int_{-1}^1 e^{x(t^3-2t^2)}\,dt$

11 $\displaystyle\int_0^e e^{-xe^t}\,dt$

12 Show that if $\sum_{n=1}^\infty (a_n - 1)$ converges, then $\prod_{n=1}^\infty a_n$ converges at a similar rate.
Hint: $\ln\left(\prod_{n=1}^\infty a_n\right) = \sum_{n=1}^\infty \ln(a_n)$.

13 Show that the infinite product in Euler's formula for $\Gamma(z)$ (equation 2.10) will always converge, except for when z is a negative integer. (The only reason that negative integers cause a problem is that one of the factors becomes undefined.) You can use the result from problem 12.

14 Show that Euler's formula for $\Gamma(z)$ (equation 2.10) satisfies $\Gamma(z+1) = z\Gamma(z)$.
Hint: Express $\Gamma(z+1)/\Gamma(z)$ as a single infinite product, which telescopes.

15 Find a "triplication" identity for the gamma function similar to example 2.13. That is, express $\Gamma(3n)$ in terms of $\Gamma(n)$, $\Gamma(n+\frac{1}{3})$, and $\Gamma(n+\frac{2}{3})$.

16 Use Euler's formula for $\Gamma(z)$ (equation 2.10) to prove the *Wallis product*:

$$\frac{\pi}{2} = \frac{2}{1}\cdot\frac{2}{3}\cdot\frac{4}{3}\cdot\frac{4}{5}\cdot\frac{6}{5}\cdot\frac{6}{7}\cdot\frac{8}{7}\cdots = 2\prod_{n=1}^{\infty}\left(\frac{2n}{2n+1}\frac{2n+2}{2n+1}\right).$$

Hint: Plug in $z = 1/2$, and square both sides.

17 Find the exact value of $\Gamma(\frac{1}{14})\Gamma(\frac{9}{14})\Gamma(\frac{11}{14})$.
Hint: Plug $n = 1/14$ in equation 2.11, then keep applying equation 2.11 to expand the "even" factor.

18 Find the exact value of $\Gamma(\frac{3}{14})\Gamma(\frac{5}{14})\Gamma(\frac{13}{14})$.

19 Find the exact value of $\Gamma(\frac{1}{3})\Gamma(\frac{1}{4})/\Gamma(\frac{1}{12})$. Note: you will need the result of problem 15.

20 Show that the function

$$f(x) = \frac{2^{2n-1}\Gamma(x)\Gamma(x+\frac{1}{2})}{\Gamma(2x)}$$

is periodic with period 1. That is, show that $f(x+1) = f(x)$.

21 Use Sterling's formula to find the limit

$$\lim_{x\to\infty}\frac{2^{2x-1}\Gamma(x)\Gamma(x+\frac{1}{2})}{\Gamma(2x)}.$$

Since the only way for a periodic function to have a limit is for the function to be constant, this proves equation 2.11 is valid for all positive x.

22 Use Euler's formula for $\Gamma(z)$ (equation 2.10) to prove the reflection formula (equation 2.12). You may use the infinite product formula for $\sin(x)$:

$$\sin(x) = x\prod_{n=1}^{\infty}\left(1-\frac{x^2}{n^2\pi^2}\right) = x\left(1-\frac{x^2}{\pi^2}\right)\left(1-\frac{x^2}{4\pi^2}\right)\left(1-\frac{x^2}{9\pi^2}\right)\cdots.$$

Hint: First show that $1/(\Gamma(z)\Gamma(-z)) = -z\sin(\pi z)/\pi$, then use $\Gamma(1-z) = -z\Gamma(-z)$.

23 Equation 2.1 introduces the constant γ. Show that $\Gamma'(1) = -\gamma$.
Hint: First use the recursion property of $\Gamma(x)$ to show that, for n an integer > 1,

$$\frac{\Gamma'(n)}{\Gamma(n)} = 1 + \frac{1}{2} + \frac{1}{3} + \cdots + \frac{1}{n-1} + \Gamma'(1).$$

Then use Stirling's formula 2.7 to show that

$$\frac{\Gamma'(x)}{\Gamma(x)} \sim \ln x - \frac{1}{2x} + \cdots \text{ as } x \to \infty.$$

24 Show that we can represent the constant γ using an improper integral

$$\gamma = \int_0^\infty -e^{-t} \ln t \, dt. \tag{2.16}$$

Hint: Use the fact that $\Gamma'(1) = -\gamma$. Differentiate equation 2.8 with respect to z under the integral, and plug in $z = 1$.

25 Use integration by parts and equation 2.16 to show that

$$\lim_{x \to 0} \left(\int_x^\infty \frac{e^{-t}}{t} \, dt + \ln x \right) = -\gamma.$$

This verifies that the constant C_3 from example 2.4 is indeed $-\gamma$.

26 A *Gaussian integral* is of the form

$$\int_0^\infty t^a e^{-bt^c} \, dt,$$

where a, b, and c are constants. Assuming the integral converges, evaluate this integral in terms of the gamma function.

27 Find the first three terms for the asymptotic series as $x \to \infty$ for the integral

$$\int_{-1}^2 \frac{e^{-xt^4}}{t+2} \, dt.$$

Note that you will have to evaluate some Gaussian integrals not in the form of equations 2.4 through 2.6. See problem 26.

28 Use Watson's lemma to find the asymptotic series as $x \to -\infty$ for the integral

$$\int_0^\pi e^{x \sin t} \, dt.$$

Note that the maximums of the exponent now occur at 0 and π. By symmetry, we can concentrate on the maximum at 0, then double the result.

For problems **29** through **36**: Use Watson's lemma to find the first three terms of the asymptotic series as $x \to \infty$ for the following integrals. You may leave the answer in terms of the gamma function.

29 $\displaystyle\int_0^\infty e^{-xt} \ln(t+1) \, dt$

30 $\displaystyle\int_0^1 \frac{e^{-xt}}{\sqrt{t+1}} \, dt$

31 $\displaystyle\int_0^e \frac{e^{-x\sqrt{t}}}{t+1} \, dt$

32 $\displaystyle\int_0^\pi \frac{e^{-xt^2}}{t^2+1} \, dt$

33 $\displaystyle\int_0^{\pi/2} e^{-x \tan t} \, dt$

34 $\displaystyle\int_0^\pi e^{x \cos t} \, dt$

35 $\displaystyle\int_0^1 \frac{e^{-xt^3}}{t+1} \, dt$

36 $\displaystyle\int_0^\infty e^{-xt^4} \ln(t+1) \, dt$

2.4 Review of Complex Numbers

Before we proceed further, we need to understand how asymptotics can be extended to the complex plane. In particular, we need to review the properties of complex numbers.

A complex number is a number of the form $z = x + iy$, where $i = \sqrt{-1}$. Simple arithmetic is governed by the property that $i^2 = -1$:

$$\begin{aligned}(x_1 + iy_1) + (x_2 + iy_2) &= (x_1 + x_2) + i(y_1 + y_2),\\ (x_1 + iy_1) - (x_2 + iy_2) &= (x_1 - x_2) + i(y_1 - y_2),\\ (x_1 + iy_1) \cdot (x_2 + iy_2) &= (x_1x_2 - y_1y_2) + i(x_1y_2 + x_2y_1).\end{aligned}$$

The complex conjugate of $z = x + iy$ is $\overline{z} = x - iy$. Division of complex numbers is accomplished by multiplying the numerator and denominator by the conjugate of the denominator.

$$\frac{x_1 + iy_1}{x_2 + iy_2} = \frac{(x_1 + iy_1)(x_2 - iy_2)}{(x_2 + iy_2)(x_2 - iy_2)} = \frac{x_1x_2 + y_1y_2}{x_2^2 + y_2^2} + i\frac{x_2y_1 - x_1y_2}{x_2^2 + y_2^2}.$$

Each complex number can be thought of as a point on the complex plane, with a horizontal real axis and vertical imaginary axis. Adding two complex numbers is geometrically equivalent to adding the two corresponding vectors in the plane. The product of two complex numbers also has a geometrical interpretation. We define the *argument* of a complex number $z = x + iy$ to be the angle (in radians) from the vector $\langle x, y\rangle$ to the positive x-axis. Then the argument of a complex product z_1z_2 will be the sums of the arguments of z_1 and z_2, as shown in figure 2.5.

Raising a number to a complex power is more complicated. Let us first determine how to raise e to a complex power. Using the law of exponents, we have $e^{x+iy} = e^x e^{iy}$, so we only have to determine how to exponentiate a purely complex number. This can be done by taking advantage of the Maclaurin series for e^x.

$$e^x = 1 + x + \frac{x^2}{2} + \frac{x^3}{3!} + \frac{x^4}{4!} + \frac{x^5}{5!} + \frac{x^6}{6!} + \frac{x^7}{7!} + \cdots.$$

Replacing x with iy, we get

$$e^{iy} = 1 + iy + \frac{-y^2}{2} + \frac{-iy^3}{3!} + \frac{y^4}{4!} + \frac{iy^5}{5!} + \frac{-y^6}{6!} + \frac{-iy^7}{7!}\cdots.$$

Grouping the real and complex parts together produces

$$e^{iy} = \left(1 - \frac{y^2}{2} + \frac{y^4}{4!} - \frac{y^6}{6!} + \cdots\right) + i\left(y - \frac{y^3}{3!} + \frac{y^5}{5!} - \frac{y^7}{7!} + \cdots\right).$$

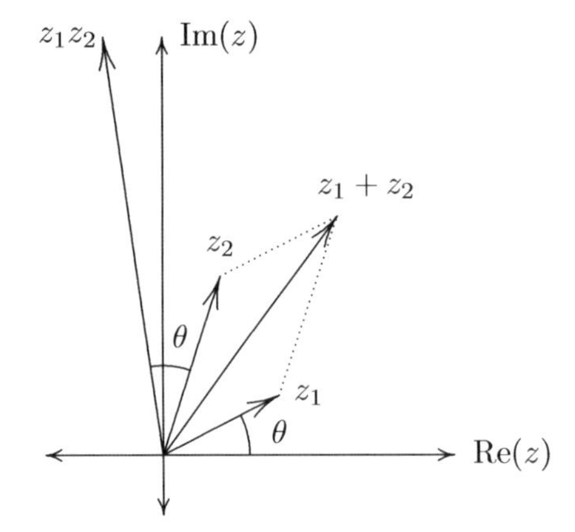

FIGURE 2.5: This illustrates the relationship of the sum and product of complex numbers in the plane to the original numbers. The sum is represented by the diagonal of a parallelogram, whereas the angle of the product vector is the sum of the original vector angles.

These two series are the series for $\cos(y)$ and $\sin(y)$. Thus,

$$\boxed{e^{iy} = \cos(y) + i\sin(y).} \tag{2.17}$$

This result is known as *Euler's formula*, which gives us a way to exponentiate any complex number:

$$e^{x+iy} = e^x \cos(y) + ie^x \sin(y). \tag{2.18}$$

Example 2.18
Find the complex exponential $e^{2+\pi i}$.
SOLUTION: Using equation 2.18, we have $e^{2+\pi i} = e^2 \cos(\pi) + ie^2 \sin(\pi) = -e^2$. Thus, the complex exponential can be real and negative. □

Having understood the complex exponential function, let us consider the inverse function, the complex logarithm. In order to compute $\ln(x + iy)$, we will need to find a complex number $a + bi$ such that $e^a \cos b = x$ and $e^a \sin b = y$. It is easy to solve for a, since $x^2+y^2 = e^{2a}\cos^2 b+e^{2a}\sin^2 b = e^{2a}$. Hence, $a = \ln(x^2 + y^2)/2$.

To solve for b, we can divide y/x to eliminate the exponentials. So $y/x = \tan b$, giving $b = \tan^{-1}(y/x)$. However, we must make sure that we are in the correct quadrant. The best way to do this is to convert the point (x, y) to polar coordinates, (r, θ). Then $b = \theta$, and in fact, $a = \ln(r)$. So we have

$$\boxed{\ln(x + iy) = \ln(r) + i\theta, \text{ where } (r, \theta) \text{ are the polar coordinates of } (x, y).} \tag{2.19}$$

There is one problem with this definition. There are, in fact, an infinite number of polar angles for a given point, since we can add or subtract 2π from θ. So the complex logarithm is in fact multi-valued. We can add $2\pi i$ to any solution and get another solution. There are several ways to fix this problem.

One solution is to insist that we pick the polar coordinates such that $-\pi < \theta \leq \pi$. Then $\ln(z)$ will be well defined, but it will be discontinuous for negative real numbers. For example, $\ln(-4 + .001i) \approx 1.38629 + 3.14134i$, but $\ln(-4 - .001i) \approx 1.38629 - 3.14134i$. Even though $-4 + .001i$ is close to $-4 - .001i$, their logarithms are far apart. The negative real axis becomes a *branch cut* because it separates two branches of the multi-valued logarithm function. It is possible to move the branch cut, say to the negative imaginary axes, but it is impossible to eliminate it altogether.

Another solution is to treat the two polar points (r, θ) and $(r, \theta + 2\pi)$ as different points. That is, a "point" not only includes the location in the complex plane, but also the number of times one has traveled around the origin to reach it. Instead of defining the function on the complex plane, we are defining the function on a surface such as figure 2.6, which is analogous to a winding staircase or a parking garage. If one begins at a positive real number, and travels counterclockwise around a circle centered at the origin, instead of ending at the original location, one is one level above the starting point. This corkscrew surface is called the *Riemann surface* for the logarithm function. The different levels are called the *sheets* of the Riemann surface. Each point on the Riemann surface (except the origin) has a unique polar coordinate (r, θ) with $r > 0$, so the complex logarithm will be uniquely defined. In fact, the complex logarithm is a 1-to-1 and onto mapping from this Riemann surface to the entire complex plane.

Many of the functions we have encountered, especially those with divergent asymptotic series, are also multi-valued, and so a Riemann surface is needed to fully understand such functions. For example, the Stieltjes integral function,

$$S(z) = \int_0^\infty \frac{e^{-t}}{1 + zt}\, dt$$

has a branch cut along the negative real axis. This can be seen by evaluating $S(-4+.001i) \approx -0.105519 - 0.611626i$, whereas $S(-4-.001i) \approx -0.105519 + 0.611626i$. By using the same Riemann surface as in figure 2.6, $S(z)$ becomes a continuous, well defined function. To find the value of $S(z)$ on the other sheets, example 2.23 will show that we can add $2\pi i e^{1/x}/x$ to the value of $S(x)$ from the previous sheet. So on the sheet with $\pi < \theta < 3\pi$, $S(-4 + .001i) \approx -0.105289 - 1.83496i$, and $S(-4 - .001i) \approx -0.105748 - 0.611711i$. Note that the value of the function just below the negative real axis on this sheet is very close to the value just above the negative real axis on the previous sheet, with $-\pi < \theta < \pi$.

Now that the complex logarithm is defined, raising a complex number to a power becomes easy. We define $z_1^{z_2}$ as $e^{\ln(z_1)z_2}$. Because of the multi-valued

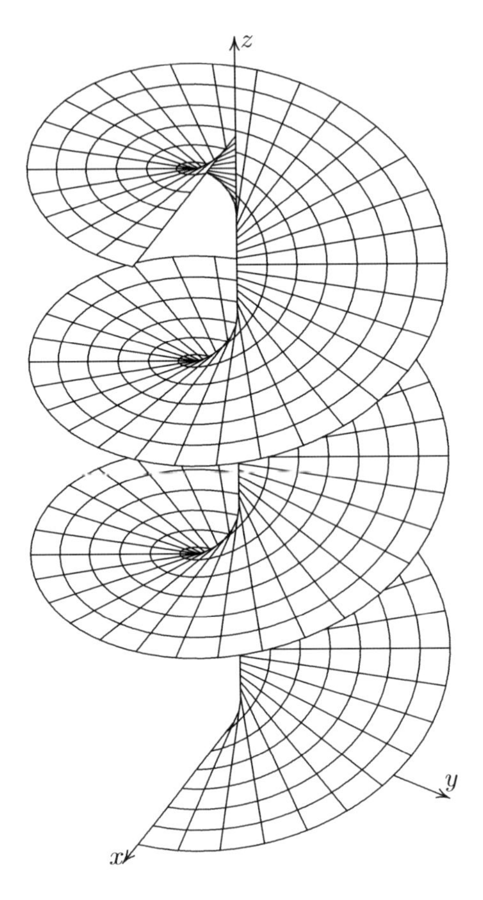

FIGURE 2.6: A three-dimensional representation of the Riemann surface for the complex logarithm. By traveling around the origin, one moves to another sheet of the surface. Although each sheet is locally flat, this concept is easiest to understand via a three dimensional model.

logarithm, there will usually be an infinite number of answers, but if z_2 is rational, there will only be a finite number of results.

Example 2.19
Find all possible values of $(2i)^{2/3}$.
SOLUTION: We rewrite this as $e^{\ln(2i)(2/3)}$. One value for $\ln(2i)$ is $\ln 2 + \pi i/2$, but we can add a multiple of $2\pi i$ to this, giving $\ln 2 + \pi i/2 + 2k\pi i$, where k is an integer. Next, we multiply by $2/3$, producing $(2\ln 2)/3 + \pi i/3 + 4k\pi i/3$. The expression $2\ln 2$ simplifies to $\ln 4$. So when we exponentiate this, we get

$$e^{(\ln 4)/3 + \pi i/3 + 4k\pi i/3} = e^{(\ln 4)/3}(\cos((4k+1)\pi/3) + i\sin((4k+1)\pi/3)).$$

Plugging in $k = 0$, $k = 1$, and $k = 2$ produces the results $\sqrt[3]{4}(1/2 + \sqrt{3}i/2)$, $\sqrt[3]{4}(1/2 - \sqrt{3}i/2)$, and $-\sqrt[3]{4}$. However, using $k = 3$ starts repeating the values. So there are 3 solutions, given by $\sqrt[3]{4}(1/2 + \sqrt{3}i/2)$, $\sqrt[3]{4}(1/2 - \sqrt{3}i/2)$, and $-\sqrt[3]{4}$. □

2.4.1 Analytic Functions

Now that we understand the basic arithmetic of complex numbers, let us work on the calculus of complex functions. Given a complex function $f(z)$, we define the derivative in the usual way:

$$f'(z) = \lim_{\Delta z \to 0} \frac{f(z + \Delta z) - f(z)}{\Delta z}. \tag{2.20}$$

It is not hard to show that the derivative of the complex function z^2 is $2z$. The derivative of other polynomials behave in the standard way.

But this limit really has two parts: a real part and the complex part. If we break the function $f(z)$ into its real and complex parts, we can write $f(x + iy) = u(x, y) + iv(x, y)$ for some functions u and v. Then

$$\frac{f(z + \Delta z) - f(z)}{\Delta z} = \frac{(u(x + \Delta x, y + \Delta y) - u(x, y)) + i(v(x + \Delta x, y + \Delta y) - v(x, y))}{\Delta x + i\Delta y} = \frac{[(u(x + \Delta x, y + \Delta y) - u(x, y)) + i(v(x + \Delta x, y + \Delta y) - v(x, y))](\Delta x - i\Delta y)}{\Delta x^2 + \Delta y^2}.$$

In order for the complex limit to exist, both the real and imaginary parts must approach a limit as $(\Delta x, \Delta y) \to (0, 0)$. These are multi-variable limits, which are more difficult to converge than single variable limits because of the two path rule. The same limit must be achieved for all paths to the origin, including curved and spiral paths. In order to get both multi-variable limits to exist, several things must happen. For starters, $u(x, y)$ and $v(x, y)$

must satisfy the partial differential equations, known as the *Cauchy-Riemann equations:*

$$\frac{\partial u}{\partial x} = \frac{\partial v}{\partial y}, \qquad \text{and} \qquad \frac{\partial u}{\partial y} = -\frac{\partial v}{\partial x}. \tag{2.21}$$

We say that a complex function $f(z)$ is *analytic* at a point z_0 if there is a open set containing z_0 for which the derivative exists. Any point that is not analytic is called a *singularity*. If a point is not analytic, but there is an open set containing the point for which the function is analytic except at that point, we say the point is an *isolated singularity.* What is surprising is that if one derivative exists on an open set, then *all* derivatives will exist on that set.

THEOREM 2.2
If any of these five statements are true, they all are:

1) $f(z)$ is analytic at a point z_0.

2) There is an open set containing z_0 for which the derivative exists.

3) The component functions $u(x, y)$ and $v(x, y)$ both have continuous first partial derivatives in an open set containing z_0, and the Cauchy-Riemann equations 2.21 hold in this set.

4) $f(z)$ has *all* derivatives existing at z_0.

5) $f(z)$ has a Taylor series at z_0:

$$f(z) = f(z_0) + f'(z_0)(z - z_0) + \frac{f''(z_0)}{2!}(z - z_0)^2 + \frac{f'''(z_0)}{3!}(z - z_0)^3 + \cdots$$

which converges within a circle of positive radius centered at z_0.

For a proof, see [6].

If $f(z)$ is analytic at z_0, there will be a value ρ, called the *radius of convergence*, for which the Taylor series converges if $|z - z_0| < \rho$, and diverges if $|z - z_0| > \rho$. If the series converges for all finite z, we say that $\rho = \infty$, and the function $f(z)$ is said to be *entire*. The boundary $|z - z_0|$ describes a circle of radius ρ, hence the terminology *radius* of convergence. In fact, the radius of convergence is easy to predict from the function $f(z)$.

THEOREM 2.3
If $f(z)$ is analytic at a point z_0, then the radius of convergence of the Taylor series at z_0 is the distance from z_0 to the nearest singularity of $f(z)$. If $f(z)$ has no singularities in the complex plane, then the Taylor series will converge for all finite z.

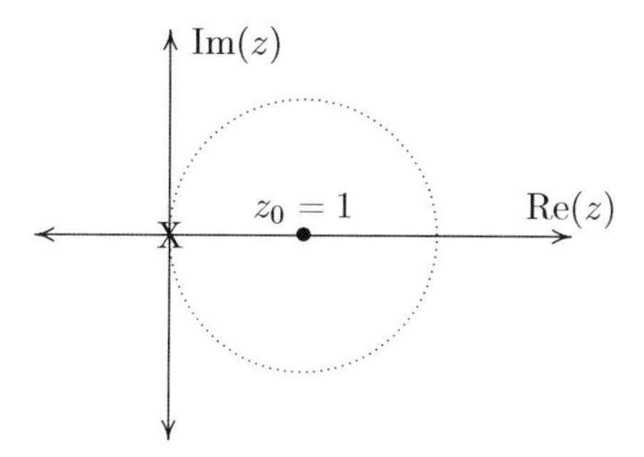

FIGURE 2.7: The Taylor series for $\ln(z)$ centered at $z_0 = 1$ will converge inside of the circle. This is the largest circle which does not contain singularities. The singularity at 0 keeps the circle from getting larger.

See [6, p. 138] for a proof.

For example, the exponential function e^z has no singularities, so this function is entire, and its Taylor series converges for all z. However, the complex logarithm $\ln(z)$ has a singularity at $z = 0$. The Taylor series for $\ln(z)$ centered at $z_0 = 1$,

$$\ln(z) = (z-1) - \frac{(z-1)^2}{2} + \frac{(z-1)^3}{3} - \frac{(z-1)^4}{4} + \cdots, \tag{2.22}$$

will have a radius of convergence of 1, shown in figure 2.7.

Within the circle of convergence of a Taylor series, the function will be analytic. So given a series that converges for a finite radius of convergence, we can pick a point near the edge of this circle and construct the Taylor series about this new point. If the point is sufficiently far from any singularities, the new region of convergence will go beyond the original circle of convergence. The new Taylor series will give us an *analytic continuation* of the original function.

For example, the point $(1+i)/\sqrt{2}$ is inside the circle in figure 2.7. So we can use the series in equation 2.22 to find a new Taylor series about $z_0 = (1+i)/\sqrt{2}$:

$$\ln(z) = \frac{\pi i}{4} + \frac{1-i}{\sqrt{2}}\left(z - \frac{1+i}{\sqrt{2}}\right) + \frac{i}{2}\left(z - \frac{1+i}{\sqrt{2}}\right)^2$$
$$- \frac{1+i}{3\sqrt{2}}\left(z - \frac{1+i}{\sqrt{2}}\right)^3 + \frac{1}{4}\left(z - \frac{1+i}{\sqrt{2}}\right)^4 + \cdots.$$

Since the only singularity is at $z = 0$, this new series will also have a radius of convergence of 1, so the new series will converge at points where the first series diverged. We can then repeat this process, as shown in figure 2.8.

The analytical continuation of an analytic function will be unique, provided we do not travel *around* a singularity. In figure 2.8, the circles of convergence

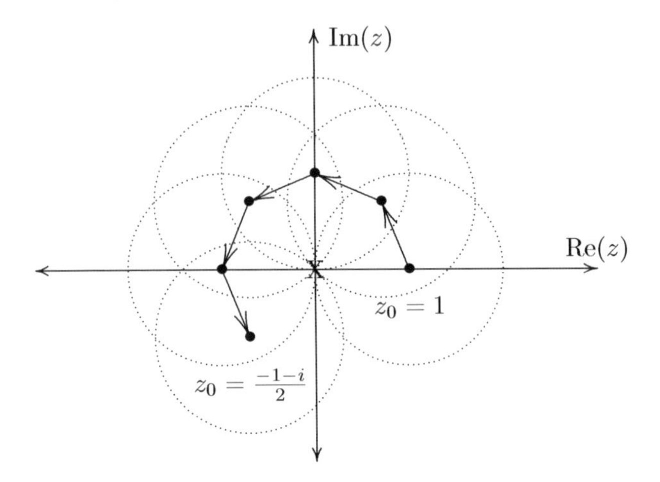

FIGURE 2.8: The Taylor series centered at $z = 1$ can be analytically continued to produce values of the function outside of the original region of convergence.

centered at $z_0 = 1$ and $z_0 = (-1 - i)/\sqrt{2}$ overlap slightly, but the functions will not agree on this overlap, since we have traveled around the singularity. Another way of thinking about this is that these two circles are on different sheets of the Riemann surface, so if we evaluated the functions in the region of the apparent overlap, the results will differ by exactly $2\pi i$.

If a function has more that one singularity, the analytic continuations of the function can be used to define the Riemann surface for this function. The Riemann surface can be fairly complicated when there are multiple singularities.

Sometimes, the singularity can be removed simply by redefining the value at the singularity. An example is $f(z) = (\sin z)/z$. Obviously, the function is undefined at $z = 0$, but if we redefine $f(0)$ to be 1, then the function can be expressed by a convergent Maclaurin series

$$f(x) = 1 - \frac{x^2}{3!} + \frac{x^4}{5!} - \frac{x^6}{7!} + \cdots.$$

We say that $f(z)$ has a *removable singularity* at z_0 if the function can be made analytic by redefining $f(z_0)$. Removable singularities do not affect the radius of convergence of Taylor series, so the series for $(\sin z)/z$ converges for all z.

We say that a singularity z_0 is a *pole* of $f(z)$ if

$$(z - z_0)^n f(z)$$

has a removable singularity at z_0 for some positive integer n. The *order* of the pole is the smallest n for which the singularity of $(z - z_0)^n f(z)$ is removable.

A pole will not alter the Riemann surface, that is, if we analytically continue a function around a pole, we will end up with the same function as we started with, unlike singularity of the logarithm function.

Analytical continuation allows us to define a complex function beyond its original definition. For example, the integral for the $\Gamma(z)$, equation 2.8, only converges when the real part of z is positive. But we can analytically continue this function to produce a function defined on the entire complex plane, except for the first order poles at the non-negative integers.

Example 2.20

The *Riemann zeta function,* $\zeta(z)$, is defined by the series

$$\zeta(z) = \sum_{n=1}^{\infty} \frac{1}{n^z} = 1 + \frac{1}{2^z} + \frac{1}{3^z} + \frac{1}{4^z} + \cdots. \tag{2.23}$$

This series converges as long as the real part of $z > 1$. However, this function can be analytically continued to form a well defined function for the entire complex plane, except for a first order pole at $z = 1$. See problems 16 through 18. ▯

2.4.2 Contour Integration

Having defined the derivative for an analytical function, how can we integrate a complex function? Because a complex function is two-dimensional, the integral will be a contour integral, involving a path C from z_1 to z_2 in the complex plane. If a complex function $f(z)$ can be written as $f(x+iy) = u(x,y) + iv(x,y)$, then we can define the contour integral of $f(z)$ over the curve C in the natural way.

$$\begin{aligned}\int_C f(z)\,dz &= \int_C (u(x,y) + iv(x,y)) \cdot (dz + idy) \\ &= \int_C u(x,y)\,dx - v(x,y)\,dy + i\int_C v(x,y)\,dx + u(x,y)\,dy. \end{aligned}\tag{2.24}$$

Thus, a complex contour integral can be calculated by evaluating two real contour integrals. Recall that a line integral

$$\int_C M(x,y)\,dx + N(x,y)\,dy$$

is *independent of path* if the vector field $\langle M(x,y), N(x,y)\rangle$ is *conservative.* A conservative field is also known as a *gradient field,* since there is a potential function ϕ such that $\nabla\phi = \langle M(x,y), N(x,y)\rangle$. The vector field is conservative if the cross partials test

$$\frac{\partial M}{\partial y} = \frac{\partial N}{\partial x}$$

is satisfied on a *simply connected* open set, that is, without holes. In order for the two real integrals in equation 2.24 to be independent of path, we need to have

$$\frac{\partial u}{\partial y} = \frac{\partial(-v)}{\partial x} \quad \text{and} \quad \frac{\partial v}{\partial y} = \frac{\partial u}{\partial x}$$

on a simply connected set. But these are precisely the Cauchy-Riemann equations 2.21. So for analytic functions defined on a simply connected set, the complex integral will only depend on the endpoints of the contour C. In fact, there will be potential functions ϕ_1 and ϕ_2 such that

$$\int_C f(z)\,dz = \int_{z_1}^{z_2} f(z)\,dz = (\phi_1(z_2) - \phi_1(z_1)) + i(\phi_2(z_2) - \phi_2(z_1)).$$

Then $F(z) = \phi_1(z) + i\phi_2(z)$ will be an *anti-derivative* of $f(z)$, for it is not hard to show that $F(z)$ will be analytic, and indeed $F'(z) = f(z)$. (See problems 20 and 21.) Any two anti-derivatives will differ by a constant.

Another result stemming from the independence of path deals with *closed curves*, that is, curves that end at the same point where they start. If a function is analytic both on a closed curve and *inside* the curve, then the line integral along the closed curve will always be 0. This is because the region inside of a closed curve will be simply connected, hence the integral will be $F(z_1) - F(z_1) = 0$. Because of the independence of the path, we can *deform the contour* to a different contour with the same endpoints, provided that the function is analytic in the region between the two contours.

But this will not be true if there is a singularity inside of the curve!

Example 2.21
Evaluate

$$\oint_C \frac{1}{z}\,dz$$

where the contour C goes counter-clockwise around the origin and returns to where it started. Here, the notation $\oint$ emphasizes that the contour is closed and is going counter-clockwise.

SOLUTION: We can no longer say that $F(z_1) - F(z_1) = 0$, since the ending point will be on a different sheet of the Riemann surface than the starting point. If z_1 is the starting point, then the ending point will be $\tilde{z}_1$, the corresponding point on the next sheet. Then $\ln(\tilde{z}_1) - \ln(z_1) = 2\pi i$. ☐

We can generalize this example to handle any simple pole, that is, an isolated pole of order 1. If $f(z)$ has a simple pole at z_0, then we define the *residue* of the pole to be

$$R = \lim_{z \to z_0} (z - z_0) f(z).$$

Then $g(z) = f(z) - (R/(z - z_0))$ will have a removable singularity at z_0, so we can write

$$f(z) = g(z) + \frac{R}{z - z_0}$$

where $g(z)$ is analytic at z_0. Then any closed contour integral going counter-clockwise around z_0, but does not go around any other singularity of $f(z)$, will evaluate to $2\pi Ri$.

Example 2.22

Evaluate

$$\oint_C \frac{z+2}{z^3+z}\, dz$$

where C goes counter-clockwise around a circle of radius 1 centered at $i/2$.

SOLUTION: There are three singularities in the integrand, at 0, i, and $-i$. However, only 0 and i are inside of C, so these are the only two poles we have to consider. The residues of these poles are easy to compute:

$$\lim_{z \to 0} z \frac{z+2}{z^3+z} = \lim_{z \to 0} \frac{z+2}{z^2+1} = 2,$$

$$\lim_{z \to i} (z - i) \frac{z+2}{z^3+z} = \lim_{z \to i} \frac{z+2}{z(z+i)} = i/2 - 1.$$

Thus,

$$\frac{z+2}{z^3+z} = g(z) + \frac{2}{z} + \frac{i/2-1}{z-i}$$

where $g(z)$ is analytic at both 0 and i. Hence,

$$\oint_C \frac{z+2}{z^3+z}\, dz = 2\pi i(2 + i/2 - 1) = -\pi + 2\pi i.$$

□

Example 2.23

The Stieltjes integral function $S(z)$ has the same Riemann surface as the complex logarithm function. What is the relationship of $S(z)$ on two consecutive sheets? That is, if we start at a point z, travel counter-clockwise around the origin, back to z, we will be on a different sheet of the Riemann surface, so we can call this new point $\tilde{z}$. Is there a relationship between $S(z)$ and $S(\tilde{z})$?

SOLUTION: The integral for $S(z)$ is given by

$$S(z) = \int_0^\infty \frac{e^{-t}}{1+zt}\, dt.$$

The integrand has a pole at $t = -1/z$. As long as z is not a negative real number, we can use the contour of real numbers from 0 to ∞ for the integral. In order to understand what happens to $S(z)$ as we travel around the origin,

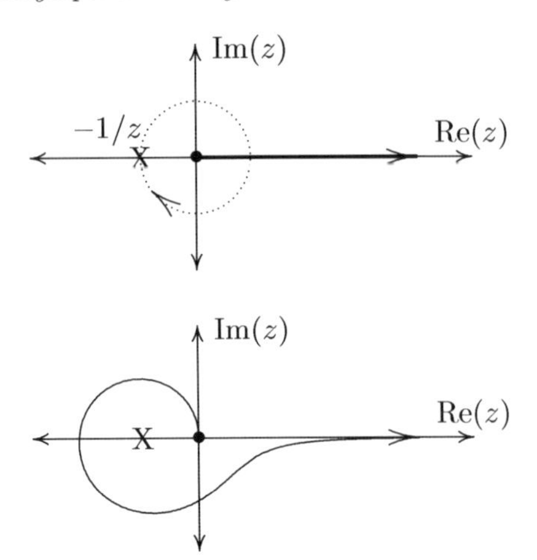

FIGURE 2.9: The integral for the Stieltjes integral function has a pole at $-1/z$. If z travels counterclockwise around to origin to reach the next sheet of the Riemann surface, $-1/z$ will travel clockwise around the origin. But the contour must avoid the singularity, so the contour is distorted in the process. The result is that the contour goes counterclockwise around the singularity.

we need to see what happens to the pole $-1/z$. It will travel clockwise around the origin, forcing the contour to be deformed. See figure 2.9. The result is that the new contour will have an extra loop going counterclockwise around the pole at $-1/z$, so we would add $2\pi i$ times the residue of the pole to $S(z)$ to determine $S(\tilde{z})$. We can easily calculate the residue of the pole:

$$\lim_{t\to -1/z} (t + 1/z)\frac{e^{-t}}{1+zt} = \lim_{t\to -1/z} \frac{e^{-t}}{z} = \frac{e^{1/z}}{z}.$$

So we have $S(\tilde{z}) = S(z) + e^{1/z}2\pi i/z$. □

2.4.3 Gevrey Asymptotics

Now that we have some understanding of complex functions, let us consider how complex functions relate to the definition of asymptotic series. Asymptotic relations are ultimately defined in terms of a limit, and for a limit to exist in the complex plane as $z \to z_0$, the limit must be independent of the path taken to get to z_0. Yet in most of the asymptotic relations, we considered only one sided limits, such as $x \to \infty$ or $x \to 0^+$. Very few asymptotic relations as $z \to z_0$ are valid over *all* paths to z_0. Hence, we will have to make some kind of restriction on the paths which are allowed.

Let us illustrate with a simple example. Since $\cosh(x) = (e^x + e^{-x})/2$, it is clear that as $x \to \infty$ along the real axis, then $\cosh x \sim e^x/2$. But this is just one path to ∞. For the complex limit as $z \to \infty$, we have to consider any path for which $|z|$ is increasing and unbounded. That is, the path can extend outward in any direction, or even spiral outward. If we choose the direction of the negative real axis, then $e^x \ll e^{-x}$, and so $\cosh(x) \sim e^{-x}/2$ in this direction.

The naive approach would be to restrict the paths to the half plane $\mathrm{Re}(z) > 0$. This would preclude the possibility that $e^x \ll e^{-x}$, but something interesting happens if we consider the path along the positive imaginary axis $z = iy$. Then

$$\cosh(iy) = \frac{e^{iy} + e^{-iy}}{2} = \frac{(\cos y + i \sin y) + (\cos y - i \sin y)}{2} = \cos y.$$

In this case, e^{iy} and e^{-iy} are of the same size, order 1, so we cannot say that $\cosh z \sim e^z/2$ along this line.

Of course, this path is not in the half plane $\mathrm{Re}(z) > 0$, but we can consider a parallel path $z = 1+iy$. Then $\cosh(1+iy) = \cosh(1)\cos(y) + i \sinh(1)\sin(y)$, which still is of order 1 as $y \to \infty$. So in order to have the relation $\cosh(z) \sim e^z/2$ hold, we can only consider those paths in which the real part of z increases to infinity.

In general, the region of validity for an asymptotic relation is a *sector*. A sector is a region in the complex plane for which $\alpha < \arg(z - z_0) < \beta$, and $z \neq z_0$. See figure 2.10. The *opening* of the sector is the angle $\beta - \alpha$. It is possible (but rare) for the opening to be greater than 2π, so when $z_0 = 0$ the sector represents a portion of the Riemann surface shown in figure 2.6. However, it is insufficient to consider all paths that stay within this sector, for we must exclude those paths that run parallel to the sides of the sector. Hence, we have the following definition.

DEFINITION 2.1 We say that $f(z)$ is *Gevrey asymptotic* to $g(z)$ as $z \to z_0$ in the sector $\alpha < \arg(z - z_0) < \beta$, if, for every path which lies entirely inside $\alpha + \epsilon < \arg(z - z_0) < \beta - \epsilon$ for some $\epsilon > 0$, we have

$$\lim_{z \to z_0} \frac{f(z)}{g(z)} = 1.$$

Likewise, we say that $f(z)$ has a Gevrey asymptotic series $\sum_{i=1}^{n} g_i(z)$ as $z \to z_0$ in the sector $\alpha < \arg(z - z_0) < \beta$ if for every n,

$$\lim_{z \to z_0} \frac{1}{g_n(z)} \left(f(z) - \sum_{i=1}^{n-1} g_i(z) \right) = 1$$

for all such paths. For the case where $z_0 = \infty$, use the same sectors as with $z_0 = 0$, that is, $\alpha < \arg(z) < \beta$.

The presence of the ϵ in the definition eliminates the paths that run parallel to a side of the sector. See figure 2.11. For example, we can say that $\cosh x$

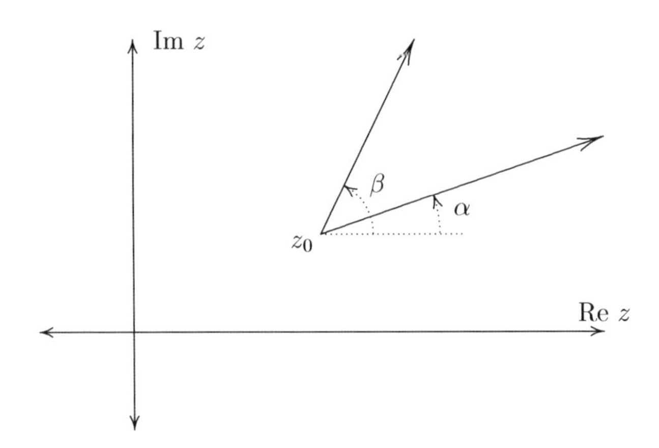

FIGURE 2.10: This shows a typical sector for which an asymptotic relation is valid.

is Gevrey asymptotic to $e^x/2$ as $x \to \infty$ in the sector $-\pi/2 < \arg(z) < \pi/2$. From now on, if a sector is mentioned with the asymptotic relation or series, it will be understood to be a Gevrey asymptotic relation.

One can often determine the sector for which the asymptotic formula holds by carefully examining the logic that produced that formula.

Example 2.24

Determine the sector of validity for problem 8 of section 2.2,

$$\int_0^x e^{t^2}\,dt \qquad \text{as } x \to \infty.$$

SOLUTION: When x is real and positive, we see that the maximum of the exponential function occurs when $t = x$, so we would change the endpoints to

$$\int_{x-\epsilon}^x e^{t^2}\,dt \qquad \text{as } x \to \infty.$$

When x is complex, we treat the original integral as a contour integral along any path from 0 to x, so we can have the path go through $x - \epsilon$, where ϵ is real and positive. But will

$$\int_0^{x-\epsilon} e^{t^2}\,dt \qquad \text{be subdominant to} \qquad \int_{x-\epsilon}^x e^{t^2}\,dt?$$

If x is a number with argument $\pm\pi/4$, then x^2 will have argument $\pm\pi/2$, that is, will be purely imaginary. Then e^{x^2} will be of order 1, and $e^{(x-\epsilon)^2}$ will be

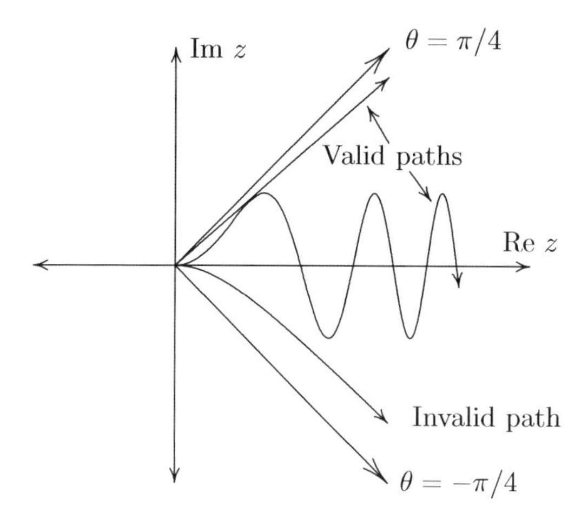

FIGURE 2.11: The sector of validity for the asymptotic series of $\int_0^x e^{t^2}\,dt$ as $x \to \infty$. Shown are two valid paths for the Gevrey asymptotics. One path is invalid, since its asymptote is parallel to a side of the sector, hence the path is not contained is a slightly smaller sector.

of the same order, so the argument for Laplace's method breaks down. Thus, the sector for validity for the asymptotic series is from $-\pi/4 < \arg(x) < \pi/4$, as shown in figure 2.11. ▯

The reason why the asymptotic relations break down near the sides of the sector is solely because of the property of the exponential function. When the exponent is purely imaginary, the exponential function has a magnitude of 1, so no component of the function can be considered as subdominant to another. We define the *Stokes lines* as being the lines in the complex plane where the exponent of the controlling exponential function is purely imaginary. In the last example, the Stokes lines were at $\theta = \pm\pi/4$ and $\pm 3\pi/4$. As a path crosses a Stokes line, the dominant and subdominant components of a function will often switch places, resulting in a totally different asymptotic behavior on the other side of the Stokes line. This exchange of dominance and subdominance is known as the *Stokes phenomenon*.

If the real part of the exponent increases as the path crosses a Stokes line, then the asymptotic expansion will still be valid past this Stokes line. For example, equation 2.8, the series for $\operatorname{erf}(x)$ as $x \to \infty$, has the controlling exponential function e^{-x^2}. The exponent will be purely complex when $\theta = \pm\pi/4$ and $\pm 3\pi/4$. But when a path crosses the Stokes line at $\theta = \pi/4$, the real part of the exponent changes from negative to positive. There may be other components to the function on this side of the line, but they will be subdominant to e^{-x^2}, which is now growing exponentially. Hence, the series in equation 2.8 will still be valid. But as we cross the Stokes line $\theta = 3\pi/4$, the real part of the exponent changes back to negative, and any component of $\text{erf}(x)$ which was subdominant will become dominant. So the series in equation 2.8 is valid in the sector $-3\pi/4 < \theta < 3\pi/4$.

Example 2.25

Find the sector of validity for the asymptotic series for the Stieltjes integral function

$$S(z) \sim 1 - z + 2z^2 - 3!z^3 + 4!z^4 + \cdots \qquad \text{as} \quad z \to \infty.$$

SOLUTION: In the definition

$$S(z) = \int_0^\infty \frac{e^{-t}}{1+zt}$$

the exponential function does not change with z, so at first there doesn't seem to be any Stokes lines. However, we saw in example 2.23 that if we go into the next sheet of the Riemann surface, the contour will be distorted, causing the t to take on values that have a negative real component. In fact, the next sheet of the surface evaluates to $2\pi i e^{1/z}/z$ more than the first sheet. As long as the real part of z is negative, this new part will be subdominant to the series, but as the real part of z becomes positive again after going around the origin, the additional term will dominate the behavior of $S(z)$. So the Stokes line does appear, but in the next sheet of the Riemann surface. Thus, the sector of validity is $-3\pi/2 < \theta < 3\pi/2$. This example is interesting because the opening of the sector is more than 2π, meaning that the same series is valid on two different sheets for a portion of the plane. This is possible because the difference of the function on these two sheets is subdominant to the series. □

2.4.4 Asymptotics for Oscillatory Functions

There is a subtle problem with asymptotic relations as $x \to \infty$ for functions that oscillate, such as $\sin(x)$ and $\cos(x)$. Because such functions have an infinite sequence of zeros, we cannot say that $f(x) \sim f(x)$ as $x \to \infty$. For example, the limit

$$\lim_{x\to\infty} \frac{\sin x}{\sin x}$$

does not exist, since the fraction is undefined whenever x is a multiple of π. Technically, for $\lim_{x\to\infty} g(x)$ to exist, $g(x)$ must be defined for sufficiently large x. This problem can be remedied by redefining the limit so that only values in the domain of $g(x)$ are used, but there is another problem that occurs with oscillating functions.

Consider, for example, the function $f(x) = \sin((x^2+1)/(x+1))$. Since

$$\frac{x^2+1}{x+1} \sim x - 1 + \frac{2}{x} - \frac{2}{x^2} + \cdots \qquad \text{as } x \to \infty,$$

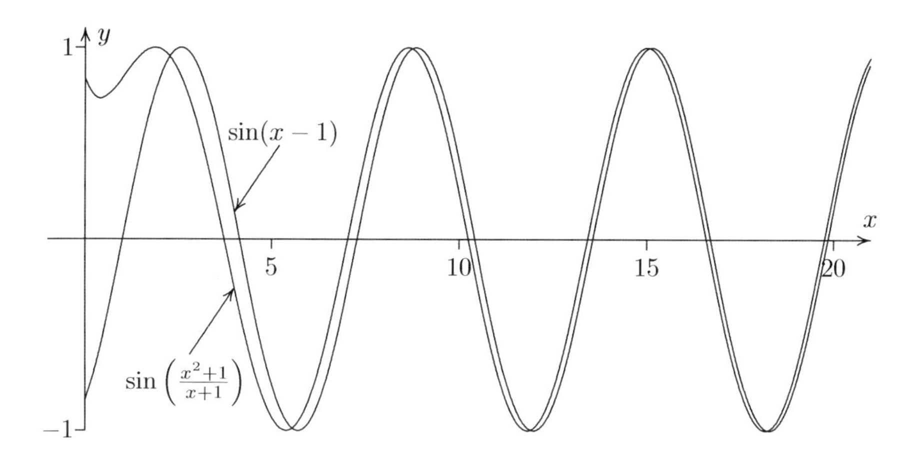

FIGURE 2.12: The graphs of $\sin((x^2+1)/(x+1))$ and $\sin(x-1)$ become nearly identical as $x \to \infty$, but the zeros of the functions are not coincident. Thus, these functions are not asymptotic as $x \to \infty$.

one would expect that this function is asymptotic to $g(x) = \sin(x-1)$ as $x \to \infty$. Indeed, the graphs of the two functions become nearly identical when x is large, as shown in figure 2.12. However, the ratio of two functions $f(x)/g(x)$ will have an infinite number of asymptotes, occurring whenever $g(x)$ is zero, and $f(x)$ is non-zero. So the limit of the ratio $f(x)/g(x)$ cannot exist.

The way to work around this complication is to express the oscillating function in terms of standard trig functions and unknown functions which are not oscillating. We can then find the asymptotic series of the unknown functions. For functions that behave like $\sin(x)$ or $\cos(x)$, we can first set $f(x) = A(x)\cos(\omega(x)x) + B(x)\sin(\omega(x)x)$, where $\omega(x)$ is the frequency of the function, possibly changing as $x \to \infty$. We can then find asymptotic approximations for $A(x)$, $B(x)$, and $\omega(x)$.

Example 2.26

Find the asymptotic behavior of $f(x) = \sin((x^2+1)/(x+1))$ as $x \to \infty$.

SOLUTION: We can rewrite

$$\frac{x^2+1}{x+1} = x - 1 + \frac{2}{x+1}.$$

Using the trig identities

$$\sin(x+y) = \sin(x)\cos(y) + \cos(x)\sin(y), \tag{2.25}$$

$$\cos(x+y) = \cos(x)\cos(y) - \sin(x)\sin(y), \tag{2.26}$$

we have that

$$f(x) = \sin(1 + 2/(x+1))\cos(x) + \cos(1 + 2/(x+1))\sin(x).$$

Thus, $\omega(x) = 1$, $A(x) = \sin(1 + 2/(x+1))$, and $B(x) = \cos(1 + 2/(x+1))$. These three functions are not oscillating, so we can find the asymptotic series for $A(x)$ and $B(x)$ as $x \to \infty$. However, in this case it is more convenient if we first add a *phase shift constant* δ. That is, we let

$$f(x) = A(x)\cos(\omega(x)x - \delta) + B(x)\sin(\omega(x)x - \delta).$$

Then $\omega(x) = 1$, $\delta = 1$,

$$A(x) = \sin\left(\frac{2}{x+1}\right) \sim \frac{2}{x} - \frac{2}{x^2} + \frac{2}{3x^3} + \frac{2}{x^4} + \cdots \text{ as } x \to \infty,$$

and

$$B(x) = \cos\left(\frac{2}{x+1}\right) \sim 1 - \frac{2}{x^2} + \frac{4}{x^3} - \frac{16}{3x^4} + \cdots \text{ as } x \to \infty.$$

□

Example 2.27

Find the asymptotic series for the integral

$$\int_x^\infty \frac{\cos(t^2)}{t}\, dt.$$

SOLUTION: Integrating by parts, we begin with $u = 1/t^2$, $dv = t\cos(t^2)$, so $v = \sin(t^2)/2$. Thus we have

$$\begin{aligned}
\int_x^\infty \frac{\cos(t^2)}{t}\, dt &= \frac{\sin(t^2)}{2t^2}\bigg|_x^\infty + \int_x^\infty \frac{\sin(t^2)}{t^3}\, dt \\
&= -\frac{\sin(x^2)}{2x^2} - \frac{\cos(t^2)}{2t^4}\bigg|_x^\infty + \int_x^\infty \frac{-2\cos(t^2)}{t^5}\, dt \\
&= -\frac{\sin(x^2)}{2x^2} + \frac{\cos(x^2)}{2x^4} - \frac{\sin(t^2)}{t^6}\bigg|_x^\infty - \int_x^\infty \frac{6\sin(t^2)}{t^7}\, dt \\
&= -\frac{\sin(x^2)}{2x^2} + \frac{\cos(x^2)}{2x^4} + \frac{\sin(x^2)}{x^6} + \frac{3\cos(t^2)}{t^8}\bigg|_x^\infty \\
&\qquad + \int_x^\infty \frac{24\cos(t^2)}{t^9}\, dt \ldots.
\end{aligned}$$

The remainder integral is decreasing in size as $x \to \infty$, but to display the asymptotic relationship, we say that

$$\int_x^\infty \frac{\cos(t^2)}{t}\, dt = A(x)\cos(x^2) + B(x)\sin(x^2),$$

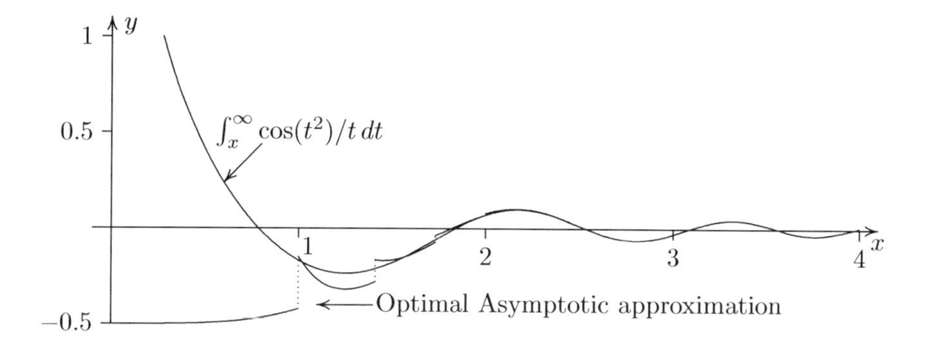

FIGURE 2.13: Comparing the true graph of $\int_x^\infty \cos(t^2)/t\,dt$ with its optimal asymptotic approximation. Note that for $x > 2$, the curves are indistinguishable.

where

$$A(x) \sim \frac{1}{2x^4} - \frac{3}{x^8} + O(x^{-12}), \qquad B(x) \sim -\frac{1}{2x^2} + \frac{1}{x^6} + O(x^{-10}) \quad \text{as } x \to \infty.$$

Note that in this example, the frequency changes with x. ▯

Another way to handle the asymptotics of an oscillating function is to consider it as the real (or imaginary) component of a complex function that has no zeros. For example, the integral in example 2.27 can be viewed as the real part of

$$I(x) = \int_x^\infty \frac{e^{it^2}}{t}\,dt,$$

since $e^{it^2} = \cos(t^2) + i\sin(t^2)$. Although the complex function still has an oscillating behavior, $I(x)$ will not have any zeros on the real axis, so we can find an asymptotic series for $I(x)$. Integrating by parts, we have

$$\int_x^\infty \frac{e^{it^2}}{t}\,dt \sim e^{ix^2}\left(\frac{i}{2x^2} + \frac{1}{2x^4} - \frac{i}{x^6} - \frac{3}{x^8} + \frac{12i}{x^{10}} + \cdots\right).$$

Figure 2.13 shows the real component of the optimal asymptotic approximation of this series verses the original integral. Note that for complex asymptotic series such as this one, the optimal asymptotic approximation is based upon the point in the series where the absolute value of the complex terms start to increase. That is, we stop just before the term with the smallest absolute value.

Problems for §2.4

For problems **1** through **10**: Compute the following complex arithmetic operations. For multiply defined functions, find all possible solutions.

1 $(3+5i)/(2-3i)$

2 $(6-i)/(5+3i)$

3 $e^{(3-2i)}$

4 $e^{(1+\pi i/2)}$

5 2^{1+3i}

6 $\ln(2-2i)$

7 $\ln(-\sqrt{3}+i)$

8 $(-1)^{1/4}$

9 $64^{1/6}$

10 $(1-i\sqrt{3})^{1/2}$

11 Use the properties $\sin(a+b) = \cos a \sin b + \cos b \sin a$, $\sin(iy) = i\sinh(y)$, and $\cos(iy) = \cosh(y)$ to derive a formula for $\sin(x+iy)$.

12 Use the technique of problem 11 to find a formula for $\cos(x+iy)$.

13 Show that in order for the two dimensional limit in equation 2.20 to exist, the Cauchy-Riemann equations 2.21 must be satisfied.

Hint: Consider the two paths $\Delta y = 0, \Delta x \to 0$, and $\Delta x = 0, \Delta y \to 0$.

14 Show that the function $e^z = e^{x+iy} = e^x \cos y + ie^x \sin y$ satisfies the Cauchy-Riemann equations 2.21.

15 Decompose the complex function $z^3 = (x+iy)^3$ into its real and complex parts. Then show that these parts satisfy the Cauchy-Riemann equations 2.21.

16 Although the series for the Riemann zeta function (equation 2.23) only converges for $\mathrm{Re}z > 1$, its alternating counterpart,

$$\eta(z) = \sum_{n=1}^{\infty} \frac{(-1)^{n+1}}{n^z} = 1 - \frac{1}{2^z} + \frac{1}{3^z} - \frac{1}{4^z} + \cdots$$

converges if $\mathrm{Re}(z) > 0$. Show that when $\mathrm{Re}(z) > 1$, the series converges to $\zeta(z)(2^z-2)/2^z$.

17 Show how we can use the series in problem 16 to analytically continue the zeta function to the region $0 < \mathrm{Re}(z) < 1$.

18 Problem 17 allows us to define the Riemann zeta function in the strip $0 < \mathrm{Re}(z) < 1$. Within this strip, $\zeta(z)$ satisfies the *reflection property*

$$\zeta(1-z) = 2^{1-z}\pi^{-z}\cos(\pi z/2)\Gamma(z)\zeta(z).$$

Use the reflexion property to analytically continue $\zeta(z)$ to the entire plane, except for $z = 1$.

19 Use the result of problem 16 to find the residue of the pole at $z = 1$ for the Riemann zeta function $\zeta(z)$.

20 Show that if $F(z) = \phi_1(z) + i\phi_2(z)$ is the potential function for the integrals in equation 2.24, then $F(z)$ satisfies the Cauchy-Riemann equations 2.21.

21 Show that if $F(z) = \phi_1(z) + i\phi_2(z)$ is the potential function for the integrals in equation 2.24, then $F'(z) = u(x, y) + iv(x, y) = f(z)$.

22 Show that if $f(z)$ has an isolated second order pole at z_0, then $f(z)$ can be written as

$$f(z) = \frac{A}{(z - z_0)^2} + \frac{R}{(z - z_0)} + g(z),$$

where $g(z)$ is analytic. Show that the contour integral $\oint_C f(z)\,dz$, where C goes counter-clockwise around z_0, and encloses no other singularities of $f(z)$, will evaluate to $2\pi Ri$. That is, R is still the residue of the pole.

23 Generalize problem 22 for a third order pole. That is, if $f(z)$ can be written as

$$f(z) = \frac{A}{(z - z_0)^3} + \frac{B}{(z - z_0)^2} + \frac{R}{(z - z_0)} + g(z),$$

find $\oint_C f(z)\,dz$, where C goes counter-clockwise around z_0, and encloses no other singularities of $f(z)$.

24 Use problem 23 to find the residue of the function $\cot(\pi z)/z^2$ at the third order pole $z = 0$.

25 The function $\cot(\pi z)/z^2$ has simple poles at all non-zero integers. Find the residues of these simple poles.

26 It can be shown, by integrating over larger and larger squares with corners at $(\pm 1 \pm i)(N + \frac{1}{2})$, that the sum of *all* of the residues of the function $\cot(\pi z)/z^2$ is zero. Use this, along with the results of problems 24 and 25, to show that $\zeta(2) = \pi^2/6$.

For problems **27** through **31**: Find the sector of validity for the following asymptotic relations.

27 $$\int_0^\pi e^{x \sin t}\,dt \sim 2e^x \sum_{n=0}^\infty \frac{(2n)!\Gamma(n + 1/2)}{8^n \sqrt{2}(n!)^2 x^{n+1/2}} \qquad \text{as } x \to \infty.$$

28 $$\int_0^1 \frac{e^{xt}}{1 + t^2}\,dt \sim \frac{\pi}{4} + \frac{\ln 2}{2}x + \frac{4 - \pi}{8}x^2 + \cdots \qquad \text{as } x \to 0.$$

29 $$\int_0^x e^{t^3}\,dt \sim e^{x^3}\left(\frac{1}{3x^2} + \frac{2}{9x^5} + \frac{10}{27x^8} + \cdots\right) \qquad \text{as } x \to \infty.$$

30 $$\int_x^\infty e^{-t^3}\,dt \sim e^{-x^3}\left(\frac{1}{3x^2} - \frac{2}{9x^5} + \frac{10}{27x^8} + \cdots\right) \qquad \text{as } x \to \infty.$$

31 $$\int_0^1 e^{-xt^2}\,dt \sim \frac{1}{2}\sqrt{\frac{\pi}{x}} + e^{-x}\left(-\frac{2}{x} + \frac{4}{x^2} - \frac{3}{8x^3} + \cdots\right) \qquad \text{as } x \to \infty.$$

2.5 Method of Stationary Phase

We have seen that asymptotic expansions are valid in sectors, and the boundary of the sector, called the Stoke's line, occurs where the exponent in the controlling exponential function is purely imaginary. But what happens if we consider the path along the Stoke's line? For example, consider the problem

$$\int_0^\pi e^{x \sin t}\, dt.$$

We have an asymptotic series valid as $x \to \infty$, but this series breaks down along the Stoke's line $\arg(z) = \pi/2$. Traveling along this Stoke's line can be accomplished by replacing x with iy, and consider the real variable $y \to \infty$. Thus, we must analyze

$$\int_0^\pi e^{iy \sin t}\, dt.$$

This is an example of a *generalized Fourier integral.* Any integral of the form

$$\int_a^b f(t) e^{iy\phi(t)}\, dt$$

is a generalized Fourier integral if $f(t)$, $\phi(t)$, y, a, and b are all real. For the particular case where $\phi(t) = t$, we have an *ordinary Fourier integral.* Although the i in the exponent causes the integral to be complex, it is easy to break up a Fourier integral into real and complex parts.

$$\int_a^b f(t) e^{iy\phi(t)}\, dt = \int_a^b f(t) \cos(y\phi(t))\, dt + i \int_a^b f(t) \sin(y\phi(t))\, dt.$$

To understand the asymptotic behavior as $y \to \infty$, consider the graphs of $\cos(y\phi(t))$ and $\sin(y\phi(t))$. When y is large and $\phi'(t) \neq 0$, the phase angle inside the trig function, $y\phi(t)$, will be changing quickly, causing the function to oscillate rapidly. When this rapid oscillation is integrated, the areas above and below the axis tend to cancel out. The exception is for points where $\phi'(t)$ is close to 0, causing the phase angle $y\phi(t)$ to be nearly stationary. See figure 2.14 for an example. Hence, the leading behavior of the integral occurs near the points where $\phi'(t) = 0$. To prove this formally, we will need to use the Riemann-Lebesgue lemma.

LEMMA 2.2: Riemann-Lebesgue Lemma
If $\int_a^b |f(t)|\, dt$ is finite, then

$$\lim_{y \to \infty} \int_a^b f(t) e^{iyt}\, dt = 0.$$

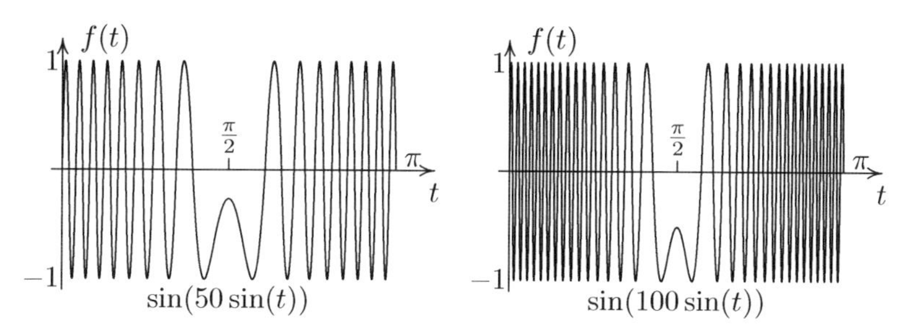

FIGURE 2.14: Plot of $\sin(y\sin(t))$ for $y = 50$ and $y = 100$. As y increases, the oscillations get tighter, causing cancellation in the integral. The exception is when the derivative of $\sin(t)$ is near 0, for the phase becomes stationary momentarily. But as $y \to \infty$, the region of this non-oscillatory behavior becomes thiner. By the Riemann-Lebesgue lemma, the integral approaches 0 as $y \to \infty$.

PROOF: First consider the case where $f(t)$ is continuously differentiable on the interval $a \le t \le b$. In this case, we can integrate by parts and find that

$$\int_a^b f(t)e^{iyt}\,dt = \frac{f(b)e^{iyb} - f(a)e^{iya}}{iy} - \frac{1}{iy}\int_a^b f'(t)e^{iyt}\,dt.$$

As $y \to \infty$, it is clear that this expression goes to 0 provided that $\int_a^b f'(t)e^{iyt}\,dt$ is bounded. But

$$\left|\int_a^b f'(t)e^{iyt}\,dt\right| \le \int_a^b |f'(t)e^{iyt}|\,dt = \int_a^b |f'(t)|\,dt.$$

Since $f(t)$ is continuously differentiable on the interval $a \le x \le b$, then $f'(t)$ will be continuous, so the last integral will be finite.

For an arbitrary function $f(t)$ for which $\int_a^b |f(t)|\,dt$ is finite, for any $\epsilon > 0$ we can approximate the function with a continuously differentiable function $g(t)$ such that

$$\int_a^b |f(t) - g(t)|\,dt < \epsilon.$$

Then we can express

$$\int_a^b f(t)e^{iyt}\,dt = \int_a^b g(t)e^{iyt}\,dt + \int_a^b (f(t) - g(t))e^{iyt}\,dt.$$

The first of these integrals will approach zero, and the second integral can be estimated by

$$\left|\int_a^b (f(t) - g(t))e^{iyt}\,dt\right| \le \int_a^b |(f(t) - g(t))e^{iyt}|\,dt = \int_a^b |f(t) - g(t)|\,dt < \epsilon.$$

Since ϵ was arbitrary, we have

$$\lim_{y\to\infty}\int_a^b f(t)e^{iyt}\,dt = 0.$$

∎

The main usefulness of this lemma is to prove the following corollary.

COROLLARY 2.1

Let $\phi(t)$ is continuously differentiable for $a \le t \le b$, such that $\phi'(x) = 0$ at only a finite number of points on the interval. If $\int_a^b |f(t)|\,dt$ converges, then

$$\lim_{y\to\infty}\int_a^b f(t)e^{iy\phi(t)}\,dt = 0.$$

PROOF: Since $\phi(t)$ is not constant on any subinterval, $\phi'(t) = 0$ at only a finite number of points on the interval $a \le t \le b$. Thus, we can divide the interval into subintervals, where $\phi'(t)$ can only be 0 at the endpoint of a subinterval. Thus, $\phi(t)$ will be monotonic within each subinterval.

Let us consider the case where $\phi(t)$ is increasing on $c < t < d$. Then $\phi(t)$ is invertible on this interval, and we can substitute $u = \phi(t)$, so that $du = \phi'(t)\,dt$. This gives us

$$\int_c^d f(t)e^{iy\phi(t)}\,dt = \int_{\phi(c)}^{\phi(d)} \frac{f(\phi^{-1}(u))}{\phi'(\phi^{-1}(u))}e^{iyu}\,du. \tag{2.27}$$

Even though this new integrand has the potential of being undefined at the endpoints, note that

$$\int_{\phi(c)}^{\phi(d)} \left|\frac{f(\phi^{-1}(u))}{\phi'(\phi^{-1}(u))}\right|\,du = \int_c^d |f(t)|\,dt < \infty.$$

Hence, we can apply the Riemann-Lebesgue lemma to equation 2.27, so

$$\lim_{y\to\infty}\int_c^d f(t)e^{iy\phi(t)}\,dt = 0.$$

The case where $\phi(t)$ is decreasing on the interval $c < t < d$ is similar. Since we have proven that the integral approaches 0 on each of the subintervals, we can combine these together to show that $\int_a^b f(t)e^{iy\phi(t)}\,dt$ will approach 0 as $y \to \infty$. ∎

We are now ready to use the Riemann-Lebesgue lemma to find the asymptotic approximations to integrals of the form

$$\int_a^b f(t)e^{iy\phi(t)}\,dt$$

as $y \to \infty$. If $\phi'(t)$ is non-zero on the interval, then we can integrate by parts and show that the integral decays like $1/y$:

$$\int_a^b \frac{f(t)}{iy\phi'(t)} iy\phi'(t)e^{iy\phi(t)}\,dt = \frac{f(t)}{iy\phi'(t)}e^{iy\phi(t)}\Big|_a^b - \frac{1}{iy}\int_a^b \left(\frac{f(t)}{\phi'(t)}\right)' e^{iy\phi(t)}\,dt.$$

By the corollary 2.1, the final integral is going to 0, so the last term decays faster than $1/y$. As long as $f(a)$ and $f(b)$ exist, and are not both zero, the first term of the asymptotic series is easily calculated.

If $\phi'(t) = 0$ for some point on the interval, then the leading behavior will be determined by the small portion of the integral near that point. Hence, we can proceed in the same manner as Laplace's method. By breaking the integral into pieces if necessary, each piece can have the point of stationary phase, that is, where $\phi'(t) = 0$, at one of the endpoints, say a. Then we can decompose the integral into two terms

$$\int_a^{a+\epsilon} f(t)e^{iy\phi(t)}\,dt + \int_{a+\epsilon}^b f(t)e^{iy\phi(t)}\,dt.$$

We have already seen that the second integral will have order $1/y$, since $\phi'(t) \neq 0$ on the interval $a + \epsilon \le t \le b$, so let us consider the first integral. Since ϵ is small, we can replace $f(t)$ with $f(a)$, and $\phi(t)$ with its Taylor series up to its first non-constant term. That is, $\phi(t) \approx \phi(a) + k(t - a)^n$, where $k \neq 0$ and $n > 0$. If $f(a) \neq 0$, this will give us the leading behavior for the first integral.

$$\int_a^{a+\epsilon} f(t)e^{iy\phi(t)}\,dt \sim \int_a^{a+\epsilon} f(a)e^{iy\phi(a)}e^{iyk(t-a)^n}\,dt.$$

We can make the substitution $u = t-a$, and in fact change the upper endpoint to ∞ which will introduce a term that decays like $1/y$. So

$$\int_a^{a+\epsilon} f(t)e^{iy\phi(t)}\,dt \sim f(a)e^{iy\phi(a)}\int_0^\infty e^{iyku^n}\,du.$$

This last integral can be computed exactly by deforming the contour. If $k > 0$, consider the contour shown in figure 2.15. Because the integrand is analytic everywhere, the integral is independent of the path, so instead of integrating along the real axis from 0 to R, we can integrate along an angle of $\pi/2n$, and then through an arc denoted by C_2. Inside this sector, $\mathrm{Re}(iku^n) < 0$, so the integral along C_2 approaches 0 exponentially as $R \to \infty$. Hence, the value of the integral is unchanged by integrating along the ray $u = se^{\pi i/2n}$, with s going from 0 to ∞ along the real axis.

The advantage of integrating along this angle is that the exponent becomes real and negative.

$$\int_0^\infty e^{iyku^n}\,du = e^{\pi i/2n}\int_0^\infty e^{-yks^n}\,ds.$$

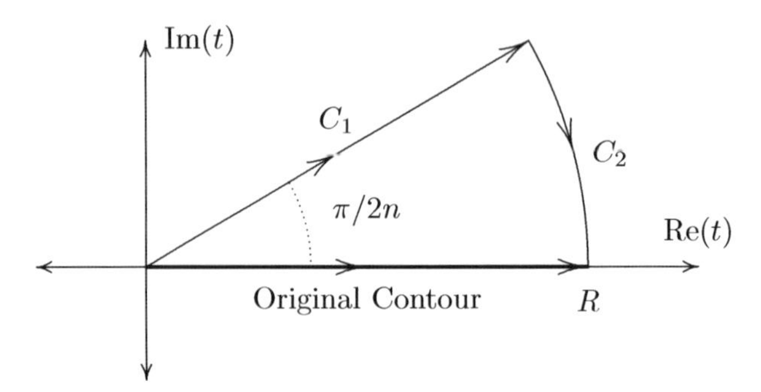

FIGURE 2.15: The contour along the real axis from 0 to R can be deformed into an angular ray C_1, plus an arc C_2. The integral for C_2 quickly goes to 0 as $R \to \infty$, so we get the same result integrating along the angular ray.

This integral can be converted to a gamma integral with a substitution $v = yks^n$, with $ds = v^{-1+1/n}y^{-1/n}k^{-1/n}dv/n$

$$\int_0^\infty e^{iyku^n}\, du = e^{\pi/2n}\frac{1}{ny^{1/n}k^{1/n}}\int_0^\infty v^{-1+1/n}e^{-v}\, dv$$
$$= e^{\pi/2n}\frac{1}{ny^{1/n}k^{1/n}}\Gamma(1/n).$$

If $k < 0$, we use a mirror image of figure 2.15, so that the new ray is turned $\pi/2n$ clockwise from the real axis. Using a similar argument,

$$\int_0^\infty e^{iyku^n}\, du = e^{-\pi i/2n}\frac{1}{ny^{1/n}(-k)^{1/n}}\Gamma(1/n).$$

These two cases can be combined, giving the case where the stationary phase occurs at the lower endpoint:

$$\boxed{\text{If } f(a) \neq 0, \int_a^b f(t)e^{iy\phi(t)}\, dt \sim f(a)e^{iy\phi(a)}e^{|k|\pi i/(2kn)}\frac{\Gamma(1/n)}{ny^{1/n}\,|k|^{1/n}}.} \qquad (2.28)$$

The case where $f(a) = 0$ must be dealt with in a different way, since the dominant behavior will usually not come from the stationary phase. The method of steepest descents, covered in the next section, would have to be used.

Example 2.28
Find the leading behavior as $y \to \infty$ of

$$\int_0^\pi e^{iy\sin(t)}\, dt.$$

SOLUTION: By symmetry, $\int_0^{\pi/2} e^{iy\sin(t)}\,dt = \int_{\pi/2}^{\pi} e^{iy\sin(t)}\,dt$, so we can consider the leading behavior of

$$2\int_{\pi/2}^{\pi} e^{iy\sin(t)}\,dt.$$

We now have the stationary phase occurring at the lower endpoint. Here, $f(a) = 1$, and the Taylor series for $\phi(t) = \sin(t)$ at $a = \pi/2$ is

$$1 - (t - \pi/2)^2/2 + \cdots.$$

So $n = 2$ and $k = -1/2$, and equation 2.28 gives us

$$2e^{iy}e^{-\pi i/4}\frac{\Gamma(1/2)}{2\sqrt{y/2}} = (\cos y + i\sin y)(1 - i)\sqrt{\frac{\pi}{y}}.$$

∎

Example 2.29
Find the leading behavior as $y \to \infty$ of

$$\int_0^1 \cos(yt^3 + t)\,dt.$$

SOLUTION: We can think of this integral as the real part of

$$\int_0^1 e^{iyt^3+it}\,dt = \int_0^1 e^{it}e^{iyt^3}\,dt.$$

So $f(t) = e^{it}$ and $\phi(t) = t^3$. The stationary phase occurs at $t = 0$, which is already at the lower endpoint of integration. It is clear that $n = 3$ and $k = 1$, so equation 2.28 produces

$$\int_0^1 e^{iyt^3+it}\,dt \sim e^0 e^{\pi i/6}\frac{\Gamma(1/3)}{3\sqrt[3]{y}}.$$

Taking the real part, we get

$$\int_0^1 \cos(yt^3 + t)\,dt \sim \frac{\Gamma(1/3)}{2\sqrt{3}\sqrt[3]{y}} \qquad \text{as} \quad y \to \infty.$$

∎

Problems for §2.5

For problems **1** through **6**: Use integration by parts to find the first three terms of the asymptotic series as $x \to \infty$ for the following complex integrals.

1 $\displaystyle\int_0^1 \ln(t+1)e^{ixt}\,dt$

2 $\displaystyle\int_0^1 \frac{e^{ixt}}{2+t}\,dt$

3 $\displaystyle\int_0^3 \sqrt{t+1}\cos(xt)\,dt$

4 $\displaystyle\int_1^2 e^{ixt^2}\,dt$

5 $\displaystyle\int_1^2 \ln(t)e^{ix(t^2-1)}\,dt$

6 $\displaystyle\int_1^2 \frac{\sin(xt^2)}{t}\,dt$

For problems **7** through **16**: Use the method of stationary phase to find the leading behavior as $y \to \infty$ of the following integrals.

7 $\displaystyle\int_0^1 \cos(t)e^{iyt^2}\,dt$

8 $\displaystyle\int_0^1 \sin(t)e^{iy(t^3-3t)}\,dt$

9 $\displaystyle\int_0^2 \frac{1}{t+2}e^{iy(\sin(t)-t)}\,dt$

10 $\displaystyle\int_0^1 \Gamma(t+1)e^{iy(2\cos(t)+t^2)}\,dt$

11 $\displaystyle\int_{-1}^1 \ln(t+3)e^{iy(\tan(t)-t)}\,dt$

12 $\displaystyle\int_{-1}^1 \frac{1}{t+2}e^{iy(t^3-3t)}$

13 $\displaystyle\int_0^\pi \cos(t+yt^2)\,dt$

14 $\displaystyle\int_0^{\pi/2} \sin(2y+yt^3)\,dt$

15 $\displaystyle\int_0^\pi \cos(y\cos t)\,dt$

16 $\displaystyle\int_0^\pi \sin(y\cos(t)+yt^2/2)\,dt$

17 When n is an integer, the *Bessel function of order* n, denoted by $J_n(x)$, can be expressed by the integral

$$J_n(x) = \frac{1}{\pi}\int_0^\pi \cos(x\sin(t) - nt)\,dt.$$

Find the leading behavior of $J_0(x)$, $J_1(x)$, and $J_2(x)$ as $x \to \infty$.

For problems **18** through **23**: Equation 2.28 cannot be used if $f(a) = 0$. Instead, we can keep the first term in the series for $f(t)$, and perform the same analysis that was done to derive equation 2.28. However, if the leading term is not larger than $1/y$, then integration by parts should be used. Use this technique to find the leading term as $y \to \infty$ for the following integrals.

18 $\displaystyle\int_0^\pi \sin(t)e^{iyt^2}\,dt$

19 $\displaystyle\int_0^1 \tan(t)e^{iyt^3}\,dt$

20 $\displaystyle\int_0^{\pi/2} \sin^2(t)e^{iy(\sin t - t)}\,dt$

21 $\displaystyle\int_0^1 \sinh(t)e^{iy(2\cosh t - t^2)}\,dt$

22 $\displaystyle\int_0^\pi (\cos t - 1)e^{iy(\cos t + t^2/2)}\,dt$

23 $\displaystyle\int_0^\pi (\sin t - t)e^{iy\sin^4(t/3)}\,dt$

2.6 Method of Steepest Descents

The problem with the method of stationary phase is that it only provides the leading behavior of the integral. In order to obtain more terms of the asymptotic series, we need a more powerful method akin to Laplace's method. The method of steepest descents gives us such a technique. Given a contour integral of the form

$$\int_C f(z)e^{x\phi(z)}\,dz$$

where $f(z)$ and $\phi(z)$ are analytic, we can deform the contour C to a new contour C' for which we can apply Laplace's method. The new contour will mainly follow curves for which $\text{Im}(\phi(z))$ is constant.

Let us demonstrate this method for the simple case of $\phi(z) = iz$. This produces the Fourier integral

$$\int_a^b f(z)e^{ixz}\,dz$$

where $b > a$. We begin by finding the curves for which $\text{Im}(\phi(z))$ is constant, which pass through the endpoints of the integral. In this case, these curves will be the straight lines $z = a+is$ and $z = b+is$, where s is a real parameter.

The next step is to determine which direction along these curves causes the real part of $\phi(z)$ to *decrease*. This direction is called the *direction of steepest descent*, and can be thought of as the "downstream" direction. In this example, $\phi(z) = iz = ia - s$, so the direction of steepest decent is the direction in which the parameter s is increasing. See figure 2.16. The direction of steepest descents is shown by arrows beside the curves.

The plan is to construct a new contour with the same endpoints as the original, but following the paths of steepest descent downstream from these two endpoints, labeled C_1 and C_3. Very rarely will these two steepest descent curves intersect, so we have to add a "bridge" C_2 that connects these two curves. In this case, the bridge is a horizontal line $z = s + Ri$, where R is a large constant. As the bridge goes further downstream, the contour integral for the bridge C_2 will generally approach 0. The new contour will start at a, go downstream along C_1, cut across the bridge C_2, and then go upstream along C_3 ending at b. As long as $f(t)$ is analytic both along and inside of this contour, the integral will be independent of the path. Finally, we let the C_2 contour go further and further downstream, causing the contribution from C_2 to go to 0, so we have two integrals along C_1 and C_3 for which Laplace's method can be applied.

In this example, the integral along C_1 is computed by the substitution $z = a + is$, producing

$$ie^{iax}\int_0^R f(a+is)e^{-sx}\,ds.$$

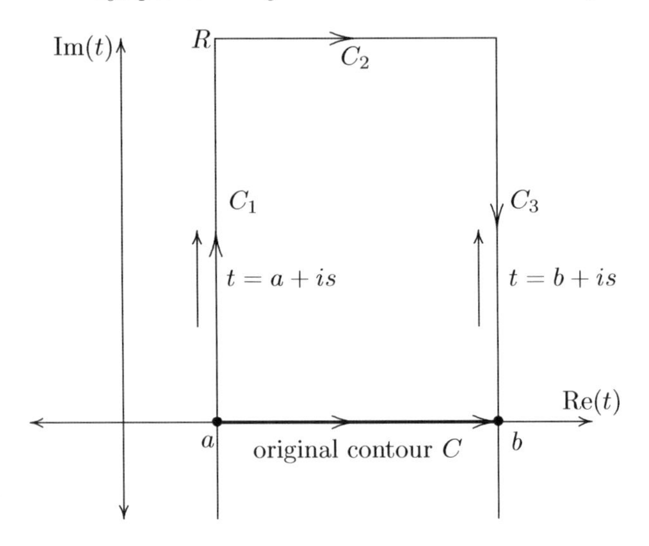

FIGURE 2.16: The paths of steepest descents for the integral $\int_a^b f(t)e^{ixt}\,dt$.

The asymptotic series as $x \to \infty$ can easily be found using Watson's lemma on the Taylor series of $f(a + is)$. Note that the asymptotic series does not depend on R.

The integral along C_3 is similar, but since this is going upstream, we add a negative sign. So this produces

$$-ie^{ibx} \int_0^R f(b + is)e^{-sx}\,ds.$$

Finally, the integral along C_2 will be

$$\int_a^b f(s + Ri)e^{-Rx}e^{isx}\,ds.$$

As long as $f(z)$ does not grow faster than an exponential function, that is, $|f(z)| < Ce^{a|z|}$ for some constants C and a, then for sufficiently large R, this will be subdominant to the other two asymptotic series as $x \to \infty$. Hence, we have

$$\boxed{\int_a^b f(t)e^{ixt}\,dt = ie^{iax} \int_0^\infty f(a + is)e^{-sx}\,ds - ie^{ibx} \int_0^\infty f(b + is)e^{-sx}\,ds.} \tag{2.29}$$

This assumes that $f(t)$ is analytic for all z with real part between a and b, and positive imaginary part. If there are poles in this region, we have to adjust this equation by adding $2\pi i$ times the sum of the residues within this region.

Example 2.30
Find the full asymptotic series for the integral

$$\int_0^1 \frac{1}{t+1} e^{ixt}\, dt.$$

SOLUTION: The function $f(t) = 1/(t+1)$ has only one singularity at $t = -1$, which is outside of the range of integration, so we can use equation 2.29 to obtain

$$\int_0^1 \frac{1}{t+1} e^{ixt}\, dt = i \int_0^\infty \frac{1}{1+is} e^{-sx}\, ds - ie^{ix} \int_0^\infty \frac{1}{2+is} e^{-sx}\, ds.$$

Using the two Maclaurin series

$$\frac{1}{1+is} = \sum_{n=0}^\infty (-i)^n s^n,$$

$$\frac{1}{2+is} = \sum_{n=0}^\infty \frac{1}{2}\left(\frac{-i}{2}\right)^n s^n,$$

we get the full asymptotic series as $x \to \infty$:

$$\int_0^1 \frac{1}{t+1} e^{ixt}\, dt \sim \sum_{n=0}^\infty \left(\frac{e^{ix}}{2^{n+1}} - 1\right) \frac{n!(-i)^{n+1}}{x^{n+1}} \qquad \text{as } x \to \infty.$$

□

Example 2.31
Find the full asymptotic series for the integral

$$\int_0^1 e^{ixt^3}\, dt.$$

SOLUTION: Because the exponent is not of the form ixt, we cannot directly use equation 2.29. But we can still find the steepest descent curves, that is, the curves for which the exponent ixt^3 has constant imaginary part. If we let $t = x + iy$, these curves can be represented by $x^3 - 3xy^2 = k$ for real constant k. Figure 2.17 shows some of these curves, along with the direction of steepest descent.

This figure also reveals our strategy for deforming the contour from 0 to 1 to a contour that is mostly along steepest descent lines. We can travel downstream along the 30° line C_1, across a small bridge C_2, and back to 1 along the curve $x^3 - 3xy^2 = 1$. It is clear that as the bridge goes further downstream, it gets smaller in length, but we also have to show that the integrand is decreasing. This is not hard, since the real part of the exponent is decreasing as we go further downstream, so the integrand is going to 0. Thus, C_2 will not contribute to the integral.

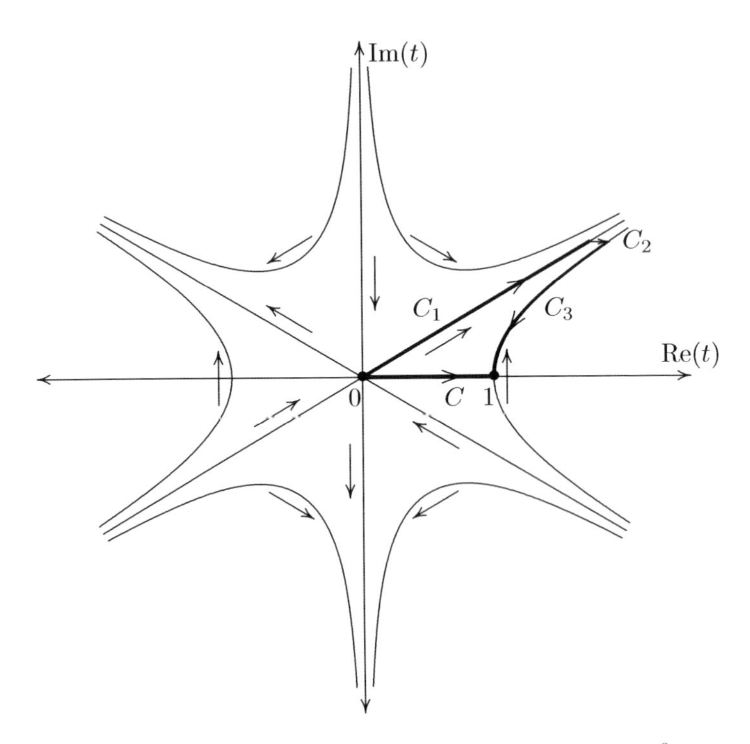

FIGURE 2.17: This shows the steepest descent lines for e^{ixt^3}. The arrows beside the lines show the downstream direction. The original contour C from 0 to 1 can be replaced by the contour C_1 going downstream, going across the bridge C_2, and finally going upstream along C_3 back to 1. As the bridge C_2 goes further downstream, its contribution approaches 0.

For C_1, we can make the substitution $xt^3 = iu$, so $t = e^{\pi i/6}u^{1/3}x^{-1/3}$, $dt = e^{\pi i/6}x^{-1/3}u^{-2/3}/3$. The integral for C_1 can be computed exactly using the gamma integral.

$$\int_0^\infty \frac{e^{\pi i/6}}{3x^{1/3}} u^{-2/3} e^{-u}\, du = \frac{e^{\pi i/6}\Gamma(1/3)}{3x^{1/3}} = \frac{(\sqrt{3}+i)\Gamma(1/3)}{6x^{1/3}}.$$

For C_3, the curve $x^3 - 3xy^2 = 1$ makes for a very inconvenient substitution. So instead, we will use the fact that $i\phi(t)$ will have constant imaginary part on this curve, so we will substitute $it^3 = -u + ik$ for some k. Since 1 is on C_3, we have that $k = 1$. If $t^3 = 1 + iu$, then $t = \sqrt[3]{1+iu}$, so $dt = i(1+iu)^{-2/3}/3\, du$. Then

$$\int_{C_3} e^{ixt^3} = -\int_0^\infty e^{ix(1+iu)} i(1+iu)^{-2/3}/3\, du = \frac{-ie^{ix}}{3}\int_0^\infty (1+iu)^{-2/3} e^{-xu}\, du.$$

We can use the binomial series (equation 1.31) to find the Maclaurin series for $(1+x)^{-2/3}$.

$$(1+x)^{-2/3} = 1 - \frac{2x}{3} + \frac{5x^2}{9} - \frac{40x^3}{81} + \cdots = \sum_{n=0}^\infty (-1)^n \frac{\Gamma(2/3+n)}{n!\Gamma(2/3)} x^n.$$

So

$$(1+iu)^{-2/3} = \sum_{n=0}^\infty (-1)^n \frac{\Gamma(2/3+n)}{n!\Gamma(2/3)} i^n u^n.$$

Applying Watson's lemma, we have

$$\int_{C_3} e^{ixt^3} = \frac{-ie^{ix}}{3}\int_0^\infty (1+iu)^{-2/3} e^{-xu}\, du \sim \frac{-ie^{ix}}{3}\sum_{n=0}^\infty \frac{\Gamma(2/3+n)(-i)^n}{n!\Gamma(2/3)x^{n+1}}.$$

Adding C_1 to C_3, we get the full asymptotic series for the integral as $x \to \infty$.

$$\int_0^1 e^{ixt^3}\, dt \sim \frac{(\sqrt{3}+i)\Gamma(1/3)}{6x^{1/3}} + \frac{-ie^{ix}}{3}\sum_{n=0}^\infty \frac{\Gamma(2/3+n)(-i)^n}{n!\Gamma(2/3)x^{n+1}}.$$

□

2.6.1 Saddle Points

As powerful as the least descents method is, there are times in which the two downstream paths for the endpoints go in different directions, so a simple bridge between the two is impossible.

Example 2.32

Find the behavior as $x \to \infty$ of

$$\int_0^\pi e^{ix\sin t}\, dt.$$

SOLUTION: From problem 11 of section 2.4,

$$\sin(x+iy) = \sin(x)\cosh(y) + i\cos(x)\sinh(y).$$

Thus, the exponent will have constant imaginary part on the curves

$$\sin(x)\cosh(y) = k.$$

At the endpoints 0 and π, k will be 0, so the steepest descent curves through the two endpoints will be vertical. However, the path of steepest descent from $t=0$ is along the positive imaginary axis, whereas the path of steepest descent from $t=\pi$ goes in the direction of the negative real axis, as shown in figure 2.18. The reason is because $\sin(\pi+iy) = -i\sinh(y)$. Since the downstream directions are opposite, no simple bridge exists between the two that is going further and further downstream.

The solution is to add more curves into the path from 0 to π. But in order to add a new curve, there must be a point on this curve for which we have *two* different directions of steepest descent. This can only happen if $\phi'(t) = 0$, where the integral involves $e^{ix\phi(t)}$. A point where $\phi'(t) = 0$ is called a *saddle point*. At a saddle point, there will be two or more paths of steepest descent coming from the point. For example, figure 2.17 shows three paths of steepest descent emulating from the origin, along with three paths of steepest ascent.

Once we have found a saddle point, we can find the contour integral downstream from the saddle point via Laplace's method. By doing this for two of the downstream paths, we have part of a connected path that can be used to deform the contour.

In the example $\int_0^\pi e^{ix\sin t}\,dt$, a saddle point occurs, among other places, at $t=\pi/2$. There are two paths of steepest descent from this saddle point, so we can use Laplace's method to find the contour integral along these paths. Thus, we can deform the contour from 0 to π to one that is going downstream R units along the positive imaginary axis, cuts across a short bridge C_2, then goes upstream to the saddle point at $\pi/2$, then goes downstream in the opposite direction, cuts across another short bridge C_5, and finally goes vertically upwards to π, as indicated in figure 2.18.

It is clear that as $R\to\infty$, the contribution from C_2 and C_5 go to zero, not because the bridges are getting shorter, but because the integrand is decaying exponentially as we travel downstream.

To compute the integral along the contours, we use a substitution of the form $i\phi(t) = -u + ik$, so that the exponential function will become $e^{-xu+ixk}$. For C_1 and C_6, k will be 0, since $\sin(0) = \sin(\pi) = 0$. Letting $\sin(t) = iu$, $t = \sin^{-1}(iu)$, so $du = \pm i/\sqrt{1-u^2}$. Since du will be going in the direction of downstream, we will use the plus sign for C_1 and the minus sign for C_6. But for C_6, the endpoints for u will go from ∞ to 0, so

$$\int_{C_1} e^{ix\sin t}\,dt = \int_{C_6} e^{ix\sin t}\,dt = \int_0^\infty \frac{i}{\sqrt{1+u^2}}e^{-ux}\,du.$$

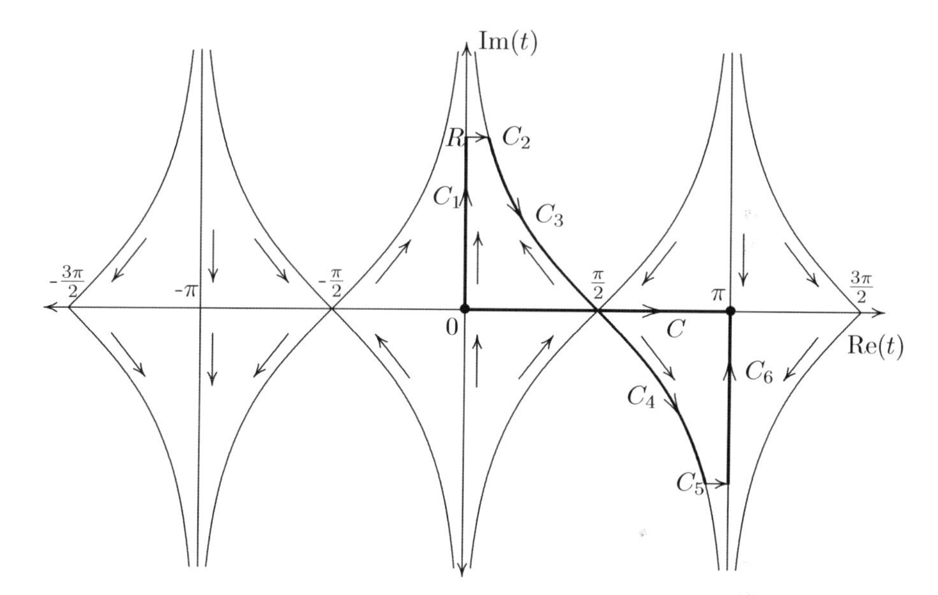

FIGURE 2.18: This shows the steepest descent lines for $ix \sin t$. The original contour C from 0 to π is deformed to the contour C_1 going downstream, going across the bridge C_2, going upstream along C_3 to the saddle point at $\pi/2$, then downstream along C_4, across another bridge C_5, and finally upstream along C_6 to π.

In order to integrate along the contour C_3, we will again use the substitution $i\phi(t) = -u + ik$ for some constant k, with u ranging from ∞ down to 0. In this case $\phi(t) = \sin(t)$, and evaluating at $\pi/2$ gives us $k = 1$. So $t = \sin^{-1}(1 + iu)$, $dt = \pm i\, du/\sqrt{1-(1+iu)^2} = \pm i\, du/\sqrt{u}\sqrt{u-2i}$. For C_3, the du has an argument close to $5\pi/4$ when u is close to 0, so we choose the plus sign. We can use the same substitution for C_4, only this time the du has an argument close to $-\pi/4$, so we choose the minus sign, and u will range from 0 to ∞. Thus,

$$\int_{C_3} e^{ix \sin t}\, dt = \int_{C_4} e^{ix \sin t}\, dt = e^{ix} \int_0^\infty \frac{-i}{\sqrt{u}\sqrt{u-2i}} e^{-xu}\, du$$
$$= e^{ix} \int_0^\infty \frac{1-i}{2\sqrt{u}\sqrt{1+(iu/2)}} e^{-xu}\, du.$$

Putting the pieces together, we get

$$\int_0^\pi e^{ix \sin t}\, dt = \int_0^\infty \frac{2i}{\sqrt{1+u^2}} e^{-ux}\, du + e^{ix} \int_0^\infty \frac{1-i}{\sqrt{u}\sqrt{1+(iu/2)}} e^{-xu}\, du.$$

We can now find the asymptotic series as $x \to \infty$ for both of these series, using Watson's lemma. The Maclaurin series for both of the series begins with equation 1.32,

$$\frac{1}{\sqrt{1-4x}} \sim \sum_{n=0}^\infty \frac{(2n)!x^n}{(n!)^2}.$$

Replacing x with $-u^2/4$, and multiplying by $2i$ produces

$$\frac{2i}{\sqrt{1+u^2}} \sim \sum_{n=0}^\infty \frac{2i(-1)^n (2n)! u^{2n}}{4^n (n!)^2}.$$

Or, replacing x with $-iu/8$ and multiplying by $(1-i)/\sqrt{u}$ gives us

$$\frac{1-i}{\sqrt{u}\sqrt{1+(iu/2)}} \sim \sum_{n=0}^\infty \frac{(1-i)(2n)! u^{n-1/2} (-i)^n}{8^n (n!)^2}.$$

So by Watson's lemma,

$$\int_0^\pi e^{ix \sin t}\, dt \sim \sum_{n=0}^\infty e^{ix} \frac{(1-i)(2n)!\Gamma(n+1/2)(-i)^n}{8^n (n!)^2 x^{n+1/2}} + \sum_{n=0}^\infty \frac{2i(-1)^n (2n)!^2}{4^n (n!)^2 x^{2n+1}}.$$

When there are two sums in an asymptotic series, it is understood that the terms are to be sorted in decreasing order as $x \to \infty$. Thus,

$$\int_0^\pi e^{ix \sin t} dt \sim e^{ix} \frac{(1-i)\sqrt{\pi}}{x^{1/2}} + \frac{2i}{x} - e^{ix} \frac{(1+i)\sqrt{\pi}}{8x^{3/2}} + e^{ix} \frac{(9i-9)\sqrt{\pi}}{128 x^{5/2}} - \frac{2i}{x^3} + \cdots. \quad \square$$

Although the substitution $i\phi(t) = -u + ik$ converts each piece of the contour integral to a form where Watson's lemma can immediately be used, the function $\phi(t)$ must be easily inverted in order to use this substitution. If $\phi(t)$ cannot be easily inverted, it is best if we use a polygonal contour, that is, a contour consisting of straight lines, each tangent to the path of steepest decent at the endpoint or saddle point.

Example 2.33

Find the behavior as $x \to \infty$ of

$$\int_0^3 \cos(x(t^3 - 3t^2))\, dt.$$

SOLUTION: At first this doesn't look like an exponential integral, but we can view this as the real part of

$$\int_0^3 e^{ix(t^3-3t^2)}\, dt.$$

Since $\phi(t) = t^3 - 3t^2$ is hard to invert, we will not use the actual curves of steepest descent, but straight lines that are tangent to these curves. So we must first find the direction of these tangent lines.

We begin with the endpoint at 0, which is also a saddle point since $\phi'(0) = 0$. When t is close to 0, $\phi(t) \sim -3t^2$, so the steepest descent lines will come out at a direction of $\theta = -\pi/4$ and $3\pi/4$. At the other endpoint, since $\phi'(3) = 9$, $\phi(t) \sim 9(3 - t)$ as $t \to 3$. Hence, the direction of steepest descent is $\theta = \pi/2$.

Since these directions do not intersect, we need a saddle point between the two. In fact, $\phi'(2) = 0$, so 2 is a saddle point. Since $\phi(t) = -4+3(t-2)^2+(t-2)^3$, the steepest descent curves come out at a direction for which $i3(t - 2)^2$ is real and negative, so $\theta = \pi/4$ or $-3\pi/4$.

We can now construct a polygonal path that will be tangent to the paths of steepest descent. Starting with $t = 0$, we go straight to $t = 1 - i$, then go to $t = 2$, continue in this direction to $t = 3 + i$, and finally straight down to $t = 3$. See figure 2.19.

The substitutions for each of the segments is fairly straightforward, but it is easier if we have increasing u be always in the downstream direction. For C_1, we let $t = (1-i)u$, for $-C_2$ we let $t = 2-(1+i)u$, for C_3 we let $t = 2+(1+i)u$, and for $-C_4$ we let $u = 3 + iu$. Each of these integrals will have the same range: $0 \le u \le 1$. Note that C_2 and C_4 are actually going upstream, so we parameterized the curves going in the opposite direction, denoted $-C_2$ and $-C_4$. This reverse of direction will change the sign of the integral for that segment, which is easy to correct. So we have the integral broken into pieces:

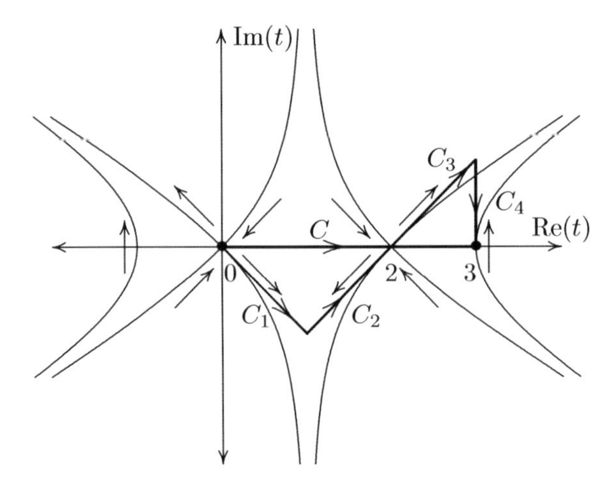

FIGURE 2.19: Rather than following the steepest descent curves, we can follow a polygonal contour that is tangent to the steepest descent curves at the endpoints and saddle points. Here, the contour from 0 to 3 is deformed into 4 pieces, each a straight line.

$$\begin{aligned}\int_0^3 e^{ix(t^3-3t^2)}\,dt = &\int_0^1 (1-i)e^{-6u^2x+(2-2i)u^3x}\,du\\ &-\int_0^1 -(1+i)e^{-4ix-6u^2x+(2+2i)u^3x}\,du\\ &+\int_0^1 (1+i)e^{-4ix-6u^2x-(2+2i)u^3x}\,du\\ &-\int_0^1 ie^{-9ux-6iu^2x+u^3x}\,du.\end{aligned}$$

The middle two integrals can be combined, using $e^{(2+2i)u^3x}+e^{-(2+2i)u^3x} = 2\cosh((2+2i)u^3x)$. So we now have the integrals

$$\begin{aligned}(1-i)\int_0^1 e^{-6u^2x}e^{(2-2i)u^3x}\,du + (2+2i)e^{-4ix}\int_0^1 e^{-6u^2x}\cosh((2+2i)u^3x)\,du\\ -i\int_0^1 e^{-9ux}e^{-6iu^2x+u^3x}\,du.\end{aligned}$$

All three of these integrals are now in the form where we can use Laplace's method. For the first two, we can make the substitution $s = 6u^2x$, so $u = \sqrt{s/6x}$, $du = 1/\sqrt{24sx}$. We can also change the upper endpoint to ϵ since this will only change subdominant terms. So the first two integrals become

$$\int_0^\epsilon e^{-s}\frac{e^{(1-i)s^{3/2}/3\sqrt{6x}}}{\sqrt{24sx}}\,ds \\ + (2+2i)e^{-4ix}\int_0^\epsilon e^{-s}\frac{\cosh((1+i)s^{3/2}/3\sqrt{6x})}{\sqrt{24sx}}\,ds.$$

Using the series expansions

$$\frac{e^{(1-i)s^{3/2}/3\sqrt{6x}}}{\sqrt{24sx}} = \sum_{n=0}^{\infty}\frac{(1-i)^n s^{(3n-1)/2}}{3^n 2\sqrt{6x}^{n+1}n!}$$

and

$$\frac{\cosh((1+i)s^{3/2}/3\sqrt{6x})}{\sqrt{24sx}} = \sum_{n=0}^{\infty}\frac{(2i)^n s^{3n-(1/2)}}{54^n x^{n+(1/2)}\sqrt{24}(2n)!},$$

we can use Watson's lemma to find the asymptotic series as $x \to \infty$ for the first two integrals.

$$\sum_{n=0}^{\infty}\frac{(1-i)^{n+1}\Gamma\left(\frac{3}{2}n+\frac{1}{2}\right)}{3^n 2\sqrt{6x}^{n+1}n!} + \sum_{n=0}^{\infty}\frac{(1+i)e^{-4ix}(2i)^n\Gamma\left(3n+\frac{1}{2}\right)}{54^n x^{n+(1/2)}\sqrt{6}(2n)!}. \tag{2.30}$$

The last integral is a bit trickier, since we have to expand an exponential function involving two terms in its exponent. We can make the substitution $s = 9ux$, so $u = s/9x$.

$$-i\int_0^1 e^{-9ux}e^{-6iu^2x+u^3x}\,du \sim \int_0^\epsilon \frac{-i}{9x}e^{-s}\exp\left(\frac{-2is^2}{27x}+\frac{s^3}{729x^2}\right)ds.$$

By expanding out the terms in

$$1+\left(\frac{-2is^2}{27x}+\frac{s^3}{729x^2}\right)+\frac{1}{2}\left(\frac{-2is^2}{27x}+\frac{s^3}{729x^2}\right)^2+\frac{1}{6}\left(\frac{-2is^2}{27x}+\frac{s^3}{729x^2}\right)^3+\cdots,$$

and collecting common powers of x, we can find the series for

$$\exp\left(\frac{-2is^2}{27x}+\frac{s^3}{729x^2}\right) = 1-\frac{2is^2}{27x}+\frac{s^3-2s^4}{729x^2}+\frac{2i(-3s^5+2s^6)}{59049x^3}+\cdots.$$

So we can get several terms for the last integral.

$$\int_0^\infty \frac{-i}{9x}e^{-s}\exp\left(\frac{-2is^2}{27x}+\frac{s^3}{729x^2}\right)ds \sim \frac{-i}{9x}-\frac{4}{243x^2}+\frac{14i}{2187x^3}+\frac{80}{19683x^4}+\cdots. \tag{2.31}$$

We are now ready to put all of the pieces together. But recall that for the original integral, we only wanted the real part of the exponential to obtain the cosine function. We can assemble the parts as

$$\int_0^3 \cos(x(t^3-3t^2))\,dt = w_1(x)\cos(4x-\pi/4)+w_2(x)\sin(4x-\pi/4)+w_3(x),$$

where

$$w_1(x) \sim \sum_{n=0}^{\infty} \frac{(-1)^n \Gamma\left(6n+\frac{1}{2}\right)}{3^{6n} x^{2n+(1/2)} \sqrt{3}(4n)!}$$
$$\sim \frac{\sqrt{\pi}}{\sqrt{3x}} - \frac{385\sqrt{\pi}}{41472\sqrt{3}x^{5/2}} + \frac{37182145\sqrt{\pi}}{10319560704\sqrt{3}x^{9/2}} + \cdots,$$

$$w_2(x) \sim \sum_{n=0}^{\infty} \frac{(-1)^n \Gamma\left(6n+\frac{7}{2}\right)}{3^{6n+3} x^{2n+(3/2)} \sqrt{3}(4n+2)!}$$
$$\sim \frac{5\sqrt{\pi}}{144\sqrt{3}x^{3/2}} - \frac{85085\sqrt{\pi}}{17915904\sqrt{3}x^{7/2}} + \frac{5391411025\sqrt{\pi}}{1486016741376\sqrt{3}x^{11/2}} + \cdots,$$

and, by combining the first sum of equation 2.30 with equation 2.31,

$$w_3 \sim \frac{\sqrt{\pi}}{2\sqrt{6x}} - \frac{5\sqrt{\pi}}{288\sqrt{6}x^{3/2}} - \frac{2}{81x^2} - \frac{385\sqrt{\pi}}{82944\sqrt{6}x^{5/2}}$$
$$+ \frac{85085\sqrt{\pi}}{35831808\sqrt{6}x^{7/2}} + \frac{40}{6561x^4} + \cdots. \tag{2.32}$$

□

We have now encountered many different ways of computing the asymptotic behavior of an integral, from the simplest technique of repeating integrating by parts to the complicated method of steepest descents. Many of these methods produced series that actually diverged, yet using optimal truncation, we could still exploit these series to produce fairly accurate approximations. In the next chapter, we will find ways to milk even more information from the divergent asymptotic series, to obtain even more precise approximations.

Problems for §2.6

For problems **1** through **8**: Use equation 2.29 to find the asymptotic series as $x \to \infty$ up to the $1/x^3$ term for the following integrals.

1 $I(x) = \int_0^1 \frac{1}{t+2} e^{ixt}\, dt$

2 $I(x) = \int_0^1 \ln(1+t) e^{ixt}\, dt$

3 $I(x) = \int_0^\pi \cos(t) e^{ixt}\, dt$

4 $I(x) = \int_0^3 \frac{1}{\sqrt{t+1}} e^{ixt}\, dt$

5 $I(x) = \int_0^4 \sqrt{t} e^{ixt}\, dt$

6 $I(x) = \int_{-1}^1 \frac{i}{t+i} e^{ixt}\, dt$

7 $I(x) = \int_{-1}^1 \frac{i}{t-i} e^{ixt}\, dt$

8 $I(x) = \int_{-1}^1 \frac{1}{1+t^2} e^{ixt}\, dt$

9 Show that example 2.31 can be done via straight line contours, with C_1 going from 0 to $1+i/\sqrt{3}$, and C_2 going down from $1+i/\sqrt{3}$ to 1.

10 Find the full asymptotic series as $x \to \infty$ for

$$\int_0^1 e^{ixt^2}\,dt.$$

For problems **11** through **18**: Use the method of steepest descents to find the asymptotic series as $x \to \infty$ up to the $1/x^3$ term for the following integrals. Note that some of these are best done via polygonal paths.

11 $I(x) = \int_0^1 e^{ixt^4}\,dt$

12 $I(x) = \int_0^1 te^{ixt^3}\,dt$

13 $I(x) = \int_0^1 te^{ixt^4}\,dt$

14 $I(x) = \int_0^{\ln 2} e^{ixe^t}\,dt$

15 $I(x) = \int_0^{\pi} e^{ix\cos(t)}\,dt$

16 $I(x) = \int_0^{\cosh^{-1}(2)} e^{ix\cosh(t)}\,dt$

17 $I(x) = \int_0^1 e^{ix(t+t^2)}\,dt$

18 $I(x) = \int_0^{\tanh^{-1}(1/2)} e^{ix\tanh t}\,dt$

For problems **19** through **24**: Use the method of steepest descents through a saddle point to find the asymptotic series as $x \to \infty$ up to the $1/x^3$ term for the following integrals. The position of the saddle point is given.

19 $I(x) = \int_0^2 e^{ix(t^2-2t)}\,dt,$ saddle point at $t = 1$.

20 $I(x) = \int_0^{\sinh^{-1}(1)} e^{ix\sinh t}\,dt,$ saddle point at $t = \pi i/2$.

21 $I(x) = \int_0^{\sqrt{3}} e^{ix(t^3-3t)}\,dt,$ saddle point at $t = 1$.

22 $I(x) = \int_0^{\infty} e^{ix(t^3+3t)}\,dt,$ saddle point at $t = i$.

23 $I(x) = \int_{-1}^3 e^{ix(t^3-3t^2)}\,dt,$ saddle points at $t = 0$ and $t = 2$.

24 $I(x) = \int_{-2}^2 e^{ix(t^3-3t)}\,dt,$ saddle points at $t = -1$ and $t = 1$.

25 The Airy function, $\mathrm{Ai}(x)$, is important in the study of differential equations. It can be represented by the integral

$$\mathrm{Ai}(x) = \frac{1}{\pi}\int_0^{\infty} \cos(t^3/3 + xt)\,dt.$$

Use the method of steepest descents through a saddle point to find the full asymptotic series as $x \to \infty$ for this function.

Hint: Substituting $t = u\sqrt{x}$, and then replacing $x^{3/2}$ with $3y$ produces

$$\mathrm{Ai}(x) = \frac{\sqrt{x}}{\pi} \mathrm{Re}\left(\int_0^\infty e^{iy(u^3+3u)}\, du \right).$$

So the problem is reduced to that of problem 22. Note that the dominant part of the integral is purely complex, so subdominant terms must be kept!

Chapter 3

Speeding Up Convergence

We have seen many instances in which we can solve a difficult problem in terms of a series. Sometimes the series is slowly converging, so that it takes many terms of the series to get a decent amount of accuracy to the problem. Furthermore, the terms of the series may not follow a simple pattern, so it takes a major effort just to get a handful of terms. To make matters worse, many of the asymptotic series turned out to be divergent. Although the optimal truncation rule allowed us to get a few places of accuracy from the series, the results were limited. The goal of this section is to milk the most out of what few terms of the series we have to get as much accuracy as possible.

3.1 Shanks Transformation

For each series $\sum_{n=0}^{\infty} a_n$, we can form the sequence of partial sums

$$A_n = \sum_{k=0}^{n} a_n.$$

Then the sum of the series is the limit of the sequence A_n. If the true sum of the series is S, then the error is $S - A_n$. Depending on the how the error relates to n, we can speed up the convergence by extrapolating the sequence.

For example, suppose that the error in A_n decays exponentially as n increases, at least approximately. Then we can express $A_n \approx S + bq^n$, for some $q < 1$. This involves 3 unknowns, so if we use three consecutive terms of the sequence, we should be able to solve for S. If we assume that the error is exactly exponential, then

$$\begin{aligned} A_{n-1} &= S + bq^{n-1} \\ A_n &= S + bq^n \\ A_{n+1} &= S + bq^{n+1}. \end{aligned}$$

Since $A_{n+1} - A_n = bq^n(q-1)$, and $A_n - A_{n-1} = bq^{n-1}(q-1)$, we can quickly solve for q:

$$q = \frac{A_{n+1} - A_n}{A_n - A_{n-1}}.$$

Then $bq^n = (A_{n+1} - A_n)/(q-1)$, and replacing q with its known value gives us

$$bq^n = \frac{(A_{n+1} - A_n)(A_n - A_{n-1})}{A_{n+1} - 2A_n + A_{n-1}}.$$

Finally, we can solve for $S = A_n - bq^n$ to get

$$S_n = \frac{A_{n+1}A_{n-1} - A_n^2}{A_{n+1} + A_{n-1} - 2A_n}. \tag{3.1}$$

This new sequence, called the *Shanks transformation* of the series, will usually converge faster than the original series. It is denoted by $S(A_n)$, and works particular well on alternating series.

Example 3.1

One series that converges to π is the following:

$$\pi = \sum_{n=0}^{\infty} \frac{4(-1)^n}{2n+1} = 4 - \frac{4}{3} + \frac{4}{5} - \frac{4}{7} + \frac{4}{9} - \frac{4}{11} + \cdots.$$

Use Shanks transformation on this series.

SOLUTION: The original series converges agonizingly slow, since $\{A_n\} =$

$$\{4, 2.666666667, 3.466666667, 2.895238095, 3.339682540, 2.976046176, \ldots\}.$$

It takes 65 terms just to get 1% accuracy. But we can apply equation 3.1 to every combination of three consecutive terms of A_n. This produces the sequence

$$\{S(A_n)\} = \{3.166666667, 3.133333333, 3.145238095, 3.139682540, \ldots\}$$

which already has more than 1% accuracy.

We can take the Shanks transformation of the Shanks transformation, denoted by $S^2(A_n)$. Table 3.1 shows how the iterated Shanks transformation quickly gives an impressive amount of accuracy. Although summing 10 terms of the original series gives a poor estimate of $\pi \approx 3.0418$, the third order iterated Shanks transformation can produce the approximation $\pi \approx 3.1415924$ using the same terms of the original series.

Figure 3.1 plots the absolute errors of the Shanks transformation on a logarithmic graph. Although there is some increase in accuracy as we increase the number of terms used in the series, the big jumps in accuracy occur as we increase the number of iterations, each gaining about 2 places of accuracy. Thus, with sufficient iterations, we can in theory compute the sum to any desired accuracy.

□

TABLE 3.1: Iterated Shanks transformation for the series $\sum_{n=0}^{\infty} 4(-1)^n/(2n+1)$. This series converges to π slowly, but the Shanks transformation speeds up the convergence considerably.

n	A_n	$S(A_n)$	$S^2(A_n)$	$S^3(A_n)$
1	4			
2	2.666666667	3.166666667		
3	3.466666667	3.133333333	3.142105263	
4	2.895238095	3.145238095	3.141450216	3.141599357
5	3.339682540	3.139682540	3.141643324	3.141590860
6	2.976046176	3.142712843	3.141571290	3.141593231
7	3.283738484	3.140881341	3.141602842	3.141592438
8	3.017071817	3.142071817	3.141587321	
9	3.252365935	3.141254824		
10	3.041839619			

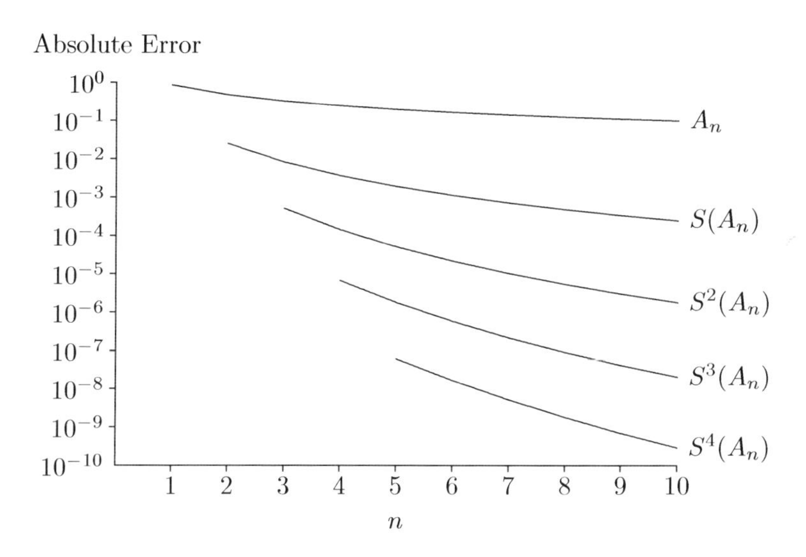

FIGURE 3.1: Plot of absolute errors of iterated Shanks transformation for the series $\sum_{n=0}^{\infty} 4(-1)^n/(2n+1)$. Note that each iteration gains about 2 decimal places when $n = 10$.

TABLE 3.2: Iterated Shanks transformation for the series $\sum_{n=1}^{\infty}(.9)^n/n = \ln(10) \approx 2.302585093$. Shanks transformation improves convergence even though this is not an alternating series.

n	A_n	$S(A_n)$	$S^2(A_n)$	$S^3(A_n)$
1	0.9			
2	1.305	1.9125		
3	1.548	2.052692308	2.245159713	
4	1.712025	2.133803571	2.268413754	2.296053112
5	1.830123	2.184417	2.281042636	2.298856749
6	1.9186965	2.217632063	2.288432590	2.300349676
7	1.987024629	2.240240634	2.292993969	2.301192122
8	2.040833030	2.256066635	2.295924710	
9	2.083879751	2.267394719		
10	2.118747595			

The Shanks transformation also works for positive series, if the error can be shown to be decreasing geometrically, not algebraically. For example, the series

$$-\ln(1-x) = x + \frac{x^2}{2} + \frac{x^3}{3} + \frac{x^4}{4} + \cdots$$

converges for $x = 9/10$, but very slowly. The true sum is $-\ln(.1) = \ln(10)$, but it takes 33 terms to get just 3 places of accuracy. Taylor's remainder theorem shows that the error is

$$\frac{(1-c_n)^{-n-1}(.9)^{n+1}}{n+1}$$

where c_n is some number between 0 and .9. Thus, the error is dominated by a geometric function, so the Shanks transformation should improve the convergence. Table 3.2 shows that indeed, the convergence improves significantly.

3.1.1 Generalized Shanks Transformation

The principle of the Shanks transformation is to eliminate a geometric error, bq^n, from the sequence of numbers. But it is possible to eliminate several such errors at the same time. That is, we will assume that

$$A_n = S + \sum_{i=1}^{m} b_i q_i^n.$$

This gives us $2m+1$ unknowns, so we will need $2m+1$ consecutive terms, A_{n-m} through A_{n+m}. In spite of the non-linearity in the system of equations, there is a closed form expression for S, expressed in terms of the determinants of $m+1$ by $m+1$ matrices:

$$\frac{\begin{vmatrix} A_{n-m} & A_{n-m+1} & \cdots & A_n \\ A_{n-m+1}-A_{n-m} & A_{n-m+2}-A_{n-m+1} & \cdots & A_{n+1}-A_n \\ A_{n-m+2}-A_{n-m+1} & A_{n-m+3}-A_{n-m+2} & \cdots & A_{n+2}-A_{n+1} \\ \vdots & \vdots & \ddots & \vdots \\ A_n-A_{n-1} & A_{n+1}-A_n & \cdots & A_{n+m}-A_{n+m-1} \end{vmatrix}}{\begin{vmatrix} 1 & 1 & \cdots & 1 \\ A_{n-m+1}-A_{n-m} & A_{n-m+2}-A_{n-m+1} & \cdots & A_{n+1}-A_n \\ A_{n-m+2}-A_{n-m+1} & A_{n-m+3}-A_{n-m+2} & \cdots & A_{n+2}-A_{n+1} \\ \vdots & \vdots & \ddots & \vdots \\ A_n-A_{n-1} & A_{n+1}-A_n & \cdots & A_{n+m}-A_{n+m-1} \end{vmatrix}}$$

We will define the *generalized Shanks transformation*, or $S_m(A_n)$, by this expression. Note that $S_1(A_n)$ reproduces that standard Shanks transformation, but $S_2(A_n) =$

$$\frac{A_{n+2}A_nA_{n-2} + 2A_{n+1}A_nA_{n-1} - A_{n-2}A_{n+1}^2 - A_{n+2}A_{n-1}^2 - A_n^3}{\begin{array}{c} A_{n-2}(A_{n+2} - 2A_{n+1} + A_n) + 2A_{n-1}(A_n + A_{n+1} - A_{n+2}) \\ +A_n(A_{n+2} + 2A_{n+1} - 3A_n) - A_{n-1}^2 - A_{n+1}^2 \end{array}}. \tag{3.2}$$

Example 3.2

Compare the generalized Shanks transformation with the iterated Shanks transformation for the series from example 3.1.

SOLUTION: Table 3.3 shows the results of using the generalized Shanks transformation. By comparing these results to table 3.1, we see that there is basically the same amount of places of accuracy in the generalized Shanks transformation as in the iterated Shanks transformation. □

Since the generalized Shanks transformation is not better than the iterated Shanks transformation, what is the point of using the more complicated formulas? There is one aspect of the iterated Shanks transformation that we have not yet addressed. Computers and calculators have a limited amount of precision in the way the numbers are stored in memory. The denominator of equation 3.1, $A_{n+1} + A_{n-1} - 2A_n$, requires subtracting two very close numbers, which causes a huge loss in precision, known as the *loss of significance*. This loss of significance effects the amount of precision in our calculation of $S(A_n)$. A general rule is that A_n is known to k digits of precision, then $S(A_n)$ is only accurate to $k/2$ digits. Iterating the Shanks transformation amplifies this loss of precision. For example, if A_n is calculated to 16 digits of precision, then the calculation of $S^3(A_n)$ would only have 2 digits of precision, even if theoretically $S^s(A_n)$ should give 5 or 6 digits of accuracy. Thus, one would need the original series calculated to 40 places of accuracy to see the 5 places of accuracy in $S^3(A_n)$. Cancellation errors also occur in the generalized Shanks transformation, but the effect is not amplified by iteration. So

TABLE 3.3: Generalized Shanks transformation for the series $\sum_{n=0}^{\infty} 4(-1)^n/(2n+1)$. Accuracy is about the same as the iterated Shanks transformation, given in table 3.1.

n	A_n	$S_1(A_n)$	$S_2(A_n)$	$S_3(A_n)$
1	4			
2	2.666666667	3.166666667		
3	3.466666667	3.133333333	3.142342342	
4	2.895238095	3.145238095	3.141391941	3.141614907
5	3.339682540	3.139682540	3.141662738	3.141587302
6	2.976046176	3.142712843	3.141563417	3.141594274
7	3.283738484	3.140881341	3.141606504	3.141592073
8	3.017071817	3.142071817	3.141585436	
9	3.252365935	3.141254824		
10	3.041839619			

having A_n known to 16 digits will allow us to calculate the generalized Shanks transformation $S_3(A_n)$ to 8 places.

Problems for §3.1

For problems **1** through **15**:

Part a: First find A_1 through A_7 for the following sequences. Note that some sums begin at $m = 0$, causing A_1 to be the sum of two terms. Then apply the iterated Shanks transformation to find $S^2(A_n)$ for $n = 3$ to $n = 5$. How many digits of precision does $S^2(A_n)$ give in comparison to the given exact limit?

Part b: Do the same as part a, but use the generalized Shanks transformation $S_2(A_n)$ given by equation 3.2.

1 $A_n = \sum_{m=1}^{n} \frac{(-1)^{m+1}}{m}$, $\qquad \lim_{n\to\infty} A_n = \ln(2)$.

2 $A_n = \sum_{m=1}^{n} \frac{(-1)^{m+1}}{m^2}$, $\qquad \lim_{n\to\infty} A_n = \pi^2/12$.

3 $A_n = \sum_{m=0}^{n} (-.9)^m + (-.8)^m$, $\qquad \lim_{n\to\infty} A_n = 185/171$.

4 $A_n = \sum_{m=0}^{n} \frac{(-1)^m}{3m+1}$, $\qquad \lim_{n\to\infty} A_n = \pi/(3\sqrt{3}) + \ln(2)/3$.

5 $A_n = \sum_{m=0}^{n} \frac{(-1)^m}{3m+2}$, $\qquad \lim_{n\to\infty} A_n = \pi/(3\sqrt{3}) - \ln(2)/3$.

6 $A_n = \sum_{m=1}^{n} \frac{(-1)^{m+1}}{\sqrt{m}}$, $\lim_{n\to\infty} A_n = (1-\sqrt{2})\zeta(1/2) \approx 0.604898643422.$

7 $A_n = \sum_{m=1}^{n} \frac{(-1)^{m+1}}{m^{1/4}}$, $\lim_{n\to\infty} A_n = (1-2^{3/4})\zeta(1/4) \approx 0.554487385914.$

8 $A_n = \sum_{m=0}^{n} \frac{1}{m!}$, $\lim_{n\to\infty} A_n = e.$

9 $A_n = \sum_{m=0}^{n} \frac{3^m}{m!}$, $\lim_{n\to\infty} A_n = e^3.$

10 $A_n = \sum_{m=1}^{n} \left(2^{2^{-m}} - 1\right)$, $\lim_{n\to\infty} A_n \approx 0.781838631839.$

11 $A_n = \left(1 + \frac{1}{2^n}\right)^{2^n}$, $\lim_{n\to\infty} A_n = e.$

12 $A_n = 2^{\left(\frac{n2^n+1}{2^n}\right)} - 2^n$, $\lim_{n\to\infty} A_n = \ln 2.$

13 $A_n = \prod_{m=1}^{n} \left(1 + \frac{1}{2^m}\right)$, $\lim_{n\to\infty} A_n \approx 2.38423102903.$

14 $A_n = \prod_{m=1}^{n} \left(1 - \frac{1}{2^m}\right)$, $\lim_{n\to\infty} A_n \approx 0.288788095087.$

15 $A_n = \prod_{m=1}^{n} \left(1 + \frac{(-1)^m}{2^m}\right)$, $\lim_{n\to\infty} A_n \approx 0.568698946265.$

3.2 Richardson Extrapolation

Although the Shanks transformation speeds up the convergence for a large number of series, it does not work for every series. For example, consider the series for the Riemann zeta function,

$$\zeta(x) = \sum_{n=1}^{\infty} \frac{1}{n^x}. \tag{3.3}$$

The integral test shows that the series converges for $x > 1$.

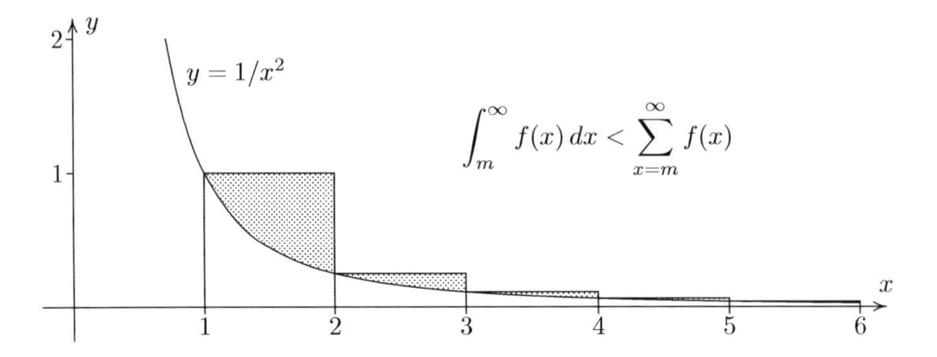

FIGURE 3.2: This shows the classical diagram proving the integral test. For a function that is positive and decreasing, the integral will always be less than the sum. Furthermore, the difference, shown by the shaded region, will be less than the first term of the sum.

In particular, consider the case with $x = 2$. To find the error in the partial sum approximation A_n, note that

$$\zeta(2) - A_n = \sum_{x=n+1}^{\infty} \frac{1}{x^2}.$$

The diagram that is typically used to prove the integral test (figure 3.2) also gives us a way to estimate this sum. In particular,

$$\frac{1}{n} = \int_n^\infty \frac{1}{x^2}\,dx > \sum_{x=n+1}^{\infty} \frac{1}{x^2} > \int_{n+1}^\infty \frac{1}{x^2}\,dx = \frac{1}{n+1}.$$

Thus, the error in the partial sum A_n will be between $1/n$ and $1/(n+1)$. Problem 1 shows that the Shanks transformation would only decrease the error by a factor of 2, which is not that impressive.

A better approach for this series would be to use *Richardson extrapolation.* Like Shanks transformation, a series must first be converted to the sequence of partial sums A_n before applying Richardson extrapolation.

If we can show that a sequence converges to S, and the error is an *algebraic* function of n, that is,

$$A_n \sim S + \frac{Q}{n^k}$$

for some k which is known, then the plan is to use both A_n and A_{2n} to solve for S and Q. Since the equations are linear, it is a trivial matter to solve for S:

$$S = \frac{2^k A_{2n} - A_n}{2^k - 1}. \tag{3.4}$$

TABLE 3.4: Richardson extrapolation for $\sum_{n=0}^{\infty} 1/n^2$. This series converges to $\pi^2/6 \approx 1.644934067$ slowly.

n	A_n	A_{2n}	$2A_{2n} - A_n$
1	1.	1.25	1.5
2	1.25	1.42361	1.59722
3	1.36111	1.49139	1.62167
4	1.42361	1.52742	1.63123
5	1.46361	1.54977	1.63592
6	1.49139	1.56498	1.63856
7	1.51180	1.57600	1.64019
8	1.52742	1.58435	1.64127
9	1.53977	1.59089	1.64202
10	1.54977	1.59616	1.64256
20	1.59616	1.62024	1.64432

Thus, we define the Richardson extrapolation $R(A_n)$ to be S from equation 3.4.

Example 3.3
Apply Richardson extrapolation for the p-series with $p = 2$.
SOLUTION: We have already shown that $k = 1$ from the error estimates. Thus, Richardson extrapolation simplifies to $R(A_n) = 2A_{2n} - A_n$. Table 3.4 shows that Richardson extrapolation on the sequence of partial sums gains us about one more place of accuracy. □

Example 3.4
One of the classic ways to estimate a definite integral is the trapezoid rule. To approximate

$$\int_a^b f(x)\,dx,$$

let

$$T_n = \Delta x(f(x_0) + 2f(x_1) + 2f(x_2) + \cdots + 2f(x_{n-1}) + f(x_n))/2,$$

where $x_i = a + i\Delta x$, with $\Delta x = (b - a)/n$. The amount of error is bounded by

$$\frac{M(b-a)^3}{12n^2},$$

where M is a constant given by the maximum of $|f''(x)|$ on the interval $a \le x \le b$. Discover what happens if we take the Richardson extrapolation of this method.

SOLUTION: This error term shows that the trapezoid rule is a perfect candidate for Richardson extrapolation, with $k = 2$. Richardson extrapolation with $k = 2$ yields $(4T_{2n} - T_n)/3$. However, to expand this, we need to adjust the definitions so that both T_n and T_{2n} use the same Δx.

If we let $\Delta x = (b - a)/(2n)$, and $x_i = a + i\Delta x$, then

$$T_n = (2\Delta x)(f(x_0) + 2f(x_2) + 2f(x_4) + \cdots + 2f(x_{2n-2}) + f(x_{2n}))/2,$$

and

$$T_{2n} = \Delta x(f(x_0) + 2f(x_1) + 2f(x_2) + 2(x_3) + \cdots 2f(x_{2n-1}) + f(x_{2n}))/2.$$

Thus, $R(T_n)$ becomes

$$\Delta x(f(x_0)+4gf(x_1)+2f(x_2)+4f(x_3)+\cdots+2f(x_{2n-2})+4f(x_{2n-1})+f(x_{2n}))/3.$$

This is recognizable as Simpson's approximation for indefinite integrals. The error is of order $1/n^4$, which is a huge improvement over the trapezoid rule. So we can think of Simpson's rule as being the Richardson's extrapolation of the trapezoid rule. □

3.2.1 Generalized Richardson Extrapolation

Iterating Richardson extrapolation is not as effective as iterating Shanks transformation, since it is hard to predict the value of k for each iteration. But another solution that does not require knowledge of the value of k ahead of time is to generalize the concept. If the error is an algebraic function of n, we can assume that

$$A_n \sim S + \frac{Q_1}{n} + \frac{Q_2}{n^2} + \frac{Q_3}{n^3} + \cdots \frac{Q_{m-1}}{n^{m-1}}.$$

Since there are now m unknowns, we will use the terms A_n, A_{2n}, $A_{3n}, \ldots A_{mn}$ to give us m equations with m unknowns. Fortunately, there is a nice closed form solution for this system of equations, giving

$$S = \sum_{j=1}^{m} \frac{(-1)^{m+j} A_{jn} j^{m-1}}{(j-1)!(m-j)!}.$$

We define this sum as the *generalized Richardson extrapolation* of the sequence, denoted $R_m(A_n)$. For example, when $m = 2$, this produces $R_2(A_n) = 2A_{2n} - A_n$, the same as the standard Richardson extrapolation with $k = 1$. It is routine to calculate

$$R_3(A_n) = (9A_{3n} - 8A_{2n} + A_n)/2,$$
$$R_4(A_n) = (64A_{4n} - 81A_{3n} + 24A_{2n} - A_n)/6,$$
$$R_5(A_n) = (625A_{5n} - 1024A_{4n} + 486A_{3n} - 64A_{2n} + A_n)/24,$$

TABLE 3.5: Generalized Richardson extrapolation for $\sum_{n=1}^{\infty} 1/n^2$. The true limit is $\pi^2/6 \approx 1.644934066848$.

	$n = 1$	$n = 2$	$n = 3$	$n = 4$
A_n	1.0	1.25	1.361111111	1.423611111
$R_2(A_n)$	1.5	1.597222222	1.621666667	1.631232993
$R_3(A_n)$	1.625	1.641805556	1.643954790	1.644512220
$R_4(A_n)$	1.643518519	1.644863001	1.644923476	1.644931428
$R_5(A_n)$	1.644965278	1.644939144	1.644934957	1.644934302
$R_6(A_n)$	1.644951389	1.644934439	1.644934095	1.644934071
$R_7(A_n)$	1.644935185	1.644934059	1.644934066	1.644934067
$R_8(A_n)$	1.644933943	1.644934065	1.644934067	1.644934067

$$\begin{aligned}
R_6(A_n) &= (7776A_{6n} - 15625A_{5n} + 10240A_{4n} - 2430A_{3n} \\
&\qquad +160A_{2n} - A_n)/120, \\
R_7(A_n) &= (117649A_{7n} - 279936A_{6n} + 234375A_{5n} - 81920A_{4n} + 10935A_{3n} \\
&\qquad -384A_{2n} + A_n)/120, \\
R_8(A_n) &= \frac{131072}{315}A_{8n} - \frac{823542}{720}A_{7n} + \frac{5832}{5}A_{6n} - \frac{78125}{144}A_{5n} + \frac{1024}{9}A_{4n} \\
&\qquad -\frac{729}{80}A_{3n} + \frac{8}{45}A_{2n} - \frac{A_n}{5040}.
\end{aligned}$$

Example 3.5

Apply the generalized Richardson extrapolation to the sum $\sum_{n=1}^{\infty} 1/n^2$ from example 3.3, which converges to $\pi^2/6$.

SOLUTION: The results are given in table 3.5. One can obtain 10 digits of accuracy using only $R_8(A_3)$, which requires only the first 24 terms of the original series. ∎

Example 3.6

Equation 2.1 defined the constant γ, but the sequence in this definition converges very slowly. Approximate this constant by using the generalized Richardson extrapolation on the sequence

$$b_n = -\ln(n) + \sum_{i=1}^{n} \frac{1}{i}$$

from the definition.

SOLUTION: The results are shown in table 3.6. In spite of the slow convergence of the original sequence, the Richardson extrapolation can easily obtain 9 places of accuracy. ∎

TABLE 3.6: Generalized Richardson extrapolation for $\lim_{n\to\infty}(\sum_{i=1}^{n} 1/i) - \ln(n)$. The true limit is $\gamma \approx 0.5772156649015$. Notice that Richardson extrapolation can obtain 9 places of accuracy with only the first 8 terms of the sequence.

	$n=1$	$n=2$	$n=3$	$n=4$
b_n	1.0	0.8068528194	0.7347210447	0.6970389722
$R_2(b_n)$	0.6137056389	0.5872251250	0.5817600169	0.5797922301
$R_3(b_n)$	0.5788334232	0.5773529093	0.5772449441	0.5772252093
$R_4(b_n)$	0.5770928784	0.5771996661	0.5772119510	0.5772144162
$R_5(b_n)$	0.5771907408	0.5772148565	0.5772155789	0.5772156484
$R_6(b_n)$	0.5772150314	0.5772156984	0.5772156699	0.5772156660
$R_7(b_n)$	0.5772159193	0.5772156690	0.5772156651	0.5772156649
$R_8(b_n)$	0.5772156950	0.5772156649	0.5772156649	0.5772156649

Problems for §3.2

1 Show that using the Shanks transformation on the sequence $a_n = 1/n$ yields the sequence $S_n = 1/(2n)$.

2 Show that if $f(x)$ is a positive decreasing function, and $\sum_{n=1}^{\infty} f(n)$ converges, then the error $S - A_n$ satisfies

$$S - A_n \sim \int_n^{\infty} f(x)\,dx.$$

For problems **3** through **14**:

First determine the value of k so that the error between A_n and the limit is proportional to $1/n^k$. Then use Richardson extrapolation to estimate the limit using only A_4 and A_8. What percentage of error is there in the Richardson extrapolation as opposed to using A_8?

3 $A_n = \sum_{m=1}^{n} \frac{1}{m^4}$, $\qquad \lim_{n\to\infty} A_n = \pi^4/90.$

4 $A_n = \sum_{m=1}^{n} \frac{1}{m^3}$, $\qquad \lim_{n\to\infty} A_n = \zeta(3) \approx 1.20205690316.$

5 $A_n = \sum_{m=1}^{n} \frac{1}{m^{3/2}}$, $\qquad \lim_{n\to\infty} A_n = \zeta(3/2) \approx 2.61237534869.$

6 $A_n = \sum_{m=1}^{n} \frac{1}{m^2+m}$, $\qquad \lim_{n\to\infty} A_n = 1.$

7 $A_n = \sum_{m=1}^{n} \frac{1}{2m^2+m}$, $\qquad \lim_{n\to\infty} A_n = 2 - 2\ln 2.$

8 $A_n = \sum_{m=1}^{n} \frac{1}{4m^2 + m}$, $\lim_{n \to \infty} A_n = 4 - 3\ln 2 - \pi/2.$

9 $A_n = \sum_{m=1}^{n} \frac{1}{4m^3 - m}$, $\lim_{n \to \infty} A_n = 2\ln 2 - 1.$

10 $A_n = \sum_{m=1}^{n} \frac{1}{m^3 + m^2}$, $\lim_{n \to \infty} A_n = \pi^2/6 - 1.$

11 $A_n = \sum_{m=1}^{n} \frac{1}{4m^4 - m^2}$, $\lim_{n \to \infty} A_n = 2 - \pi^2/6.$

12 $A_n = \sum_{m=1}^{n} \frac{1}{9m^4 - m^2}$, $\lim_{n \to \infty} A_n = 9/2 - \sqrt{3}\pi/2 - \pi^2/6.$

13 $A_n = \left(1 + \frac{1}{2} + \frac{1}{3} + \cdots + \frac{1}{n}\right) - \ln(n)$, $\lim_{n \to \infty} A_n = \gamma.$
Hint: We are essentially summing the shaded areas of figure 2.1.

14 $A_n = \prod_{m=1}^{n} \left(1 + \frac{1}{m^2}\right)$, $\lim_{n \to \infty} A_n = \sinh(\pi)/\pi.$
Hint: See problem 12 of section 2.3.

15 We have seen that the Richardson extrapolation of the Trapezoid rule reproduces the Simpson's rule. What happens if we do the Richardson extrapolation on Simpson's rule? Instead of the familiar pattern for the coefficients, (1, 4, 2, 4, 2, ... 4, 1), what will the new pattern be? Hint: The estimated error for Simpson's rule is $O(1/n^4)$.

For problems **16** through **27**:

Find A_1 through A_6 for the following sequences, and then use these to find the generalized Richardson extrapolation $R_6(A_1)$. How many places of accuracy do we get?

16 $A_n = \left(1 + \frac{1}{n}\right)^n$, $\lim_{n \to \infty} A_n = e.$

17 $A_n = \left(1 + \frac{2}{n}\right)^n$, $\lim_{n \to \infty} A_n = e^2.$

18 $A_n = \frac{n!}{n^n e^{-n} \sqrt{n}}$, $\lim_{n \to \infty} A_n = \sqrt{2\pi}.$

19 $A_n = \sum_{m=1}^{n} \frac{1}{m^3}$, $\lim_{n \to \infty} A_n = \zeta(3) \approx 1.20205690316.$

20 $A_n = \sum_{m=1}^{n} \frac{1}{m^4}$, $\lim_{n\to\infty} A_n = \zeta(4) = \pi^4/90.$

21 $A_n = \sum_{m=1}^{n} \frac{1}{m^{3/2}}$, $\lim_{n\to\infty} A_n = \zeta(3/2) \approx 2.61237534869.$

22 $A_n = \sum_{m=1}^{n} \frac{1}{m^2+m}$, $\lim_{n\to\infty} A_n = 1.$

23 $A_n = \sum_{m=1}^{n} \left(\frac{1}{2m-1} - \frac{1}{2m}\right)$, $\lim_{n\to\infty} A_n = \ln(2).$

Note that this is the alternating harmonic series disguised as a positive series.

24 $A_n = \prod_{m=1}^{n} \left(1 - \frac{1}{4m^2}\right)$, $\lim_{n\to\infty} A_n = 2/\pi.$

25 $A_n = \prod_{m=1}^{n} \left(1 - \frac{1}{9m^2}\right)$, $\lim_{n\to\infty} A_n = 3\sqrt{3}/(2\pi).$

26 $A_n = \prod_{m=1}^{n} \left(\sqrt[4]{1 + \frac{1}{m}} \frac{4m}{4m+1}\right)$, $\lim_{n\to\infty} A_n = \Gamma(5/4) \approx 0.906402477055.$

27 $A_n = \prod_{n=1}^{\infty} \left(\frac{2m}{2m+1} \cdot \frac{2m+2}{2m+1}\right)$, $\lim_{n\to\infty} A_n = \pi/4.$

See problem 16 from section 2.3.

3.3 Euler Summation

We now turn our attention to *divergent* series. We have already seen many asymptotic series that turn out to be divergent, yet we were able to get some numerical information out of the series though optimal asymptotic truncation rule. This raises the question of whether we can assign a reasonable value to certain kinds of divergent series.

To understand how we can assign a value to a series, we first must understand what causes many series to diverge. For example, the series

$$\sum_{n=0}^{\infty} (-1)^n = 1 - 1 + 1 - 1 + \cdots \tag{3.5}$$

diverges, but this series represents plugging $x = 1$ into the power series

$$f(x) = \sum_{n=0}^{\infty} (-1)^n x^n = 1 - x + x^2 - x^3 + \cdots . \tag{3.6}$$

Note that this is a geometric series, so the sum is $f(x) = 1/(x+1)$ for $|x| < 1$. Also $f(x)$ has a singularity at $x = -1$, which prevents the series from converging at $x = 1$. Since the region of convergence for a Taylor series is always inside a circle, a singularity at one point on this circle often causes the Taylor series to diverge elsewhere on the circle, even though the function $f(x)$ may be defined at this point.

This motivates the definition of the Euler summation technique. As long as the terms of a series $\sum_{n=0}^{\infty} a_n$ grow like a power of n, then

$$f(x) = \sum_{n=0}^{\infty} a_n x^n$$

will converge for $|x| < 1$. We then can define the *Euler sum* of the series to be $\lim\limits_{x \to 1^-} f(x)$, whenever this limit is finite. For example, the Euler sum of the series in equation 3.5 is $1/2$, since we already saw that the sum of equation 3.6 is $1/(x+1)$.

Example 3.7
Find the Euler sum of

$$1 + 0 - 1 + 1 + 0 - 1 + 1 + 0 - 1 + \cdots .$$

SOLUTION: We first convert it to a power series,

$$f(x) = 1 + 0 - x^2 + x^3 + 0 - x^5 + x^6 + 0 - x^8 + \cdots .$$

This is not a geometric series, but it is the sum of two geometric series:

$$f(x) = 1 + x^3 + x^6 + \cdots + (-x^2) + (-x^5) + (-x^8) + \cdots .$$

The sum of a geometric series is $a/(1-r)$ where a is the first term, and r is the common ratio. Thus

$$f(x) = \frac{1}{1-x^3} - \frac{x^2}{1-x^3} = \frac{1-x^2}{1-x^3}.$$

We can now take the limit as $x \to 1^-$ to get the Euler sum to be $2/3$. □

The Euler sum requires being able to find the exact sum of a given power series, producing a function $f(x)$. Most calculus texts emphasize the other direction, that is, converting a given function into a power series, either through series manipulations or by Taylor's theorem. However, there is a straightforward procedure in the case where a_n is a rational function of n.

The first step is to completely factor the numerator and denominator of the coefficient a_n into linear factors of n. Then we set $f(x)$ equal to the power series, and then perform the same operations to both sides of the equation until the series on one side of the equation is geometric, hence summable to $a/(1-r)$, where a is the first of the geometric series, and r is the common ratio between terms. Then we must undo the operations on both sides to solve for $f(x)$. The operations involved are multiplying/dividing by a power of x, integrating, and differentiating.

To eliminate a factor of the form $(n+b)$ from the *denominator*, first multiply by a power of x so that the series involves x^{n+b}, and then *differentiate* both sides. To eliminate a factor of the form $(n+b)$ in the *numerator*, first multiply by a power of x so that the series involves x^{n+b-1}, then *integrate* both sides of the equation. This is best illustrated with an example.

Example 3.8

Find the Euler sum of the series

$$\sum_{n=0}^{\infty}(-1)^n\frac{n+2}{n+1} = 2 - \frac{3}{2} + \frac{4}{3} - \frac{5}{4} + \cdots.$$

SOLUTION: The series clearly diverges, since the terms do not go to 0. Yet we can find the Euler sum by setting

$$f(x) = \sum_{n=0}^{\infty}\frac{(n+2)(-1)^n x^n}{n+1}$$

and solving for $f(x)$. To get rid of the $n+1$ in the denominator of the sum, we must first multiply by x, to get the power of x to be $n+1$, setting up the derivative.

$$xf(x) = \sum_{n=0}^{\infty}\frac{(n+2)(-1)^n x^{n+1}}{n+1}$$

$$[xf(x)]' = \sum_{n=0}^{\infty}(n+2)(-1)^n x^n$$

Now to eliminate the $n+2$, we multiply both sides of the equation by x, so that the power of x is one less than the factor to be eliminated. This sets up the integration.

$$x[xf(x)]' = \sum_{n=0}^{\infty}(n+2)(-1)^n x^{n+1}$$

$$\int x[xf(x)]'\,dx = \sum_{n=0}^{\infty}(-1)^n x^{n+2}$$

This finally is a geometric series, since n only appears in an exponent. This series sums to $x^2/(1+x)$, so we have

$$\int x[xf(x)]'\,dx = \frac{x^2}{1+x}.$$

Now we solve for $f(x)$, unraveling the steps that we have done. So first, we differentiate

$$\begin{aligned} x[xf(x)]' &= \frac{x^2+2x}{(x+1)^2} \\ [xf(x)]' &= \frac{x+2}{(x+1)^2} \\ xf(x) &= \int \frac{x+2}{(x+1)^2}\,dx \\ &= \ln(x+1) - \frac{1}{x+1} + C \\ f(x) &= \frac{\ln(x+1)}{x} - \frac{1}{x^2+x} + \frac{C}{x}. \end{aligned}$$

Finally, we use the fact that $\lim_{x\to 0} f(x) = 2$ to show that $C = 1$. Thus,

$$f(x) = \frac{\ln(x+1)}{x} + \frac{1}{x+1}.$$

Plugging in $x = 1$ gives us the Euler sum of the original series to be $1/2+\ln(2)$. ▯

Note that if the original series converges, then the Euler sum will be the same as the original convergent sum. This is obvious, since the original sum will converge to $f(1)$, and the function is continuous, so $\lim_{x\to 1^-} f(x) = f(1)$. We say that a summation method is *regular* if, whenever the method is applied to a convergent series, the method will always succeed, and produce the same sum as the original convergent sum. Thus, we can say that the Euler summation method is regular.

Problems for §3.3

1 Find the Euler sum of the series

$$1+0-1+0+1+0-1+0+1+0+-1+0+\cdots.$$

2 Find the Euler sum of the series

$$\sum_{n=0}^{\infty}(-1)^n(n+1) = 1-2+3-4+\cdots.$$

3 Find the Euler sum of the series

$$\sum_{n=0}^{\infty}(-1)^n(2n+1) = 1 - 3 + 5 - 7 + \cdots.$$

4 Find the Euler sum of the series

$$\sum_{n=0}^{\infty}(-1)^n(n+1)^2 = 1 - 4 + 9 - 16 + \cdots.$$

5 Find the Euler sum of the series

$$\sum_{n=0}^{\infty}(-1)^n\frac{(n+k)!}{n!}.$$

Hint: Look at the cases $k = 0, 1, 2, 3$ and find the pattern.

6 Show that if $\sum a_n$ and $\sum b_n$ are both Euler summable, then $\sum(a_n + b_n)$ is Euler summable, and its sum is the addition of the first two sums.

7 Find the Euler sum of the series

$$\sum_{n=0}^{\infty}(-1)^n(n^2+1) = 1 - 2 + 5 - 10 + \cdots.$$

Rather than dealing with complex powers, can we combine the results from the previous problems?

8 Find the Euler sum of the series

$$\sum_{n=0}^{\infty}(-1)^n\frac{n+1}{n+2} = \frac{1}{2} - \frac{2}{3} + \frac{3}{4} - \frac{4}{5} + \cdots.$$

9 Find the Euler sum of the series

$$\sum_{n=0}^{\infty}(-1)^n\frac{n^2}{n+1} = 0 - \frac{1}{2} + \frac{4}{3} - \frac{9}{4} + \cdots.$$

10 Is the series

$$\sum_{n=0}^{\infty}(-2)^n = 1 - 2 + 4 - 8 + 16 - 32 + \cdots$$

Euler summable? What happens if we Shank the partial sums of this series?

11 We saw in example 3.7 that the Euler sum of $1 + 0 - 1 + 1 + 0 - 1 + 1 + 0 - 1 + \cdots$ is $2/3$. What would happen if we did a repeated Shank on the sequence of partial sums $\{1, 1, 0, 1, 1, 0, 1, 1, 0, 1, 1, 0, \ldots\}$?

12 Is the harmonic series

$$\sum_{n=0}^{\infty} \frac{1}{n+1} = 1 + \frac{1}{2} + \frac{1}{3} + \frac{1}{4} + \cdots$$

Euler summable? Why or why not?

13 The series for the Riemann zeta function (equation 2.23) is not Euler summable if $\mathrm{Re}(z) < 1$. However, its alternating counterpart, seen in problem 16 of section 2.4, *is* Euler summable for all $z \neq 1$. This gives us another way to extend the Riemann zeta function to the entire plane, except for $z = 1$. Use this method to evaluate

a) $\zeta(0)$, b) $\zeta(-1)$, c) $\zeta(-2)$, d) $\zeta(-3)$.

For problems **14** through **18**:

The technique used in example 3.8 can be used to find the sum of many different power series. The goal may not be to convert the series into a geometric series, but rather any familiar series in which the sum is known. Use this technique to determine what function has the following Maclaurin series.

14

$$f(x) = \sum_{n=0}^{\infty} \frac{(n+2)x^n}{n!}$$

15

$$f(x) = \sum_{n=0}^{\infty} \frac{x^n}{(n+2)n!}$$

16

$$f(x) = \sum_{n=0}^{\infty} \frac{(n+2)(-x)^n}{(2n)!}$$

17

$$f(x) = \sum_{n=0}^{\infty} \frac{(n+1/2)(-x^2)^n}{n!}$$

18

$$f(x) = \sum_{n=0}^{\infty} \frac{(-x^2)^n}{(n+1/2)n!}$$

3.4 Borel Summation

Although the Euler summation allows us to "sum" certain divergent series

$$\sum_{n=0}^{\infty} a_n,$$

it will only work if the radius convergence of the power series

$$f(x) = \sum_{n=0}^{\infty} a_n x^n$$

is 1. This means that the coefficients a_n can grow no faster than a polynomial. If the coefficients grow faster than a polynomial, another technique is needed.

Instead of considering the original power series, suppose we instead consider the series

$$\phi(x) = \sum_{n=0}^{\infty} \frac{a_n x^n}{n!}. \tag{3.7}$$

With the extra $n!$ in the denominator, this series will be much more likely to converge, as least for some non-zero radius of convergence. If so, and if the function $\phi(x)$ can be analytically continued to include the positive real axis, then we can define the *Borel sum* of $f(x)$ to be

$$B(x) = \int_0^{\infty} e^{-t}\phi(xt)\,dt \tag{3.8}$$

provided this integral converges. In particular, the Borel sum of the original series is $B(1)$.

To understand the logic in this definition, let us compute the asymptotic series for this integral. If we substitute $u = xt$, then

$$B(x) = \int_0^{\infty} e^{-u/x}\phi(u)/x\,du,$$

and we can use Watson's lemma (2.1) since $1/x$ is going to ∞. Since the Taylor series of $\phi(u)$ is given by equation 3.7, we have

$$B(x) \sim \frac{1}{x}\sum_{n=0}^{\infty} \frac{a_n}{n!}\frac{n!}{(1/x)^{n+1}} \sim \sum_{n=0}^{\infty} a_n x^n.$$

Thus, we have found a function with the same asymptotic series as our original series.

Example 3.9

Find the Borel sum of the series $0! - 1! + 2! - 3! + 4! - 5! + \cdots$.

SOLUTION: Obviously, this series diverges. If we add powers of x to the series,

$$1 - x + 2x - 6x^2 + 24x^3 - 120x^4 + \cdots,$$

we get a series that has a zero radius of convergence. Yet if we divide the nth term by $n!$, we get

$$\phi(x) = 1 - x + x - x + x - x + \cdots.$$

This geometric series converges to $1/(1 + x)$, at least for $|x| < 1$. This can be analytically continued so that ϕ is defined for the positive real numbers. Hence,

$$B(x) = \int_0^\infty \frac{e^{-t}}{1 + xt}\, dt.$$

In particular, the Borel sum of the original series is

$$\int_0^\infty \frac{e^{-t}}{1 + t}\, dt \approx 0.596347362323.$$

□

Example 3.10
Find the Borel sum of

$$\sum_{n=0}^{\infty} (n + 1)(-2)^n = 1 - 4 + 12 - 32 + 80 + \cdots.$$

SOLUTION: This is not Euler summable, since the series with the x^n added,

$$\sum_{n=0}^{\infty} (n + 1)(-2x)^n = 1 - 4x + 12x^2 - 32x^3 + 80x^4 + \cdots$$

has a radius of convergence of only $1/2$. However, if we divide the nth term by $n!$, we get

$$\phi(x) = \sum_{n=0}^{\infty} \frac{(n + 1)(-2)^n x^n}{n!}.$$

This converges for all x, but to what function? Let us work to simplify the sum. To get rid of the $n + 1$, we need to have the power of x to be n, (which it already is,) and then integrate both sides of the equation. Thus,

$$\int \phi(x)\, dx = \sum_{n=0}^{\infty} \frac{(-2)^n x^{n+1}}{n!} = x \sum_{n=0}^{\infty} \frac{(-2x)^n}{n!} = xe^{-2x}.$$

Taking the derivative of both sides, we get

$$\phi(x) = (1 - 2x)e^{-2x}.$$

So the Borel sum of the series is

$$B(x) = \int_0^\infty e^{-t}(1-2xt)e^{-2xt}\,dt = e^{-(2x+1)t}\frac{4tx^2+2tx-1}{(1+2x)^2}\Big|_0^\infty = \frac{1}{(1+2x)^2}.$$

Since the integral converges, the Borel sum of the series is $B(1) = 1/9$. □

3.4.1 Generalized Borel Summation

In the case where not only does the original series diverge, but also the series

$$\phi(x) = \sum_{n=0}^\infty \frac{a_n x^n}{n!}$$

diverges for all non-zero x, it may be possible to iterate the Borel summation. That is, we could apply the Borel sum to $\phi(x)$, and then apply equation 3.8 to get back to $B(x)$.

In general, if there is a positive integer k such that the series

$$\phi(x) = \sum_{n=0}^\infty \frac{a_n x^n}{(n!)^k}$$

converges for a non-zero radius of convergence, and if $\phi(x)$ can be analytically continued to include the positive real numbers, then we define the *generalized Borel sum* of the series $\sum_{n=0}^\infty a_n x^n$ to be the multiple integral

$$B(x) = \int_0^\infty\int_0^\infty\cdots\int_0^\infty e^{-t_1-t_2-\cdots-t_k}\phi(xt_1t_2\ldots t_k)\,dt_1\,dt_2\cdots dt_k.$$

In particular, the sum of $\sum_{n=0}^\infty a_n = B(1)$. We say that a series is k-Borel summable if the generalized Borel sum exists for this value of k.

Example 3.11

Find the generalized Borel sum to the series

$$\sum_{n=0}^\infty (-1)^n(n!)^2 = 1 - 1 + 4 - 36 + 576 - 14400 - \cdots.$$

SOLUTION: Multiplying the nth term by x^n gives

$$\sum_{n=0}^\infty (-x)^n(n!)^2 = 1 - x + 4x^2 - 36x^3 + 576x^4 - 14400x^5 + \cdots.$$

This series has zero radius of convergence and is not even Borel summable. But if we divide the nth term by $(n!)^2$, we get

$$\phi(x) = \sum_{n=0}^\infty (-x)^n = 1 - x + x^2 - x^3 + \cdots.$$

Thus, the series is 2-Borel summable. Since $\phi(x)$ sums to $1/(1+x)$, we find that the generalized Borel sum is the double integral

$$B(x) = \int_0^\infty \int_0^\infty \frac{e^{-t-s}}{1+xts}\, dt\, ds.$$

In particular, the generalized Borel sum of the original series is

$$B(1) = \int_0^\infty \int_0^\infty \frac{e^{-t-s}}{1+ts}\, dt\, ds \approx 0.668091326378.$$

□

Example 3.12
Find the generalized Borel sum to the series

$$\sum_{n=0}^{\infty} (-1)^n (n!)^3 = 1 - 1 + 8 - 216 + 13824 - 1728000 + \cdots.$$

SOLUTION: This time, after multiplying the nth term by x^n, we have to divide each term by $(n!)^3$ before it converges. Thus, $k = 3$, and once again

$$\phi(x) = \sum_{n=0}^{\infty} (-x)^n = 1 - x + x^2 - x^3 + \cdots = 1/(1+x).$$

So we have the generalized Borel sum to be the triple integral

$$B(x) = \int_0^\infty \int_0^\infty \int_0^\infty \frac{e^{-t-s-r}}{1+xtsr}\, dt\, ds\, dr.$$

In particular, the generalized Borel sum of the original series is

$$B(1) = \int_0^\infty \int_0^\infty \int_0^\infty \frac{e^{-t-s-r}}{1+tsr}\, dt\, ds\, dr \approx 0.723749944182.$$

□

Example 3.13
Find the generalized Borel sum of the series

$$\sum_{n=0}^{\infty} (-1)^n (2n)! = 1 - 2 + 24 - 720 + 40320 - 3628800 + \cdots.$$

SOLUTION: At first glance, this looks like it is growing faster than the last example, but we only have to divide each term by $(n!)^2$ to get the series to have a non-zero radius of convergence. In fact, by comparing this series to equation 1.32, we find that

$$\sum_{n=0}^{\infty} \frac{(-1)^n (2n)! x^n}{(n!)^2} = \frac{1}{\sqrt{1+4x}}.$$

Thus, the generalized Borel sum for the series is

$$B(x) = \int_0^\infty \int_0^\infty \frac{e^{-t-s}}{\sqrt{1+4xts}}\, dt\, ds.$$

The first integral can in fact be computed in terms of the $\operatorname{erf}(x)$ function. However, the second integral becomes tricky. *Mathematica* computes the final result to be

$$B(x) = \frac{2\sin(1/\sqrt{x})\operatorname{Ci}(1/\sqrt{x}) + \cos(1/\sqrt{x})(\pi - 2\operatorname{Si}(1/\sqrt{x}))}{2\sqrt{x}}.$$

Here, $\operatorname{Ci}(x)$ and $\operatorname{Si}(x)$ are as defined in problems 1 and 2 of section 2.2. In particular, the generalized Borel sum of the original series is

$$B(1) = \int_0^\infty \int_0^\infty \frac{e^{-t-s}}{\sqrt{1+4ts}}\, dt\, ds \approx 0.621449624236.$$

∎

There is an alternative way of computing the generalized Borel sum that doesn't involve multiple integrals [3].

PROPOSITION 3.1

If a series

$$\sum_{n=0}^{\infty} a_n x^n \tag{3.9}$$

is k-Borel summable, then the series

$$\psi(x) = \sum_{n=0}^{\infty} \frac{a_n x^n}{(kn)!}$$

converges for a non-zero radius of convergence, and the generalized Borel sum can be computed via the single integral

$$B(x) = \int_0^\infty e^{-t} \psi(xt^k)\, dt. \tag{3.10}$$

PROOF: Note that when $k = 1$, this new formula simplifies to the definition of the standard Borel summation. Thus, the proposition is true for $k = 1$.

We can use induction on k, and assume that the proposition is true for the previous k. If (3.9) is k-Borel summable, then the series

$$\sum_{n=0}^{\infty} \frac{a_n x^n}{n!} \tag{3.11}$$

is $(k-1)$-Borel summable. Hence, by induction, we can find the generalized Borel sum of this series to be

$$\phi(x) = \int_0^\infty e^{-u} \varphi(xu^{k-1})\, du$$

where $\varphi(x)$ is the analytical extension of the series

$$\sum_{n=0}^{\infty} \frac{a_n x^n}{n!(kn-n)!}.$$

Again, by induction, this series will have a non-zero radius of convergence, and its analytical extension will be defined for all positive x.

Having established the generalized Borel sum of (3.11), we can now compute the generalized Borel sum of (3.9) to be

$$B(x) = \int_0^\infty e^{-t} \int_0^\infty e^{-u} \varphi(xtu^{k-1})\, du\, dt.$$

All we have to do is show that this is equivalent to (3.10).

The next step is to introduce the substitution $s = t + u$ for the outside integral. Then $ds = dt$, and s will still range from 0 to ∞, but u will only range from 0 to s. Hence, we have

$$B(x) = \int_0^\infty \int_0^s e^{-s} \varphi(x(s-u)u^{k-1})\, du\, ds.$$

So the task is to study the function

$$\theta(s, x) = \int_0^s \varphi(x(s-u)u^{k-1})\, du.$$

Since $\varphi(x)$ is analytic for all $x > 0$, then $\theta(s, x)$ will also be analytic for $x > 0$ and $s > 0$. In fact, it is not hard to find the Taylor expansion in x. Replacing φ with its Taylor series gives us

$$\theta(s, x) = \int_0^s \sum_{n=0}^{\infty} \frac{a_n x^n (s-u)^n u^{kn-n}}{n!(kn-n)!}\, du.$$

If x is sufficiently small, the series will be uniformly convergent over the range of the integral. Thus, we can interchange the sum and the integral to obtain

$$\theta(s, x) = \sum_{n=0}^{\infty} \frac{a_n x^n}{n!(kn-n)!} \int_0^s (s-u)^n u^{kn-n}\, du.$$

We can evaluate this integral using the Eulerian beta integral [24, p. 253]:

$$\int_0^1 x^m(1-x)^n\,dx = \beta(m+1,n+1) = \frac{\Gamma(m+1)\Gamma(n+1)}{\Gamma(m+n+2)}. \tag{3.12}$$

From this, we can obtain (if m and n are non-negative integers)

$$\int_0^b x^m(b-x)^n\,dx = b^{m+n+1}\frac{m!\,n!}{(m+n+1)!}.$$

So we have

$$\theta(s,x) = \sum_{n=0}^{\infty} \frac{a_n x^n}{n!(kn-n)!}\,\frac{s^{kn+1}n!(kn-n)!}{(kn+1)!} = \sum_{n=0}^{\infty}\frac{a_n x^n s^{kn+1}}{(kn+1)!}.$$

Although this expression involves both x and s, we can take the partial derivative of $\theta(s,x)$ with respect to s to get

$$\theta_s(s,x) = \sum_{n=0}^{\infty}\frac{a_n x^n (kn+1)s^{kn}}{(kn+1)!} = \sum_{n=0}^{\infty}\frac{a_n x^n s^{kn}}{(kn)!} = \psi(xs^k).$$

There is still an issue of whether this series still converges. But by Sterling's formula, $n! \sim n^n e^{-n}\sqrt{2\pi n}$, it is easy to show that

$$\frac{(n!)^k}{(nk)!} \sim \frac{(2\pi n)^{k-1}}{k^{kn}\sqrt{k}}.$$

Since we are given that the original series (3.9) is k-Borel summable, then the series

$$\sum_{n=0}^{\infty}\frac{a_n x^n}{(n!)^k}$$

has a positive radius of convergence. So

$$\sum_{n=0}^{\infty}\frac{a_n x^n}{(kn)!}$$

will have an even larger radius of convergence, by a factor of k^k. So the series for ψ converges.

The final step is to integrate by parts the integral

$$B(x) = \int_0^{\infty} e^{-s}\theta(s,x)\,ds.$$

Using $u = \theta(s,x)$, $dv = e^{-s}\,ds$, we get

$$B(x) = -e^{-s}\theta(s,x)\Big|_0^{\infty} + \int_0^{\infty} e^{-s}\theta_s(s,x)\,ds.$$

Because the integral converges, $\lim_{s\to\infty} e^{-s}\theta(s,x) = 0$. It is also clear from the series that $\theta(0,x) = 0$. Thus,

$$B(x) = \int_0^\infty e^{-s}\theta_s(s,x)\,ds = \int_0^\infty e^{-t}\varphi(xt^k)\,dt.$$

□

Example 3.14
Use this new method to find the generalized Borel sum of

$$\sum_{n=0}^{\infty}(-1)^n(2n)! = 1 - 2 + 24 - 720 + 40320 - 3628800 + \cdots.$$

SOLUTION: With $k = 2$, we find

$$\psi(x) = \sum_{n=0}^{\infty}(-1)^n x^n = \frac{1}{1+x}.$$

Thus, the generalized Borel sum is

$$\int_0^\infty \frac{e^{-t}}{1+xt^2}\,dt.$$

In particular, the Borel sum of the original series is

$$\int_0^\infty \frac{e^{-t}}{1+t^2}\,dt \approx 0.621449624236.$$

□

3.4.2 Stieltjes Series

There is one subtle question that has yet to be addressed. We have managed to find a sum for a divergent asymptotic series to produce a function, yet which function have we found? Recall that there are *many* functions that are asymptotic to a series, since we can always add a subdominant function. So how does the Borel sum pick one function to converge to?

In the case of a Taylor series that converges for at some non-zero radius of convergence,

$$f(x) \sim \sum_{n=0}^{\infty} a_n(x-x_0)^n,$$

there will be a unique function that is analytic at x_0, so there is no problem with uniqueness. But if the radius of convergence is 0, then no function with this asymptotic series representation will be analytic at x_0. So how else does the Borel sum pick a unique function to converge too?

This can be a very difficult question to answer in general, but there is a large class of series for which the answer is well understood. We say that a series

$$f(x) \sim \sum_{n=0}^{\infty} a_n x^n$$

is a *Stieltjes series* if there is a non-negative function $\rho(t)$, defined for all $t > 0$, such that

$$a_n = (-1)^n \int_0^{\infty} t^n \rho(t)\, dt. \tag{3.13}$$

Here, all improper integrals must converge, but the function $\rho(t)$ is allowed to be a distribution. For example, the series

$$\sum_{n=0}^{\infty} (-1)^n n! x^n$$

is a Stieltjes series, because letting $\rho(x) = e^{-t}$ causes

$$(-1)^n \int_0^{\infty} t^n e^{-t}\, dt = (-1)^n n! = a_n.$$

Unfortunately, it usually is very hard to determine $\rho(t)$, so this definition makes it almost impossible to determine if a series is Stieltjes. Furthermore, there may be more than one $\rho(t)$ which satisfies equation 3.13. However, there is a result, known as Carleman's condition, which allows us to determine if the $\rho(t)$ is unique [1].

$$\boxed{\text{If } \sum_{n=1}^{\infty} |a_n|^{(-1/2n)} = \infty, \text{ then there can be at most one possible } \rho(t).} \tag{3.14}$$

There are fortunately other ways to determine if a series is Stieltjes. For example, the series is Stieltjes if, and only if, it is alternating (starting with a positive constant term) and

$$\sum_{i=0}^{n} \sum_{j=0}^{n} |a_{i+j}| v_i v_j \geq 0$$

for all real vectors $\{v_0, v_1, v_2, \ldots v_n\}$. Another way of saying this is that all of the symmetric matrices of the form

$$\begin{pmatrix} a_0 & -a_1 & a_2 & \cdots & (-1)^n a_n \\ -a_1 & a_2 & -a_3 & \cdots & (-1)^{n+1} a_{n+1} \\ a_2 & -a_3 & a_4 & \cdots & (-1)^{n+2} a_{n+2} \\ \vdots & \vdots & \vdots & \ddots & \vdots \\ (-1)^n a_n & (-1)^{n+1} a_{n+1} & (-1)^{n+2} a_{n+2} & \cdots & a_{2n} \end{pmatrix} \tag{3.15}$$

are *positive semi-definite*. This means that all of the symmetrical submatrices must have non-negative determinant.

Example 3.15
Test whether

$$\sum_{n=0}^{\infty}(-2x)^n = 1 - 2x + 4x^2 - 8x^3 + \cdots$$

is Stieltjes.
SOLUTION: We need to determine whether all submatrices of

$$\begin{pmatrix} 1 & 2 & 4 & \cdots & 2^n \\ 2 & 4 & 8 & \cdots & 2^{n+1} \\ 4 & 8 & 16 & \cdots & 2^{n+2} \\ \vdots & \vdots & \vdots & \ddots & \vdots \\ 2^n & 2^{n+1} & 2^{n+2} & \cdots & 2^{2n} \end{pmatrix}$$

have non-negative determinant. Clearly the 1×1 submatrices will have positive determinant, but note that any two rows are multiples of each other. The same will be true for any submatrix 2×2 or larger, so all of these determinants will be 0. Hence, this matrix is positive semi-definite, and so the series is Stieltjes.

Ironically, in this easy example, it is actually quite tricky to find the $\rho(t)$ that satisfies equation 3.13. The distribution $\rho(t)$ is given by $\delta(t-2)$, where $\delta(t)$ is the Dirac delta function. This "spike function" satisfies the property

$$\boxed{\text{If } f(x) \text{ is continuous at } c\text{, then } \int_{-\infty}^{\infty} f(t)\delta(t-c)\,dt = f(c).} \tag{3.16}$$

So naturally,

$$(-1)^n \int_0^{\infty} t^n \delta(t-2)\,dt = (-1)^n 2^n = a_n.$$

Technically, the Dirac delta function is a distribution instead of a function, but $\rho(t)$ is allowed to be a distribution. See problems 18 to 21 for further discussion on the Dirac delta function. ∎

Another way to test whether an asymptotic series is Stieltjes is through the function $f(x)$ that it is asymptotic to.

DEFINITION 3.1 We say that the complex function $f(z)$ is a *Stieltjes function* if the following four properties hold:

1.) $f(z)$ is analytic in the cut complex plane, where the cut is along the negative real axis. That is, $f(z)$ is analytic if either $\text{Im}(z) \neq 0$ or $\text{Re}(z) > 0$.

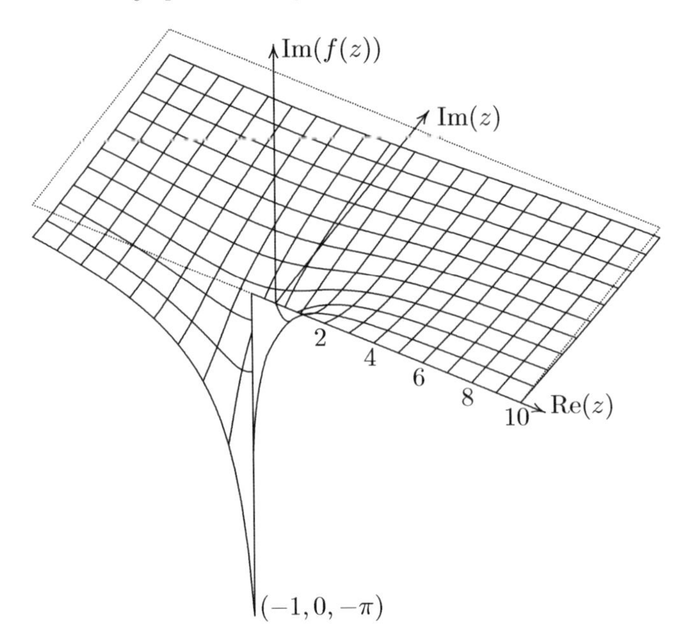

FIGURE 3.3: Graphical demonstration that the imaginary part of $f(z) = \ln(z+1)/z$ is negative whenever $\text{Im}(z)$ is positive. The xy-plane is shown by the dotted lines. This graph gives evidence that, at least in the vicinity of the origin, $\text{Im}(\ln(z+1)/z) < 0$ whenever $\text{Im}(z) > 0$.

2.) $f(z)$ is real when z is a positive real number, and $\lim\limits_{z\to\infty} f(z)$ is a non-negative constant.

3.) $f(z) \sim \sum\limits_{n=0}^{\infty} a_n z^n$ as $x \to 0$ is an alternating series, with $(-1)^n a_n > 0$.

4.) $\text{Im}(f(z)) < 0$ whenever $\text{Im}(z) > 0$.

For example, the function

$$\frac{\ln(x+1)}{x} \sim 1 - \frac{x}{2} + \frac{x^2}{3} - \frac{x^3}{4} + \cdots$$

is Stieltjes. The first 3 properties are easy to verify, but the fourth is a bit tricky. Figure 3.3 shows a *Mathematica* plot of $\text{Im}(f(z))$ for $\text{Im}(z) > 0$. Although this only shows a portion of the complex plane, the trend is easy to visualize. The entire graph for the half plane is below the xy-plane.

The connection between Stieltjes series and Stieltjes functions is as follows: Because of subdominance, a given Stieltjes series will have many possible functions asymptotic to that series. But if there is a function that stands out:

$$f(x) = \int_0^{\infty} \frac{\rho(t)}{1+xt}\, dt,$$

where $\rho(t)$ is any function that satisfies equation 3.13. Using the procedure of section 2.1, we have

$$\int_0^\infty \frac{\rho(t)}{1+xt}\,dt \sim \int_0^\infty \sum_{n=0}^\infty \rho(t)(-xt)^n\,dt \tag{3.17}$$

$$\sim \sum_{n=0}^\infty (-1)^n x^n \int_0^\infty t^n \rho(t)\,dt \sim \sum_{n=0}^\infty a_n x^n. \tag{3.18}$$

Furthermore, this function can be shown to satisfy all four properties for the Stieltjes function (see problem 22). If, in fact, there is only one $\rho(t)$ that satisfies equation 3.13, the integral in equation 3.17 is called the *Stieltjes sum* of the series. This property works in both directions, so the asymptotic series of any Stieltjes function will be a Stieltjes series.

We can now give a partial answer to the question of which function the Borel sum selects. If the series is Stieltjes, and is also Borel summable, then Carleman's condition, equation 3.14, will hold. This means that there will be a unique $\rho(t)$ satisfying equation 3.13. The Borel sum of the series will be identical to the Stieltjes sum, that is, the unique Stieltjes function which is asymptotic to that series.

Problems for §3.4

1 Find the Borel sum of the series $1 - 2 + 4 - 8 + 16 - 32 + \cdots$. Compare this with problem 10 of section 3.3.

2 Find the Borel sum of

$$\sum_{n=0}^\infty (-1)^n n\,n! = 0 - 1 + 4 - 18 + 96 - 600 + \cdots$$

in terms of an integral. Then use a calculator to approximate this integral numerically.

3 Find the Borel sum of

$$\sum_{n=0}^\infty (-1)^n (n+2)n! = 2 - 3 + 8 - 30 + 144 - 840 + \cdots$$

in terms of an integral. Then find an anti-derivative that will compute the integral exactly.

4 Find the Borel sum of

$$\sum_{n=0}^\infty a_n = 0 + 1 - 1 + 2 - 6 + 24 - 120 + \cdots,$$

where

$$a_n = \begin{cases} 0 & \text{if } n = 0, \\ (-1)^{n-1}(n-1)! & \text{if } n \geq 1. \end{cases}$$

How does adding a leading 0 effect the Borel sum? Compare with example 3.9.

5 Show that, in general, inserting a leading 0 in the sum, as was done in problem 4, does not affect the Borel sum.

6 Show that Borel summation is *regular*. That is, if the original series converges, then the Borel sum of the series is the same as the original sum.
Hint: Use Watson's lemma.

7 Find the Borel sum of the divergent series

$$f(x) \sim \sum_{n=0}^{\infty} \frac{(-1)^n (2n)! x^n}{n!} \sim 1 - 2x + 12x^2 - 120x^3 + 1680x^4 - 30240x^5 + \cdots.$$

Express the answer in terms of an integral. Find the numerical evaluation of $f(1)$.

8 Show that the Borel sum of the series

$$\sum_{n=0}^{\infty} (-1)^n (2n)! x^n = 1 - 2x + 24x^2 - 720x^3 + \cdots$$

is not Borel summable, but the series

$$\sum_{n=0}^{\infty} (-1)^n (2n)! x^{2n} = 1 - 2x^2 + 24x^4 - 720x^6 + \cdots$$

is Borel summable. Find the Borel sum of this series in terms of an integral, and find the numerical approximation at $x = 1$. How does this relate to the generalized Borel sum of the original series, done in example 3.14?

9 Use proposition 3.1 to find the generalized Borel sum of the series

$$\sum_{n=0}^{\infty} (-1)^n (2n+1)! x^n = 1 - 6x + 120x^2 - 5040x^3 + \cdots$$

in terms of a single integral. Find the numerical evaluation of $f(1)$.

10 Use proposition 3.1 to find the generalized Borel sum of the series

$$\sum_{n=0}^{\infty} \frac{(-1)^n (3n)! x^n}{n!} = 1 - 6x + 360x^2 - 60480x^3 + \cdots$$

in terms of a single integral. Find the numerical evaluation of $f(1)$.
(Hint: Even though $k = 2$ will do, use $k = 3$.)

11 Use the Eulerian beta integral (equation 3.12) to show that

$$\int_0^b x^m (b-x)^n \, dx = b^{m+n+1} \frac{m!\, n!}{(m+n+1)!}$$

if m and n are positive integers.

12 Show that the series

$$\sum_{n=0}^{\infty} (-1)^n 2^{(n^2)}$$

is *not* generalized Borel summable.

13 Show that the series

$$\sum_{n=0}^{\infty} (-1)^n (2n)! x^n$$

is Stieltjes.

14 Show that for the series in problem 13, the Carleman's condition (equation 3.14) is satisfied, showing that $\rho(t)$ is unique.

15 Show that the series $\sum_{n=0}^{\infty} (-1)^n (n+1) x^n = 1 - 2x + 3x^2 - 4x^3 + \cdots$ is *not* Stieltjes.

16 It was shown graphically that $\ln(x+1)/x$ is a Stieltjes function, so it's Taylor series must be a Stieltjes series. Hence, there is a positive function $\rho(x)$ such that $\int_0^\infty t^n \rho(t)\, dt = 1/(n+1)$. Find such a function.
Hint: Think piecewise defined functions.

17 Show that the sum of two Stieltjes functions is a Stieltjes function.

18 Example 3.15 introduced the Dirac delta function, which is really a distribution instead of a function. It can be described as an infinitely thin spike centered at $x = 0$ such that the area under the spike is 1. This can be denoted mathematically by

$$\delta(x) = 0 \text{ for all } x \neq 0, \qquad \int_{-\infty}^{\infty} \delta(x)\, dx = 1. \tag{3.19}$$

Show that the key property of the Dirac delta function, equation 3.16, can be derived from this definition.

19 Although the Dirac delta function is not really a function, there are several ways to express it as the limit of a sequence of functions. One such sequence is

$$f_\epsilon(x) = \begin{cases} 0 & \text{if } |x| > \epsilon, \\ 1/(2\epsilon) & \text{if } |x| \leq \epsilon. \end{cases}$$

Show that $\lim\limits_{\epsilon \to 0+} f_\epsilon(x)$ satisfies equation 3.19 from problem 18.

20 Another typical way to express the Dirac delta function is

$$\delta(x) = \lim_{\epsilon \to 0^+} \frac{e^{-x^2/\epsilon}}{\sqrt{\pi\epsilon}}.$$

Show that this limit satisfies equation 3.19 from problem 18.

21 One final way to express the Dirac delta function is

$$\delta(x) = \lim_{\epsilon \to 0^+} \frac{\epsilon}{\pi(x^2 + \epsilon^2)}.$$

Show that this limit satisfies equation 3.19 from problem 18.

22 Show that if $\rho(t) \geq 0$ for all $t > 0$, and $\rho(t)$ is not identically 0, then the function

$$f(z) = \int_0^\infty \frac{\rho(t)}{1 + zt}\, dt$$

satisfies $\mathrm{Im}(f(z)) < 0$ whenever $\mathrm{Im}(z) > 0$. Assume that the integral converges.

3.5 Continued Fractions

Many times a divergent asymptotic series

$$f(x) \sim \sum_{n=0}^{\infty} a_n x^n \sim a_0 + a_1 x + a_2 x^2 + \cdots$$

can be expressed in a form which actually converges. One such form is a *continued fraction*, or *C-fraction*,

$$f(x) \sim \cfrac{c_0}{1 + \cfrac{c_1 x}{1 + \cfrac{c_2 x}{1 + \cfrac{c_3 x}{1 + \cfrac{c_4 x}{1 + \cdots}}}}}.$$

To understand why convergence is more likely with a continued fraction, consider what it means for the original series to converge. For a particular value x_0, the sequence

$$a_0, \quad a_0 + a_1 x_0, \quad a_0 + a_1 x_0 + a_2 x_0^2, \quad a_0 + a_1 x_0 + a_2 x_0^2 + a_3 x_0^3, \ldots$$

must converge, meaning that we have to approximate the function $f(x)$ with *polynomials*, which cannot have singularities. On the contrary, the continued fraction converges at x_0 if the sequence

$$F_0 = c_0, \quad F_1 = \frac{c_0}{1 + c_1 x_0}, \quad F_2 = \frac{c_0}{1 + \dfrac{c_1 x_0}{1 + c_2 x_0}}, \quad F_3 = \frac{c_0}{1 + \dfrac{c_1 x_0}{1 + \dfrac{c_2 x_0}{1 + c_3 x_0}}},$$

converges, so we are approximating $f(x)$ with a sequence of *rational functions*, which are allowed to have singularities. Even singularities other than poles, such as branch cuts, can be approximated by a grouping of poles created by the rational functions.

Although we are mainly interested in asymptotic power series, a continued fraction representation can be made for any asymptotic series for $f(x)$. We can let $f_0(x) = f(x)$, and then recursively define

$$g_n(x) = \text{ the leading term of } f_n(x), \qquad (3.20)$$
$$f_{n+1}(x) = \frac{g_n(x)}{f_n(x)} - 1.$$

Then

$$f(x) \sim g_0(x)/(1 + g_1(x)/(1 + g_2(x)/(1 + g_3(x)/(1 + g_4(x)/(\cdots.$$

Notice that we made the continued fraction notation more compact by introducing open parenthesis. The procedure in equation 3.20 can be used to find the continued fraction representation of a function using a symbolic manipulator such as *Mathematica*® or *Maxima*.

Example 3.16

Convert

$$f(x) = e^x \sim 1 + x + \frac{x^2}{2} + \frac{x^3}{6} + \frac{x^4}{24} + \frac{x^5}{120} + \frac{x^6}{720} + \frac{x^7}{5040} + \cdots \qquad \text{as} \quad x \to 0$$

to a continued fraction.

SOLUTION: The leading term as $x \to 0$ is obviously $g_0(x) = 1$, so $f_1(x) = 1/f_0(x) - 1$, so

$$f_1(x) = e^{-x} - 1 \sim -x + \frac{x^2}{2} - \frac{x^3}{6} + \frac{x^4}{24} - \frac{x^5}{120} + \frac{x^6}{720} - \frac{x^7}{5040} + \cdots.$$

This time, the leading term is $g_1(x) = -x$, and so $f_2(x) = -x/(e^{-x} - 1) - 1$, which starts getting complicated. *Mathematica* can produce the results

$$f_2(x) \sim \frac{x}{2} + \frac{x^2}{12} - \frac{x^4}{720} + \frac{x^6}{30240} + O(x^7).$$

$$f_3(x) \sim -\frac{x}{6} + \frac{x^2}{36} - \frac{x^3}{540} - \frac{x^4}{6480} + \frac{x^5}{27216} + O(x^6).$$

$$f_4(x) \sim \frac{x}{6} + \frac{x^2}{60} - \frac{x^4}{8400} + O(x^5).$$

$$f_5(x) \sim -\frac{x}{10} + \frac{x^2}{100} - \frac{x^3}{3500} + O(x^4).$$

$$f_6(x) \sim \frac{x}{10} + \frac{x^2}{140} + O(x^3).$$

$$f_7(x) \sim -\frac{x}{14} + O(x^2).$$

Thus, $g_2(x) = x/2$, $g_3(x) = -x/6$, $g_4(x) = x/6$, $g_5(x) = -x/10$, $g_6(x) = x/10$, and $g_7(x) = -x/14$. So the continued fraction expansion of e^x as $x \to 0$ is

$$e^x \sim 1\Big/\Big(1 - x\Big/\Big(1 + \tfrac{x}{2}\Big/\Big(1 - \tfrac{x}{6}\Big/\Big(1 + \tfrac{x}{6}\Big/\Big(1 - \tfrac{x}{10}\Big/\Big(1 + \tfrac{x}{10}\Big/\Big(1 - \tfrac{x}{14}\Big/\Big(\cdots.$$

The pattern is starting to be apparent: $c_0 = 1$, $c_1 = -1$, $c_2 = 1/2$, $c_{2n-1} = -1/(4n-2)$, and $c_{2n} = 1/(4n-2)$ for $n > 1$. If we plug in $x = 1$, we get the sequence

$$F_0 = 1, F_1 = \infty, F_2 = 3, F_3 \approx 2.6666667, F_4 \approx 2.7142957, F_5 = 2.71875,$$

$$F_6 \approx 2.718309859, F_7 \approx 2.71827957, \ldots.$$

The sequence is clearly converging to e, but the rate of convergence is only slightly better than the Taylor series. (The sum of the Taylor series up to the x^7 term is 2.718253968.) The reason is that the function e^x has no singularities, so it can be approximated with a polynomial as easy as it can be approximated with a rational function. □

DEFINITION 3.2 A useful and concise notation for continued fractions is the "Big K" notation. A generalized continued fraction can be written

$$\mathop{\mathrm{K}}_{n=0}^{\infty} \frac{a_n}{b_n} = \cfrac{a_0}{b_0 + \cfrac{a_1}{b_1 + \cfrac{a_2}{b_3 + \cfrac{a_3}{b_3 + \cfrac{a_4}{b_4 + \cdots}}}}}.$$

We can combine this notation with the open parenthesis notation to get

$$\mathop{\mathrm{K}}_{n=0}^{\infty} a_n \Big/\!\Big(b_n = a_0/(b_0 + a_1/(b_1 + a_2/(b_2 + a_3/(b_3 + a_4/(b_4 \cdots.$$

With this new notation, we can express the continued fraction expansion of e^x as

$$e^x = 1\Big/\Big(1 - x\Big/\Big(1 + \frac{x}{2}\Big/\Big(1 + \mathop{\mathrm{K}}_{n=2}^{\infty} \frac{-x}{4n-2}\Big/\Big(1 + \frac{x}{4n-2}\Big/\Big(1.$$

Notice how we can include two "terms" at once, since c_{2n-1} has a different pattern than c_{2n}.

A continued fraction representation of $f(x)$ is called *regular* if $g_n(x) = c_n x$, for all $n \geq 1$. If there is a regular continued fraction representation of $f(x)$, it is easy to compute from the series

$$f(x) \sim a_0 + a_1 x + a_2 x^2 + a_3 x^3 + \cdots.$$

We begin by letting $b_{-3} = -1$, $b_{-2} = 1$, $b_{-1} = 1$,

$$b_0 = a_0, \qquad b_1 = a_1, \qquad b_2 = \begin{vmatrix} a_0 & a_1 \\ a_1 & a_2 \end{vmatrix}, \qquad b_3 = \begin{vmatrix} a_1 & a_2 \\ a_2 & a_3 \end{vmatrix}, \tag{3.21}$$

$$b_4 = \begin{vmatrix} a_0 & a_1 & a_2 \\ a_1 & a_2 & a_3 \\ a_2 & a_3 & a_4 \end{vmatrix}, \qquad b_5 = \begin{vmatrix} a_1 & a_2 & a_3 \\ a_2 & a_3 & a_4 \\ a_3 & a_4 & a_5 \end{vmatrix}, \ldots.$$

As long as all $b_n \neq 0$, we can let $c_n = -b_{n-3}b_n/(b_{n-1}b_{n-2})$, giving

$$f(x) \sim c_0\Big/\Big(1 + \mathop{\mathrm{K}}_{n=1}^{\infty} c_n x\Big/\Big(1.$$

Example 3.17

Find the continued fraction expansion of $\ln(x+1)$.

SOLUTION: The Taylor series for $\ln(x+1)$ centered at $x_0 = 0$ is

$$\ln(x+1) \sim \sum_{n=1}^{\infty} (-1)^{n-1}\frac{x^n}{n} \sim x - \frac{x^2}{2} + \frac{x^3}{3} - \frac{x^4}{4} + \frac{x^5}{5} - \frac{x^6}{6} + \cdots.$$

Since the leading term is x instead of a constant, this will not quite a regular continued fraction representation. But we can instead consider $\ln(x+1)/x$, for which $a_n = (-1)^n/(n+1)$.

Then

$$b_0 = 1, \quad b_1 = -\frac{1}{2}, \quad b_2 = \begin{vmatrix} 1 & -1/2 \\ -1/2 & 1/3 \end{vmatrix} = \frac{1}{12}, \quad b_3 = \begin{vmatrix} -1/2 & 1/3 \\ 1/3 & -1/4 \end{vmatrix} = \frac{1}{72},$$

$$b_4 = \begin{vmatrix} 1 & -1/2 & 1/3 \\ -1/2 & 1/3 & -1/4 \\ 1/3 & -1/4 & 1/5 \end{vmatrix} = \frac{1}{2160}, \quad b_5 = \begin{vmatrix} -1/2 & 1/3 & -1/4 \\ 1/3 & -1/4 & 1/5 \\ -1/4 & 1/5 & -1/6 \end{vmatrix} = \frac{-1}{43200}.$$

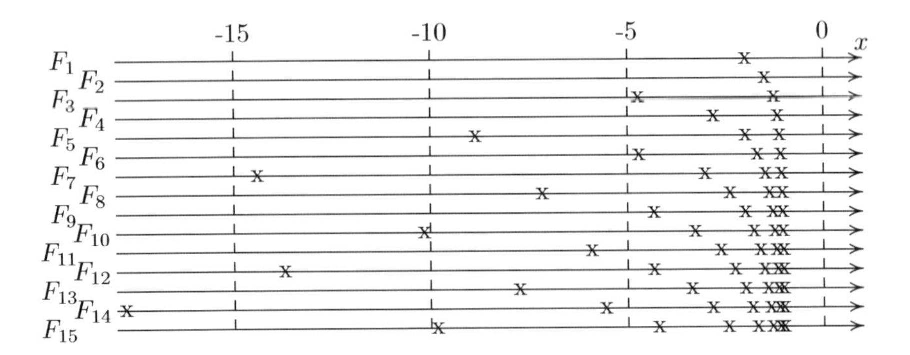

FIGURE 3.4: This shows how the singularities of the continued fraction approximations for $\ln(x+1)$ fill up the branch cut $x \leq -1$. The x's show the location of the poles for each F_n. As n increases, the poles become more dense.

This allows us to compute $c_0 = 1$, $c_1 = 1/2$, $c_2 = 1/6$, $c_3 = 1/3$, $c_4 = 1/5$, and $c_5 = 3/10$. The pattern may not be obvious, but it does exist: $c_{2n-1} = n/(4n-2)$ and $c_{2n} = n/(4n+2)$, for $n \geq 1$. Once again, the even and odd coefficients follow different patterns, so we can express this as

$$\frac{\ln(x+1)}{x} \sim 1\Big/\Big(1+\mathop{\mathbf{K}}_{n=1}^{\infty}\frac{nx}{4n-2}\Big/\Big(1+\frac{nx}{4n+2}\Big/\Big(1.$$

So $\ln(x+1) \sim$

$$x\Big/\Big(1+\frac{x}{2}\Big/\Big(1+\frac{x}{6}\Big/\Big(1+\frac{2x}{6}\Big/\Big(1+\frac{2x}{10}\Big/\Big(1+\frac{3x}{10}\Big/\Big(1+\frac{3x}{14}\Big/\Big(1+\frac{4x}{14}\Big/\Big(\cdots.$$

The original Taylor series only converges for $-1 < x \leq 1$, but the continued fraction representation converges for all complex numbers except for negative real numbers ≤ -1. At $x = 2$, we get the sequence

$$F_0 = 2, \quad F_1 = 1, \quad F_2 = 1.142857, \quad F_3 = 1.090909,$$

$$F_4 = 1.101449, \quad F_5 = 1.098039, \quad F_6 = 1.098805646, \ldots$$

which is converging to $\ln 3 \approx 1.098612289$. □

This example demonstrates how a logarithmic branch cut can be approximated by a conglomeration of poles. All of the poles for each approximation $F_n(x)$ lie on the negative real axis. Figure 3.4 illustrates how these poles fill up the branch cut $x \leq -1$.

If $f(x)$ is an even function (or odd), then every leading term of $g_n(x)$ in equation 3.20 would be of order x^2 for $n \geq 1$. So it makes sense to first let $t = x^2$, so $x = \sqrt{t}$.

Example 3.18
Find the continued fraction representation of

$$\tan^{-1}(x) = x - \frac{x^3}{3} + \frac{x^5}{5} - \frac{x^7}{7} + \frac{x^9}{9} - \frac{x^{11}}{11} + \cdots.$$

SOLUTION: We first replace x with $\sqrt{t}$, and consider the series for

$$\frac{\tan^{-1}(\sqrt{t})}{\sqrt{t}} = 1 - \frac{t}{3} + \frac{t^2}{5} - \frac{t^3}{7} + \frac{t^4}{9} - \frac{t^5}{11} + \cdots.$$

This produces $b_0 = 1, b_1 = -1/3$, $b_2 = 4/45$, $b_3 = 4/525$, $b_4 = 256/496125$, and $b_5 = -256/22920975$. Since all of these are non-zero, we can compute $c_0 = 1$, $c_1 = 1/3$, $c_2 = 4/15$, $c_3 = 9/35$, $c_4 = 16/63$, and $c_5 = 25/99$. So

$$\frac{\tan^{-1}(\sqrt{t})}{\sqrt{t}} = 1\Big/\Big(1 + \frac{t}{3}\Big/\Big(1 + \frac{4t}{15}\Big/\Big(1 + \frac{9t}{35}\Big/\Big(1 + \frac{16t}{63}\Big/\Big(1 + \frac{25t}{99}\Big/\Big(1 + \cdots.$$

This time there is a straight-forward pattern: $c_n = n^2/(4n^2 - 1)$ for $n \geq 1$. Replacing t with x^2, and multiplying by x gives

$$\tan^{-1} x \sim x\Big/\Big(1 + \mathop{\mathbf{K}}_{n=1}^{\infty} \frac{n^2 x^2}{4n^2 - 1}\Big/\Big(1.$$

The original Taylor series only converged for $|x| \leq 1$, but the continued fraction representation converges for all real x. There are branch cuts, but they are on the imaginary axis above i and below $-i$. □

Of course the real test of the continued fraction representation is to try it on an asymptotic series that diverges for all x.

Example 3.19
In example 2.7, we found the Stieltjes integral function to have the asymptotic series

$$S(x) \sim \sum_{n=0}^{\infty} (-1)^n n! x^n \sim 1 - x + 2x^2 - 6x^3 + 24x^4 - 120x^5 \text{ as } x \to 0^+.$$

This sum diverges for all x. Find the continued fraction representation for this series.

TABLE 3.7: The Stieltjes integral function vs. its continued fraction representation, which converges to the appropriate value, in spite of the fact that the asymptotic series diverges for all x.

	$x = 1$	$x = 1/2$	$x = 1/3$
F_1	0.5	0.666667	0.75
F_2	0.666667	0.75	0.8
F_3	0.571429	0.714286	0.782609
F_4	0.615385	0.727273	0.787879
F_5	0.588235	0.72093	0.785714
F_6	0.60274	0.723684	0.786517
F_7	0.593301	0.722222	0.786151
F_8	0.598802	0.72293	0.786305
F_9	0.595084	0.722531	0.786229
F_{10}	0.597383	0.722739	0.786264
F_{11}	0.595783	0.722617	0.786246
$\int_0^\infty e^{-t}/(1+xt)\,dt$	0.596347	0.722657	0.786251

SOLUTION: $b_0 = 1$, $b_1 = -1$, $b_2 = 1$, $b_3 = 2$, $b_4 = 4$, and $b_5 = -24$, so $c_0 = c_1 = c_2 = 1$, $c_3 = c_4 = 2$, and $c_5 = 3$. In fact, $c_{2n-1} = c_{2n} = n$ for all $n \geq 1$. So this divergent series has a continued fraction representation

$$S(x) = 1/(1 + x/(1 + x/(1 + 2x/(1 + 2x/(1 + 3x/(1 + 3x/(1 + 4x/(\cdots.$$

This can be expressed using the big K notation as

$$S(x) = 1\Big/\Big(1 + \mathop{\mathrm{K}}_{n=1}^{\infty} nx\Big/\Big(1 + nx\Big/\Big(1.$$

This representation actually *converges* for all $x > 0$! Table 3.7 shows this representation evaluated at $x = 1/3$, $x = 1/2$, and $x = 1$. It is clear that the continued fraction converges in all three cases, although convergence is faster when x is closer to 0. This is certainly an improvement over the original asymptotic series, which diverges for all x. □

This example shows that not only does the continued fraction representation converge, but it converges to the correct function. This raises the same question that was asked about Borel summation in section 3.4.2. There are many functions whose asymptotic series is a given divergent series, differing by subdominant terms. Which of these functions does the continued fraction representation converge to?

There is an elegant result for the case where all of the coefficients c_n are positive. First of all, a series is a Stieltjes series if, and only if, all of the coefficients c_n are non-negative. (Note that if one of the c_n is zero, then the continued fraction truncates, so the series converges to a rational function.)

We call the continued fraction representation is an *S-fraction* if this is the case. Furthermore, if an S-fraction does converge, then there will be a *unique* $\rho(t)$ that satisfies equation 3.13, and the S-fraction converges to the Stieltjes sum [4, p. 408]. For example, we found a pattern of the continued fraction coefficients for $\tan^{-1}(\sqrt{x})/\sqrt{x}$ to be $c_1 = 1$, and $c_n = n^2/(4n^2-1)$ for $n \geq 1$. Since these are all positive, the series is Stieltjes, hence $\tan^{-1}(\sqrt{x})/\sqrt{x}$ is a Stieltjes function. (Note that $\tan^{-1}(x)$ by itself is *not* a Stieltjes function, since some of the coefficients would be 0.)

Problems for §3.5

For problems **1** through **7**: Determine c_0 through c_5 for the continued fraction approximation as $x \to 0$ of the series. Use these to form the F_5 approximation to the function. Note that one may have to divide by x and/or substitute $t = x^2$ before using equation 3.21.

1

$$\sqrt{x+1} \sim 1 + \frac{x}{2} - \frac{x^2}{8} + \frac{x^3}{16} - \frac{5x^4}{128} + \frac{7x^5}{256} + \cdots.$$

2

$$\frac{1}{\sqrt{4x+1}} \sim \sum_{n=0}^{\infty} \frac{(-1)^n (2n)! x^n}{(n!)^2} \sim 1 - 2x + 6x^2 - 20x^3 + 70x^4 - 252x^5 + \cdots.$$

3

$$\cos(x) \sim 1 - \frac{x^2}{2} + \frac{x^4}{24} - \frac{x^6}{720} + \frac{x^8}{40320} - \frac{x^{10}}{3628800} + \cdots.$$

4

$$\sin x \sim x - \frac{x^3}{6} + \frac{x^5}{120} - \frac{x^7}{5040} + \frac{x^9}{362880} - \frac{x^{11}}{39916800} + \cdots.$$

5

$$\tan x \sim x + \frac{x^3}{3} + \frac{2x^5}{15} + \frac{17x^7}{315} + \frac{62x^9}{2835} + \frac{1382x^{11}}{155925} + \cdots.$$

6

$$\sec x \sim 1 + \frac{x^2}{2} + \frac{5x^4}{24} + \frac{61x^6}{720} + \frac{277x^8}{8064} + \frac{50521x^{10}}{3628800} + \cdots.$$

7

$$\tanh x \sim x - \frac{x^3}{3} + \frac{2x^5}{15} - \frac{17x^7}{315} + \frac{62x^9}{2835} - \frac{1382x^{11}}{155925} + \cdots.$$

8 In example 2.8, we derived an asymptotic series for the function $\text{erf}(x)$ as $x \to \infty$

$$\text{erf}(x) \sim 1 - \frac{e^{-x^2}}{\sqrt{\pi}} \left(\frac{1}{x} - \frac{1}{2x^3} + \frac{3}{4x^5} - \frac{15}{8x^7} + \frac{105}{16x^9} - \frac{945}{32x^{11}} + \cdots \right).$$

Find a continued fraction representation of this series. Can you determine the pattern in the coefficients?

9 Find the continued fraction representation for the divergent series

$$f(x) \sim \sum_{n=0}^{\infty} \frac{(-1)^n (2n)! x^n}{n!} \sim 1 - 2x + 12x^2 - 120x^3 + 1680x^4 - 30240x^5 + \cdots.$$

Investigate whether this representation converges at $x = 1$.

10 The binomial series for $(x+1)^k$ about $x = 0$ is given by

$$1 + kx + \frac{k(k-1)}{2!}x^2 + \frac{k(k-1)(k-2)}{3!}x^3 + \frac{k(k-1)(k-2)(k-3)}{4!}x^4 + \cdots.$$

Find the c_0 through c_5 expressed in terms of k. Can you find a pattern for c_{2n} and c_{2n+1}? Notice that when k is an integer, one of the c_n will be 0.

11 Determine the Taylor series up to the x^6 term for the continued fraction representation

$$f(x) \sim 1 \Big/ \Big(1 + x \Big/ \Big(1 + x^2 \Big/ \Big(1 + x^3 \Big/ \Big(1 + x^4 \Big/ \Big(1 + \cdots.$$

12 Determine what function the continued fraction representation

$$f(x) = \mathop{\mathrm{K}}_{n=0}^{\infty} x \Big/ 1$$

$$= x \Big/ \Big(1 + x \Big/ \Big(1 + x \Big/ \Big(1 + x \Big/ \Big(1 + x \Big/ \Big(1 + \cdots$$

converges to.

13 Find the continued fraction representation for the divergent series

$$f(x) \sim \sum_{n=0}^{\infty} (-1)^n 2^{n^2} \sim 1 - 2x + 16x^2 - 512x^3 + 65536x^4 - 33554432x^5 + \cdots.$$

By observing the pattern of the coefficients c_n, show that this is a Stieltjes series.

14 Show that the series in problem 13 does not satisfy Carleman's condition (equation 3.14). Thus, there may be more that one Stieltjes function that has this asymptotic series.

15 Show that if we let

$$\rho(t) = \frac{2^{-(\log_4 t)^2}}{2t\sqrt{\pi}\ln 2},$$

then $\int_0^\infty t^n \rho(t)\, dt = 2^{n^2}$.

16 Show that if we let

$$\rho(t) = \frac{2^{-(\log_4 t)^2}}{2t\sqrt{\pi}\ln 2}(1 + \sin(\pi \log_4 t)),$$

then $\int_0^\infty t^n \rho(t)\,dt$ will also be 2^{n^2}. Thus, there are two distinct Stieltjes functions which have the series in problem 13 as their asymptotic series.

17 Show that the continued fraction representation of the series in problem 13 does *not* converge at $x = 1$.

18 Show that if we used the $\rho(t)$ from problem 15, then the "Stieltjes sum" for the series in problem 13, given by

$$\int_0^\infty \frac{\rho(t)}{1 + xt}\,dt$$

is equal to 1/2 when $x = 1$. You may use the fact that

$$\int_{-\infty}^\infty \frac{e^{-t^2}}{1 + e^{at}}\,dt = \frac{\sqrt{\pi}}{2}$$

for all a.

For problems **19** through **22**: Equation 3.20 can be generalized to consider *continued-function* representations. Let $h(x)$ be a simple function such that $h(0) = 1$. Then we let

$$g_n(x) = \text{ the leading term of } f_n(x), \qquad (3.22)$$
$$f_{n+1}(x) = h^{-1}\left(\frac{f_n(x)}{g_n(x)}\right).$$

Finally, we can represent

$$f(x) \sim g_0(x) \cdot h\Big(g_1(x) \cdot h\Big(g_2(x) \cdot h\Big(g_3(x) \cdot h\Big(g_4(x) \cdots.$$

For example, if $h(x) = 1/(1 + x)$, then $h^{-1}(x) = 1/x - 1$, so we get continued fractions. If we let $h(x) = 1 + x$, we reproduce the classic Taylor series. But if we try other functions such as $h(x) = e^x$, so that $h^{-1}(x) = \ln(x)$, we get continued exponential representations:

$$f(x) \sim g_0(x) \exp\Big(g_1(x) \exp\Big(g_2(x) \exp\Big(g_3(x) \exp\Big(g_4(x) \cdots.$$

If $h(x) = \sqrt{1 + x}$, so that $h^{-1} = x^2 - 1$, we get the continued root representation

$$f(x) \sim g_0(x)\sqrt{1 + g_1(x)\sqrt{1 + g_2(x)\sqrt{1 + g_3(x)\sqrt{1 + g_4(x)\cdots}}}}.$$

These may converge faster than the original power series.

19 Find the order 3 continued root representation (that is, up to $g_3(x)$) for the divergent series

$$f(x) \sim \sum_{n=0}^{\infty} n!x^n \sim 1 + x + 2x^2 + 6x^3 + \cdots.$$

20 Find the order 3 continued root representation (that is, up to $g_3(x)$) for the divergent series

$$f(x) \sim \sum_{n=0}^{\infty} (2n)!x^n/(n!) \sim 1 + 2x + 12x^2 + 120x^3 + \cdots.$$

21 Find the order 3 continued exponential representation (that is, up to $g_3(x)$) for the divergent series

$$f(x) \sim \sum_{n=0}^{\infty} n!x^n \sim 1 + x + 2x^2 + 6x^3 + \cdots.$$

22 Find the order 3 continued exponential representation (that is, up to $g_3(x)$) for the divergent series

$$f(x) \sim \sum_{n=0}^{\infty} (2n)!x^n/(n!) \sim 1 + 2x + 12x^2 + 120x^3 + \cdots.$$

3.6 Padé Approximants

Continued fractions allow us to convert a divergent asymptotic series into a format that converges. However, all of the determinants b_n in equation 3.21 must be non-zero or else the continued fraction representation breaks down. Although there are ways to get around this problem, a better approach is to approximate a series with a sequence of *rational functions*. After all, if we expand each of the partial fraction approximations F_n, we will produce a rational function.

The plan is to find a rational function $A(x)/B(x)$ that best approximates the asymptotic series $\sum_{n=0}^{\infty} a_n x^n$, where the degree of the polynomial $A(x)$ is at most N, and the degree of $B(x)$ is at most M. Because we can multiply the numerator and denominator by a constant without affecting the fraction, we can assume that the constant term of $B(x)$ is 1. Thus,

$$\frac{A(x)}{B(x)} = \frac{A_0 + A_1 x + A_2 x^2 + \cdots A_N x^N}{1 + B_1 x + B_2 x^2 + \cdots B_M x^M}.$$

There are $M + N + 1$ unknowns in this expression, so we should be able to get the Taylor series of this function to agree with the first $M + N + 1$ terms of the asymptotic series $\sum_{n=0}^{\infty} a_n x^n$. The trick is to multiply through by the denominator $B(x)$, so we have

$$\begin{aligned}(1 + B_1x + B_2x^2 + \cdots B_Mx^M)(a_0 + a_1x + a_2x^2 + \cdots a_{M+N}x^{M+N}) \\ = A_0 + A_1x + A_2x^2 + \cdots A_Nx^N + O(x^{M+N+1}).\end{aligned} \tag{3.23}$$

Once the coefficients for $A(x)$ and $B(x)$ are determined, we can form the rational function $A(x)/B(x)$, called the N, M*-Padé approximant*, denoted by $P_M^N(x)$.

Example 3.20
Determine the formula for the $1, 1-$Padé.
SOLUTION: Equation 3.23 becomes

$$\begin{aligned}A_0 + A_1x &= (1 + B_1x)(a_0 + a_1x + a_2x^2) + O(x^3) \\ &= a_0 + (a_0B_1 + a_1)x + (a_1B_1 + a_2)x^2 + O(x^3).\end{aligned}$$

So $A_0 = a_0$, $a_0B_1 + a_1 = A_1$, and $a_1B_1 + a_2 = 0$. We can solve for $B_1 = -a_2/a_1$, and $A_1 = (a_1^2 - a_0a_2)/a_1$. Thus,

$$P_1^1(x) = \frac{a_0 + (a_1^2 - a_0a_2)x/a_1}{1 - a_2x/a_1}.$$

□

The general Padé approximant can be computed via matrices. The first step is to solve for the denominator coefficients $B_1, B_2, \ldots B_M$ by solving the matrix equation

$$\begin{bmatrix} a_N & a_{N-1} & a_{N-2} & \cdots & a_{N-M+1} \\ a_{N+1} & a_N & a_{N-1} & \cdots & a_{N-M+2} \\ a_{N+2} & a_{N+1} & a_N & \cdots & a_{N-M+3} \\ \vdots & \vdots & \vdots & \ddots & \vdots \\ a_{N+M-1} & a_{N+M-2} & a_{N+M-3} & \cdots & a_N \end{bmatrix} \cdot \begin{bmatrix} B_1 \\ B_2 \\ B_3 \\ \vdots \\ B_M \end{bmatrix} = \begin{bmatrix} -a_{N+1} \\ -a_{N+2} \\ -a_{N+3} \\ \vdots \\ -a_{N+M} \end{bmatrix}. \tag{3.24}$$

Note that any negative subscripts, such as a_{-1}, are defined to be 0. This equation can be solved either by taking the inverse of the M by M matrix, or by using Cramer's rule. Either way, once the denominator coefficients are found, the numerator coefficients $A_1, A_2, \ldots A_N$ are quickly found using

$$A_i = \sum_{j=0}^{i} a_{i-j}B_j,$$

where $B_j = 0$ for $j > N$.

Example 3.21
Determine the formula for the $2, 2-$Padé.

SOLUTION: The matrix equation to solve is

$$\begin{bmatrix} a_2 & a_3 \\ a_1 & a_2 \end{bmatrix} \cdot \begin{bmatrix} B_1 \\ B_2 \end{bmatrix} = \begin{bmatrix} -a_3 \\ -a_4 \end{bmatrix}.$$

Solving this produces

$$B_1 = \frac{a_3 a_4 - a_3 a_2}{a_2^2 - a_1 a_3} \qquad B_2 = \frac{a_1 a_3 - a_2 a_4}{a_2^2 - a_1 a_3}.$$

Then we can easily solve for

$$A_0 = a_0, \qquad A_1 = a_1 + a_0 \frac{a_3 a_4 - a_3 a_2}{a_2^2 - a_1 a_3},$$

$$A_2 = a_2 + a_1 \frac{a_3 a_4 - a_3 a_2}{a_2^2 - a_1 a_3} + a_0 \frac{a_1 a_3 - a_2 a_4}{a_2^2 - a_1 a_3}.$$ □

Padé approximants are a generalization of continued fractions. When $M = N$ or $M = N + 1$, then the Padé approximants are identical to the regular continued fraction approximants. That is, $P_N^N = F_{2N}$ and $P_{N+1}^N = F_{2N+1}$, so the sequence of continued fractions can be written as $P_0^0(x)$, $P_1^0(x)$, $P_1^1(x)$, $P_2^1(x)$, $P_2^2(x)$, $P_3^2(x)\ldots$. In the case of continued fractions, if one of the b_n was 0, the continued fraction was irregular, and none of the coefficients c_i could be computed for $i \geq n$. But for Padé approximants, if one of the matrices in equation 3.24 has a zero determinant, then only one of the Padé approximants is undefined. The rest of the Padé approximates can still be used as approximations to the function.

Example 3.22

Find the Padé approximants for the function $f(x) = \sqrt{1 + 2x + 3x^2}$.

SOLUTION: The series for the function $f(x) = \sqrt{1 + 2x + 3x^2}$ has the Maclaurin series

$$\sqrt{1 + 2x + 3x^2} = 1 + x + x^2 - x^3 + \frac{x^4}{2} + \frac{x^5}{2} - \frac{3x^6}{2} + \frac{3x^7}{2} + \frac{3x^8}{8} + \cdots.$$

If we try to find the continued fraction expansion for this function, we find that $b_2 = 0$, so the continued fraction expansion is irregular. Indeed, the matrix for computing $P_2^1(x)$ has a zero determinant, so this particular Padé approximant is undefined. However, the next Padé approximant, P_2^2 is well defined:

$$P_2^2(x) = \frac{1 + 7x/4 + 2x^2}{1 + 3x/4 + x^2/4}.$$

In fact, all future Padé approximants are defined, and approach the function fairly quickly, as seen in table 3.8. □

TABLE 3.8: $\sqrt{1+2x+3x^2}$ vs. its Padé approximants, which eventually converges for all real x, in spite of the fact that the Maclaurin series has a radius of convergence less than 1. Note that unlike the continued fraction representation, Padé approximants can recover from the "bump" created by an undefined approximant.

	$x = 1$	$x = 2$	$x = 3$
$P_0^0(x)$	1	1	1
$P_1^0(x)$	∞	-1	-0.5
$P_1^1(x)$	∞	-1	-0.5
$P_2^1(x)$	undefined	undefined	undefined
$P_2^2(x)$	2.375	3.57143	4.40909
$P_3^2(x)$	2.53125	6.11111	-216.5
$P_3^3(x)$	2.45946	4.26829	6.39568
$P_4^3(x)$	2.41975	3.22472	2.50312
$P_4^4(x)$	2.44856	4.10345	5.74864
$P_5^4(x)$	2.451	4.19034	6.28595
$P_5^5(x)$	2.44959	4.12632	5.84717
$\sqrt{1+2x+3x^2}$	2.44949	4.12311	5.83095

We have seen the connection between Padé approximants and continued fractions, but there is also a connection between the Padé approximants and the generalized Shanks transformation. If we let A_n be the sequence of partial sums created by summing the terms up to x^n for a power series evaluated at x_0, then for $n \geq k$, the generalized Shanks transformation $S_k(A_n)$ is identical to $P_k^n(x_0)$.

Example 3.23

In the last series for $\sqrt{1+2x+3x^2}$, plugging in $x_0 = 1$ gives the series $1 + 1 + 1 - 1 + \frac{1}{2} + \cdots$. So the sequence of partial sums is $A_0 = 1$, $A_1 = 2$, $A_2 = 3$, $A_3 = 2$, and $A_4 = 5/2$. We can then use equation 3.2 to find $S_2(A_2) = 19/8$. This agrees with the entry for $P_2^2(1) = 2.375$ in table 3.8. □

This can be used as a shortcut for computing a Padé approximant if we only need to evaluate the approximant at one point. Since the generalized Shanks transformation only involves two $M+1$ by $M+1$ determinants as opposed to solving a system of M linear equations, it may be computationally easier to compute $S_k(A_n)$. In the case where $n < k$, the generalized Shanks could still be used, but any A_n with negative subscripts must be treated as 0.

3.6.1 Two-point Padé

If a function has a known asymptotic power series at two different points, it is possible to incorporate both of the series into one rational function, called a two-point Padé. That is, we find a rational function that matches the first i terms of one series, and the first j terms of the other series, where $i+j = N+M+1$. Typically, the two points are chosen to be 0 and ∞, but any two points will do. Since the rational function must be close to the original function at two different places, the results are usually very impressive.

Example 3.24
One example of a function that has an asymptotic power series both at 0 and ∞ is related to the error function,

$$f(x) = e^{x^2} \int_x^\infty e^{-t^2}\, dt.$$

Find the two-point Padé approximates for this function.
SOLUTION: In example 2.8, we had that

$$\int_x^\infty e^{-t^2}\, dt \sim e^{-x^2} \left(\frac{1}{2x} - \frac{1}{4x^3} + \frac{3}{8x^5} - \frac{15}{16x^7} + \frac{105}{32x^9} - \cdots \right) \qquad \text{as } x \to \infty.$$

This is not an asymptotic power series, because of the exponential function in front. But $f(x)$ multiplies this series by e^{x^2}, so this will have an asymptotic power series at ∞. Also, the error function can be expressed in terms of $f(x)$, namely,

$$\text{erf}(x) = 1 - \frac{2e^{-x^2}}{\sqrt{\pi}} f(x).$$

But $f(x)$ also has a convergent power series at 0, noting that

$$f(x) = e^{x^2} \left(\frac{\sqrt{\pi}}{2} - \frac{\sqrt{\pi}}{2} \text{erf}(x) \right) = e^{x^2} \left(\frac{\sqrt{\pi}}{2} - \int_0^x e^{-t^2}\, dt \right).$$

The Maclaurin series is found simply by integrating the series for e^{-t^2} term by term, and finally multiplying by the series for e^{x^2}, producing

$$f(x) = \frac{\sqrt{\pi}}{2} - x + \frac{\sqrt{\pi}x^2}{2} - \frac{2x^3}{3} + \frac{\sqrt{\pi}x^4}{4} - \frac{4x^5}{15} + \cdots \qquad \text{as } x \to 0.$$

Combining this with

$$f(x) \sim \frac{1}{2x} - \frac{1}{4x^3} + \frac{3}{8x^5} - \frac{15}{16x^7} + \frac{105}{32x^9} - \cdots \qquad \text{as } x \to \infty,$$

we have the makings for a two-point Padé. Since the leading term for series at infinity is $O(1/x)$, it is natural to choose a rational function in which the

TABLE 3.9: Comparing $e^{x^2}\int_x^\infty e^{-t^2}\,dt$ with the two-point Padé approximants. Note how fast the two-point Padé approximants converge uniformly over all positive numbers.

	$x = 1/2$	$x = 1$	$x = 2$
$P_1^0(x)$	0.469841095731	0.319654347056	0.194993382396
$P_2^1(x)$	0.539515358736	0.372254966969	0.222561479155
$P_3^2(x)$	0.545246791735	0.378293334779	0.225911056204
$P_4^3(x)$	0.545619189548	0.378880177078	0.226292172514
$P_5^4(x)$	0.545640234986	0.378931574733	0.226333673714
$P_6^5(x)$	0.545641308082	0.378935737413	0.226338032937
$e^{x^2}\int_x^\infty e^{-t^2}\,dt$	0.545641360765	0.378936078071	0.226338524991

denominator has one more degree than the numerator. Since P_{N+1}^N has $2N+2$ unknowns, it is natural to force the rational function to match the first $N+1$ terms in both series. The P_1^0 case is fairly trivial (see problem 15), so consider calculating the two-point P_2^1 Padé. Since we need

$$\frac{A_0 + A_1 x}{1 + B_1 x + B_2 x^2} \sim \sqrt{\pi}/2 - x + O(x^2) \qquad \text{as } x \to 0,$$

we have

$$(1 + B_1 x + B_2 x^2)(\sqrt{\pi}/2 - x) = A_0 + A_1 x + O(x^2).$$

Also, we need

$$\frac{A_0 + A_1 x}{1 + B_1 x + B_2 x^2} = \frac{A_1 x^{-1} + A_0 x^{-2}}{B_2 + B_1 x^{-1} + x^{-2}} \sim x^{-1}/2 + O(x^{-3}) \qquad \text{as } x \to \infty,$$

or

$$(B_2 + B_1 x^{-1} + x^{-2})(x^{-1}/2) = A_1 x^{-1} + A_0 x^{-2} + O(x^{-3}).$$

Thus, we need

$$\sqrt{\pi}/2 = A_0, \quad B_1\sqrt{\pi}/2 - 1 = A_1, \quad B_2/2 = A_1, \quad \text{and } B_1/2 = A_0.$$

The four linear equations are simple enough to solve for the four coefficients to produce

$$P_2^1 = \frac{\sqrt{\pi}/2 + (\pi/2 - 1)x}{1 + \sqrt{\pi}x + (\pi - 2)x^2}.$$

Figure 3.5 shows that the graph of this rational function is almost indistinguishable from the original $f(x)$. Table 3.9 shows how the two-point P_{N+1}^N Padé quickly converge to the function. It is possible to find a rational function that produces 12 digits of accuracy for all positive numbers, giving a convenient method of evaluating the error function. □

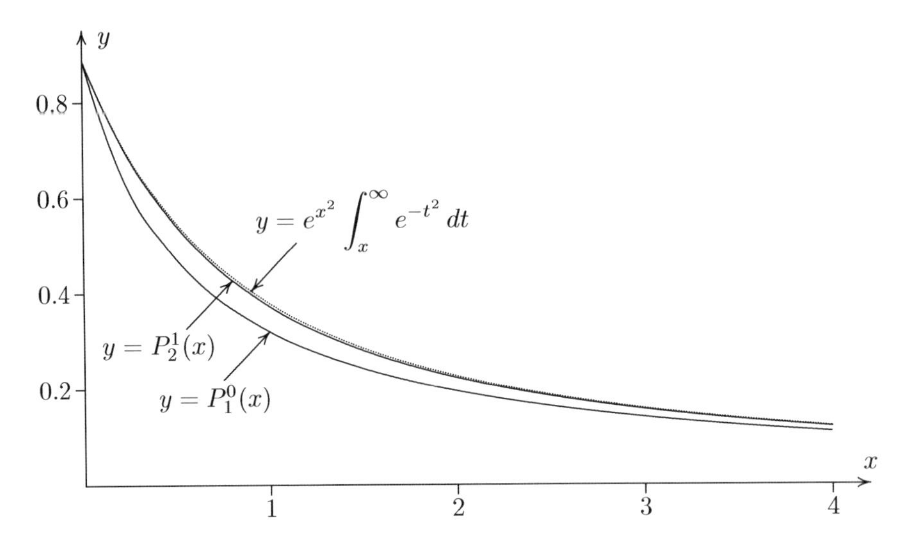

FIGURE 3.5: Plot of the two-point Padé approximants $P_1^0(x)$ and $P_2^1(x)$ for the function $e^{x^2}\int_x^\infty e^{-t^2}\,dt$. Note that even $P_2^1(x)$, in spite of its low order, is almost indistinguishable from the exact solution, shown with a dotted line. The reason is that we are forcing agreement in the curves both near 0 and near ∞.

Problems for §3.6

For problems **1** through **7**: Find the P_2^2 Padé of the function. Note that unlike the continued fraction representation, we do not have to divide by x and/or substitute $t = x^2$ first.

1

$$\sqrt{x+1} \sim 1 + \frac{x}{2} - \frac{x^2}{8} + \frac{x^3}{16} - \frac{5x^4}{128} + \frac{7x^5}{256} + \cdots.$$

2

$$\frac{1}{\sqrt{4x+1}} \sim \sum_{n=0}^{\infty} \frac{(-1)^n (2n)! x^n}{(n!)^2} \sim 1 - 2x + 6x^2 - 20x^3 + 70x^4 - 252x^5 + \cdots.$$

3

$$e^x \sim 1 + x + \frac{x^2}{2} + \frac{x^3}{6} + \frac{x^4}{24} + \frac{x^5}{120} + \cdots.$$

4

$$\cos(x) \sim 1 - \frac{x^2}{2} + \frac{x^4}{24} - \frac{x^6}{720} + \frac{x^8}{40320} - \frac{x^{10}}{3628800} + \cdots.$$

5

$$\sec x \sim 1 + \frac{x^2}{2} + \frac{5x^4}{24} + \frac{61x^6}{720} + \frac{277x^8}{8064} + \frac{50521x^{10}}{3628800} + \cdots.$$

6

$$\ln(x+1) \sim x - \frac{x^2}{2} + \frac{x^3}{3} - \frac{x^4}{4} + \frac{x^5}{5} + \cdots.$$

7

$$\frac{\sin(x)}{x} \sim 1 - \frac{x^2}{6} + \frac{x^4}{120} - \frac{x^6}{5040} + \frac{x^8}{362880} - \frac{x^{10}}{39916800} + \cdots.$$

8 Table 3.8 mentions that the Maclaurin series for $\sqrt{1+2x+3x^2}$ has a radius of convergence less than 1. Find the radius of convergence.

For problems **9** through **13**: Show that the continued fraction representation of the function is irregular, and find the P_2^2 Padé of the function in spite of the irregularity.

9

$$\sqrt{1-2x+3x^2} = 1 - x + x^2 + x^3 + \frac{x^4}{2} - \frac{x^5}{2} - \frac{3x^6}{2} - \frac{3x^7}{2} + \frac{3x^8}{8} + \frac{29x^9}{8} + \cdots.$$

10

$$e^{x^2/2-x} = 1 - x + x^2 - \frac{2}{3}x^3 + \frac{5}{12}x^4 - \frac{13}{60}x^5 + \frac{19}{180}x^6 + \cdots.$$

11

$$e^{x^2/2+x} = 1 + x + x^2 + \frac{2}{3}x^3 + \frac{5}{12}x^4 + \frac{13}{60}x^5 + \frac{19}{180}x^6 + \cdots.$$

12

$$(1+2x-3x^2)^{-5/2} = 1 - 5x + 25x^2 - 105x^3 + 420x^4 - 1596x^5 + 5880x^6 + \cdots.$$

13

$$\frac{\ln(1+2x+x^2-x^3/6)}{x} = 2 - x + \frac{x^2}{2} - \frac{x^3}{6} - \frac{x^4}{10} + \frac{23x^5}{72} - \frac{31x^6}{63} + \frac{11x^7}{18} + \cdots.$$

14 Use the generalized Shanks transformation to evaluate the P_2^2 Padé at $x_0 = 2$ for the function $\sqrt{1+2x+3x^2}$, as was done in example 3.23.

15 Find the two-point Padé P_1^0 for the function $f(x)$ in example 3.24.

16 The Dawson integral is defined by $f(x) = e^{-x^2}\int_0^x e^{t^2}\,dt$. The asymptotic series are given by

$$f(x) \sim x - \frac{2x^3}{3} + \frac{4x^5}{15} - \frac{8x^7}{105} + \frac{16x^9}{945} + \cdots \text{ as } x \to 0,$$

$$f(x) \sim \frac{1}{2x} + \frac{1}{4x^3} + \frac{3}{8x^5} + \frac{15}{16x^7} + \frac{105}{32x^9} + \cdots \text{ as } x \to \infty.$$

Find the two-point P_2^1 Padé for this function.

17 Find the two-point P_4^3 Padé for the Dawson integral of problem 16.

Hint: Because both the series at 0 and at ∞ involve only odd powers of x, we can assume that the two-point P_4^3 Padé will be an odd function. Hence, it will be of the form $(A_1 x + A_3 x^3)/(1 + B_2 x^2 + B_4 x^4)$.

18 Using the two series

$$\sqrt{1+x^2} \sim 1 + \frac{x^2}{2} - \frac{x^4}{8} + \cdots \quad \text{as } x \to 0,$$

$$\sqrt{1+x^2} \sim x + \frac{1}{2x} - \frac{1}{8x^3} + \cdots \quad \text{as } x \to \infty,$$

find the two-point P_1^2 Padé for $\sqrt{1+x^2}$.

19 Find the two-point P_2^3 Padé for $\sqrt{1+x^2}$. See problem 18.

20 Using the two series

$$\sqrt{1+6x+x^2} \sim 1 + 3x - 4x^2 + 12x^3 + \cdots \quad \text{as } x \to 0$$

$$\sqrt{1+6x+x^2} \sim x + 3 - \frac{4}{x} + \frac{12}{x^2} + \cdots \quad \text{as } x \to \infty,$$

find the two-point P_1^2 Padé for $\sqrt{1+6x+x^2}$.

21 Find the two-point P_2^3 Padé for $\sqrt{1+6x+x^2}$. See problem 20.

Chapter 4

Differential Equations

One of the most important applications of asymptotics is to approximate the solutions to an ordinary differential equation. We will not go into the techniques for finding the exact solution to the equations, except for the simplest cases. Our focus will be to find the approximate solutions, either through a Taylor series or an asymptotic series. However, some theory will be necessary to understand the nature of the solutions to the differential equations.

4.1 Classification of Differential Equations

We begin by explaining the concept of a differential equation.

DEFINITION 4.1 An *ordinary differential equation* is an equation involving a single function of one variable, and at least one of its derivatives. The *order* of the equation is the degree of the highest derivative that appears in the equation.

For example, in the equation

$$xy''(x) = y'(x)y(x), \tag{4.1}$$

y is the function of x, and since the second derivative appears in the equation, but no higher derivatives, this is a second order equation. Usually the equation can be solved for the highest derivative, putting the equation into the form

$$y^{(n)} = F[y^{(n-1)}, y^{(n-2)}, \ldots, y'(x), y(x), x]. \tag{4.2}$$

Here, we used the notation $y^{(k)}$ to denote the kth derivative of y. For equation 4.1, we have $F[y', y, x] = y'(x)y(x)/x$. Notice that F is a multi-variable function in the variables x, y, and y', so it is possible to take the partial derivative of F with respect to y', producing $y(x)/x$.

The *solution* to a differential equation is a function that yields an identity when substituted into the equation. For example, if $y(x) = -1-2/\ln(x)$, then $y'(x) = 2/(x(\ln x)^2)$, and $y'' = -4/(x^2(\ln x)^3) - 2/(x^2(\ln x)^2)$, so $y'(x)y(x)$

is the same as $xy''(x)$. Thus, $-1 - 2/\ln(x)$ is a solution to equation 4.1. However, there will usually be an infinite number of solutions to a differential equation. In fact, we can specify the values of $y(x_0), y'(x_0), \ldots y^{(n-1)}(x_0)$, and we will still typically have a solution to the differential equation. These additional constraints are called *initial conditions.* The differential equation, along with the initial conditions, is called an *initial value problem.*

THEOREM 4.1

For the initial value problem

$$y^{(n)} = F[y^{(n-1)}, y^{(n-2)}, \ldots, y'(x), y(x), x]$$

with the initial conditions

$$y(x_0) = k_0, \quad y'(x_0) = k_1, \quad y''(x_0) = k_2, \quad \ldots \quad y^{(n-1)}(x_0) = k_{n-1},$$

if all of the functions F, $\partial F/\partial y, \partial F/\partial y', \partial F/\partial y'', \ldots, \partial F/\partial y^{(n-1)}$ are continuous in a neighborhood of the initial conditions, then there exists a unique solution to the initial value problem, which is valid on some open set containing x_0.

For a proof, see [17, p. 64].

For the example in equation 4.1, we have $\partial F/\partial y = y'/x$ and $\partial F/\partial y' = y/x$. These will be continuous in a neighborhood of the initial conditions provided that $x_0 \neq 0$. Thus, if we fix x_0 away from 0, there will be a unique solution for each combination of k_0 and k_1.

Since one can specify n values in the initial condition, we expect there to be a family of solutions involving n arbitrary constants. An *n-parameter family of solutions* is a solution involving n arbitrary constants, for which the same set of solutions cannot be expressed using fewer constants. For example, $y = c_0^2 - c_1^2$ is a solution to equation 4.1 involving two arbitrary constants, but the same set of solutions can be expressed using only one constant: $y = d$. Thus, this is only a one-parameter family of solutions.

For an nth degree equation, an n-parameter family of solutions is called a *general solution* to the equation. Most of the time, the general solution will not be expressible in closed form, but will rely on a Taylor series or asymptotic series. However, equation 4.1 actually does have a general solution expressible in closed form:

$$y(x) = -1 + 2b\tan(a + b\ln(x)). \tag{4.3}$$

This can be verified by taking two derivatives.

$$y'(x) = \frac{2b^2\sec^2(a + b\ln(x))}{x},$$

$$y''(x) = \frac{-2b^2\sec^2(a + b\ln(x))}{x^2} + \frac{4b^3\sec^2(a + b\ln(x))\tan(a + b\ln(x))}{x^2},$$

so indeed, $xy''(x)$ is the same thing as $y'(x)y(x)$. This gives us a two-parameter family of solutions, and since equation 4.1 is second order, we have a general solution.

Note, however, that the general solution does not contain *all* of the solutions! We already observed that $y(x) = -1 - 2/\ln(x)$ is a solution, yet this is not in the family of solutions given by equation 4.3. Furthermore, equation 4.3 left out the one parameter family of solutions, $y(x) = d$. These solution are called *special solutions*. Often certain assumptions are made in the process of finding the general solution, such as assuming that an expression is non-zero before dividing by that expression. Some special solutions may fall through the cracks for the cases where the assumption is false.

Once we have a general solution, if we are given initial conditions for the problem, we can solve n equations with n unknowns to find the unique solution to the initial value problem.

Example 4.1

Solve the initial value problem

$$xy''(x) = y'(x)y(x), \qquad y(1) = 1, \quad y'(1) = 2.$$

SOLUTION: We begin with the known general solution,

$$y(x) = -1 + 2b\tan(a + b\ln(x)),$$

and use the initial conditions to solve for the arbitrary constants. Since $y(1) = 1$, we have $1 = -1 + 2b\tan(a)$, or $b\tan(a) = 1$. The derivative of the general solution is $y'(x) = 2b^2\sec^2(a + b\ln(x))/x$, so we can utilize the second part of the initial condition to show that $2 = 2b^2\sec^2(a)$, or $b\sec(a) = \pm 1$.

Since $\tan(a) = \sin(a)\sec(a)$, we can divide these two equations to prove that $\sin(a) = \pm 1$. But now we have a problem, since the only angles for which the sine function is ± 1 are the poles of the tangent function. Hence, we cannot solve for a and b in the general solution.

However, theorem 4.1 guarantees that there *is* a unique solution. Apparently, the solution to the initial value problem is a *special* solution not covered in the general solution. Clearly $y = d$ will not satisfy both of the initial conditions, but there is another one-parameter family of solutions,

$$y(x) = -1 - \frac{2}{c + \ln(x)}.$$

See problem 1. Can we find a value of c that satisfies both of the initial conditions?

Since $y(1) = 1$, we have $1 = -1 - 2/c$, so $c = -1$. But we also must have $y'(1) = 2$. Since

$$y'(x) = \frac{2}{x(c + \ln(x))^2},$$

we find that $2 = 2/c^2$. Indeed, $c = -1$ will solve this as well, so we have the unique solution

$$y(x) = -1 + \frac{2}{1 - \ln(x)}. \tag{4.4}$$

□

There is another quirk to the solutions of equation 4.1. We expect the solutions to have a singularity at $x = 0$, since this would cause $F = y'(x)y(x)/x$ to be undefined. However, the solution in equation 4.4 also has a singularity at $x = e$, which is not predictable by the original equation. In fact, the general solution will have a singularity whenever $a + b \ln x$ is an odd multiple of $\pi/2$. Such singularities are called *spontaneous singularities* or *movable singularities*, since their locations depend on the initial conditions. Because of the spontaneous singularities, it is impossible to predict ahead of time the domain of the solution to the differential equation.

4.1.1 Linear vs. Non-Linear

The reason why the solutions of equation 4.1 had so many complications is because the equation was *non-linear*. There is an important class of equations for which the solutions are much more predictable than the solutions to equation 4.1.

DEFINITION 4.2 An nth degree differential equation is *linear* if it can be put in the form

$$y^{(n)} + p_{n-1}(x)y^{(n-1)} + \cdots + p_2(x)y'' + p_1(x)y' + p_0(x)y = g(x). \tag{4.5}$$

A differential equation that cannot be put in this form is called *non-linear*.

Note that for a linear equation, each term can only involve y or one of its derivatives, multiplied by a function of x. Thus, equation 4.1 is non-linear, since one term involves both y and y'. On the other hand,

$$xy'' = y' + y$$

is a linear equation.

In general, linear equations are much easier to solve than non-linear ones. For starters, there will be no spontaneous singularities. In fact, the domain of the solutions can be determined by the equation, as determined by the following result:

THEOREM 4.2

If the coefficient functions $p_0(x), p_1(x), \ldots p_{n-1}(x)$ and $g(x)$ are analytic functions in an open simply connected set S containing x_0, then the solution to the initial value problem

$$y^{(n)} + p_{n-1}(x)y^{(n-1)} + \cdots + p_2(x)y'' + p_1(x)y' + p_0(x)y = g(x),$$

with the initial conditions

$$y(x_0) = k_0, \quad y'(x_0) = k_1, \quad y''(x_0) = k_2, \quad \ldots, y^{(n-1)}(x_0) = k_{n-1},$$

will be analytic in the set S.

The original proof was done by Fuchs in 1865 [9].

This theorem shows that the only singularities in a solution must occur at a point where there is a singularity in one of the coefficient functions.

Example 4.2

Determine the possible locations of the singularities in the solutions to the equation

$$xy'' = y' + y.$$

SOLUTION: Putting this equation into standard form, we have

$$y'' - \frac{1}{x}y' - \frac{1}{x}y = 0.$$

There is only one singularity in the coefficient functions, at $x = 0$. Thus, this is the only possible singularity point of all of the solutions. Note that this is not a guarantee that the solutions will have a singularity at $x = 0$. In fact, one solution,

$$y(x) = \sum_{n=2}^{\infty} \frac{x^n}{n!(n-2)!} = \frac{x^2}{2} + \frac{x^3}{6} + \frac{x^4}{48} + \frac{x^5}{720} + \cdots,$$

is analytic in the entire plane. See problem 8. □

Theorem 4.2 shows that if all of the coefficient functions

$$p_0(x), p_1(x), \ldots p_{n-1}(x) \text{ and } g(x)$$

are analytic at x_0, then all solutions will have a convergent Taylor series centered at x_0. This gives us one way to approximate the solution to the equation. We can assume that

$$y = \sum_{n=0}^{\infty} a_n(x - x_0)^n.$$

Differentiating can be done term-wise, so we have

$$y' = \sum_{n=0}^{\infty} na_n(x - x_0)^{n-1}.$$

We can then plug this series into the differential equation, and determine what the coefficients a_n must be in order for the series to solve the equation. We will cover this technique in depth in section 4.3.

4.1.2 Homogeneous vs. Inhomogeneous

Another important feature of linear equations is the lack of special solutions. That is, once we have a general solution, we have *all* of the solutions. This can be demonstrated by understanding the form of the general solution. We will first cover equations for which $g(x) = 0$, which have a special designation.

DEFINITION 4.3 A linear equation is *homogeneous* if the $g(x)$ term is 0. That is, the equation is of the form

$$y^{(n)} + p_{n-1}(x)y^{(n-1)} + \cdots + p_2(x)y'' + p_1(x)y' + p_0(x)y = 0.$$

If the $g(x)$ term is non-zero, the equation is said to be *inhomogeneous*.

For example, the equation $xy'' = y' + y$ is homogeneous, since every term involves y or one of its derivatives. A homogeneous equation will always have $y = 0$ as a solution, but a more important property is that the sum of two solutions to a homogeneous equation will again be a solution to the equation. Also, any constant times a known solution will be a solution. Thus, if we can manage to find n non-zero solutions $y_1(x), y_2(x), \ldots y_n(x)$ to the homogeneous equation, then the combination

$$y(x) = c_1y_1(x) + c_2y_2(x) + \cdots + c_ny_n(x) \tag{4.6}$$

will be a solution to the differential equation. But will this be an n-parameter family of solutions, or will there be a way to simplify this expression to a form with fewer constants? We say that the set of n solutions is a *fundamental set of solutions* if equation 4.6 gives the general solution.

To determine whether a set of solutions forms a fundamental set, we will utilize the *Wronskian* of the functions $y_1, y_2, \ldots, y_n$.

DEFINITION 4.4 Given a set of n functions $y_1(x), y_2(x), \ldots, y_n(x)$, the *Wronskian* of the set of functions, denoted by $W(y_1, y_2, \ldots, y_n)$, is the n by n determinant

$$\begin{vmatrix} y_1(x) & y_2(x) & y_3(x) & \cdots & y_n(x) \\ y_1'(x) & y_2'(x) & y_3'(x) & \cdots & y_n'(x) \\ y_1''(x) & y_2''(x) & y_3''(x) & \cdots & y_n''(x) \\ \vdots & \vdots & \vdots & \ddots & \vdots \\ y_1^{(n-1)}(x) & y_2^{(n-1)}(x) & y_3^{n-1}(x) & \cdots & y_n^{(n-1)}(x) \end{vmatrix}.$$

THEOREM 4.3
If the functions $y_1(x), y_2(x), \dots y_n(x)$ are n analytical solutions to an nth order homogeneous linear differential equation, and if the Wronskian of these functions is non-zero at any point, then the functions form a fundamental set of solutions to the equation. That is, the general solution to the differential equation can be given by

$$y(x) = c_1 y_1(x) + c_2 y_2(x) + \cdots c_n y_n(x).$$

Furthermore, *all* analytical solutions to the differential equation will be of this form.

PROOF: Suppose that $W(y_1, y_2, \dots y_n) \neq 0$ at a point x_0. Theorem 4.1 shows that for each combination of $k_1, k_2, \dots, k_n$, there will be a unique solution to the differential equations which satisfies

$$y(x_0) = k_0, \quad y'(x_0) = k_1, \quad y''(x_0) = k_2, \quad \dots, y^{(n-1)}(x_0) = k_{n-1}. \qquad (4.7)$$

Since every solution would have corresponding values of k_i, it is sufficient to show that for every combination of the k_i, there is a solution in the form of equation 4.6 which satisfies equation 4.7. This amounts to solving the equations

$$\begin{aligned}
c_1 y_1(x_0) + c_2 y_2(x_0) + c_3 y_3(x_0) + \cdots + c_n y_n(x_0) &= k_0, \\
c_1 y_1'(x_0) + c_2 y_2'(x_0) + c_3 y_3'(x_0) + \cdots + c_n y_n'(x_0) &= k_1, \\
c_1 y_1''(x_0) + c_2 y_2''(x_0) + c_3 y_3''(x_0) + \cdots + c_n y_n''(x_0) &= k_2, \\
&\cdots \\
c_1 y_1^{(n-1)}(x_0) + c_2 y_2^{(n-1)}(x_0) + c_3 y_3^{(n-1)}(x_0) + \cdots + c_n y_n^{(n-1)}(x_0) &= k_{n-1}.
\end{aligned}$$

This can be expressed in matrix form.

$$\begin{bmatrix}
y_1(x_0) & y_2(x_0) & y_3(x_0) & \cdots & y_n(x_0) \\
y_1'(x_0) & y_2'(x_0) & y_3'(x_0) & \cdots & y_n'(x_0) \\
y_1''(x_0) & y_2''(x_0) & y_3''(x_0) & \cdots & y_n''(x_0) \\
\vdots & \vdots & \vdots & \ddots & \vdots \\
y_1^{(n-1)}(x_0) & y_2^{(n-1)}(x_0) & y_3^{(n-1)}(x_0) & \cdots & y_n^{(n-1)}(x_0)
\end{bmatrix} \cdot \begin{bmatrix} c_1 \\ c_2 \\ c_3 \\ \vdots \\ c_n \end{bmatrix} = \begin{bmatrix} k_0 \\ k_1 \\ k_2 \\ \vdots \\ k_{n-1} \end{bmatrix}.$$

From linear algebra, we know that this will have a unique solution for c_1, c_2, $\cdots c_n$ if, and only if, the determinant of the square matrix is non-zero. But this determinant is precisely the Wronskian of the functions at x_0, which we know is non-zero. Thus, the set y_1, y_2, y_3, $\dots$ y_n form a fundamental set of solutions. ∎

Example 4.3
Both $\cos(x)$ and $\sin(x)$ are particular solutions to the equation $y'' = -y$. Use these to find the general solution to this equation.

SOLUTION: We already have two solutions to a second order equation, so we only need to show that these form a fundamental set of solutions. This is accomplished by showing that the Wronskian is non-zero. Since

$$W(\cos(x), \sin(x)) = \begin{vmatrix} \cos(x) & \sin(x) \\ -\sin(x) & \cos(x) \end{vmatrix} = \cos^2(x) + \sin^2(x) = 1,$$

we see that all of the solutions can be put in the form

$$y = c_1 \cos(x) + c_2 \sin(x).$$

□

We can now turn our attention to inhomogeneous equations. One might think that this would be a much more difficult problem, but if we can find just *one* solution to an inhomogeneous equation, we can reduce the problem to that of a homogeneous equation. For a given inhomogeneous equation,

$$y^{(n)} + p_{n-1}(x)y^{(n-1)} + \cdots + p_2(x)y'' + p_1(x)y' + p_0(x)y = g(x), \tag{4.8}$$

we define the *associated homogeneous equation* as the similar equation without the inhomogeneous term, that is,

$$y^{(n)} + p_{n-1}(x)y^{(n-1)} + \cdots + p_2(x)y'' + p_1(x)y' + p_0(x)y = 0.$$

We can now consider the following theorem.

THEOREM 4.4

If y_p is a single solution to the inhomogeneous equation 4.8, and the functions $u_1, u_2, u_3, \ldots u_n$ are a fundamental set of solutions to the associated homogeneous equation, then all solutions to equation 4.8 are given by

$$y(x) = y_p(x) + c_1 u_1(x) + c_2 u_2(x) + \cdots + c_n u_n(x). \tag{4.9}$$

PROOF: We begin by making the substitution $y(x) = u(x) + y_p(x)$. This would produce

$$(u^{(n)} + y_p^{(n)}) + p_{(n-1)}(x)(u^{(n-1)} + y_p^{(n-1)}) + \cdots$$
$$+ p_2(x)(u'' + y_p'') + p_1(x)(u' + y_p') + p_0(x)(u + y_p) = g(x).$$

We can rearrange this to produce

$$u^{(n)} + p_{n-1}(x)u^{(n-1)} + \cdots + p_2(x)u'' + p_1(x)u' + p_0(x)u$$
$$+ y_p^{(n)} + p_{n-1}(x)y_p^{(n-1)} + \cdots + p_2(x)y_p'' + p_1(x)y_p' + p_0(x)y_p = g(x).$$

Now we can use the fact that y_p is a solution to the original equation, so that all of the terms on the second line will cancel. Thus, we have the homogeneous equation for $u(x)$.

$$u^{(n)} + p_{n-1}(x)u^{(n-1)} + \cdots + p_2(x)u'' + p_1(x)u' + p_0(x)u = 0.$$

Since $u_1, u_2, \ldots u_n$ form a fundamental set of solutions to this homogeneous equation, we have that

$$u(x) = c_1 u_1(x) + c_2 u_2(x) + \cdots + c_n u_n(x).$$

Finally, letting $y(x) = u(x) + y_p(x)$ gives us equation 4.9. $\square$

Example 4.4

Solve the equation $y'' + y = x^2$.

SOLUTION: Since this is an inhomogeneous equation, our task is to find just one solution to this equation. Since $g(x)$ is a polynomial, it makes sense to try polynomial solutions. If we try $y(x) = Ax^2 + Bx + C$, then $y'' + y = Ax^2 + Bx + (C + 2A)$. This can equal x^2 if $A = 1$, $B = 0$, and $C = -2$. So we have one solution, $y = x^2 - 2$.

Having one solution reduces the problem to a homogeneous equation, $y'' + y = 0$. The solution was found in example 4.3 to be $c_1 \cos x + c_2 \sin(x)$. Adding this to the particular solution gives us the complete solution,

$$y = x^2 - 2 + c_1 \cos(x) + c_2 \sin(x). \qquad \square$$

There is a method which will always find a particular solution to an inhomogeneous equation, given that we have solved the associated homogeneous equation. The method is known as *variation of parameters*. With this technique, an inhomogeneous equation is no more difficult to solve than a homogeneous equation.

THEOREM 4.5

If $y_1(x)$ and $y_2(x)$ are two solutions to the associated homogeneous equation $y'' + p_1(x)y' + p_0(x)y = 0$ with $W(y_1, y_2) \neq 0$, then a particular solution to the inhomogeneous equation

$$y'' + p_1(x)y' + p_0(x)y = g(x)$$

is given by

$$\boxed{y_p = -y_1(x) \int \frac{g(x)y_2(x)}{W(y_1, y_2)}\, dx + y_2(x) \int \frac{g(x)y_1(x)}{W(y_1, y_2)}\, dx.} \tag{4.10}$$

PROOF: See [10] for a traditional method of deriving this formula. For our purposes, we only need to show that equation 4.10 solves the differential equation, which is easy to do.

If we calculate y_p' using the product rule, two of the terms will cancel, so

$$y_p' = -y_1'(x) \int \frac{g(x)y_2(x)}{W(y_1, y_2)}\, dx + y_2'(x) \int \frac{g(x)y_1(x)}{W(y_1, y_2)}\, dx.$$

Taking another derivative, two terms will combine to produce $g(x)$, so

$$y_p'' = -y_1''(x)\int \frac{g(x)y_2(x)}{W(y_1,y_2)}\,dx + y_2''(x)\int \frac{g(x)y_1(x)}{W(y_1,y_2)}\,dx + g(x).$$

Finally, we can plug y_p into the equation to get

$$\begin{aligned} y_p'' + p_1(x)y_p' + p_0(x)y_p = \\ -(y_1'' + p_1(x)y_1' + p_0(x)y_1)\int \frac{g(x)y_2(x)}{W(y_1,y_2)}\,dx \\ + (y_2'' + p_1(x)y_2' + p_0(x)y_2)\int \frac{g(x)y_1(x)}{W(y_1,y_2)}\,dx + g(x). \end{aligned}$$

Since $y_1(x)$ and $y_2(x)$ satisfy the homogeneous equations, this simplifies to $g(x)$, hence y_p will satisfy the inhomogeneous equation. □

Although this eliminates the guesswork from determining a particular solution, the integrals are not always easy to evaluate.

Example 4.5
Use variation of parameters to solve $y'' + y = x^2$.
SOLUTION: In example 4.3, the solution to the associated homogeneous equation was $c_1 \cos x + c_2 \sin(x)$. Computing the Wronskian, we get

$$W(\cos(x), \sin(x)) = \cos^2(x) - (-\sin^2(x)) = 1.$$

Thus, equation 4.10 gives us

$$y_p = -\cos(x)\int x^2 \sin(x)\,dx + \sin x \int x^2 \cos(x)\,dx.$$

Evaluating the integrals, we obtain

$$y_p = -\cos(x)(2x\sin(x) + (2 - x^2)\cos(x)) + \sin x(2x\cos(x) + (x^2 - 2)\sin(x)),$$

which simplifies to

$$y_p = (\cos^2(x) + \sin^2(x))(x^2 - 2) = x^2 - 2.$$

Thus, we get the same result as example 4.4. □

Equation 4.10 can be generalize to an nth order inhomogeneous equation. If y_1, y_2, $\ldots y_n$ form a fundamental set of solutions to the associated homogeneous equation, then a particular solution is given by

$$\boxed{y_p = \sum_{i=1}^{n} (-1)^{i+n} y_i \int \frac{g(x)W_i(x)}{W(x)}\,dx,} \tag{4.11}$$

where $W(x)$ is the Wronskian of all n solutions y_1, y_2, $\dots y_n$, and $W_i(x)$ is the Wronskian of $n-1$ solutions with y_i left out, that is,

$$W_i(x) = W(y_1, y_2, \dots y_{i-1}, y_{i+1}, \dots y_n).$$

See problem 18 for the proof.

Because of variation of parameters, an inhomogeneous equation is no more difficult to solve than a homogeneous equation. In fact, the difficulty often lies in first solving the associated homogeneous equation.

4.1.3 Initial Conditions vs. Boundary Conditions

In solving initial value problems, linear equations are much easier than non-linear equations. This is because the solutions will be in the form of equation 4.9, so that in solving for the n arbitrary constants, we have to solve a system of n *linear* equations. Recall that in example 4.1 the system of equations were non-linear, and hence much harder to solve.

Example 4.6
Find the solution to the initial value problem $y'' + y = 0$, with $y(\pi/3) = 1$ and $y'(\pi/3) = -1$.
SOLUTION: We have already seen that the general solution to the differential equation is $y = c_1 \cos(x) + c_2 \sin(x)$. Plugging in $x = \pi/3$ gives $c_1/2 + c_2\sqrt{3}/2 = 1$. If we take the derivative of the general solution, we get $y'(x) = -c_1 \sin(x) + c_2 \cos(x)$, which we can plug in $\pi/3$ to get a second linear equation $-c_1\sqrt{3}/2 + c_2/2 = -1$. Thus, we have two linear equations with two unknowns, which will have a unique solution, $c_1 = (1+\sqrt{3})/2$ and $c_2 = (\sqrt{3}-1)/2$. Thus, the unique solution is

$$y = (1+\sqrt{3})\cos x/2 + (\sqrt{3}-1)\sin x/2. \qquad \square$$

There is, however, another type of problem for which we specify n pieces of information at two (or more) values of x. This type of problem is known as a *boundary value problem*. Although theorem 4.1 guarantees that initial values problems have a unique solution under appropriate conditions, there is no guarantee of a unique solution to a boundary value problem, even in the case of a linear homogeneous equation.

Example 4.7
If possible, solve the boundary value problem $y''+y = 0$, $y(0) = 1$, $y'(\pi/2) = 2$.
SOLUTION: Again, we have the general solution as $y = c_1 \cos(x) + c_2 \sin(x)$. Plugging in $x = 0$ produces $y(0) = c_1 = 1$. But the derivative of the general solution is $y'(x) = -c_1 \sin(x) + c_2 \cos(x)$, and plugging in $x = \pi/2$ gives us $y'(\pi/2) = -c_1 = 2$. Since c_1 cannot be both 1 and -2, there is no solution.

Notice that neither boundary condition allowed us to determine the value of c_2. ∎

Just as there are cases where there is no solution to a boundary value problem, there are times when there are an infinite number of solutions. In fact, many important problems in physics and engineering stem from determining which differential equations will have a non-unique solution as a boundary value problem. The differential equation will involve a parameter, called the *eigenvalue*, and only certain choices of the eigenvalue will allow for a non-unique solution. The solutions are called the *eigenfunctions* corresponding to the eigenvalue.

Example 4.8
It is routine to show that $\cos(x\sqrt{E})$ and $\sin(x\sqrt{E})$ solve the equation $y'' + Ey = 0$ for $E > 0$. Use this information to find the values of E for which the boundary value problem

$$y'' + Ey = 0, \qquad y(0) = 0, \text{ and } y(\pi) = 0$$

has a non-unique solution.
SOLUTION: Note that $y = 0$ will always satisfy both the differential equation and the initial conditions. Thus, a non-unique solution is simply a non-zero solution. The general solution is $y = c_1 \cos(x\sqrt{E}) + c_2 \sin(x\sqrt{(E)})$. Plugging in $x = 0$ gives us $y(0) = c_1 = 0$. In order to have a non-zero solution, we must have $c_2 \neq 0$. Thus, plugging in $x = \pi$ produces $\sin(\pi\sqrt{E}) = 0$. This is only possible if $\pi\sqrt{E}$ is a multiple of π, which happens if $E = 0, 1, 4, 9, 16, \ldots$. Actually, $E = 0$ is not an eigenvalue, since this would cause the equation to be $y'' = 0$, which has a totally different solution, $y = c_1 + c_2 x$. When $E = n^2$, for n is a positive integer, then the corresponding eigenfunction is $\sin(nx)$. ∎

Many second order eigenvalue problems can be expressed in the form

$$y''(x) + p_1(x)y'(x) + p_0(x)y(x) = Eh(x)y(x), \qquad y(a) = y(b) = 0. \quad (4.12)$$

Such problems can always be converted into a *Sturm-Liouville eigenvalue problem,* which has the form

$$\frac{d}{dx}\left(\mu(x)y'(x)\right) + [q(x) + Er(x)]y(x) = 0, \qquad y(a) = y(b) = 0. \quad (4.13)$$

See problem 19. Sturm-Liouville equations and their solutions possess many special mathematical properties. For example, if $\mu(x) > 0$, $q(x) \leq 0$ and $r(x) > 0$ in the interval $[a, b]$, then there will be an infinite number of eigenvalues, all of which are real and positive. Furthermore, the eigenfunctions $y_n(x)$ will satisfy

$$\int_a^b y_n(x)y_m(x)r(x)\,dx = 0 \qquad \text{if } n \neq m. \quad (4.14)$$

By scaling the eigenfunctions appropriately, we can also get

$$\int_a^b y_n(x)^2 r(x)\, dx = 1. \tag{4.15}$$

Equations 4.14 and 4.15 together form the *orthonormal property* of the eigenfunctions with respect to the weight function $r(x)$. [10]

Example 4.9

Scale the solutions of example 4.8 so that they have the orthonormal property.

SOLUTION: Example 4.8 is an example of a Strum-Liouville problem, with $\mu(x) = 1$, $q(x) = 0$, and $r(x) = 1$. Thus, equation 4.14 states that, when $n \neq m$,

$$\int_0^\pi \sin(nx)\sin(mx)\, dx = 0. \tag{4.16}$$

See problem 20 to see how this can be verified. Before equation 4.15 will be verified, we will have to scale the eigenfunctions by a constant. If $y_n(x) = k_n \sin(nx)$, then

$$\begin{aligned}\int_0^\pi y_n(x)^2\, dx &= k_n^2 \int_0^\pi \sin^2(nx)\, dx = k_n^2 \int_0^\pi \frac{1-\cos(2nx)}{2}\, dx \\ &= k_n^2\left(\frac{x}{2} - \frac{\sin(2nx)}{4n}\right)\Big|_0^\pi = \frac{k_n^2\pi}{2}.\end{aligned}$$

For this to equal 1, we need $k_n = \sqrt{2/\pi}$ for all n. Thus, $y_n = \sqrt{2/\pi}\sin(nx)$ gives an orthonormal set of eigenfunctions. ▯

4.1.4 Regular Singular Points vs. Irregular Singular Points

For a linear homogeneous equation

$$y^{(n)} + p_{n-1}(x)y^{(n-1)} + \cdots + p_2(x)y'' + p_1(x)y' + p_0(x)y = 0,$$

if all of the functions $p_0(x)$, $p_1(x)$, $p_2(x)$, $\ldots$ $p_{n-1}(x)$ are analytic at a point x_0, then theorem 4.2 states that all of the solutions will be analytic at x_0. We will say that x_0 is an *ordinary point* of the differential equation for this situation. If one of the functions $p_0(x)$, $p_1(x)$, $p_2(x)$, $\ldots$ p_{n-1} is not analytic at x_0, we say that x_0 is a *singular point* of the differential equation. However, we want to distinguish between two different kinds of singular points, since the type of singular point will determine the behavior of the solutions as $x \to x_0$.

DEFINITION 4.5 For the differential equation

$$y^{(n)} + p_{n-1}(x)y^{(n-1)} + \cdots + p_2(x)y'' + p_1(x)y' + p_0(x)y = 0,$$

if one of the functions $p_0(x)$, $p_1(x)$, $p_2(x)$, ... p_{n-1} is not analytic at a finite point x_0, yet the functions

$$(x - x_0)p_{n-1}(x),\ (x - x_0)^2 p_{n-2}(x),\ \ldots\ (x - x_0)^{n-1}p_1(x),\ (x - x_0)^n p_0(x) \tag{4.17}$$

all are analytic or have a removable singularity at x_0, we say that x_0 is a *regular singular point* of the differential equation. On the other hand, if one of the functions in equation 4.17 has a singularity that is not removable, then the differential equation has an *irregular singular point* at x_0.

Example 4.10

Classify the finite points of the differential equation

$$(x^3 - 2x^2)y'' + (x + 3)y' + (x - 2)y = 0.$$

SOLUTION: If we put the equation into standard form we have

$$y'' + \frac{x+3}{x^2(x-2)}y' + \frac{1}{x^2}y = 0.$$

Note that we were able to simplify $p_0(x)$, since it had a removable singularity. The only places where $p_1(x)$ or $p_0(x)$ is non-analytic is at 0 or 2. so all other finite points are ordinary points. To determine whether 0 is a regular singular point or irregular singular point, we consider the functions

$$x\frac{x+3}{x^2(x-2)} = \frac{x+3}{x(x-2)}, \qquad \text{and} \qquad x^2\frac{1}{x^2} = 1.$$

Although one of these has a removable singularity at $x = 0$, the other does not, so 0 is an irregular singular point of the equation. To test the point 2, we consider the functions

$$(x-2)\frac{x+3}{x^2(x-2)} = \frac{x+3}{x^2}, \qquad \text{and} \qquad (x-2)^2\frac{1}{x^2} = \frac{(x-2)^2}{x^2}.$$

Both of these have a removable singularities at $x = 2$, so 2 is a regular singular point of the equation. □

The classification of the singular points is important because we can determine the type of asymptotic series the solution will have as $x \to x_0$. For ordinary points, all solutions will have a Taylor series of the form

$$y = \sum_{n=0}^{\infty} a_n (x - x_0)^n.$$

If x_0 is a regular singular point, there will be an least one solution of the form

$$y = (x - x_0)^k \sum_{n=0}^{\infty} a_n (x - x_0)^n$$

for some k. Such a series is called a *Frobenius series*. The other solutions will either be Frobenius series, or will involve a Frobenius series multiplied by $\ln(x - x_0)^n$. In either case, the series will converge for all x within a positive radius of convergence of x_0. Since the form of the series is known ahead of time, the strategy will be to plug the series into the differential equation, and solve for the coefficients a_n. This will be covered in depth in section 4.4.

For irregular singular points, the nature of the asymptotic series as $x \to x_0$ cannot be predicted ahead of time. Often it will involve exponential functions, and usually it will diverge for all $x \neq x_0$. However, methods such as Borel summation and continued fractions can utilize these divergent series.

We also want to consider the asymptotic series as $x \to \infty$, which means we must classify the point at ∞ as well as the finite points. This is easy to do, using a substitution $x = 1/t$. This will interchange the points at 0 and infinity, so we can classify the point at 0 in the new equation, and we define this to be the classification of the point at ∞ for the original differential equation. Note that if $x = 1/t$, then by the chain rule,

$$y'(x) = \frac{dy}{dt}\frac{dt}{dx} = \frac{dy}{dt}\frac{-1}{x^2} = -t^2 y'(t). \tag{4.18}$$

Continuing with higher derivatives, we have

$$y''(x) = t^4 y''(t) + 2t^3 y'(t),$$

$$y'''(x) = -t^6 y'''(t) - 6t^5 y''(t) - 6t^4 y'(t),$$

$$y^{(4)}(x) = t^8 y^{(4)}(t) + 12t^7 y'''(t) + 36t^6 y''(t) + 24t^5 y'(t), \ldots.$$

Example 4.11

Classify the point at ∞ for the differential equation

$$(x^3 - 2x^2)y'' + (x + 3)y' + (x - 2)y = 0.$$

SOLUTION: Substituting $x = 1/t$, $y(x) = y(t)$, $y'(x) = -t^2 y'(t)$, and $y''(x) = t^4 y''(t) + 2t^3 y'(t)$ produces

$$(1/t^3 - 2/t^2)(t^4 y''(t) + 2t^3 y'(t)) + (3 + 1/t)(-t^2 y'(t)) + ((1/t) - 2)y(t) = 0.$$

This simplifies to

$$(t - 2t^2)y''(t) + (2 - 5t - 3t^2)y'(t) + (1 - 2t)y(t)/t = 0.$$

We now classify the point at $t = 0$ for this new equation. Dividing by $t - 2t^2$, we have

$$y'' + \frac{2 - 5t - 3t^2}{t(1 - 2t)}y' + \frac{1}{t^2}y(t) = 0.$$

Clearly the coefficient functions are not analytic at $t = 0$, but the functions

$$t\frac{2 - 5t - 3t^2}{t(1-2t)} = \frac{2 - 5t - 3t^2}{1 - 2t}, \qquad \text{and} \qquad t^2\frac{1}{t^2} = 1$$

have removable singularities at $t = 0$. So $t = 0$ is a regular singular point of the new equation, and hence $x = \infty$ is a regular singular point of the original differential equation. □

This section has only touched on the methods of solving differential equations. A much more comprehensive treatment can be found in a typical differential equations textbook, such as [5]. The purpose of this section is to introduce the theory of differential equations, so that we can interpret the approximate solutions. For example, equation 4.6 gives us the form of the general solution to a linear homogeneous equation, so if we find n approximate solutions to the equation, we know that we can approximate every solution. Many of the strategies for solving the equation exactly were left out, since the focus will be on finding the asymptotic series of the solutions, not the exact solutions.

Problems for §4.1

1 Show that $y(x) = -1 - 2/(c + \ln(x))$ is a one-parameter family of solutions to equation 4.1 that is not covered in the general solution given in equation 4.3.

2 Show that $y = ae^{bx}$ gives a general solution to the non-linear equation

$$y''(x)y(x) = [y'(x)]^2.$$

3 Show that $y = a/(x+b)$ gives a general solution to the non-linear equation

$$y''(x)y(x) = 2[y'(x)]^2.$$

4 Use equation 4.3 to solve the initial value problem

$$xy''(x) = y'(x)y(x), \qquad y(1) = 0, \quad y'(1) = 1.$$

5 Use equation 4.3 to solve the initial value problem

$$xy''(x) = y'(x)y(x), \qquad y(1) = 0, \quad y'(1) = -1.$$

Note that you will have to use complex arithmetic temporarily, but the answer will come out real.

6 Use the result of problem 2 to solve the initial value problem

$$y''(x)y(x) = [y'(x)]^2, \qquad y(0) = 3, \quad y'(0) = 2.$$

7 Use the result of problem 3 to solve the initial value problem

$$y''(x)y(x) = 2[y'(x)]^2, \qquad y(0) = 3, \quad y'(0) = 2.$$

8 Show that the series

$$y(x) = \sum_{n=2}^{\infty} \frac{x^n}{n!(n-2)!}$$

solves the equation $xy'' = y' + y$.

Hint: differentiate the series term by term, but shift the index on one of the series so they can be compared.

9 Consider the linear initial value problem,

$$y^{(n)} + p_{n-1}(x)y^{(n-1)} + \cdots + p_2(x)y'' + p_1(x)y' + p_0(x)y = g(x), \qquad \text{with}$$

$$y(x_0) = k_0, \quad y'(x_0) = k_1, \quad y''(x_0) = k_2, \quad \ldots \quad y^{(n-1)}(x_0) = k_{n-1}.$$

Show that if the coefficient functions $p_0(x)$, $p_1(x)$, $\ldots$ $p_{n-1}(x)$ and $g(x)$ are continuous at x_0, then there will be a unique solution to the initial value problem.

Hint: Use theorem 4.1.

10 Show that both $y = e^x$ and $y = e^{-x}$ satisfy the equation $y''(x) = y(x)$. From this information, find the general solution to the equation.

11 Show that both $y = x^2$ and $y = 1/x$ satisfy the equation $x^2y''(x) = 2y(x)$. From this information, find the general solution to the equation.

12 Show that $y = 1$, $y = \sin^2(x)$, and $y = \cos^2(x)$ all satisfy the third order equation $y'''(x) + 4y'(x) = 0$. Can we say that the general solution to the equation is $y = c_1 + c_2 \sin^2(x) + c_3 \cos^2(x)$?

13 Find a particular solution to the equation

$$xy'' - y' - y = x^2.$$

Hint: Try plugging in a polynomial.

14 Using the results of example 4.3, find the general solution to $y'' + y = x^3$.

15 Use equation 4.10 to find a particular solution to $y'' + y = \sin x$.

16 Using problem 10, apply variation of parameters to find a particular solution to $y'' - y = e^x$.

17 Using problem 11, apply variation of parameters to find a particular solution to $x^2y'' - 2y = x^2$.

Note: The equation must first be put into standard form.

18 Use the technique of theorem 4.5 to prove that y_p in equation 4.11 satisfies equation 4.5.

Hint: First establish by expanding by minors along the bottom row, that

$$\begin{vmatrix} y_1 & y_2 & y_3 & \cdots & y_n \\ y_1' & y_2' & y_3' & \cdots & y_n' \\ y_1'' & y_2'' & y_3'' & \cdots & y_n'' \\ \vdots & \vdots & \vdots & \ddots & \vdots \\ y_1^{(n-2)} & y_2^{(n-2)} & y_3^{(n-2)} & \cdots & y_n^{(n-2)} \\ y_1^{(k)} & y_2^{(k)} & y_3^{(k)} & \cdots & y_n^{(k)} \end{vmatrix} = \sum_{i=1}^{n} y_i^{(k)}(-1)^{i+n} W_i(x).$$

Thus, the left hand side is 0 if $k < n-1$, and $W(x)$ if $k = n-1$.

19 Show that a second order eigenvalue problem in the form of equation 4.12 can be transformed into a Sturm-Liouville problem as in equation 4.13.

Hint: Multiply equation 4.12 by the "integrating factor" $\mu(x) = e^{\int p_1(x)\,dx}$.

20 Use the trig identity

$$\sin(\alpha)\sin(\beta) = \frac{1}{2}(\cos(\alpha-\beta)) - \cos(\alpha+\beta))$$

to verify equation 4.16.

21 Show that $y = \sqrt{x}\sin(\sqrt{E-1/4}\ln(x))$ and $y = \sqrt{x}\cos(\sqrt{E-1/4}\ln(x))$ are solutions to the equation

$$x^2y''(x) + Ey(x) = 0.$$

22 Use the result of problem 21 to solve the eigenvalue problem

$$x^2y''(x) + Ey(x) = 0, \qquad y(1) = y(e) = 0.$$

Find both the eigenvalues and their associated eigenfunctions. You do not have to scale them to be orthonormal.

23 The equation in problem 22 can be scaled to put it into Sturm-Liouville form: $(y'(x))' + Ey(x)/x^2 = 0$. Scale the eigenfunctions from this problem so that they will be orthonormal with respect to the weight $1/x^2$.

24 Show that $y = \sin(\sqrt{E-1}\ln(x))/x$ and $y = \cos(\sqrt{E-1}\ln(x))/x$ solve the equation

$$x^2y''(x) + 3xy'(x) + Ey(x) = 0.$$

25 Use the result of problem 24 to solve the eigenvalue problem

$$x^2y''(x) + 3xy'(x) + Ey(x) = 0, \qquad y(1) = y(2) = 0.$$

Find both the eigenvalues and their associated eigenfunctions. You do not have to scale them to be orthonormal.

26 The equation in problem 24 is not currently in Sturm-Liouville form. Convert this equation to Sturm-Liouville form, and determine the weight function $r(x)$ for which the solutions will be orthonormal. See the hint for problem 19.

For problems **27** through **36**: Find all singular points (possibly including ∞), and state whether each singular point is regular or irregular.

27 $xy'' + y = 0$
28 $x^3 y''' = y$
29 $x^3 y'' + x^2 y' + y = 0$
30 $x^5 y'' + 2x^4 y' - y = 0$
31 $xy'' + \sin(x) y' + \tan(x) y = 0$
32 $(1 - x^2) y'' - 2xy' + a(a+1) y = 0$
33 $(x^2 + 1) y'' + 2xy' - a(a+1) y = 0$
34 $(x^3 + x^2) y'' + 2xy' + y = 0$
35 $(x^4 - x^3) y'' + 2x^3 y' + 4y = 0$
36 $(x^4 - x^2) y'' + 2x^3 y' + 2y = 0$

4.2 First Order Equations

Until now, we have covered the theoretical aspects of differential equations, but have not shown any techniques for *solving* a differential equation, except for special cases. Because we will mainly focus on the approximate solutions to the equation, most of the standard tricks for solving differential equations would not apply. However, there will be times that we will have to know how to solve some simple first order equations in order to obtain the approximate solutions to a higher order equation. Hence, this section will focus on solving first order equations exactly.

4.2.1 Separable Equations

The easiest differential equations to solve are first order equations

$$y' = F(x, y)$$

in which $F(x, y)$ *factors* into a function of x, times a function of y.

$$y' = g(x)h(y).$$

Such differentiable equations are called *separable equations*. The strategy for solving such equations is to rewrite $y'(x)$ into its Leibniz notation: dy/dx. Then we will multiply both sides of the equation by dx, and divide both sides by $h(y)$, to obtain

$$\frac{dy}{h(y)} = g(x)\, dx.$$

We now add an integral sign to both sides, resulting in an integral with respect to y on the left, and an integral with respect to x on the right.

$$\boxed{\int \frac{1}{h(y)}\,dy = \int g(x)\,dx + C.} \tag{4.19}$$

In doing the integrals, we only need to add an arbitrary constant to one side, since adding two different constants on both sides would allow the constants to be combined. After integrating, we should, if possible, solve for y.

Example 4.12
Solve the equation $y' = xy^2 - 1 + x - y^2$.
SOLUTION: The equation factors into

$$\frac{dy}{dx} = (x-1)(y^2+1),$$

so multiplying both sides by dx, dividing by $y^2 + 1$, and adding an integral sign, we get

$$\int \frac{1}{y^2+1}\,dy = \int x - 1\,dx.$$

Integrating both sides, we have

$$\tan^{-1}(y) = \frac{x^2}{2} - x + C.$$

This can easily be solved for y, to produce $y = \tan(x^2/2 - x + C)$. □

Since a separable equation is usually non-linear, there is a possibility of special solutions. Since we divided by $h(y)$ in the process of solving the equation, any real solution to $h(y) = 0$ will become a constant special solution.

Example 4.13
Find all of the solutions to the equation

$$y' = x\sqrt{1-y^2}.$$

SOLUTION: Separating the variables, we get

$$\int \frac{1}{\sqrt{1-y^2}}\,dy = \int x\,dx.$$

Thus, we have $\sin^{-1}(y) = x^2/2 + C$, or $y = \sin(x^2/2 + C)$. Even though this is a general solution, it does not give *all* of the solutions to the equation.

Since we divided by $\sqrt{1-y^2}$ in the process of solving the equation, we have to consider the possibility that $\sqrt{1-y^2}=0$. This indeed happens if $y=\pm 1$. So there are two special solutions to the equation. ☐

Note that in the last example, there are regions in the (x, y) plane in which the conditions of theorem 4.1 are *not* met. The functions $F(x,y)=x\sqrt{1-y^2}$ and $\partial F/\partial y = -xy/\sqrt{1-y^2}$ are only both continuous if $1 < y < 1$. If we pick an initial condition such as $y(0)=1$, we find that $\partial F/\partial y$ is not continuous at this initial condition, so we are not guaranteed a unique solution. Indeed, there are two solutions for the initial value problem, $y=\sin((x^2+\pi)/2)$ and $y=1$.

Given an initial value problem, we can use the initial condition to solve for the arbitrary constant from the general solution. But for non-linear equations, there may be more than one set of solutions, so care must be taken to determine which solution satisfies the initial condition.

Example 4.14

Solve the initial value problem

$$y' = \frac{x+5}{y-3}, \qquad y(0)=0.$$

SOLUTION: We begin by finding the general solution to the differential equation. Multiplying both sides by $(y-3)\,dx$ and adding an integral sign, we get

$$\int (y-3)\,dy = \int (x+5)\,dx.$$

This can quickly be integrated, to produce

$$\frac{y^2}{2} - 3y = \frac{x^2}{2} + 5x + C.$$

This is a quadratic equation in y, so we can either use the quadratic equation or complete the square to solve for y. In this case, completing the square is fairly easy if we double both sides, followed by adding 9 to both sides.

$$y^2 - 6y + 9 = x^2 + 10x + 9 + 2C.$$

A common trick for simplifying expressions involving an arbitrary constant is to replace a constant expression with another constant. In this case, $9+2C$ can be replaced by a new constant, K.

$$(y-3)^2 = x^2 + 10x + K.$$

This is now easy to solve for y, giving us $y = 3 \pm \sqrt{x^2+10x+K}$.

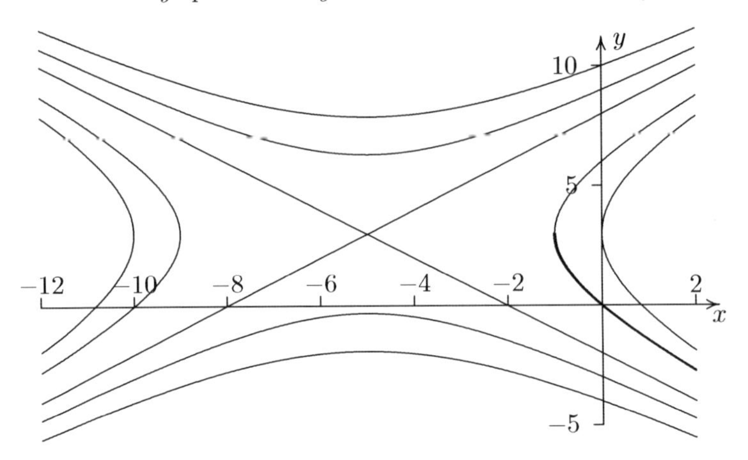

FIGURE 4.1: This shows the solutions to the equation $y' = (x+5)/(y-3)$. The solution which passes through the origin is shown in bold. Except for the two straight lines, the solutions are hyperbolas.

We now can use the fact that $y(0) = 0$ to solve for K. Since $0 = 3 \pm \sqrt{K}$, we need $K = 9$, *and* we need to use the minus sign in the $\pm$. Thus,

$$y = 3 - \sqrt{x^2 + 10x + 9}.$$

Figure 4.1 shows the plot of this curve, along with other solutions to the differential equation. Note that this solution is only defined for $x \geq -1$, because of a spontaneous singularity in the square root. □

4.2.2 First Order Linear Equations

We now turn our attention to linear first order equations. Such equations can be expressed in the form

$$y' + p(x)y = g(x). \tag{4.20}$$

The technique for first order linear equations is to multiply both sides of the equation by a function $\mu(x)$, known as the *integrating factor*, which will cause the left side to be the exact derivative of a product, via the product rule. To illustrate, we multiply both sides of equation 4.20 with $\mu(x)$, we get

$$\mu(x)y'(x) + \mu(x)p(x)y(x) = \mu(x)g(x).$$

Now the left hand side is almost the derivative of $\mu(x)y(x)$, which would be $\mu(x)y'(x) + \mu'(x)y(x)$. In fact, it would be this derivative if $\mu'(x)$ happened to be $\mu(x)p(x)$. But we can solve the equation $\mu'(x) = \mu(x)p(x)$, for this is a separable equation.

$$\int \frac{1}{\mu(x)}\, d\mu = \int p(x)\, dx \tag{4.21}$$

We only need one function μ that will work for an integrating factor, so we can ignore the arbitrary constant at this time. Integrating equation 4.21, we get $\ln(\mu(x)) = \int p(x)\, dx$, so $\mu(x) = e^{\int p(x)\, dx}$.

Having established the function μ, we can return to solving the first order equation. Since $(\mu(x)y(x))' = \mu(x)y'(x) + \mu(x)p(x)y(x)$, the equation becomes

$$(\mu(x)y(x))' = \mu(x)g(x).$$

Integrating both sides, and dividing by $\mu(x)$, we solve the equation to produce

$$\boxed{y(x) = \frac{\int \mu(x)g(x)\, dx + C}{\mu(x)}, \text{ where } \mu(x) = e^{\int p(x)\, dx}.} \tag{4.22}$$

Example 4.15
Solve the equation

$$xy' + 2y = x^4.$$

SOLUTION: First, we divide by x to get the equation into the form of equation 4.20. This produces

$$y' + \frac{2}{x}y = x^3.$$

Since $p(x) = 2/x$, $\int p(x)\, dx = 2\ln x$, hence $\mu(x) = e^{2\ln x} = (e^{\ln(x)})^2 = x^2$. Thus,

$$y = \frac{\int x^2 x^3\, dx + C}{x^2} = \frac{x^6/6 + C}{x^2} = \frac{x^4}{6} + \frac{C}{x^2}. \qquad \square$$

Note that in this example, $p(x)$ had a singularity at $x = 0$, so by theorem 4.2, the solution is only guaranteed to be analytic if $x \neq 0$. Indeed, most of the solutions are discontinuous at $x = 0$, although there is one solution, $y = x^4/6$, which is analytic everywhere.

If we add an initial condition, $y(x_0) = y_0$, we will be able to solve for the constant C, producing a unique solution.

Example 4.16
Solve the initial value problem

$$y' + 2xy = x^3, \qquad y(0) = 1.$$

SOLUTION: Since $p(x) = 2x$, $\mu(x) = e^{\int 2x\, dx} = e^{x^2}$. Hence,

$$y = \frac{\int e^{x^2} x^3\, dx + C}{e^{x^2}}.$$

We can integrate by parts, using $u = x^2$, $dv = xe^{x^2}$, $v = e^{x^2}/2$, to get

$$\int e^{x^2} x^3 \, dx = \frac{x^2 e^{x^2}}{2} - \frac{e^{x^2}}{2} + C.$$

Thus, we have

$$y = \frac{x^2}{2} - \frac{1}{2} + Ce^{-x^2}.$$

To find the value of C, we use the fact that when $x = 0$, $y = 1$. Plugging these into the solution, we get

$$1 = 0 - \frac{1}{2} + Ce^0,$$

which shows that $C = 3/2$. Thus,

$$y = \frac{x^2}{2} - \frac{1}{2} + \frac{3}{2}e^{-x^2}. \qquad \square$$

This only gives a small sample of the techniques for solving differential equations. But this will give the tools necessary for finding the appropriate solutions to more complicated problems.

Problems for §4.2

For problems **1** through **8**: Solve the following separable equations. Don't forget to consider special solutions.

1 $y' = y^2 x$

2 $y' = \sqrt{1 - y^2}$

3 $y' = y \ln x$

4 $y' = xy + x + y + 1$

5 $y' = e^{x+y}$

6 $y' = xy^2 - x$

7 $y' = \sqrt{xy}$

8 $y' = (x + 1)/(y - 1)$

For problems **9** through **16**: Solve the separable initial value problems.

9 $y' = x^2 y^2$, $y(1) = 2$

10 $y' = xy^2 + x$, $y(0) = 1$

11 $y' = xy^3$, $y(1) = -\frac{1}{2}$

12 $y' = \sqrt{x - xy^2}$, $y(0) = \frac{1}{2}$

13 $y' = (y + 1)/(x - 1)$, $y(0) = 1$

14 $y' = (x + 1)/(y - 2)$, $y(1) = 1$

15 $y' = xy - x + y - 1$, $y(0) = 0$

16 $y' = (xy^2 + y^2)/(xy^2 + x)$, $y(1) = 1$

For problems **17** through **24**: Solve the following linear first order equations.

17 $y' + 2xy = 2x$

18 $y' + \tan(x) y = \sin(x)$

19 $xy' + 2y = x^2$

20 $xy' - 2y = x^2$

21 $y' + 2y = x^2$

22 $xy' + 3y = e^x$

23 $(x^2 + 1)y' + 2xy = e^x$

24 $(x^2 - 1)y' + xy = x$

For problems **25** through **32**: Solve the following linear initial value problems.

25 $xy' - 3y = x^2$, $y(1) = 2$

26 $xy' + 2y = x^2$, $y(2) = 3$

27 $y' + 3y = x$, $y(0) = 1$

28 $y' - y = x^3$, $y(0) = 2$

29 $(1 - x^2)y' - xy = 2$, $y(0) = 1$

30 $x^2 y' + 2y = 1$, $y(2) = 0$

31 $xy' + (1 + x)y = x^2$, $y(1) = 3$

32 $xy' + (1 + x^2)y = x$, $y(2) = 1$

4.3 Taylor Series Solutions

Although we have learned how to solve some first order equations exactly, our focus will be on finding general methods that can be used to *approximate* solutions to higher order equations. Our main tool is the asymptotic approximation to the solutions as $x \to x_0$, where x_0 can either be a finite number or ∞. Our strategy will depend on whether x_0 is an ordinary point, a regular singular point, or an irregular singular point of the equation. This section will cover the case of a regular singular point, section 4.4 will cover the situation for a regular singular point, and chapter 5 will deal with the most complicated case, the irregular singular point.

If x_0 is a regular singular point for a homogeneous linear equation, then theorem 4.2 guarantees that all solutions will be analytic at x_0, hence can be expressed by a convergent Taylor series

$$y(x) = \sum_{n=0}^{\infty} a_n (x - x_0)^n.$$

In the case where x_0 is ∞, we consider the Taylor series centered at ∞ to be

$$y(x) = \sum_{n=0}^{\infty} a_n x^{-n} = a_0 + \frac{a_1}{x} + \frac{a_2}{x^2} + \frac{a_3}{x^3} + \cdots.$$

This series will converge absolutely whenever $|x| > R$ for some R.

We will take advantage of the fact that a Taylor series can be differentiated term by term. Thus, if x_0 is finite,

$$y'(x) = \sum_{n=0}^{\infty} n a_n (x - x_0)^{n-1}, \qquad y''(x) = \sum_{n=0}^{\infty} n(n-1) a_n (x - x_0)^{n-2},$$

$$y'''(x) = \sum_{n=0}^{\infty} n(n-1)(n-2) a_n (x - x_0)^{n-3}, \text{ etc.}$$

Thus, we can plug the *series* into the differential equation, and use the properties of Taylor series to determine the coefficients a_n for the series.

The easiest case to consider is if $x_0 = 0$, and the coefficient functions $p_i(x)$ are rational functions of x. By multiplying by the common denominator, we can express the equation in the form

$$P_n(x) y^{(n)} + P_{n-1}(x) y^{(n-1)} + \cdots + P_1(x) y' + P_0(x) y = 0,$$

where $P_i(x)$ are polynomials in x. We can then distribute each polynomial over y and its derivatives, so that every term will involve a power of x, times y or one of its derivatives.

Consider for example

$$y'' + \frac{2x}{1+x^2}y' - xy = 0.$$

We can multiply by the common denominator $1 + x^2$ to produce

$$(1 + x^2)y'' + 2xy' - (x + x^3)y = 0.$$

Distributing each polynomial over the derivatives of y, we get

$$y'' + x^2y'' + 2xy' - xy - x^3y = 0.$$

The purpose of distributing the polynomials is so that each term can be expressed as a series of the form

$$\sum_{n=0}^{\infty} b(n)a_n x^{n+k}$$

for some function $b(n)$ involving only n, and some integer k. Note that a power of x can be brought inside of the sum, and combined with the power of x that is currently there.

One final technique required is *shifting the index* of a sum. This amounts to making a linear substitution $n = m+k$ into the sum, where k is an integer. For example, if we substitute $n = m + 2$ into the sum $\sum_{n=0}^{\infty} n(n-1)a_n x^{n-2}$, we get

$$\sum_{m+2=0}^{\infty} (m+2)((m+2)-1)a_{m+2}x^{(m+2)-2}.$$

Note that the substitution is also made in the starting point of the sum. Solving $m+2 = 0$, we get $m = -2$, so the new starting point is different than the original sum. However, the terms for $m = -2$ and $m = -1$ evaluate to zero, so in this case we can start the sum at $m = 0$. It doesn't matter which variable we used for m, so in fact we can replace m with n. This may seem to contradict the first substitution of $n = m + 2$, but the new n will not be the same as the old n. The sum can be rewritten as

$$\sum_{n=0}^{\infty}(n+2)(n+1)a_{n+2}x^n = 2a_2 + 6a_3x + 12a_4x^2 + 20a_5x^3 + \cdots.$$

Notice that the list of terms is exactly the same as the original series, except for the initial zero terms.

$$\sum_{n=0}^{\infty} n(n-1)a_n x^{n-2} = 0 + 0 + 2a_2 + 6a_3x + 12a_4x^2 + 20a_5x^3 + \cdots.$$

We now have all the tools required to determine the Taylor series for the solution of a linear equation.

Example 4.17

Find the Maclaurin series for the solutions to the linear differential equation

$$(x^2 - 1)y'' + 2xy' - 6y = 0.$$

SOLUTION: We first distribute the polynomial over y'', to produce the equation $x^2y'' - y'' + 2xy' - 6y = 0$. Then we substitute

$$y = \sum_{n=0}^{\infty} a_n x^n, \quad y' = \sum_{n=0}^{\infty} n a_n x^{n-1}, \text{ and } y'' = \sum_{n=0}^{\infty} n(n-1)a_n x^{n-2}$$

into this equation, to produce

$$\sum_{n=0}^{\infty} n(n-1)a_n x^n - \sum_{n=0}^{\infty} n(n-1)a_n x^{n-2} + 2\sum_{n=0}^{\infty} n a_n x^n - 6\sum_{n=0}^{\infty} a_n x^n = 0.$$

The second sum involves x^{n-2} instead of x^n, so we can shift the index on this sum, replacing n with $n+2$. Thus, we have

$$\sum_{n=0}^{\infty} n(n-1)a_n x^n - \sum_{n=-2}^{\infty} (n+2)(n+1)a_{n+2} x^n + 2\sum_{n=0}^{\infty} n a_n x^n - 6\sum_{n=0}^{\infty} a_n x^n = 0.$$

Now all of the powers of x are the same in all of the sums, but the starting points are different. However, the terms for $n = -2$ and $n = -1$ in the second sum would evaluate to 0, so we can change the starting point of the second sum to $n = 0$ without affecting the equation. Now that all of the starting points are the same, we can combine the series into one.

$$\sum_{n=0}^{\infty} (n(n-1)a_n - (n+2)(n+1)a_{n+2} + 2na_n - 6a_n)x^n = 0.$$

Since Taylor series are unique, each term of this series must be 0 to solve the equation. Thus, the coefficients a_n satisfy the *recursion formula*

$$n(n-1)a_n - (n+2)(n+1)a_{n+2} + 2na_n - 6a_n = 0.$$

The highest index appearing in this equation is a_{n+2}, so we can solve for this term, which will give us a way of finding every coefficient of the series in terms of the previous ones.

$$a_{n+2} = \frac{n^2 + n - 6}{(n+2)(n+1)} a_n. \tag{4.23}$$

Since we have a second order equation, $y(0) = a_0$ and $y'(0) = a_1$ will be the two arbitrary constants in the general solution. Hence, every term in the series can be expressed in terms of the first two terms, a_0 and a_1. Plugging in $n = 0, 1, 2, \ldots$ into the recursion formula, we find

$$\begin{aligned}
n = 0) &\qquad a_2 = -3a_0, \\
n = 1) &\qquad a_3 = -\frac{2a_1}{3}, \\
n = 2) &\qquad a_4 = 0a_2 = 0, \\
n = 3) &\qquad a_5 = \frac{6}{20}a_3 = -\frac{a_1}{5}, \\
n = 4) &\qquad a_6 = \frac{14}{30}a_4 = 0, \\
n = 5) &\qquad a_7 = \frac{24}{42}a_5 = -\frac{4a_1}{35}, \\
n = 6) &\qquad a_8 = \frac{36}{56}a_6 = 0, \\
n = 7) &\qquad a_9 = \frac{50}{72}a_7 = -\frac{5a_1}{63}, \ldots.
\end{aligned}$$

We are now ready to find the series for the solutions. Since

$$y = \sum_{n=0}^{\infty} a_n x^n = a_0 + a_1 x + a_2 x^2 + a_3 x^3 + a_4 x^4 + a_5 x^5 + \cdots,$$

we substitute the values for $a_2, a_3, a_4, \ldots$ that we have determined, to get

$$y = a_0 + a_1 x - 3a_0 x^2 - \frac{2a_1}{3}x^3 - \frac{a_1}{5}x^5 - \frac{4a_1}{35}x^7 - \frac{5a_1}{63}x^9 - \cdots.$$

We can group those terms involving a_0 and a_1 separately, to produce two different series solutions:

$$y = a_0(1 - 3x^2) + a_1\left(x - \frac{2x^3}{3} - \frac{x^5}{5} - \frac{4x^7}{35} - \frac{5x^9}{63} - \cdots\right).$$

Normally we will not be able to express the sum in closed form, but in this case we can. One solution truncates to produce a polynomial solution. It is easy to check that $y_1 = 1 - 3x^2$ is indeed an exact solution to the equation. The other solution is an infinite sum, but the pattern starts to become apparent if we rewrite $-x^5/5$ as $-3x^5/15$. Apparently, when n is odd, we have the following pattern:

$$a_n = \frac{-(n+1)a_1}{2n(n-2)}.$$

We can verify this pattern will continue by showing that the pattern satisfies equation 4.23.

$$\frac{n^2 + n - 6}{(n+2)(n+1)}a_n = \frac{-(n+1)a_1}{2n(n-2)}\frac{(n-2)(n+3)}{(n+2)(n+1)} = \frac{-(n+3)a_1}{2n(n+2)} = a_{n+2}.$$

Hence, the second solution can be expressed as the sum

$$y_2 = \sum_{n=0}^{\infty} \frac{-(n+1)x^{2n+1}}{(2n+1)(2n-1)}. \tag{4.24}$$

□

There is a shortcut that eliminates the need of keeping track of where each series starts. If we consider $a_n = 0$ whenever $n < 0$, then we can express the original sum as

$$y = \sum_{n=-\infty}^{\infty} a_n x^n.$$

Then shifting the index will not change the starting points of the sum. In fact, if we can abbreviate the sum as

$$y = \sum_n a_n x^n$$

to indicate that the sum is over all integer values of n, positive or negative.

Example 4.18

The Airy equation is $y'' = xy$. Find the Maclaurin series for two solutions.

SOLUTION: Replacing y with $\sum_n a_n x^n$, and y'' with $\sum_n n(n-1)a_n x^{n-2}$, we have

$$\sum_n n(n-1)a_n x^{n-2} = \sum_n a_n x^{n+1}.$$

The left hand sum can be shifted up by 2, and the right hand sum can be shifted down by 1 to get all sums involving x^n.

$$\sum_n (n+2)(n+1)a_{n+2}x^n = \sum_n a_{n-1}x^n.$$

Thus, we get the recursion formula $(n+2)(n+1)a_{n+2} = a_{n-1}$, which can be solved for a_{n+2}.

$$a_{n+2} = \frac{a_{n-1}}{(n+2)(n+1)}.$$

Plugging in $n = 0$ gives $a_2 = a_{-1}/2$, which by our convention is equal to 0. Other values of n produce

$$a_3 = \frac{a_0}{3\cdot 2}, \quad a_4 = \frac{a_1}{4\cdot 3}, \quad a_5 = 0, \quad a_6 = \frac{a_0}{6\cdot 5\cdot 3\cdot 2}, \quad a_7 = \frac{a_1}{7\cdot 6\cdot 4\cdot 3},$$
$$a_8 = 0, \quad a_9 = \frac{a_0}{9\cdot 8\cdot 6\cdot 5\cdot 3\cdot 2}, \quad a_{10} = \frac{a_1}{10\cdot 9\cdot 7\cdot 6\cdot 4\cdot 3}, \ldots.$$

Plugging these values into $y = \sum_n a_n x^n$ gives us

$$
\begin{aligned}
y &= a_0 + a_1 x + \frac{a_0 x^3}{3 \cdot 2} + \frac{a_1 x^4}{4 \cdot 3} + \frac{a_0 x^6}{6 \cdot 5 \cdot 3 \cdot 2} + \frac{a_1 x^7}{7 \cdot 6 \cdot 4 \cdot 3} + \cdots \\
&= a_0 \left(1 + \frac{x^3}{3 \cdot 2} + \frac{x^6}{6 \cdot 5 \cdot 3 \cdot 2} + \frac{x^9}{9 \cdot 8 \cdot 6 \cdot 5 \cdot 3 \cdot 2} + \cdots \right) \\
&+ a_1 \left(x + \frac{x^4}{4 \cdot 3} + \frac{x^7}{7 \cdot 6 \cdot 4 \cdot 3} + \frac{x^{10}}{10 \cdot 9 \cdot 7 \cdot 6 \cdot 4 \cdot 3} + \cdots \right).
\end{aligned}
$$

This time, the pattern is clear, but difficult to express in closed form. We can express the terms in closed form using the "zipper" property of the Gamma function.

$$
\begin{aligned}
\frac{x^{10}}{10 \cdot 9 \cdot 7 \cdot 6 \cdot 4 \cdot 3} &= \frac{x^{10}\Gamma(4/3)}{3^6 (10/3) \cdot 3 \cdot (7/3) \cdot 2 \cdot (4/3) \cdot 1 \cdot \Gamma(4/3)} \\
&= \frac{x^{10}\Gamma(4/3)}{3! \cdot 3^6 \Gamma(13/3)}.
\end{aligned}
$$

So with the Gamma function, we can express the sums in closed form

$$
y_1 = \sum_{n=0}^{\infty} \frac{x^{3n}\Gamma(2/3)}{9^n \, n! \Gamma(n + (2/3))}, \qquad y_2 = \sum_{n=0}^{\infty} \frac{x^{3n+1}\Gamma(4/3)}{9^n \, n! \Gamma(n + (4/3))}.
$$

∎

Series solutions will also work for inhomogeneous equations, if we have the Taylor series for the inhomogeneous term $g(x)$. However, there will rarely be a pattern in the coefficients of the solution.

Example 4.19

Find the Maclaurin series up to the x^8 term for the equation

$$y'' - xy = e^x.$$

SOLUTION: Note that the associated homogeneous equation is the Airy equation. Replacing y with $\sum_n a_n x^n$, y'' with $\sum_n n(n-1)a_n x^{n-2}$, and e^x with its Maclaurin series $\sum_n x^n/n!$, we get

$$\sum_n n(n-1)a_n x^{n-2} - \sum_n a_n x^{n+1} = \sum_n \frac{x^n}{n!}.$$

Shifting the first sum up by 2, and the second down by 1, we get

$$\sum_n (n+2)(n+1)a_{n+2} x^n - \sum_n a_{n-1} x^n = \sum_n \frac{x^n}{n!}.$$

So the recursion formula is slightly different than in example 4.18:

$$(n+2)(n+1)a_{n+2} - a_{n-1} = 1/n!.$$

Solving for a_{n+2}, we get

$$a_{n+2} = \frac{a_{n-1}}{(n+2)(n+1)} + \frac{1}{(n+2)(n+1)n!} = \frac{a_{n-1}}{(n+2)(n+1)} + \frac{1}{(n+2)!}.$$

Plugging in $n = 0, 1, 2, \ldots$ gives us

$$\begin{aligned}
a_2 &= \frac{1}{2} \\
a_3 &= \frac{a_0}{6} + \frac{1}{6} \\
a_4 &= \frac{a_1}{12} + \frac{1}{24} \\
a_5 &= \frac{a_2}{20} + \frac{1}{120} = \frac{1}{30} \\
a_6 &= \frac{a_3}{30} + \frac{1}{720} = \frac{a_0}{180} + \frac{1}{144} \\
a_7 &= \frac{a_4}{42} + \frac{1}{5040} = \frac{a_1}{504} + \frac{1}{840} \\
a_8 &= \frac{a_5}{56} + \frac{1}{40320} = \frac{5}{8064}
\end{aligned}$$

The terms involving a_0 or a_1 are the same as for example 4.18. However, there are additional terms which do not involve either a_0 or a_1. If we set both a_0 and a_1 to 0, we get the particular solution

$$y_p = \frac{x^2}{2} + \frac{x^3}{6} + \frac{x^4}{24} + \frac{x^5}{30} + \frac{x^6}{144} + \frac{x^7}{840} + \frac{5x^8}{8064} + \cdots.$$

Thus, we can express the general solution as

$$y = a_0 \sum_{n=0}^{\infty} \frac{x^{3n}\Gamma(2/3)}{9^n\, n!\Gamma(n + (2/3))} + a_1 \sum_{n=0}^{\infty} \frac{x^{3n+1}\Gamma(4/3)}{9^n\, n!\Gamma(n + (4/3))} + y_p.$$

□

So far, we have only found series centered at $x_0 = 0$. The procedure works for the general Taylor series $y = \sum_n a_n(x - x_0)^n$, but it is easier if we first make a substitution $t = x - x_0$. Then $dx = dt$, so $y'(x) = y'(t)$, $y''(x) = y''(t)$, etc. Every x is replaced with $t + x_0$.

Example 4.20

The equation

$$xy'' + y = 0$$

has a singular point at $x = 0$, so the solutions will probably not be analytic at $x = 0$. However we can find the Taylor series of the solutions centered at $x = 1$. Find the series up to the $(x - 1)^6$ term.

SOLUTION: Substituting $t = x - 1$, or $x = t + 1$ into the equation, we get $(t+1)y''(t) + y(t) = 0$. Distributing the polynomial over y'' produces

$$ty'' + y'' + y = 0.$$

We now replace y with $\sum_n a_n t^n$ and y'' with $\sum_n n(n-1)a_n t^{n-2}$ to get

$$\sum_n n(n-1)a_n t^{n-1} + \sum_n n(n-1)a_n t^{n-2} + \sum_n a_n t^n = 0.$$

Shifting the index of the first sum up by 1, and the second sum up by 2, we have

$$\sum_n n(n+1)a_{n+1} t^n + \sum_n (n+2)(n+1)a_{n+2} t^n + \sum_n a_n t^n = 0.$$

This produces the recursion formula

$$n(n+1)a_{n+1} + (n+2)(n+1)a_{n+2} + a_n = 0.$$

Solving for a_{n+2} gives us

$$a_{n+2} = -\frac{na_{n+1}}{n+2} - \frac{a_n}{(n+2)(n+1)}.$$

Plugging in $n = 0, 1, 2, \ldots$ gives us

$$\begin{aligned}
a_2 &= -\frac{a_0}{2}, \\
a_3 &= -\frac{a_2}{3} - \frac{a_1}{6} = \frac{a_0 - a_1}{6}, \\
a_4 &= -\frac{2a_3}{4} - \frac{a_2}{12} = \frac{2a_1 - a_0}{24}, \\
a_5 &= -\frac{3a_4}{5} - \frac{a_3}{20} = \frac{2a_0 - 5a_1}{120}, \\
a_6 &= -\frac{4a_5}{6} - \frac{a_4}{30} = \frac{18a_1 - 5a_0}{720}, \ldots .
\end{aligned}$$

Hence, since $y = \sum_n a_n t^n = \sum_n a_n (x-1)^n$, we have

$$\begin{aligned}
y &= a_0 + a_1(x-1) - \frac{a_0}{2}(x-1)^2 + \frac{a_0 - a_1}{6}(x-1)^3 + \frac{2a_1 - a_0}{24}(x-1)^4 \\
&\quad + \frac{2a_0 - 5a_1}{120}(x-1)^5 + \frac{18a_1 - 5a_0}{720}(x-1)^6 + \cdots \\
&= a_0\left(1 - \frac{(x-1)^2}{2} + \frac{(x-1)^3}{6} - \frac{(x-1)^4}{24} + \frac{(x-1)^5}{60} - \frac{7(x-1)^6}{720} + \cdots\right) \\
&\quad + a_1\left((x-1) - \frac{(x-1)^3}{6} + \frac{(x-1)^4}{12} - \frac{(x-1)^5}{24} + \frac{(x-1)^6}{40} + \cdots\right).
\end{aligned}$$

□

When we have a power series solution, we would like to know what the radius of convergence will be for the series. Fortunately, there is an easy way to get a lower estimate on the radius of convergence using the differential equation.

THEOREM 4.6

Let z_0 be an ordinary point for the differential equation

$$y^{(n)} + p_{n-1}(x)y^{(n-1)} + \cdots + p_2(x)y'' + p_1(x)y' + p_0(x)y = 0.$$

Then a Taylor series solution about x_0 has a radius of convergence at least as large as the distance from x_0 to the closest singular point of the equation.

PROOF: Let r be the distance from x_0 to the closest singular point of the equation. Then the coefficient functions $p_{n-1}(x)$, $p_{n-2}(x), \ldots p_0(x)$ will be analytic inside the circle of radius r centered at x_0. Since this set is simply connected, we know from theorem 4.2 that all solutions to the equation will be analytic on this set. But theorem 2.3 states that the radius of convergence ρ of a solution $y(x)$ is the distance from x_0 to the nearest non-analytic point of $y(x)$. Thus, $\rho \geq r$. □

Example 4.21

Without actually finding the Taylor series solution, determine a lower bound for the radius of convergence of these series solutions to

$$(x^2 + 1)y'' - e^x y' + \sin xy = 0$$

centered at $x_0 = 2$.

SOLUTION: The equation, expressed in standard form, is

$$y'' - \frac{e^x}{x^2+1}y' + \frac{\sin x}{x^2+1}y = 0.$$

The coefficient functions have a singularity when $x^2 + 1 = 0$, or $\pm i$. The distance from these singularities and $x_0 = 2$ is

$$|\pm i - 2| = \sqrt{(-2)^2 + (\pm 1)^2} = \sqrt{5}.$$

Thus, the radius of convergence is at least $\sqrt{5}$. □

Knowing the radius of convergence can determine how many terms are necessary to compute the sum of the series to a prescribed amount of accuracy. In the next section, we will examine series solutions for the case where we have a regular singular point at x_0.

Problems for §4.3

For problems **1** through **8**: Determine the Maclaurin series for the solutions, up to the x^6 term, for the following differential equations. Note that some of the series may truncate to produce a polynomial solution.

1 $y'' + xy' - y = 0$
2 $y'' + xy' + 2y = 0$
3 $y'' = x^2 y$
4 $(x^2 + 1)y'' - 12y = 0$
5 $y'' + (x + 1)y' - y = 0$
6 $(x^2 + 1)y'' - 6xy' + 10y = 0$
7 $(x^2 + 1)y'' - xy' + y = 0$
8 $(x + 1)y'' - y = 0$

For problems **9** through **14**: Find the series solutions centered at $x_0 = 0$ to the differential equation. For the series that do not truncate, determine the pattern for the nth term in the series, so that the sums can be expressed in closed form.

9 $(1 - x^2)y'' + 2y = 0$
10 $(x^2 - 1)y'' + 2xy' - 2y = 0$
11 $(1 - x^2)y'' + 6y = 0$
12 $(x^2 + 1)y'' + 6xy' + 6y = 0$
13 $(x^2 + 1)y'' + 6xy' + 4y = 0$
14 $(x^2 - 1)y'' + 8xy' + 12y = 0$

15 Use the technique of example 3.8 to sum the series for y_2 in equation 4.24.

For problems **16** through **21**: Find the Taylor series solutions centered at the given x_0 to the differential equation, up to the $(x - x_0)^5$ term.

16 $xy'' + y' + y = 0$, $x_0 = 1$
17 $xy'' - xy' + y = 0$, $x_0 = 1$
18 $x^2 y'' - 6y = 0$, $x_0 = -1$
19 $xy'' - y' + xy = 0$, $x_0 = 2$
20 $(x^2 - 2x)y'' - y = 0$, $x_0 = 1$
21 $xy'' + (1 - x)y' + (x + 1)y = 0$, $x_0 = -1$

For problems **22** through **27**: For the following inhomogeneous equations, find the Maclaurin series up to the x^6 term for the special solution in which $a_0 = a_1 = 0$.

22 $y'' - y = e^x$
23 $y'' + xy' - y = e^{-x}$
24 $y'' + y = \ln(x + 1)$
25 $y'' + xy' - y = \tan^{-1}(x)$
26 $y'' + xy' + 2y = \cos(x)$
27 $y'' + (x + 1)y' + y = \sin(x)$

28 In the rare case where ∞ is an ordinary point, we can make the substitution $x = 1/t$, as was done in equation 4.18. We then find the Maclaurin series of the new equation in t, and finally substitute back to get a series in $1/x$. Use this technique to find the series solutions centered at $x_0 = \infty$ for the equation

$$x^4 y'' + (2x^3 - x)y' - 2y = 0.$$

For problems **29** through **32**: *Without finding the power series solution*, find a lower bound for the radius of convergence that the power series solutions would have if expanded about $x = x_0$.

29 $(x^2 - x - 1)y'' - y = 0$, $x_0 = 0$
30 $(x^2 + 1)y'' + y' - y = 0$, $x_0 = 1$
31 $(x^2 + 2x + 10)y'' + y'' - y = 0$, $x_0 = 3$
32 $(x^3 + 1)y'' + y = 0$, $x_0 = 2$

4.4 Frobenius Method

When $x = x_0$ is a regular singular point of a differential equation, the solutions will usually not be able to be expressed as a Taylor series. However, there is a way to modify a Taylor series so that there will be a series solution, merely by multiplying the Taylor series by $(x - x_0)^k$.

DEFINITION 4.6 A *Frobenius series* centered at x_0 is a series of the form

$$\sum_{n=0}^{\infty} a_0(x - x_0)^{n+k}$$

for which the series converges within some positive radius of convergence of x_0. Here, k can be real or complex. Note that we can assume that $a_0 \neq 0$, since if $a_0 = 0$, we could rewrite the series increasing k by 1.

For regular singular points, all of the solutions of the differential equation can be expressed in terms of a Frobenius series.

THEOREM 4.7
If x_0 is a regular singular point of a differential equation, then there will be at least one solution that can be expressed by a Frobenius series about x_0

$$y_1(x) = f_1(x)$$

If the equation is of second order or higher, there will be a second solution of the form

$$y_2(x) = f_2(x) + C_1 \ln(x - x_0)y_1,$$

where $f_2(x)$ can be expressed as a Frobenius series about x_0, and C_1 is a constant. For third order equations, there will be a third solution of the form

$$y_3(x) = f_3(x) + C_2 \ln(x - x_0)y_1 + C_3 \ln(x - x_0)^2 y_1 + C_4 \ln(x - x_0)y_2,$$

where $f_3(x)$ is a Frobenius series about x_0, and C_2, C_3, and C_4 are constants. The pattern continues for higher order equations.

Usually, the constants C_i will be zero, so all of the solutions will be expressed as Frobenius series. Later we will determine the conditions for which these constants are non-zero.

Because at least one solution we have is a Frobenius series, we begin by substituting $y = \sum_n a_n(x - x_0)^{n+k}$ into the equation. We will again use the convention that $a_n = 0$ for $n < 0$. We use the fact that $a_0 \neq 0$ to determine the value of possible values of k. Since multiplying a solution by a constant produces another solution, we can in fact assume that $a_0 = 1$, and use the recursion formula to determine the remaining a_n.

Example 4.22

Find the Frobenius series solutions for the equation

$$4x^2y'' + 4xy' + (4x^2 - 9)y = 0.$$

SOLUTION: First we distribute the polynomials, to get

$$4x^2y'' + 4xy' + 4x^2y - 9y = 0.$$

Next, we substitute

$$y = \sum_n a_n x^{n+k}, \quad y' = \sum_n (n+k)a_n x^{n+k-1}, \text{ and}$$

$$y'' = \sum_n (n+k)(n+k-1)a_n x^{n+k-2}$$

into the equation to get

$$4\sum_n (n+k)(n+k-1)a_n x^{n+k} + 4\sum_n (n+k)a_n x^{n+k}$$
$$+ 4\sum_n a_n x^{n+k+2} - 9\sum_n a_n x^{n+k} = 0.$$

To get all terms to involve x^{n+k}, we need to shift the index of the third sum down by 2. Then we can combine all of the sums into one.

$$\sum_n [4(n+k)(n+k-1)a_n + 4(n+k)a_n + 4a_{n-2} - 9a_n]x^{n+k} = 0.$$

As with Taylor series, Frobenius series are unique, so the only way for the sum to be 0 for all x is for the coefficients to all be zero. So we have the recursion formula

$$4(n+k)(n+k-1)a_n + 4(n+k)a_n + 4a_{n-2} - 9a_n.$$

In order to determine k, we plug in the value of n which causes *the largest index of a to be 0*. In this case, the highest index of a is a_n, so we plug in $n = 0$ to get

$$4k(k-1)a_0 + 4ka_0 + 4a_{-2} - 9a_0 = 0.$$

By convention, $a_{-2} = 0$, and we are already assuming that $a_0 = 1$. So k must satisfy the equation

$$4k(k-1) + 4k - 9 = 0.$$

This is known as the *indicial polynomial* for the differential equation. The degree of the indicial polynomial will always be the order of the equation, and there will be a solution corresponding to each root of the indicial polynomial. In this case the indicial polynomial simplifies to $4k^2 - 9 = 0$, or $k = \pm 3/2$.

Let us find the solution corresponding to the largest root $k = 3/2$. Substituting $k = 3/2$ into the recursion formula causes it to simplify to

$$[(2n+3)(2n+1) + 2(2n+3) - 9]a_n + 4a_{n-2} = 0. \qquad (4.25)$$

Solving for the highest index of a, we get

$$a_n = -\frac{a_{n-2}}{n(n+3)}.$$

Assuming that $a_0 = 1$, and plugging in $n = 1, 2, 3, \ldots$ gives us the coefficients for the Frobenius series.

$$\begin{aligned}
n=1)\ a_1 &= -\frac{a_{-1}}{1\cdot 4} = 0\\
n=2)\ a_2 &= -\frac{a_0}{2\cdot 5} = -\frac{1}{10}\\
n=3)\ a_3 &= -\frac{a_1}{3\cdot 6} = 0\\
n=4)\ a_4 &= -\frac{a_2}{4\cdot 7} = \frac{1}{2\cdot 4\cdot 5\cdot 7} = \frac{1}{280}\\
n=5)\ a_5 &= -\frac{a_3}{5\cdot 8} = 0\\
n=6)\ a_6 &= -\frac{a_4}{6\cdot 9} = -\frac{1}{2\cdot 4\cdot 5\cdot 6\cdot 7\cdot 9} = -\frac{a_0}{15120}
\end{aligned}$$

It is clear that the terms with an odd coefficient will go to 0, but there is also a pattern for the even coefficients. Note that the denominators are always two factors away from being a factorial. Thus,

$$a_{2n} = (-1)^n \frac{3(2n+2)}{(2n+3)!}.$$

Hence, we can express one solution as a Frobenius series

$$y_1 = \sum_{n=0}^{\infty} (-1)^n \frac{6(n+1)x^{2n+(3/2)}}{(2n+3)!}.$$

To find the solution corresponding to $k = -3/2$, we proceed the same way. Substituting $k = -3/2$ into the recursion formula, we get

$$[(2n-3)(2n-5) + 2(2n-3) - 9]a_n + 4a_{n-2} = 0,$$

which produces

$$a_n = -\frac{a_{n-2}}{n(n-3)}.$$

This gives us the first few terms of the series

$$a_1 = -\frac{a_{-1}}{-2} = 0, \qquad a_2 = -\frac{a_0}{-2} = \frac{1}{2}, \qquad a_3 = -\frac{a_1}{0} = \frac{0}{0}?$$

We have run into a conflict computing the coefficient a_3. Returning to the recursion formula in equation 4.25, we find that $0a_3 + a_1 = 0$. So the recursion formula is actually satisfied for any value of a_3.

To understand what is going on, consider that the term corresponding to $n = 3$ would involve $x^{3-(3/2)} = x^{3/2}$. This is the first term of the series for the first solution y_1. Since we can add any multiple of y_1 to y_2 to get another solution, we see that a_3 is arbitrary, so we might as well set $a_3 = 0$. Continuing, we get

$$\begin{aligned} n=4) \quad a_4 &= -\frac{a_2}{4\cdot 1} = -\frac{1}{1\cdot 2\cdot 4} \\ n=5) \quad a_5 &= -\frac{a_3}{5\cdot 2} = 0 \\ n=6) \quad a_6 &= -\frac{a_4}{6\cdot 3} = \frac{1}{1\cdot 2\cdot 3\cdot 4\cdot 6}. \end{aligned}$$

Once again, the odd coefficient terms go to zero, and the denominators of the even terms are one factor shy of a factorial. It is not hard to determine the general pattern:

$$a_{2n} = (-1)^{n+1}\frac{(2n-1)}{(2n)!},$$

so the second solution can be expressed as the sum

$$y_2 = \sum_{n=0}^{\infty}(-1)^{n+1}\frac{(2n-1)x^{n-(3/2)}}{(2n)!}.$$

□

We see from this example that when two roots of the indicial polynomial differ by an integer, the solution corresponding to the larger root can interfere with the solution using the smaller root. In the last example we got lucky, since the recursion formula produced the equation $0 = 0$, which is satisfied by any a_3. But most of the time, the recursion formula will produce a contradiction, such as $0 = 1$. In this case, the solution will involve logarithms, so the constants in theorem 4.7 will be non-zero. However, we can utilize the knowledge of the exact form of the second solution, $y_2 = f_2(x) + C_1 \ln(x - x_0)y_1$. We can make the substitution

$$y(x) = u(x)+C\ln(x-x_0)y_1, \qquad y'(x) = u'(x)+C\frac{y_1}{x-x_0}+C\ln(x-x_0)y_1'(x),$$

$$y''(x) = u''(x) - C\frac{y_1}{(x-x_0)^2} + 2C\frac{y_1'}{x-x_0} + C\ln(x-x_0)y_1''(x), \text{ etc.}$$

Because y_1 solves the original equation, all of the logarithm terms will cancel. The resulting equation for $u(x)$ will be similar to the original equation, except that there will be an inhomogeneous term. At the point where the first solution interferes with the second solution, we can solve for the constant C instead of solving for a coefficient a_n. This a_n will be arbitrary because of the previous solution, so we can set this particular a_n to 0 for convenience.

Example 4.23

The Bessel equation of order v is given by

$$x^2y'' + xy' + (x^2 - v^2)y = 0.$$

In fact, example 4.22 is the Bessel equation of order 3/2. Find the series solution centered at $x_0 = 0$ for the Bessel equation of order 1.

SOLUTION: Plugging $y = \sum_n a_n x^{n+k}$, $y' = \sum_n (n+k)a_n x^{n+k-1}$, and

$$y'' = \sum_n (n+k)(n+k-1)a_n x^{n+k-2}$$

into the distributed equation $x^2y'' + xy' + x^2y - y = 0$ produces

$$\sum_n (n+k)(n+k-1)a_n x^{n+k} + \sum_n (n+k)a_n x^{n+k} + \sum_n a_n x^{n+k+2} - \sum_n a_n x^{n+k} = 0.$$

Shifting the index down 2 on the third sum allows us to combine the sums together.

$$\sum_n [(n+k)(n+k-1)a_n + (n+k)a_n + a_{n-2} - a_n]x^{n+k} = 0. \qquad (4.26)$$

Hence, we have the recursion formula

$$(n+k)(n+k-1)a_n + (n+k)a_n + a_{n-2} - a_n.$$

To find the indicial polynomial, we find the value of n so that the highest subscript of a becomes 0, which is $n = 0$. Plugging $n = 0$ into the recursion formula produces $[k(k-1)+k-1]a_0 = 0$, so the indicial polynomial is $k^2 - 1$, having ± 1 as roots.

Because the roots differ by an integer, one solution may interfere with the other. Thus, we will always start with the larger of the roots. Plugging $k = 1$ into the recursion formula, and solving for a_0, we get

$$a_n = -\frac{a_{n-2}}{n(n+2)}.$$

Plugging in various values of n into the equation, we get

$$\begin{aligned}
n = 1)\ a_1 &= -\frac{a_{-1}}{1 \cdot 3} = 0 \\
n = 2)\ a_2 &= -\frac{a_0}{2 \cdot 4} = -\frac{1}{8} \\
n = 3)\ a_3 &= -\frac{a_1}{3 \cdot 5} = 0 \\
n = 4)\ a_4 &= -\frac{a_2}{4 \cdot 6} = \frac{1}{2 \cdot 4 \cdot 4 \cdot 6} = \frac{1}{2^4 2!3!} = \frac{1}{192} \\
n = 5)\ a_5 &= -\frac{a_3}{5 \cdot 7} = 0 \\
n = 6)\ a_6 &= -\frac{a_4}{6 \cdot 8} = -\frac{1}{2 \cdot 4 \cdot 4 \cdot 6 \cdot 6 \cdot 8} = -\frac{1}{2^6 3!4!} = -\frac{1}{9216}.
\end{aligned}$$

By dividing each factor in the denominator by 2, the pattern is quickly revealed. For the odd terms we have $a_{2n+1} = 0$, and for the even terms,

$$a_{2n} = \frac{(-1)^n}{2^{2n} n!(n+1)!}.$$

So the first series solution to the Bessel equation of order 1 is

$$y_1 = \sum_{n=0}^{\infty} (-1)^n \frac{x^{2n+1}}{2^{2n} n!(n+1)!}.$$

For the second solution, we plug in $k = -1$ into the recursion formula, and solve for a_n to get $(n^2 - 2n)a_n + a_{n-2} = 0$. Plugging in $n = 1$ gives us $a_1 = a_{-1} = 0$, but plugging in $n = 2$ yields $0a_2 + a_0 = 0$, which is a contradiction. So this series will involve logarithms.

Since we know the solution will be of the form $y_2 = f_2 + C_1 y_1 \ln x$, we substitute $y(x) = u(x) + Cy_1 \ln x$ in the differential equation. The result is

$$\begin{aligned}
x^2(u'' - Cy_1/x^2 + 2Cy_1'/x + C\ln(x)y_1'') + x(u' + Cy_1/x + C\ln(x)y_1') \\
+ (x^2 - 1)(u + C\ln(x)y_1) = 0,
\end{aligned}$$

which can be rewritten as

$$x^2 u'' + xu' + (x^2 - 1)u + C\ln x(x^2 y_1'' + xy_1 + (x^2 - 1)y_1) = -2Cxy_1'.$$

Since y_1 solves the original differential equation, all of the logarithm terms cancel, so we end up with the original equation, only with the extra inhomogeneous term

$$\begin{aligned} -2Cxy_1' &= \sum_{n=0}^{\infty}(-1)^{n+1}\frac{2C(2n+1)x^{2n+1}}{2^{2n}n!(n+1)!} \\ &= C\left(-2x + \frac{3x^3}{4} - \frac{5x^5}{96} + \frac{7x^7}{4608} - \cdots\right). \end{aligned}$$

Since this Frobenius series only has odd powers of x, let us express this series as $C\sum_n b_n x^{n+1}$, where

$$b_n = \begin{cases} -(-1)^{n/2}\frac{2(n+1)}{2^n(n/2)!(n/2+1)!} & \text{if } n \text{ is even,} \\ 0 & \text{if } n \text{ is odd.} \end{cases}.$$

Adding the inhomogeneous term to equation 4.26, and plugging in $k=-1$, we get

$$\sum_n [(n-1)(n-2)a_n + (n-1)a_n + a_{n-2} - a_n]x^{n-1} = C\sum_n b_n x^{n+1}.$$

We still need to shift the index on the inhomogeneous term down by two to have all of the powers of x to match. So we get the recursion formula

$$n(n-2)a_n + a_{n-2} = Cb_{n-2}.$$

Now when we plug in $n=2$, we get $0a_2 + a_0 = Cb_0 = -2C$, so $C=-1/2$, and a_2 is arbitrary, so for simplicity we set $a_2 = 0$. The new equation for a_n becomes

$$a_n = \frac{-b_{n-2}/2 - a_{n-2}}{n(n-2)}.$$

Plugging in $n = 3, 4, 5, \ldots$, we get

$$\begin{aligned} n=3)\ a_3 &= \frac{-b_1/2 - a_1}{3\cdot 1} = 0 \\ n=4)\ a_4 &= \frac{-b_2/2 - a_2}{4\cdot 2} = \frac{-(-3/4)/2 - 0}{4\cdot 2} = -\frac{3}{64} \\ n=5)\ a_5 &= \frac{-b_3/2 - a_3}{5\cdot 3} = 0 \\ n=6)\ a_6 &= \frac{-b_4/2 - a_4}{6\cdot 4} = \frac{-(-5/96)/2 - (-3/64)}{6\cdot 4} = \frac{7}{2304} \\ n=7)\ a_7 &= \frac{-b_5/2 - a_5}{7\cdot 5} = 0 \\ n=8)\ a_8 &= \frac{-b_6/2 - a_6}{8\cdot 6} = \frac{-(7/4608)/2 - (7/2304)}{8\cdot 6} = \frac{-35}{442368}. \end{aligned}$$

There is no clear pattern for the terms this time. We can express the first several terms of the series $-\ln(x)y_1/2 + \sum_n a_n x^{n-1}$ as

$$y_2 = \frac{1}{x} - \frac{x\ln(x)}{2} + \frac{x^3\ln(x)}{16} - \frac{3x^3}{64} - \frac{x^5\ln(x)}{384} + \frac{7x^5}{2304} + \frac{x^7\ln(x)}{18432} - \frac{35x^7}{442368}.$$

∎

The case of the indicial polynomial having a double root is handled similarly, only in this case we can guarantee that the second series will require logarithms.

Example 4.24
The Bessel equation of order 0 is

$$x^2y'' + xy' + x^2y = 0.$$

Find the series for the solutions of this equation.
SOLUTION: Substituting $y = \sum_n a_n x^{n+k}$ into the equation produces

$$\sum a_n(n+k)(n+k-1)x^{n+k} + \sum a_n(n+k)x^{n+k} + \sum a_n x^{n+k+2} = 0.$$

Shifting the index on the last sum down by 2, we can combine this into a single sum, producing the recursion formula

$$[(n+k)(n+k-1) + (n+k)]a_n + a_{n-2} = 0.$$

When $n = 0$, this produces $k^2a_0 = 0$, so $k = 0$ is a double root of the indicial equation. Substitution this value of k into the recursion formula, and solving for a_n gives us

$$a_n = \frac{-a_{n-2}}{n^2}.$$

Plugging in $n = 1, 2, 3, \ldots$ produces

$$a_1 = 0, \quad a_2 = -\frac{1}{2^2}, \quad a_3 = 0, \quad a_4 = \frac{1}{2^2 4^2}, \quad a_5 = 0, \quad a_6 = -\frac{1}{2^2 4^2 6^2}, \ldots.$$

The odd coefficients are zero, and the even coefficients have a clear pattern,

$$a_{2n} = \frac{(-1)^n}{2^{2n}(n!)^2}.$$

So the first series can be written as

$$y_1 = \sum_{n=0}^{\infty} \frac{(-1)^n x^{2n}}{2^{2n}(n!)^2}.$$

For the second solution, we know that the solution will be of the form $y_2 = f_2(x) + C_1 \ln(x) y_1$, so we substitute $y = u + C \ln(x) y_1$. When simplified, the new equation will be

$$x^2 u'' + xu' + x^2 u = -2Cxy_1'.$$

The inhomogeneous term has the series

$$C \sum_{n=0}^{\infty} \frac{(-4n)(-1)^n x^{2n}}{2^{2n}(n!)^2}$$

which can be described as $C \sum_n b^n x^n$, where

$$b_n = \begin{cases} -(-1)^{n/2} \frac{2n}{2^n (n/2)!^2} & \text{if } n \text{ is even,} \\ 0 & \text{if } n \text{ is odd.} \end{cases}$$

This modifies the recursion formula to

$$n^2 a_n + a_{n-2} = Cb_n.$$

Let us explore what happens when we plug $n = 0, 1, 2, \ldots$ into this recursion formula.

$$\begin{aligned} n = 0)&\ \ 0 + a_{-2} = Cb_0 = 0, \text{which is an identity} \\ n = 1)&\ \ a_1 + a_{-1} = Cb_1 = 0, \ \text{so } a_1 = 0 \\ n = 2)&\ \ 4a_2 + a_0 = Cb_2 = C \end{aligned}$$

This last equation needs to determine a_2, but we have yet to determine C. In the case of a double root of the indicial polynomial, the two solutions will interfere with each other on the very first term. That is, a constant times the first solution can be added to the second solution, so a_0 is arbitrary. In fact, we can set $a_0 = 0$, so the second series will begin with $a_2 x^2$. By multiplying the solution by a constant, we can assume that $a_2 = 1$, so $C = 4$. We can now solve the recursion formula for a_n,

$$a_n = \frac{4b_n - a_{n-2}}{n^2}. \tag{4.27}$$

We can now continue to find the terms of the series

$$\begin{aligned} n = 3)&\ \ a_3 = (4b_3 - a_1)/9 - 0 \\ n = 4)&\ \ a_4 = (4b_4 - a_2)/16 = (4(-1/8) - 1)/16 = -3/32 \\ n = 5)&\ \ a_5 = (4b_5 - a_3)/25 = 0 \\ n = 6)&\ \ a_6 = (4b_6 - a - 4)/36 = (4(1/192) + 3/32)/36 = 11/3456. \end{aligned}$$

The pattern is not apparent, but in section 6.2 we will see a way to solve the recursion formula in equation 4.27 exactly. For now, we will list the first several terms of $y_2 = 4\ln(x) y_1 + \sum_n a_n x^n$.

$$y_2 = 4 \ln x - x^2 \ln x + x^2 + \frac{x^4 \ln x}{16} - \frac{3x^4}{32} - \frac{x^6 \ln x}{576} + \frac{11x^6}{3456} + \cdots. \qquad \square$$

Although we were able to find series solutions for ordinary points and regular singular points, the situation becomes more complicated with irregular singular points. For one thing, the series could involve a wide variety of different types of terms. Hence, there is no statement like theorem 4.7 which will predict the form of the series. Furthermore, the series will often diverge for all x, creating a series that must be "summed" using the techniques of chapter 3. We will explore the nuances of the final type of singularity in chapter 5.

Problems for §4.4

For problems **1** through **12**: Find the two Frobenius series solutions centered at $x_0 = 0$ to the differential equation, up to the $a_4{}^{\text{th}}$ term. In these equations, the two solutions will not interfere with each other.

1 $2xy'' + y' + 2y = 0$
2 $2xy'' + 3y' - 2y = 0$
3 $2x^2y'' + 3xy' + (x-1)y = 0$
4 $2x^2y'' + (2x^2 + x)y' - y = 0$
5 $9x^2y'' + (x^2 + 2)y = 0$
6 $9x^2y'' + (x^2 - 4)y = 0$
7 $2x^2y'' + xy' + (x^2 - 3)y = 0$
8 $2x^2y'' + (x^2 + x)y' - (2x+1)y = 0$
9 $3x^2y'' + (x^2 + x)y' - (3x+1)y = 0$
10 $3x^2y'' + (x^2 + 2x)y' + (x-2)y = 0$
11 $x^2y'' + xy' + (x^2 + 1)y = 0$
12 $x^2y'' + xy' + (2x^2 + 4)y = 0$

For problems **13** through **24**: Find the two Frobenius series solutions centered at $x_0 = 0$ to the differential equation, up to the $a_3{}^{\text{rd}}$ term. Some of these equations will have solutions involving logarithms.

13 $xy'' - y = 0$
14 $xy'' + 2y = 0$
15 $xy'' + xy' - 2y = 0$
16 $xy'' - xy' + 3y = 0$
17 $xy'' - 2y' + xy = 0$
18 $xy'' - y' + xy = 0$
19 $xy'' - y' + (1-x)y = 0$
20 $x^2y'' + (x-2)y = 0$
21 $x^2y'' + xy' + (x-1)y = 0$
22 $4x^2y'' + 4xy' + (x-1)y = 0$
23 $4x^2y'' + (x-3)y = 0$
24 $4x^2y'' - 2x^2y' + (x-3)y = 0$

Chapter 5

Asymptotic Series Solutions for Differential Equations

So far we have learned two ways of finding a series solution to a differential equation. If x_0 is an ordinary point, we can find a Taylor series solution to the differential equation that will converge with a positive radius of convergence. If we have a regular singular point at x_0, we can still use the Frobenius method to find the solutions expressed as series with a slightly different term. There is one case that we have yet to explore, and that is the behavior of a solution to a differential equation at an irregular singular point. If we consider the point at ∞, this is in fact the most common case that occurs.

Although a variety of behaviors to the solutions could occur at an irregular singular point, the most common behavior can only be expressed as an asymptotic series. The series will not converge, but we have seen in chapter 3 several ways to give a meaning to an asymptotic series that diverges. In this chapter we will find ways of determining the asymptotic series for several different types of differential equations. In all cases, our main tool will be the method of dominant balance, which was covered in section 1.5.

5.1 Behavior for Irregular Singular Points

The first step is to determine the typical behavior of a solution to a differential equation near an irregular singular point. Note that this involves replacing an exact differential equation with an asymptotic equation, and determining which functions solve this asymptotic relationship. However, there may be many behaviors that solve the asymptotic relation, even though these are not the behaviors of any exact solutions.

For example, consider the equation $y' = y$, for which there is an irregular singular point at ∞. If we convert the equation to an asymptotic relation, $y' \sim y$, then of course the solution to the exact equation, $y = Ce^x$, will satisfy the asymptotic relation. But xe^x and e^x/x^2 also satisfy the asymptotic relation, even though these obviously do not satisfy the original differential equation. Thus, we see that it is not enough to find a behavior which is consistent with the equation, but we must also find the next term in the asymptotic series before we have verified the behavior of the exact solution.

But how do we begin to find a possible asymptotic behavior for the solution to the differential equation? Examining a few examples where we know the solution will give us a clue. We have already seen that $y' = y$ has an exponential solution, and example 4.3 showed that $\sin(x)$ and $\cos(x)$ were solutions to $y'' + y = 0$, which also has an irregular singular point at ∞. Although the trigonometric functions seem like a radically different behavior then the exponential solution, recall that the trig functions can in fact be expressed as exponentials.

$$\cos x = \frac{e^{ix} + e^{-ix}}{2}, \qquad \sin x = \frac{e^{ix} - e^{-ix}}{2i}.$$

So in both cases, the solutions can be expressed in the form $e^{S(x)}$ for some $S(x)$.

This motivates making the following substitution for a linear homogeneous differential equation.

$$\boxed{y = e^{S(x)}, \qquad y' = S'(x)e^{S(x)}, \qquad y'' = (S''(x) + [S'(x)]^2)e^{S(x)}, \text{ etc.}} \tag{5.1}$$

This substitution actually will reduce the order of the equation, since the $e^{S(x)}$'s will cancel out, and the new equation will only involve $S'(x)$ and higher derivatives. However, this will usually produce a non-linear equation, which is more difficult to solve than a linear equation. But since our goal is to find the asymptotic behavior, instead of the exact solution, the non-linear equations have an advantage over linear equations. It will be easier to use the method of dominant balance, since we will have a clue as to which terms will be smaller in comparison to other terms. If $S(x)$ is growing algebraically, then we can usually assume that $S''(x) \ll [S'(x)]^2$ as $x \to \infty$. Hence, we can replace the exact differential equation with an asymptotic equation.

Bear in mind that it will not be sufficient finding just the leading behavior of the function $S(x)$, since this function must be exponentiated to determine the leading behavior of $y(x)$. For example, the first three terms of the asymptotic series for $S(x)$ might be $x + \ln(x) + \ln(3)$, but the exponential will be

$$y(x) = e^{S(x)} = e^x e^{\ln x} e^{\ln 3} = 3xe^x.$$

Thus, many terms of $S(x)$ may contribute to just the leading behavior of $e^{S(x)}$. In fact, any term of $S(x)$ which is either asymptotic growing or is constant will be involved in determining the leading behavior of $y(x)$.

Example 5.1

Determine the possible leading behaviors for the solutions to the Airy equation $y'' = xy$ as $x \to \infty$.

SOLUTION: We cannot apply the method of dominant balance directly to this equation, since there are only two terms to the equation, so neither can be considered small. So we will apply the substitution $y = e^S$, $y'' = (S'' + (S')^2)e^S$ to get

$$S''(x) + [S'(x)]^2 = x.$$

Since we now have three terms, we can utilize the method of dominant balance, and in fact we should begin by assuming that $S'' \ll (S')^2$ as $x \to \infty$. The resulting asymptotic relation is

$$(S'(x))^2 \sim x,$$

which is trivial to solve, $S'(x) \sim \pm\sqrt{x}$, so $S(x) \sim \pm 2x^{3/2}/3$ as $x \to \infty$.

Next, we have to test if this is consistent with our assumptions. If $S'(x) \sim \pm\sqrt{x}$, then $S''(x) \sim x^{-1/2}/2$ as $x \to \infty$, which is indeed smaller than $(S')^2$. So we have shown that our assumption is consistent, but to verify the first term, we must compute the next correction for $S(x)$.

If we make the substitution $S(x) = \pm 2x^{3/2}/3 + g(x)$, where $g(x) \ll x^{3/2}$ as $x \to \infty$, then substituting into the equation $S''(x) + [S'(x)]^2 = x$ gives the following equation for $g(x)$:

$$\pm x^{-1/2} + g'' \pm 2\sqrt{x}g' + (g')^2 = 0. \tag{5.2}$$

Since $g' \ll \sqrt{x}$ and $g'' \ll x^{-1/2}$ as $x \to \infty$, we can assume that the second term is smaller than the first, and that the fourth term is smaller than the third. This leaves us with $\pm x^{-1/2} \sim \mp 2\sqrt{x}g'$, which means in either case $g'(x) \sim -1/(4x)$. Then $g(x) \sim -\ln(x)/4 + C$, which is smaller than $x^{3/2}$ as $x \to \infty$, so we have the first two terms for $S(x) \sim \pm 2x^{3/2}/3 - \ln(x)/4$.

However, knowing that $S(x) = \pm 2x^{3/2}/3 - \ln(x)/4 + h(x)$ where $h(x) \ll \ln(x)$ as $x \to \infty$ is not quite enough information to determine the leading term for $y(x)$. Since $y(x) = e^{S(x)}$, we have

$$y(x) = e^{\pm 2x^{3/2}/3 - \ln(x)/4 + h(x)} = e^{\pm 2x^{3/2}/3} e^{-\ln x/4} e^{h(x)} = x^{-1/4} e^{\pm 2x^{3/2}/3} e^{h(x)}.$$

Notice that both of the known terms for $S(x)$ will contribute to the first term of $y(x)$. In fact, if $h(x)$ is growing as $x \to \infty$, then this will also contribute to the first term of $y(x)$. However, if $h(x)$ is decreasing to 0 or approaching a constant as $x \to \infty$, then $e^h(x)$ will approach a constant, and we will have the first term of $y(x)$. Thus, we have the following principle.

> In order to obtain the leading term of $y(x) = e^{S(x)}$, we need to find the series for $S(x)$ until we reach the first term that is not increasing.

(5.3)

This means we still have to find the next term of the series for $S(x)$. Substituting $g(x) = -\ln(x)/4 + h(x)$ into equation 5.2, we get

$$\frac{1}{4x^2} + h'' \pm 2\sqrt{x}h' + \frac{1}{16x^2} - \frac{h'}{2x} + (h')^2 = 0.$$

Clearly, $h'/(2x) \ll \sqrt{x}h'$, and since $h' \ll 1/x$, $(h')^2 \ll h'/x$ as $x \to \infty$. So we have the asymptotic equation

$$\frac{5}{16x^2} \sim \mp 2\sqrt{x}h'.$$

This produces $h'(x) \sim \mp 5x^{-5/2}/32$, or $h(x) \sim \pm 5x^{-3/2}/48 + C$. This time, the arbitrary constant is kept, since it is larger than the other term as $x \to \infty$. Since we have shown that $h(x)$ is approaching a constant, we have the first term of $y(x)$ to be

$$y \sim Kx^{-1/4}e^{\pm 2x^{3/2}/3}.$$

In the process of finding the first term, we have in fact found the first two terms for two different solutions. We can combine this information into the form

$$y \sim c_1 \frac{e^{2x^{3/2}/3}}{x^{1/4}}\left(1 + \frac{5}{48x^{3/2}} + \cdots\right) + c_2 \frac{e^{-2x^{3/2}/3}}{x^{1/4}}\left(1 - \frac{5}{48x^{3/2}} + \cdots\right). \quad (5.4)$$

□

The *Airy function of the first kind*, $\mathrm{Ai}(x)$, is defined to be the solution of the Airy equation for which $c_1 = 0$ and $c_2 = 1/(2\sqrt{\pi})$. The Airy function of the second kind, denoted by $\mathrm{Bi}(x)$, has $c_1 = 1/\sqrt{\pi}$. However, the series involving c_2 is subdominant to the series involving c_1, so this information alone does not define $\mathrm{Bi}(x)$. However, in example 4.18, we found the Maclaurin series for all solutions to the Airy equation. Thus, we can also define $\mathrm{Ai}(x)$ and $\mathrm{Bi}(x)$ in terms of their Maclaurin series.

$$\mathrm{Ai}(x) = \sum_{n=0}^{\infty} \frac{3^{-2/3}x^{3n}}{9^n\, n!\Gamma(n + (2/3))} - \sum_{n=0}^{\infty} \frac{3^{-4/3}x^{3n+1}}{9^n\, n!\Gamma(n + (4/3))}. \quad (5.5)$$

$$\mathrm{Bi}(x) = \sum_{n=0}^{\infty} \frac{3^{-1/6}x^{3n}}{9^n\, n!\Gamma(n + (2/3))} + \sum_{n=0}^{\infty} \frac{3^{-5/6}x^{3n+1}}{9^n\, n!\Gamma(n + (4/3))}. \quad (5.6)$$

Figure 5.1 compares the graphs of $\mathrm{Ai}(x)$ and $\mathrm{Bi}(x)$ to the two term asymptotic approximation given in equation 5.4.

Example 5.2

The *parabolic cylinder equation* is given by

$$y'' + (v + 1/2 - x^2/4)y = 0, \quad (5.7)$$

where v is a constant. Find the leading term of the asymptotic series as $x \to \infty$.

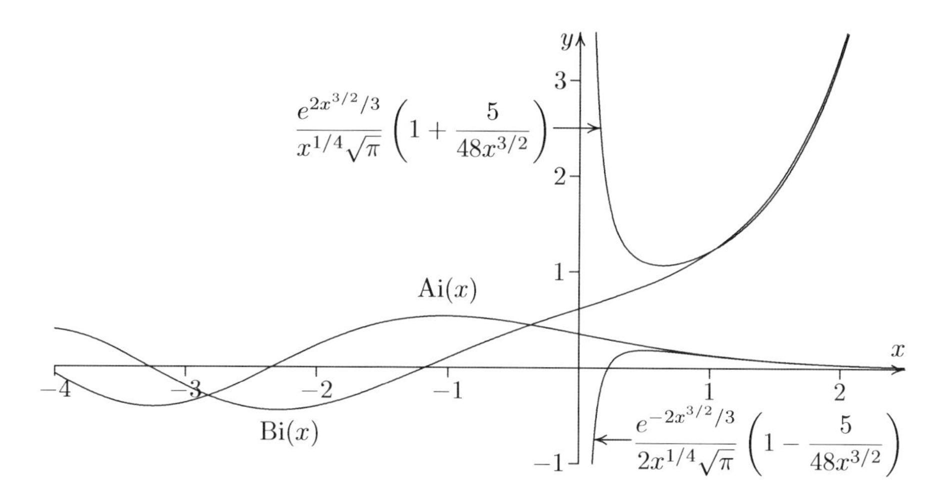

FIGURE 5.1: This shows the Airy functions $\mathrm{Ai}(x)$ and $\mathrm{Bi}(x)$, along with the first two terms of their asymptotic series as $x \to \infty$. Notice that just two terms give an impressive amount of accuracy for $x > 1$.

SOLUTION: We again begin with the substitution $y = e^S$, $y'' = (S'' + (S')^2)e^S$, which gives us the non-linear equation

$$S'' + (S')^2 + v + 1/2 - x^2/4 = 0.$$

Clearly, $v + 1/2 \ll x^2$ as $x \to \infty$, and we will first assume that $S'' \ll (S')^2$. This assumption gives us the asymptotic equation

$$(S')^2 \sim x^2/4,$$

which produces $S' \sim \pm x/2$, or $S(x) \sim \pm x^2/4$. Then $S'' \sim 1/2$, which is indeed smaller than $(S')^2$ as $x \to \infty$, so we have consistency.

However, we need the first few terms of $S(x)$, until we reach a term that is going to 0 as $x \to \infty$, before we have the leading behavior of $y(x)$. We substitute $S(x) = \pm x^2/4 + g(x)$, which creates the following equation for $g(x)$:

$$g'' \pm xg' + (g')^2 + v + 1/2 \pm 1/2 = 0. \tag{5.8}$$

We can assume that $g' \ll x/2$ and $g'' \ll 1/2$. If the constant terms happen to cancel, that is, $v + 1/2 \pm 1/2 = 0$, then $g = 0$ is an exact solution. In fact, $e^{-x^2/4}$ solves the parabolic cylinder equation for $v = 0$, and $e^{x^2/4}$ solves the equation for $v = -1$. For all other cases, we can assume that $v+1/2\pm 1/2 \neq 0$, so g'' will be small compared to a constant. Also, $(g')^2 \ll xg'$, so we get the asymptotic equation for $g(x)$ to be

$$\pm xg' = -v - 1/2 \mp 1/2.$$

This is quickly solved to get $g(x) \sim (-1/2 \mp v \mp 1/2)\ln x$, which is indeed smaller than $x^2/4$ as $x \to \infty$. But it is not going to 0, so we need yet another term.

Substituting $g = (-1/2 \mp v \mp 1/2)\ln x + h(x)$ into equation 5.8 gives us the following equation for $h(x)$:

$$(1/2 \pm v \pm 1/2)/x^2 + h'' \pm xh' + (1/2 \pm v \pm 1/2)^2/x^2 - (1 \pm 2v \pm 1)h'/x + (h')^2 = 0.$$

Clearly, $h'' \ll 1/x^2$, and $(h')^2 \ll h'/x \ll xh'$. So we have the asymptotic relation

$$(1 + v + v^2 \pm 1 \pm 2v)/x^2 \sim \mp xh',$$

which is solved by $h \sim (1 + 2v \pm 1 \pm v \pm v^2)/(2x^2) + C$. Because $h(x)$ is approaching a constant, this will only affect the second term of the asymptotic relation for $y(x)$. Thus, the possible leading terms for the solution to the parabolic cylinder equation are

$$y(x) \sim c_1 e^{-x^2/4} x^v + c_2 e^{x^2/4} x^{-1-v}.$$

The solution which is asymptotic to $x^v e^{-x^2/4}$ is called the parabolic cylinder function, denoted by $D_v(x)$. It turns out that $D_{-1-v}(ix)$ will also solve the parabolic cylinder function of order v. (See problem 12.) Hence, the general solution of the parabolic cylinder equation can be expressed by

$$y(x) = c_1 D_v(x) + c_2 D_{-1-v}(ix). \qquad \square$$

Although we usually begin the asymptotic analysis of a linear equation by making the substitution $y = e^S$, there are times that we can apply the method of dominant balance to the original equation.

Example 5.3

Find the possible leading behaviors as $x \to \infty$ to the equation

$$x^3 y'' + x^2 y' + \ln(x) y = 0.$$

SOLUTION: The point at ∞ is an irregular singular point because of the logarithm. In fact, if $\ln(x)$ is replaced with a constant, then ∞ would be a regular singular point. (See problem 29 from section 4.1.) This indicates that the solution is probably not an exponential. We can apply the method of dominant balance directly to the differential equation.

The disadvantage of beginning with the method of dominant balance is that we have no clue as to which term is small compared to the others, so we have to use trial and error. Also, if we assume that $x^2 y' \ll x^3 y''$, we get the equation $x^3 y'' + \ln(x) y = 0$, which we have no way of solving. But the other two cases can be solved fairly easily.

Case 1) $\ln(x) y \ll x^2 y'$.

This would imply that $x^3y'' \sim -x^2y'$, or that $y'' \sim -y'/x$. This is a separable equation in y', easily solved to give us $y' \sim k/x$, so $y \sim k\ln(x) + C$. Since $C \ll \ln(x)$, we have that $y \sim k\ln(x)$. This is consistent with our assumption that $\ln(x)y \ll x^2y'$.

We must test this by finding the next term in the asymptotic expansion of y. If $y = \ln(x) + g(x)$, then g will satisfy the inhomogeneous equation

$$x^2g'' + x^2g' + g\ln x = -(\ln x)^2.$$

Since $g \ll \ln(x)$, we know that $g\ln x \ll (\ln x)^2$. Rather than determining which of the remaining terms is small, we can solve the remaining equation, which is a linear equation in g'. The integrating factor is simply x, so

$$g'(x) = \frac{(\ln x)^2 + 2\ln x + 2}{x^2} + \frac{C}{x}.$$

Because $g(x) \ll \ln(x)$, we know that $g'(x) \ll 1/x$ as $x \to \infty$, so C must be 0. Then we can integrate again to get

$$g(x) = -\frac{(\ln x)^2 + 4\ln x + 6}{x} + C_2.$$

The main point is that we have found that $g(x)$ is indeed smaller than $\ln(x)$ as $x \to \infty$, so we have one solution of the form $y \sim \ln(x)$.

Case 2) $x^3y'' \ll x^2y'$.

This would imply that $x^2y' \sim -\ln(x)y$, which is a separable equation. The solution to this equation is

$$y = Ke^{\ln x/x + 1/x} \sim K\left(1 + \frac{\ln x}{x} + \frac{1}{x}\right).$$

Is this consistent with our assumption that x^2y'' is small as $x \to \infty$? If $y = 1 + \ln x/x + 1/x$, then $x^3y'' \sim 2\ln x - 1$, which is the same size as the terms that we kept. This suggests that we have the right size terms, but the wrong coefficients. If we substitute $y = 1 + a\ln x/x + b/x$ into the original equation, we get

$$(2a\ln x + 2b - 3a) + (a - a\ln x - b) + \ln x + O(\ln(x)/x).$$

Solving for a and b, we get $a = -1$ and $b = -2$, so we have a second consistent behavior, $y \sim 1 - \ln x/x - 2/x$. Thus, we have our two solutions,

$$y \sim c_1\left(\ln(x) - \frac{(\ln x)^2 + 4\ln x + 6}{x} + \cdots\right) + c_2\left(1 - \frac{\ln x}{x} - \frac{2}{x} + \cdots\right).$$

The actual solutions, along with the first two terms of the asymptotic series as $x \to \infty$, are shown in figure 5.2. □

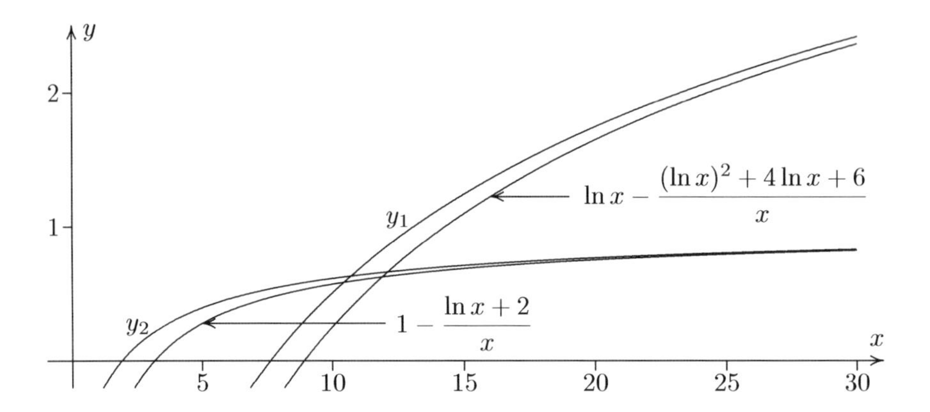

FIGURE 5.2: This shows two solutions for the equation $x^3y'' + x^2y' + \ln(x)y = 0$. Also shown are the second order asymptotic approximations to these solutions as $x \to \infty$.

If we have zero as the irregular singular point, then we proceed the same way, only doing all of our asymptotics as $x \to 0$.

Example 5.4

Find the possible leading behaviors as $x \to 0$ to the equation

$$x^3y'' - xy' + 3y = 0.$$

SOLUTION: We make the substitution $y = e^{S(x)}$ to obtain

$$x^3(S'' + (S')^2) - xS' + 3 = 0.$$

We can assume that $S'' \ll (S')^2$, but we will still have 3 terms left in the equation. Rather than trying dominant balance on the remaining terms, note that it is a quadratic equation in S', so we can use the quadratic formula to get

$$S' \sim \frac{x \pm \sqrt{x^2 - 12x^3}}{2x^3} \qquad \text{as } x \to 0. \tag{5.9}$$

If we choose the positive in the $\pm$ sign, then we can use the fact that

$$\sqrt{x^2 - 12x^3} \sim x \qquad \text{as } x \to 0,$$

and so $S' \sim 1/x^2$, hence $S(x) \sim -1/x$. Substituting $S(x) = -1/x + g(x)$ gives us the equation

$$x^3g'' + x^3(g')^2 + xg' + 1 = 0.$$

Since $g' \ll x^{-2}$ and $g'' \ll x^{-3}$ as $x \to 0$, only the last two terms are not small. Thus, $g' \sim -1/x$ as $x \to 0$, so $g \sim -\ln x$.

If we substitute $g = -\ln(x) + h(x)$, the new equation in $h(x)$ is

$$x^3 h'' + x^3(h')^2 - 2x^2 h' + xh' + 2x = 0.$$

Since $h' \ll x^{-1}$ and $h'' \ll x^{-2}$ as $x \to 0$, we quickly see that $h' \sim -2$, so $h \sim C - 2x$ as $x \to 0$. Since this is approaching a constant, we have enough information for the leading behavior, $y_1 \sim x^{-1}e^{-1/x}$.

But we still have to find the leading behavior of the second solution. If we choose the negative in the $\pm$ sign in equation 5.9, we must replace the square root with a few terms of its Taylor series.

$$S' \sim \frac{x - x(1 - 6x - 18x^2 + \cdots)}{2x^3} \sim \frac{6x^2 + 18x^3 + \cdots}{2x^3} \sim \frac{3}{x} \qquad \text{as } x \to 0.$$

Thus, $S(x) \sim 3\ln(x)$, and if we let $S(x) = 3\ln(x) + g(x)$, we get the equation

$$x^3 g'' + x^3(g')^2 + 6x^2 g' - xg' + 6x = 0.$$

Since $g' \ll x^{-1}$ and $g'' \ll x^{-2}$ as $x \to 0$, only the last two terms are not small. Hence, $g' \sim 6$, so $g \sim C + 6x$, which is approaching a constant. So the second behavior is $y_2 \sim x^3$.

Finally, we can put the two behaviors together,

$$y \sim c_1 x^{-1} e^{-1/x} + c_2 x^3 \qquad \text{as } x \to 0.$$

□

We now have the general procedure for finding the possible leading terms for a homogeneous linear equation with an irregular singular point. The usual strategy is to substitute $y = e^{S(x)}$, then assume that $S''(x) \ll (S'(x))^2$ to eliminate all terms except those that involve $S'(x)$. This becomes an algebraic equation for $S'(x)$, which can easily be solved to determine the possible behaviors of $S(x)$. However, we must peel off the possible leading behaviors to find more terms in the asymptotic series for $S(x)$, namely until we reach a term that is decaying. Finally, we exponentiate the series for $S(x)$ to find the leading term for $y(x)$.

Problems for §5.1

For problems **1** through **10**: Find the possible leading behaviors as $x \to \infty$ for the following equations which have an irregular singular point at ∞.

1 $y'' - y' - x^2 y = 0$

2 $y'' - 25\sqrt{x}y = 0$

3 $xy'' + xy' - y = 0$

4 $y'' + xy' - 4y = 0$

5 $y'' - 2xy' + (x^2 - x)y = 0$

6 $y'' + y' - xy = 0$

7 $y'' - (\ln(x))^2 y = 0$

8 $y''' - xy = 0$

9 $y''' + x^2 y = 0$

10 $y^{(4)} - x^2 y = 0$

11 Show that the possible leading behaviors of the solutions to $x^3y''-(2x^3+x^2)y'+(x^3+x^2+x+1)y=0$ are K_1xe^x and $K_2x\ln xe^x$.

Hint: Show that substitution $y=e^xu$ yields an equation that has a regular singular point at ∞.

12 Show that if $D_{-1-v}(x)$ is a solution to the parabolic cylinder equation with parameter $-1-v$, then $D_{-1-v}(ix)$ solves the parabolic cylinder equation with parameter v.

13 Show that an equation of the form

$$y''-(ax+b)y=0,$$

where $a\neq 0$, can be solved in terms of Airy functions.

Hint: Determine what linear substitution converts this to the Airy equation.

14 The generalized parabolic cylinder equation is given by

$$y''-(ax^2+bx+c)y=0.$$

Show that when $a\neq 0$, the solutions can be expressed in terms of parabolic cylinder functions $D_v(x)$.

Hint: First assume that $a>0$, and complete the square. Determine a linear substitution that converts this to the form of the parabolic cylinder equation. The solution produced will still be valid for $a<0$, but it will involve a complex value of v.

For problems **15** through **24**: Find the possible leading behaviors as $x\to 0$ for the following equations which have an irregular singular point at 0.

15 $x^3y''-y=0$
16 $x^4y''-y=0$
17 $x^4y''+x^2y'-2y=0$
18 $x^4y''-3x^2y'+2y=0$
19 $x^4y''-4x^2y'+4y=0$
20 $x^3y''-xy'+y=0$
21 $x^3y''+xy'+2y=0$
22 $x^3y''-y'-y=0$
23 $x^4y''-e^xy=0$
24 $x^3y''-e^{\sqrt{x}}y=0$

For problems **25** through **28**: Use the method of dominant balance directly on the following equations to find the possible leading behaviors as $x\to\infty$ of the solutions.

25 $x^3y''+3x^2y'+(\ln x)y=0$
26 $x^3y''+2x^2y'-(\ln x)y=0$
27 $xy''+4y'+e^{-x}y=0$
28 $xy''-2y'+e^{-x}y=0$

29 The functional relation for the parabolic cylinder function is

$$D_v(-x)=e^{-iv\pi}D_v(z)-\frac{i\sqrt{2\pi}}{\Gamma(-v)}e^{iv\pi/2}D_{-v-1}(-ix).$$

Use this to show that

$$\int_{-\infty}^{\infty}D_i(x)D_j(x)\,dx=0$$

whenever i and j are two different non-negative integers.

Hint: Consider the Sturm-Liouville eigenvalue problem

$$y'' + (E - x^2/4)y = 0, \qquad \lim_{x\to\infty} y(x) = 0, \qquad \lim_{x\to-\infty} y(x) = 0.$$

Show that the only solutions are when E is a positive half-integer.

5.2 Full Asymptotic Expansion

Although we have found a way of finding the first few terms of the asymptotic series, what we need is a technique for finding *all* of the terms of the series, similar to the methods of sections 4.3 and 4.4. To do this, we need to peel off the leading behavior from the solutions from the equation. That is, if the first term of the asymptotic series is $\tilde{y}_1$, we substitute $y(x) = \tilde{y}_1 u(x)$, and get a linear equation for $u(x)$. There is actually an easy way to do this that takes advantage of the work already done finding the leading behavior.

Our strategy for finding the leading behavior was to substitute $y = e^S$, and then find the first few terms for the asymptotic series for $S(x)$. If we let $f(x)$ denote the terms of the asymptotic series of $S(x)$ which do not approach 0 as $x \to \infty$, then we can write $S(x) = f(x) + h(x)$, where $h(x) \to 0$ as $x \to \infty$. Then $y = e^S = e^{f(x)}e^{h(x)} \sim e^{f(x)}$, since $e^{h(x)} \to 1$. So the first term of the asymptotic series for $y(x)$ is $e^{f(x)}$, and we can substitute $y = e^{f(x)}u(x)$ to peel off the leading behavior. Note that these equations imply that $u(x) = e^{h(x)}$.

In the process of finding the leading behavior, we have found an equation for which $h(x)$ satisfies, and in fact we used this equation to show that $h(x) \to 0$ as $x \to \infty$. Using this equation to find further terms is inefficient, since this is usually a non-linear equation, and the series for $h(x)$ would have to be exponentiated to produce the series for y. But if we make the substitution $h(x) = \ln(u(x))$, then the non-linear equation will become a linear equation for u, since $u(x) = e^{h(x)}$. The substitutions required are as follows.

$$\boxed{\text{If } h = \ln u, \quad h' = \frac{u'}{u}, \quad h'' = \frac{uu'' - (u')^2}{u^2}, \quad h'' + (h')^2 = \frac{u''}{u}.} \tag{5.10}$$

This substitution will *relinearize* the equation, resulting in a linear equation in u, for which some solutions will approach a constant as $x \to \infty$.

Example 5.5

Find the full asymptotic series as $x \to \infty$ for the solution to the Airy equation, $y'' = xy$.

SOLUTION: We have already found the leading behavior of the solutions in example 5.1. We let $y = e^S$, and found that $S(x) = \pm 2x^{3/2}/3 - \ln(x)/4 + h(x)$, where $h(x)$ satisfies the equation

$$h'' \pm 2\sqrt{x}h' + \frac{5}{16x^2} - \frac{h'}{2x} + (h')^2 = 0.$$

Thus, the leading term is $y \sim x^{-1/4}e^{\pm 2x^{3/2}/3}$, and if we let $u = \ln(h(x))$, then $y = x^{-1/4}e^{\pm 2x^{3/2}/3}u(x)$. Using the substitution given in equation 5.10, we get the equation

$$\frac{u''}{u} \pm 2\sqrt{x}\frac{u'}{u} - \frac{1}{2x}\frac{u'}{u} + \frac{5}{16x^2} = 0,$$

which after multiplying through by x^2u gives us the linear equation

$$x^2u'' - \frac{1}{2}xu' \pm 2x^{5/2}u' + \frac{5}{16}u = 0. \tag{5.11}$$

We also have a clue as to the nature of the series for u by observing leading behavior we found for $h(x)$ in example 5.1. We had $h(x) \sim C \pm 5x^{-3/2}/48$. This means that $e^{h(x)}$ would have a series like

$$e^{h(x)} \sim K\left(1 \pm \frac{5x^{-3/2}}{48} + a_2x^{-3} + a_3x^{-9/2} + \cdots\right)$$

where $a_2, a_3, \ldots$ would be determined by exponentiating the series. So it is natural to assume that the asymptotic series for $u(x)$ as $x \to \infty$ is of the form

$$u(x) \sim \sum_{n=0}^{\infty} a_n x^{kn}, \quad \text{with } k = -3/2. \tag{5.12}$$

Note that this is not quite in the form of a Frobenius series, since there are fractional powers of x embedded into the series. This differs from Frobenius series in another respect, since the asymptotic series about an irregular singular point will rarely converge.

Let us plug the series in equation 5.12 into equation 5.11. First of all, we have to take two derivatives of u, done term-wise.

$$u'(x) \sim \sum_n a_n(-3n/2)x^{-3n/2-1}, \qquad u''(x) \sim \sum_n a_n(9n^2 + 6n)x^{-3n/2-2}/4,$$

so we have

$$\sum_n a_n \frac{9n^2 + 6n}{4}x^{-3n/2} - \frac{1}{2}\sum_n a_n \frac{-3n}{2}x^{-3n/2} \pm 2\sum_n a_n(-3n/2)x^{-3n/2+3/2}$$
$$\sim -\frac{5}{16}\sum_n a_n x^{-3n/2}.$$

Note that we put a term on the other side of the equation, since technically we cannot have an asymptotic relation with 0 on one side. Since $x^{-3x/2+3/2} =$

$x^{-3(x-1)/2}$, we must shift this term up by one so that the sum involves $x^{-3n/2}$ like the other terms. In spite of the fractional powers in the sums, we can only shift the index by an integral amount. The new equation becomes

$$\sum_n a_n \frac{9n^2+6n}{4} x^{-3n/2} + \sum_n a_n \frac{3n}{4} x^{-3n/2} \pm 2 \sum_n a_{n+1}(-3(n+1)/2) x^{-3n/2}$$
$$\sim -\frac{5}{16} \sum_n a_n x^{-3n/2}.$$

Since all of the series now involve $x^{-3n/2}$, we can set the coefficients equal to obtain the recursion relationship.

$$a_n \frac{9n^2+6n}{4} + a_n \frac{3n}{4} \mp 3(n+1)a_{n+1} = -\frac{5}{16} a_n.$$

Since the highest index of a is a_{n+1}, we solve for a_{n+1} to obtain

$$a_{n+1} = \pm \frac{36n^2 + 36n + 5}{48(n+1)} a_n = \pm \frac{3}{4} \frac{(n+\frac{1}{6})(n+\frac{5}{6})}{n+1} a_n.$$

If we assume that $a_0 = 1$, then

$$a_1 = \pm \frac{3}{4} \frac{\frac{1}{6}\frac{5}{6}}{1}, \quad a_2 = \frac{3^2}{4^2} \frac{\frac{1}{6}\frac{5}{6}\frac{7}{6}\frac{11}{6}}{1 \cdot 2}, \quad a_3 = \pm \frac{3^3}{4^3} \frac{\frac{1}{6}\frac{5}{6}\frac{7}{6}\frac{11}{6}\frac{13}{6}\frac{17}{6}}{1 \cdot 2 \cdot 3}, \ldots.$$

To get a closed form expression for a_n, we can use the properties of the Gamma function. If we multiply the top and bottom of the fractions by $\Gamma(1/6)\Gamma(5/6)$, we get

$$a_3 = \pm \frac{3^3}{4^3} \frac{\Gamma(\frac{1}{6})\Gamma(\frac{5}{6})\frac{1}{6}\frac{5}{6}\frac{7}{6}\frac{11}{6}\frac{13}{6}\frac{17}{6}}{\Gamma(\frac{1}{6})\Gamma(\frac{5}{6}) 1 \cdot 2 \cdot 3} = \pm \frac{3^3}{4^3} \frac{\Gamma(\frac{19}{6})\Gamma(\frac{23}{6})}{\Gamma(\frac{1}{6})\Gamma(\frac{5}{6}) 3!}.$$

Here, we used the "zipper" property that $n\Gamma(n) = \Gamma(n+1)$. We now have a pattern for a_n:

$$a_n = (\pm 1)^n \frac{3^n}{4^n} \frac{\Gamma(n+\frac{1}{6})\Gamma(n+\frac{5}{6})}{\Gamma(\frac{1}{6})\Gamma(\frac{5}{6}) n!}.$$

We can further simplify this using the reflective property of the Gamma function, since $\Gamma(1/6)\Gamma(5/6) = \pi/\sin(\pi/6) = 2\pi$. This gives us the final result:

$$y \sim c_1 x^{-1/4} e^{2x^{3/2}/3} \sum_{n=0}^{\infty} \frac{3^n \Gamma(n+\frac{1}{6})\Gamma(n+\frac{5}{6}) x^{-3n/2}}{4^n 2\pi n!} \tag{5.13}$$
$$+ c_2 x^{-1/4} e^{-2x^{3/2}/3} \sum_{n=0}^{\infty} \frac{(-3)^n \Gamma(n+\frac{1}{6})\Gamma(n+\frac{5}{6}) x^{-3n/2}}{4^n 2\pi n!}.$$

□

The asymptotic series in example 5.5 will diverge for all finite x, but this has never stopped us in the past. For large x, the optimal asymptotic approximation yields amazing accuracy, as we can see in table 5.1. Also, we can

TABLE 5.1: This compares the optimal asymptotic approximation of the Airy function series given in equation 5.13 with $c_1 = 0$ and $c_2 = 1/(2\sqrt{\pi})$ to the exact Airy function $\mathrm{Ai}(x)$. Also in this comparison are the continued fraction approximants F_n for this series. Note that when x is large, the convergence is impressive.

x	1/2	1	2	4
Optimal Approximation	0.231694	0.129746	0.034840	0.0009515652
Relative Error	14.39%	4.10%	0.241%	0.000143%
F_1	0.204712	0.131169	0.034715	0.0009506583
F_2	0.241137	0.13646	0.0349608	0.0009516398
F_3	0.223456	0.13466	0.0349131	0.0009515531
F_4	0.235436	0.135538	0.034927	0.0009515653
F_5	0.228351	0.135144	0.0349231	0.0009515636
F_6	0.23343	0.13536	0.0349245	0.0009515639
F_7	0.230122	0.135249	0.034924	0.0009515638
F_8	0.232582	0.135314	0.0349242	0.0009515639
F_9	0.230882	0.135278	0.0349241	0.0009515639
F_{10}	0.23218	0.1353	0.0349241	0.0009515639
Exact $\mathrm{Ai}(x)$	0.231694	0.135292	0.0349241	0.0009515639

convert the divergent series into a continued fraction, after first substituting $t = x^{-3/2}$. The result is

$$y_1 \sim x^{-1/4} e^{2x^{3/2}/3} \Big/ \Big(1 - \frac{5}{48} x^{-3/2} \Big/ \Big(1 - \frac{67}{96} x^{-3/2} \Big/ \Big(1 - \frac{16247}{19296} x^{-3/2} \Big/ \Big(1 - \cdots.$$

$$y_2 \sim x^{-1/4} e^{-2x^{3/2}/3} \Big/ \Big(1 + \frac{5}{48} x^{-3/2} \Big/ \Big(1 + \frac{67}{96} x^{-3/2} \Big/ \Big(1 + \frac{16247}{19296} x^{-3/2} \Big/ \Big(1 + \cdots.$$

The pattern is not clear, but the sign pattern does continue. The asymptotic series for y_2 is a Stieltjes series in the variable t, and in fact the continued fraction form converges to $2\sqrt{\pi}\mathrm{Ai}(x)$. The continued fraction form for y_1 has the branch cut along the positive real axis, but it does converge on each half plane $\mathrm{Re}(x) > 0$ and $\mathrm{Re}(z) < 0$. In each case, it converges to a linear combination of $\mathrm{Ai}(x)$ and $\mathrm{Bi}(x)$.

Example 5.6

Find the full asymptotic expansion for the parabolic cylinder function $D_v(x)$ for arbitrary v.

SOLUTION: We saw in example 5.2 that the parabolic cylinder function satisfies the equation

$$y'' + (v + 1/2 - x^2/4)y = 0.$$

We also found that $D_v(x) = e^S$, where $s = -x^2/4 + v\ln(x) + h(x)$, where $h(x) \sim (v - v^2)/(2x^2)$ and satisfies the equation

$$-v/x^2 + h'' - xh' + v^2/x^2 + 2vh'/x + (h')^2 = 0.$$

But making the substitution in equation 5.10, we have $u(x) = e^{h(x)}$, where u satisfies

$$x^2u'' + (2vx - x^3)u' + (v^2 - v)u = 0.$$

To determine the series for $u(x)$, we can use the hint that $h(x)$ is of order x^{-2} as $x \to \infty$. So $e^{h(x)}$ has the form

$$u = \sum_n a_n x^{-2n}.$$

Then

$$u' = \sum_n -2na_n x^{-2n-1}, \qquad u'' = \sum_n -2n(-2n-1)a_n x^{-2n-2},$$

so substituting these sums into the equation for u produces

$$\sum_n (4n^2+2n)a_n x^{-2n} - \sum_n 4vna_n x^{-2n} + \sum_n 2na_n x^{-2n+2} \sim \sum_n (v-v^2)a_n x^{-2n}.$$

Shifting the third sum up by one gets all sums to involve x^{-2n}, so we can equate the coefficients to produce the equation

$$(4n^2 + 2n)a_n - 4vna_n + 2(n+1)a_{n+1} = (v - v^2)a_n.$$

This can be solved for a_{n+1} to produce

$$a_{n+1} = -\frac{(2n - v)(2n + 1 - v)}{2(n+1)}.$$

If $a_0 = 1$, then

$$a_1 = -\frac{(-v)(1-v)}{2 \cdot 1}, \quad a_2 = \frac{(-v)(1-v)(2-v)(3-v)}{2^2 \cdot 1 \cdot 2},$$

$$a_3 = \frac{(-v)(1-v)(2-v)(3-v)(4-v)(5-v)}{2^3 \cdot 1 \cdot 2 \cdot 3}, \ldots.$$

If v is a non-negative integer, then the series truncates, and we have an exact solution for $D_v(x)$. Otherwise, we can multiply the numerators and denominators by $\Gamma(-v)$, and use the "zipper" property of the Gamma function to obtain the closed form solution

$$a_n = (-1)^n \frac{\Gamma(2n - v)}{\Gamma(-v)2^n n!}.$$

So we have the full asymptotic expansion for $D_v(x)$:

$$D_v(x) \sim e^{-x^2/4} x^v \sum_{n=0}^{\infty} \frac{(-1)^n \Gamma(2n - v)}{\Gamma(-v) 2^n n! x^{2n}}. \tag{5.14}$$

The Stoke lines are at $\arg(x) = \pm\pi/4$ and $\arg(x) = \pm 3\pi/4$. Since this solution to the equation is subdominant to the other solution, we can pass beyond the first set of Stoke lines, so this series is valid in the region $-3\pi/4 < \arg(x) < 3\pi/4$. □

Example 5.7
Find the full asymptotic expansion as $x \to \infty$ for the solutions to the Bessel equation of order v:

$$x^2 y'' + xy' + (x^2 - v^2)y = 0.$$

SOLUTION: Unlike the previous two examples, we have not solved for the leading behavior of the solutions yet, so we must begin by finding these leading behaviors. Substituting $y = e^S$ gives us the equation

$$x^2 S'' + x^2 (S')^2 + xS' + x^2 - v^2 = 0.$$

Clearly $v^2 \ll x^2$ as $x \to \infty$, and it is natural to assume that $S'' \ll (S')^2$. The remaining terms give us a quadratic equation for S':

$$x^2 (S')^2 + xS' \sim -x^2.$$

Using the quadratic equation, we get $S' \sim (-x \pm \sqrt{x^2 - 4x^4})/(2x^2)$. The first order approximation gives us $S' \sim \pm i$, so $S(x) \sim \pm ix$. Although we are getting a complex result at this stage, we can remedy this in the end when we have the full series, and produce two real solutions.

Since $S'' = 0$, this clearly is smaller than $(S')^2$, so the assumptions are consistent. To find the next term, we let $S(x) = \pm ix + g(x)$, so the equation for $g(x)$ is

$$x^2 g'' \pm 2ix^2 g' + x^2 (g')^2 \pm ix + xg' - v^2 = 0.$$

Clearly, $v^2 \ll ix$, and since $g' \ll \pm i$, we also find $xg' \ll ix$ and $x^2(g')^2 \ll ix^2 g'$. It is also safe to assume that $g'' \ll g'$, since g' is going to 0. The remaining terms indicate that $\pm 2ix^2 g' \sim \mp ix$, or $g' \sim -1/(2x)$, which makes $g(x) \sim -\ln(x)/2$. Indeed, g'' is smaller than g', so we are ready to continue to the next term.

Substituting $g(x) = -\ln(x)/2 + h(x)$ into the equation for $g(x)$ yields the following equation for $h(x)$.

$$x^2 h'' \pm 2ix^2 h' + x^2 (h')^2 + 1/4 - v^2 = 0.$$

If $v = \pm 1/2$, then there is an exact solution with $h'(x) = 0$, or h being a constant. In this case, the exact solutions are given by $y = e^{\pm i}/\sqrt{x}$. Otherwise, there is a non-zero constant term, and since $h' \ll 1/x$ and $h'' \ll 1/x^2$, it is clear that $\pm 2ix^2 h' \sim v^2 - 1/4$. So $h \sim C \pm (4v^2 - 1)i/(8x)$. Since we have found a term that is going to zero as $x \to \infty$, we have enough to compute the leading behavior of the solutions:

$$y \sim c_1 x^{-1/2} e^{ix} + c_2 x^{-1/2} e^{-ix}.$$

We can now relinearize the equation for $h(x)$ with the substitution $h(x) = \ln(u(x))$, to obtain

$$x^2 u'' \pm 2ix^2 u' + (1/4 - v^2)u = 0.$$

Since h is of order $1/x$, we expect $e^{h(x)}$ to be a series in powers of $1/x$, so we can let

$$u = \sum_n a_n x^{-n}, \quad u' = \sum_n -n a_n x^{-n-1}, \quad u'' = \sum_n n(n+1) a_n x^{n-2}.$$

Substituting these sums into the equation for u, we get

$$\sum_n n(n+1) a_n x^{-n} \pm 2i \sum_n -n a_n x^{-n+1} \sim (v^2 - 1/4) \sum_n a_n x^{-n}.$$

The second sum must have the index shifted up by 1, so that we can equate the coefficients to obtain

$$(n^2 + n + 1/4 - v^2) a_n = \pm 2i(n+1) a_{n+1},$$

which we can solve for a_{n+1} to obtain

$$a_{n+1} = \frac{\mp i \left(n + \frac{1}{2} + v\right)\left(n + \frac{1}{2} - v\right)}{2(n+1)} a_n.$$

If $a_0 = 1$ then

$$a_1 = \frac{\mp i \left(\frac{1}{2} - v\right)\left(\frac{1}{2} + v\right)}{2 \cdot 1}, \quad a_2 = \frac{(\mp i)^2 \left(\frac{1}{2} - v\right)\left(\frac{1}{2} + v\right)\left(\frac{3}{2} - v\right)\left(\frac{3}{2} + v\right)}{2^2 \cdot 1 \cdot 2},$$

$$a_3 = \frac{(\mp i)^3 \left(\frac{1}{2} - v\right)\left(\frac{1}{2} + v\right)\left(\frac{3}{2} - v\right)\left(\frac{3}{2} + v\right)\left(\frac{5}{2} - v\right)\left(\frac{5}{2} + v\right)}{2^3 \cdot 1 \cdot 2 \cdot 3}, \ldots.$$

If v is a half-integer, then the series truncates, and we have an exact solution to the equation. Otherwise, we can multiply the numerators and denominators by $\Gamma(1/2 - v)\Gamma(1/2 + v)$, to get a closed form solution

$$a_n = \frac{(\mp i)^n \Gamma\left(\frac{1}{2} + n - v\right)\Gamma\left(\frac{1}{2} + n + v\right)}{2^n \Gamma\left(\frac{1}{2} - v\right)\Gamma\left(\frac{1}{2} + v\right) n!}.$$

Thus, we have

$$y \sim c_1 x^{-1/2} e^{ix} \sum_{n=0}^{\infty} \frac{(-i)^n \Gamma\left(\frac{1}{2}+n-v\right)\Gamma\left(\frac{1}{2}+n+v\right)}{2^n \Gamma\left(\frac{1}{2}-v\right)\Gamma\left(\frac{1}{2}+v\right) n! x^n}$$
$$+ c_2 x^{-1/2} e^{-ix} \sum_{n=0}^{\infty} \frac{(i)^n \Gamma\left(\frac{1}{2}+n-v\right)\Gamma\left(\frac{1}{2}+n+v\right)}{2^n \Gamma\left(\frac{1}{2}-v\right)\Gamma\left(\frac{1}{2}+v\right) n! x^n}.$$

The only problem with this solution is that it involves complex numbers, whereas we know that the solutions are real when x is real. But it is obvious that the two solutions are complex conjugates of each other, so if we let $c_1 = (k_1 + k_2 i)/2$, and $c_2 = (k_1 - k_2 i)/2$, the imaginary components will cancel. Using the fact that $\cos(x) = (e^{ix} + e^{-ix})/2$ and $\sin(x) = (e^{ix} - e^{-ix})/(2i)$, we get that

$$y - k_1 x^{-1/2}(\cos(x) w_1(x) + \sin(x) w_2(x)) + k_2 x^{-1/2}(\cos(x) w_2(x) - \sin(x) w_1(x)),$$

where

$$w_1(x) \sim \sum_{n=0}^{\infty} \frac{(-1)^n \Gamma\left(\frac{1}{2}+2n-v\right)\Gamma\left(\frac{1}{2}+2n+v\right)}{4^n \Gamma\left(\frac{1}{2}-v\right)\Gamma\left(\frac{1}{2}+v\right)(2n)! x^{2n}} \qquad \text{as } x \to \infty,$$

and

$$w_2(x) \sim \sum_{n=0}^{\infty} \frac{(-1)^n \Gamma\left(\frac{3}{2}+2n-v\right)\Gamma\left(\frac{3}{2}+2n+v\right)}{2 \cdot 4^n \Gamma\left(\frac{1}{2}-v\right)\Gamma\left(\frac{1}{2}+v\right)(2n+1)! x^{2n+1}} \qquad \text{as } x \to \infty.$$

Here, we used the technique introduced in subsection 2.4.4 to deal with the oscillating behavior. Typically, a phase shift is added as well. We define the *Bessel function of the first kind* J_v to be the solution with the asymptotic behavior

$$J_v(x) \sim \sqrt{2/\pi} x^{-1/2} \left(\cos\left(x - \frac{v\pi}{2} - \frac{\pi}{4}\right) w_1(x) + \sin\left(x - \frac{v\pi}{2} - \frac{\pi}{4}\right) w_2(x)\right).$$

The phase angle is added so that when v is an integer, it agrees with the integral definition from problem 17 in section 2.5. This is the solution whose Frobenius series at $x = 0$ does not involve a logarithm.

The *Bessel function of the second kind*, $Y_v(x)$, is defined as the complimentary solution with the asymptotic series

$$Y_v(x) \sim \sqrt{2/\pi} x^{-1/2} \left(\cos\left(x - \frac{v\pi}{2} - \frac{3\pi}{4}\right) w_1(x) + \sin\left(x - \frac{v\pi}{2} - \frac{3\pi}{4}\right) w_2(x)\right).$$

Figure 5.3 shows the Bessel functions of order 0, and compares them to the leading behaviors as $x \to \infty$. ☐

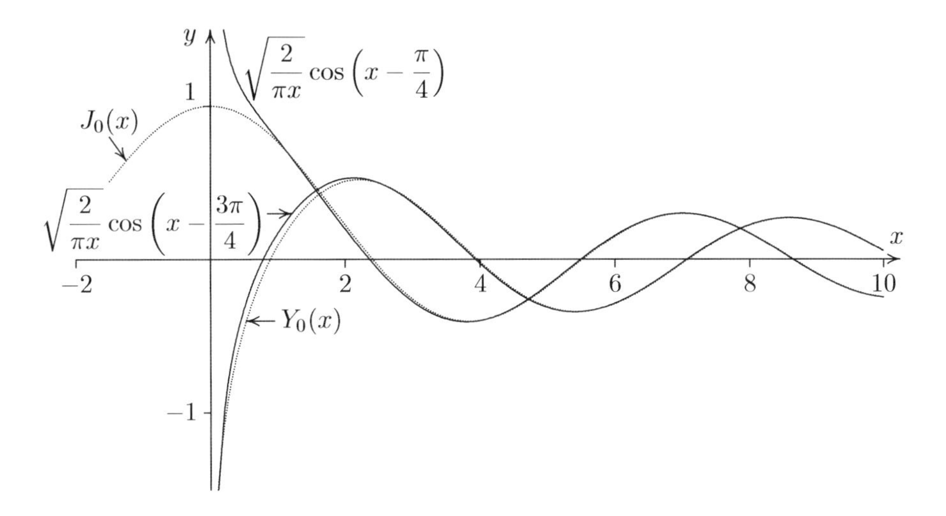

FIGURE 5.3: This shows the Bessel functions $J_0(x)$ and $Y_0(x)$, along with their leading behaviors as $x \to \infty$. Notice that even the first term of the asymptotic series gives impressive accuracy for $x > 2$.

In figure 5.1, we find that the Airy functions have an oscillatory behavior as $x \to -\infty$. We can apply the technique used in example 5.7, first finding the complex behaviors, and then converting the complex exponentials to trig functions.

Example 5.8

Find the full asymptotic behavior as $x \to -\infty$ for the Airy equation $y'' = xy$.

SOLUTION: We can actually utilize the full asymptotic behavior as $x \to \infty$, done in example 5.5. By replacing x with $-t$ in equation 5.13, and adjusting the constants, we obtain

$$y \sim k_1 t^{-1/4} e^{2it^{3/2}/3} \sum_{n=0}^{\infty} \frac{(-i)^n 3^n \Gamma(n+\frac{1}{6})\Gamma(n+\frac{5}{6}) t^{-3n/2}}{4^n 2\pi n!}$$
$$+ k_2 t^{-1/4} e^{-2it^{3/2}/3} \sum_{n=0}^{\infty} \frac{(i)^n 3^n \Gamma(n+\frac{1}{6})\Gamma(n+\frac{5}{6}) t^{-3n/2}}{4^n 2\pi n!}.$$

We now consider t to be real and positive. Again, we see that the two solutions are complex conjugates of each other, so we can let $k_1 = (C_1 + C_2 i)/2$ and $k_2 = (C_1 - C_2 i)/2$ to express the answer in terms of real functions. This gives us

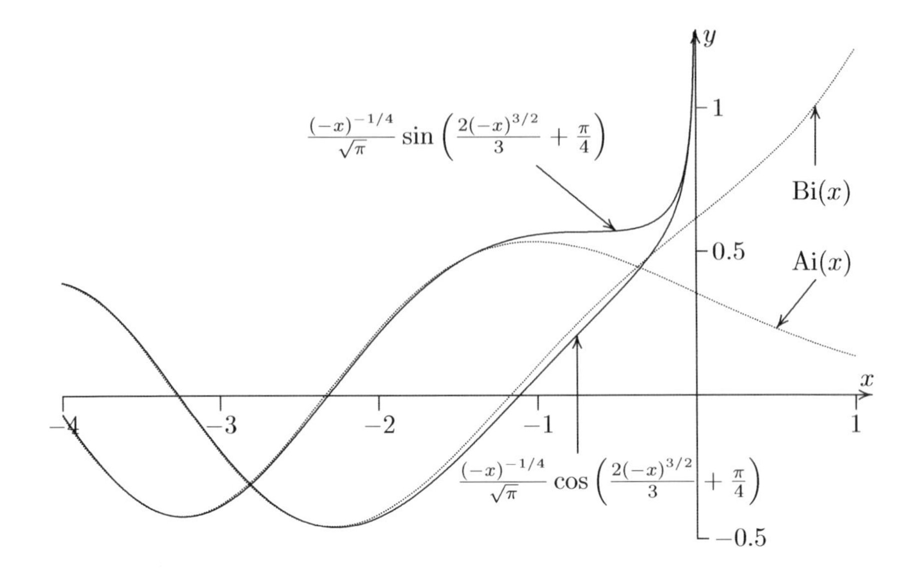

FIGURE 5.4: This compares the leading behaviors of the Airy functions as $x \to \infty$ to the actual Ai(x) and Bi(x), shown with a dotted line. For $x < -3$, the curves are almost indistinguishable.

$$y \sim C_1 t^{-1/4}\left(w_1(t)\cos\left(\frac{2}{3}t^{3/2}\right) + w_2(t)\sin\left(\frac{2}{3}t^{3/2}\right)\right)$$
$$+ C_2 t^{-1/4}\left(w_2(t)\cos\left(\frac{2}{3}t^{3/2}\right) - w_1(t)\sin\left(\frac{2}{3}t^{3/2}\right)\right),$$

where

$$w_1(t) \sim \sum_{n=0}^{\infty} \frac{(-1)^n 9^n \Gamma(2n+\frac{1}{6})\Gamma(2n+\frac{5}{6})t^{-3n}}{16^n 2\pi(2n)!},$$

and

$$w_2(t) \sim \sum_{n=0}^{\infty} \frac{(-1)^n 3^{2n+1}\Gamma(2n+\frac{7}{6})\Gamma(2n+\frac{11}{6})t^{-3n-3/2}}{4^{2n+1}2\pi(2n+1)!}.$$

As we try to relate these solutions to the Maclaurin series solutions of equations 5.5 and 5.6, we are stymied because we require a *global analysis*, which we cannot obtain from the differential equation. However, we can use an integral definition to obtain global information about the Airy function, such as in problem 25 of section 2.6. We can give the result which, not surprisingly, will involve a phase shift as seen in example 5.7. It turns out that

$$\begin{aligned}\mathrm{Ai}(x) \sim & \frac{(-x)^{-1/4}}{\sqrt{\pi}} w_1(-x) \sin\left(\frac{2}{3}(-x)^{3/2} + \frac{\pi}{4}\right) \\ & - \frac{(-x)^{-1/4}}{\sqrt{\pi}} w_2(-x) \cos\left(\frac{2}{3}(-x)^{3/2} + \frac{\pi}{4}\right) \quad \text{as } x \to -\infty,\end{aligned} \tag{5.15}$$

and

$$\begin{aligned}\mathrm{Bi}(x) \sim & \frac{(-x)^{-1/4}}{\sqrt{\pi}} w_1(-x) \cos\left(\frac{2}{3}(-x)^{3/2} + \frac{\pi}{4}\right) \\ & + \frac{(-x)^{-1/4}}{\sqrt{\pi}} w_2(-x) \sin\left(\frac{2}{3}(-x)^{3/2} + \frac{\pi}{4}\right) \quad \text{as } x \to -\infty.\end{aligned} \tag{5.16}$$

See figure 5.4 to compare this approximation to the exact solution.

Because the Stoke lines for the Airy functions occur at $\arg(x) = \pi$ and $\pm\pi/3$, and the series is computed along the Stoke line $\arg(x) = \pi$, the series will be valid for $\pi/3 < \arg(x) < 5\pi/3$. □

Problems for §5.2

For problems **1** through **10**: The first two terms of the asymptotic series for a solution to the differential equation are given. Use this as a guide to find the full asymptotic series for this solution, which will be in the form $y \sim \sum_{n=0}^{\infty} a_n x^{kn}$ for some k.

1 $x^2y'' + (3x+1)y' + y = 0,\quad y \sim 1 - x \quad \text{as } x \to 0$

2 $x^2y'' + (5x+1)y' + 3y = 0,\quad y \sim 1 - 3x \quad \text{as } x \to 0$

3 $x^3y'' + (4x^2-1)y' + 2xy = 0,\quad y \sim 1 + x^2 \quad \text{as } x \to 0$

4 $x^2y'' + (3x+\sqrt{x})y' + y = 0,\quad y \sim 1 - 2\sqrt{x} \quad \text{as } x \to 0$

5 $x^2y'' + (4x+\sqrt[3]{x})y' + 2y = 0,\quad y \sim 1 - 3x^{2/3} \quad \text{as } x \to 0$

6 $x^2y'' + (x^2-2x)y' + 2y = 0,\quad y \sim 1 + 2/x \quad \text{as } x \to \infty$

7 $x^2y'' + (x^3-2x)y' - 4y = 0,\quad y \sim 1 - 2/x^2 \quad \text{as } x \to \infty$

8 $x^2y'' + (x^{3/2}-2x)y' + 2y = 0,\quad y \sim 1 + 2/\sqrt{x} \quad \text{as } x \to \infty$

9 $x^2y'' + (x^{5/2}-3x)y' + 3y = 0,\quad y \sim 1 + 2x^{-3/2} \quad \text{as } x \to \infty$

10 $x^2y'' - x^{5/3}y' - 2y = 0,\quad y \sim 1 + 3x^{-2/3} \quad \text{as } x \to \infty$

11 Find the full asymptotic series for the solution $y_1 \sim x^{-1}e^{-1/x}$ to the equation from example 5.4.

12 Find the full asymptotic series for the solution $y_2 \sim x^3$ to the equation from example 5.4.

For problems **13** through **24**: The first two terms of the asymptotic series for one of the solutions to the differential equation is given. Find the full asymptotic series for this solution.

13 $x^3y'' - xy' - y = 0,\quad y \sim x^3 e^{-1/x}(1 - 6x)$ as $x \to 0$
14 $x^3y'' - xy' - 2y = 0,\quad y \sim x^4 e^{-1/x}(1 - 12x)$ as $x \to 0$
15 $x^3y'' - x^2y - y = 0,\quad y \sim x^{5/4} e^{2/\sqrt{x}}(1 - 15\sqrt{x}/16)$ as $x \to 0$
16 $x^3y'' - y = 0,\quad y \sim x^{3/4} e^{-2/\sqrt{x}}(1 + 3\sqrt{x}/16)$ as $x \to 0$
17 $x^4y'' + x^2y' - 2y = 0,\quad y \sim x^{2/3} e^{-1/x}(1 + 2x/27)$ as $x \to 0$
18 $x^4y'' + x^2y' - 2y = 0,\quad y \sim x^{4/3} e^{2/x}(1 + 4x/27)$ as $x \to 0$
19 $xy'' + xy' - y = 0,\quad y \sim x^{-1} e^{-x}(1 - 2/x)$ as $x \to \infty$
20 $y'' + xy' - 2y = 0,\quad y \sim x^{-3} e^{-x^2/2}(1 - 6/x^2)$ as $x \to \infty$
21 $y'' + x^2y' - xy = 0,\quad y \sim x^{-3} e^{-x^3/3}(1 - 4/x^3)$ as $x \to \infty$
22 $y'' - \sqrt{x}y = 0,\quad y \sim x^{-1/8} e^{4x^{5/4}/5}(1 + 9x^{-5/4}/160)$ as $x \to \infty$
23 $y'' - y' - x^2y = 0,\quad y \sim x^{-5/8} e^{(x-x^2)/2}(1 - 65x^{-2}/256)$ as $x \to \infty$
24 $y'' - y' - x^2y = 0,\quad y \sim x^{-3/8} e^{(x+x^2)/2}(1 + 33x^{-2}/256)$ as $x \to \infty$

For problems **25** through **30**: Find the full asymptotic series for the following differential equations as $x \to \infty$. Note that some of the series solutions may truncate.

25 $xy'' - y = 0$
26 $y'' - x^2y = 0$
27 $y'' - x^4y = 0$
28 $y'' + x^2y' - 2x^4y = 0$
29 $y'' + xy' + 5y = 0$
30 $xy'' + xy' + 3y = 0$

5.3 Local Analysis of Inhomogeneous Equations

Now that we have understood how to find the asymptotic behavior of homogeneous differential equations, let us turn our attention to inhomogeneous equations. Unfortunately, the substitution $y = e^S$ does not help with inhomogeneous equations. However, we know from equation 4.9 the form of the general solution. Since we can find the asymptotic behavior of the associated homogeneous equation, all we need is to find the asymptotic behavior of one particular solution which incorporates the inhomogeneous term.

If all of the coefficient functions are analytic at $x = x_0$, then theorem 4.2 shows that the solutions will all have a Taylor series centered at x_0. So one strategy would be to force a Taylor series to solve the equation. However, one must express the inhomogeneous term as a series, sometimes with the aid of the *Kronecker delta function*

$$\delta_i = \begin{cases} 1 & \text{if } i = 0, \\ 0 & \text{if } i \neq 0. \end{cases}$$

Example 5.9
Find the Maclaurin series for a particular solution to the inhomogeneous equation

$$y'' - xy = x^3.$$

SOLUTION: We must first express the inhomogeneous term x^3 as an infinite sum. For polynomials, this is accomplished with the help of the Kronecker delta function. We can express x^3 as

$$\sum_{n=0}^{\infty} \delta_{n-3} x^n,$$

since all but one term will be zero, namely the x^3 term. We now replace y with $\sum_n a_n x^n$ in the equation, and get

$$\sum_n n(n-1)a_n x^{n-2} - \sum_n a_n x^{n+1} = \sum_n \delta_{n-3} x^n.$$

The first sum can be shifted up by 2, and the second sum shifted down by 1, to produce

$$\sum_n (n+2)(n+1)a_{n+2} x^n - \sum_n a_{n+1} x^n = \sum_n \delta_{n-3} x^n.$$

So our recursion formula is $(n+2)(n+1)a_{n+2} - a_{n-1} = \delta_{n-3}$, which can be solved for a_{n+2} to produce

$$a_{n+2} = \frac{a_{n-1} + \delta_{n-3}}{(n+2)(n+1)}.$$

Notice the similarities between this example and example 4.18. The only change is in the δ_{n-3} term, which will only affect the computation of a_5. However, this in turn will affect further terms in the series.

$$a_2 = \frac{a_{-1}}{2} = 0, \quad a_3 = \frac{a_0}{3 \cdot 2}, \quad a_4 = \frac{a_1}{4 \cdot 3}, \quad a_5 = \frac{a_2 + 1}{4 \cdot 5} = \frac{1}{4 \cdot 5},$$
$$a_6 = \frac{a_0}{6 \cdot 5 \cdot 3 \cdot 2}, \quad a_7 = \frac{a_1}{7 \cdot 6 \cdot 4 \cdot 3}, \quad a_8 = \frac{a_5}{7 \cdot 8} = \frac{1}{4 \cdot 5 \cdot 7 \cdot 8}, \quad \ldots.$$

In order to obtain the simplest solution, we can let both a_0 and a_1 be zero. Then the non zero terms of the series are

$$a_5 = \frac{1}{4 \cdot 5} = \frac{2 \cdot 3}{5!}, \quad a_8 = \frac{1}{4 \cdot 5 \cdot 7 \cdot 8} = \frac{2 \cdot 3 \cdot 6}{8!}, \quad a_{11} = \frac{2 \cdot 3 \cdot 6 \cdot 9}{11!}, \quad \ldots.$$

So we can express a solution in closed form

$$y_p = \sum_{n=1}^{\infty} \frac{2\, 3^n\, n!\, x^{3n+2}}{(3n+2)!}.$$

□

If any of the coefficient functions of the equation have a singular point at $x = x_0$, then we must use a different approach. The simplest strategy is the direct application of dominant balance. However, we know that an inhomogeneous term must be one of terms used in the balance, lest we obtain the homogeneous equation solution.

Example 5.10
Find the first two terms for the asymptotic series as $x \to 0$ for a particular solution to

$$y'' - xy = -\frac{1}{x^2}.$$

SOLUTION: We will apply the method of dominant balance. Note that if we assume that $1/x^2 \ll xy$, then we obtain the homogeneous equation, which is analytic at $x = 0$, so $1/x^2$ will not be small. In general, we will not consider the case of throwing out the inhomogeneous term, so there are only two cases to consider.
Case 1) $y'' \ll x^{-2}$ as $x \to 0$

This would imply that $xy \sim x^{-2}$, or $y \sim x^{-3}$. But then $y'' \sim 12x^{-5}$, which is not smaller than x^{-2} as $x \to 0$.
Case 2) $xy \ll x^{-2}$ as $x \to 0$.

This would produce $y'' \sim -x^{-2}$, which is satisfied by $y \sim \ln(x)$. Adding arbitrary constants in the integrals would only produce terms which are smaller as $x \to 0$. Since $xy \sim x \ln x$ is indeed smaller than x^{-2}, at least we have consistency.

To verify that $y \sim \ln x$ is the leading behavior, we peel off this term by substituting $y = \ln x + g(x)$. The new equation in $g(x)$ is

$$g'' - xg = x \ln x.$$

Note that making an additive substitution on a linear equation only changes the inhomogeneous term. Since $g(x) \ll \ln(x)$, we have that $xg \ll x \ln x$, so $g'' \sim x \ln x$. Integrating twice, we get $g \sim (6 \ln x - 5)x^3/36$. This time, adding arbitrary constants will not be small, but the terms produced would be the terms for the homogeneous solutions, which are analytic. Because we can always add the homogeneous solutions to the particular solution, we can assume that the constants of integration are 0. Thus we have one particular solution for which

$$y_p \sim \ln x + \frac{(6 \ln x - 5)x^3}{36} + \cdots.$$

Figure 5.5 compares the first two asymptotic approximations with the exact solution.

It is clear how we can extend the pattern to obtain further terms in the series. We can write

$$y_p \sim \sum_{n=1}^{\infty} f_n(x)$$

where $f_1(x) = \ln x$, and f_{n+1} is obtained from integrating $x f_n(x)$ twice. □

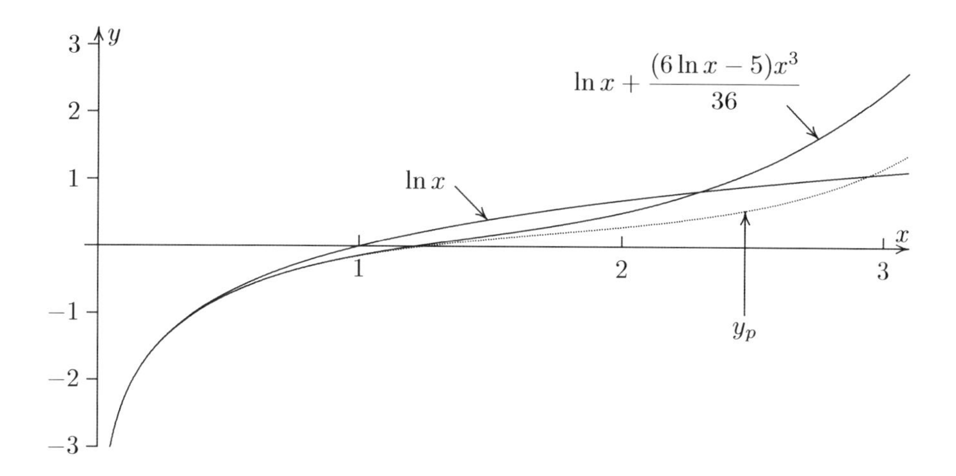

FIGURE 5.5: This shows a particular solution to the equation $y'' - xy = -1/x^2$, along with the first two asymptotic approximations. Note that the second order approximation is indistinguishable from the exact solution for $x < 1$.

Sometimes, the inhomogeneous solution will be subdominant to one of the solutions to the homogeneous solution series as $x \to a$. In this case, we are only interested in those solutions which do not have the same asymptotic behavior as a homogeneous solution.

Example 5.11
Find the asymptotic series as $x \to \infty$ for a solution to the equation

$$y'' - xy = -\frac{1}{x^2}$$

for which the limit as $x \to \infty$ is 0.
SOLUTION: Note that the condition $\lim_{x\to\infty} y_p(x) = 0$ eliminates the possibility that y_p is asymptotic to a solution of the homogeneous equation, $c_1\text{Ai}(x) + c_2\text{Bi}(x)$. Since $\text{Bi}(x) \to \infty$ as $x \to \infty$, c_2 would have to be 0, but then $x\text{Ai}(x) \ll 1/x^2$ as $x \to \infty$. Hence, there are only two cases to consider:
Case 1) $xy \ll 1/x^2$ as $x \to \infty$.
This would produce $y'' \sim -1/x^2$, which would indicate that $y \sim \ln(x) + Cx + D$ as $x \to \infty$. But then $xy \gg 1/x^2$ as $x \to \infty$, so this case is inconsistent.
Case 2) $y'' \ll 1/x^2$ as $x \to \infty$.
This gives us the equation $-xy \sim -1/x^2$, or $y \sim 1/x^3$. Then $y'' \sim 12/x^5$, which is small compared to $1/x^2$ as $x \to \infty$. So at least this case is consistent. To show that this is indeed the first term, let us peel off $1/x^3$ by letting

$y = 1/x^3 + g(x)$. The new equation for $g(x)$ is

$$g'' - xg = -\frac{12}{x^5}.$$

Since we can assume that $g(x) \ll 1/x^3$, so $g'' \ll 1/x^5$. Hence, we must have $-xg \sim -12/x^5$, or $g \sim 12/x^6$. Thus, the first two terms of the asymptotic series for y_p are

$$y_p \sim \frac{1}{x^3} + \frac{12}{x^6} + \cdots \qquad \text{as } x \to \infty.$$

To find the full asymptotic series, we can assume that the series will have the form

$$y_p \sim \sum_{n=0}^{\infty} a_n x^{-3n-3}.$$

Plugging this into the original equation, we get

$$\sum_n (3n+3)(3n+4)a_n x^{-3n-5} - \sum_n a_n x^{-3n-2} = -\sum_n \delta_n x^{-3n-2}.$$

Shifting the index down one on the first term, we get the recursion formula

$$3n(3n+1)a_{n-1} - a_n = -\delta_n.$$

So $a_0 = 1$, and for all $n > 0$, $a_n = 3n(3n+1)a_{n-1}$. Thus, $a_1 = 3 \cdot 4$, $a_2 = 3 \cdot 4 \cdot 6 \cdot 7$, etc. We can use the Gamma function to express the general term as we did in example 4.18.

$$y_p \sim \sum_{n=0}^{\infty} \frac{9^n \Gamma\left(n + \frac{4}{3}\right) n!}{\Gamma\left(\frac{4}{3}\right) x^{3n+3}} \qquad \text{as } x \to \infty.$$

We have found the unique asymptotic series for a solution to the differential equation which goes to zero as $x \to \infty$. The paradox is that there are in fact an infinite number of solutions to the differential equation with this condition. A constant times $\mathrm{Ai}(x)$ can be added to one solution in order to produce another solution. But $\mathrm{Ai}(x)$ is subdominant to the asymptotic series for y_p. Figure 5.6 shows one such solution to the equation along with the first three approximations given by the asymptotic series. In spite of the non-uniqueness of the solutions, the approximations show a great amount of accuracy for large x. ▯

The curious part of this example is that one of the homogeneous solutions, $\mathrm{Ai}(x)$, is subdominant to the asymptotic series of the inhomogeneous solution, whereas this series is subdominant to the asymptotic series of another homogeneous solution, $\mathrm{Bi}(x)$.

There are times in which many terms seem to be of the same size asymptotically. In this case, we try a three-term dominant balance.

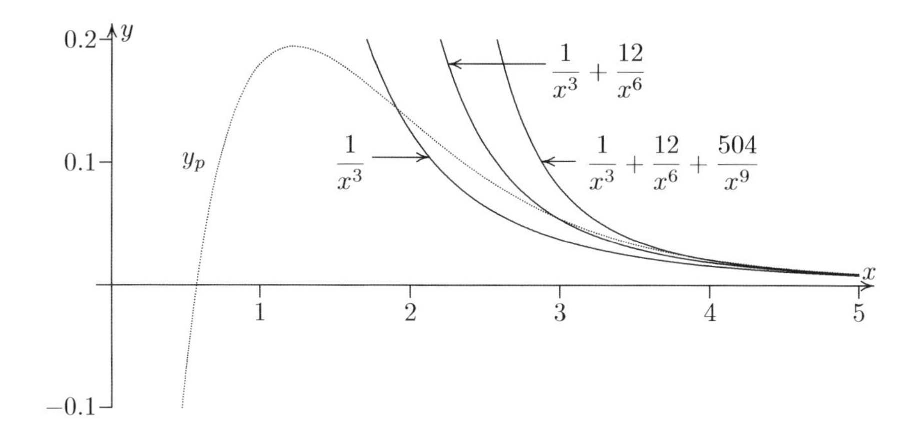

FIGURE 5.6: This shows a solution to the equation $y'' - xy = -1/x^2$, along with the first three asymptotic approximations as $x \to \infty$. Even though there are many solutions to the equation with the same asymptotic series, the differences between these solutions will be subdominant to the series.

Example 5.12

Find the leading behavior as $x \to 0$ for a particular solution to the equation

$$x^2 y'' - 2y = \frac{e^x}{x}.$$

SOLUTION: It is fruitless to assume that e^x/x is the smallest term, since this would yield a second order equation that we do not yet know how to solve. So let us consider the following two cases.

Case 1) $x^2 y'' \ll e^x/x \sim 1/x$ as $x \to 0$.

Then $-2y \sim 1/x$, so $y \sim -1/(2x)$. But then $y'' \sim 1/x^3$, so $x^2 y''$ is the same order as $1/x$, not smaller. So this case is inconsistent.

Case 2) $-2y \ll 1/x$ as $x \to 0$.

This produces $y'' \sim 1/x^3$, which would indicate that $y \sim 1/2x$. But then $-2y$ would be of the same order as $1/x$, so this case is inconsistent.

Since both of these cases are inconsistent, we will not be able to assume that any term is small. This means that all three terms are about the same order. But in both of these cases, we came up with $y \sim C/x$, so let us assume that the leading term is of this form, and try to solve for C. Then $y'' \sim 2C/x^3$, so we are solving the equation

$$\frac{2C}{x} - 2\frac{C}{x} \sim \frac{1}{x}.$$

Here we encounter a problem, since the C's cancel out. This similar problem occurred in example 4.23, because two solutions interfered with each other. In

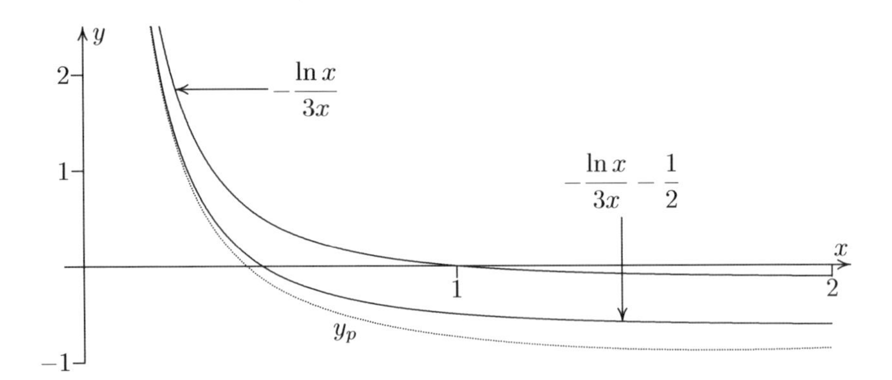

FIGURE 5.7: This shows a particular solution to the equation $x^2y'' - 2y = e^x/x$, along with its first and second order approximations as $x \to 0$. The first term involves a logarithm because of interference with a homogeneous solution.

fact, the same thing is happening here, because $y = 1/x$ is an exact solution to the homogeneous equation $x^2y'' - 2y = 0$. In example 4.23, we found the second solution involved a logarithm. This suggests that we consider multiplying our first guess, $y \sim C/x$, by $\ln x$. If $y \sim C\ln(x)/x$, then $y'' \sim (2\ln x - 3)C/x^3$. Thus, the differential equation becomes

$$\frac{2C\ln x - 3C}{x} - \frac{2C\ln x}{x} \sim \frac{1}{x}.$$

The logarithm terms cancel, and the asymptotic relation is satisfied if $C = -1/3$. Thus, having $y \sim -\ln x/(3x)$ is consistent.

If we let $y = -\ln x/(3x) + g(x)$, the new equation for $g(x)$ is

$$x^2g'' - 2g = \frac{e^x - 1}{x}.$$

Note that $(e^x - 1)/x \sim 1$ as $x \to 0$. We still have a dominant balance problem, but this time assuming $x^2g'' \ll 1$ yields the consistent solution $g \sim -1/2$. So we have the first two terms of a solution

$$y \sim -\frac{\ln x}{3x} - \frac{1}{2} + \cdots.$$

Figure 5.7 shows how the first two terms accurately approximate a particular solution to the equation. □

5.3.1 Variation of Parameters

Another way of finding the behavior of the particular solution to an inhomogeneous equation is through *variation of parameters*, given in equation 4.11.

Although this is an exact method for finding a particular solution in terms of integrals, it can also be used as an approximation method, since we can then apply asymptotical analysis on the integrals.

Recall that for a second order inhomogeneous equation, a particular solution can be given by the formula

$$y_p = -y_1 \int \frac{g(x)y_2}{y_1y_2' - y_2y_1'} + y_2 \int \frac{g(x)y_1}{y_1y_2' - y_2y_1'}. \tag{5.17}$$

Since we have techniques for finding the asymptotic series for integrals, and we can find the asymptotic series for a homogeneous equation solutions, we can use techniques such as integration by parts to find the asymptotic series for the particular solution of an inhomogeneous equation. The trickiest part is finding the asymptotic series for the Wronskian in the denominators. However, there is a shortcut for finding the exact value of the Wronskian without finding the solutions.

THEOREM 5.1

If $y_1, y_2, \ldots y_n$ are n solutions to the homogeneous equation

$$y^{(n)} + p_{n-1}(x)y^{(n-1)} + \cdots + p_2(x)y'' + p_1(x)y' + p_0(x)y = 0,$$

then the Wronskian of the solutions is given by

$$W(y_1, y_2, \ldots, y_n) = Ke^{-\int p_{n-1}(x)\,dx} \tag{5.18}$$

for some constant K. Equation 5.18 is known as *Abel's identity*, or *Abel's differential equation identity*.

PROOF: We will let $W(x)$ denote $W(y_1, y_2, \ldots, y_n)$, given by the matrix determinant

$$W(x) = \begin{vmatrix} y_1(x) & y_2(x) & y_3(x) & \cdots & y_n(x) \\ y_1'(x) & y_2'(x) & y_3'(x) & \cdots & y_n'(x) \\ y_1''(x) & y_2''(x) & y_3''(x) & \cdots & y_n''(x) \\ \vdots & \vdots & \vdots & \ddots & \vdots \\ y_1^{(n-2)}(x) & y_2^{(n-2)}(x) & y_3^{(n-2)} & \cdots & y_n^{(n-2)} \\ y_1^{(n-1)}(x) & y_2^{(n-1)}(x) & y_3^{(n-1)} & \cdots & y_n^{(n-1)} \end{vmatrix}.$$

It is sufficient to show that $W'(x) = -p_{n-1}(x)W(x)$, since the solution to this equation would yield equation 5.18. Note that every term of the determinant involves one factor from each row of the matrix. Thus, the derivative of the determinant can be expressed as a sum of n determinants, where for each of these matrices, we take the derivative of a different row. That is,

$$W'(x) = \begin{vmatrix} y_1' & y_2' & \cdots & y_n' \\ y_1' & y_2' & \cdots & y_n' \\ y_1'' & y_2'' & \cdots & y_n'' \\ \vdots & \vdots & \ddots & \vdots \\ y_1^{(n-2)} & y_2^{(n-2)} & \cdots & y_n^{(n-2)} \\ y_1^{(n-1)} & y_2^{(n-1)} & \cdots & y_n^{(n-1)} \end{vmatrix} + \begin{vmatrix} y_1 & y_2 & \cdots & y_n \\ y_1'' & y_2'' & \cdots & y_n'' \\ y_1'' & y_2'' & \cdots & y_n'' \\ \vdots & \vdots & \ddots & \vdots \\ y_1^{(n-2)} & y_2^{(n-2)} & \cdots & y_n^{(n-2)} \\ y_1^{(n-1)} & y_2^{(n-1)} & \cdots & y_n^{(n-1)} \end{vmatrix}$$

$$+ \begin{vmatrix} y_1 & y_2 & \cdots & y_n \\ y_1' & y_2' & \cdots & y_n' \\ y_1''' & y_2''' & \cdots & y_n''' \\ \vdots & \vdots & \ddots & \vdots \\ y_1^{(n-2)} & y_2^{(n-2)} & \cdots & y_n^{(n-2)} \\ y_1^{(n-1)} & y_2^{(n-1)} & \cdots & y_n^{(n-1)} \end{vmatrix} + \cdots + \begin{vmatrix} y_1 & y_2 & \cdots & y_n \\ y_1' & y_2' & \cdots & y_n' \\ y_1'' & y_2'' & \cdots & y_n'' \\ \vdots & \vdots & \ddots & \vdots \\ y_1^{(n-2)} & y_2^{(n-2)} & \cdots & y_n^{(n-2)} \\ y_1^{(n)} & y_2^{(n)} & \cdots & y_n^{(n)} \end{vmatrix}.$$

Note that all of these determinants except the last one will evaluate to 0, since these matrices have two identical rows. Thus, we have that

$$W'(x) = \begin{vmatrix} y_1 & y_2 & y_3 & \cdots & y_n \\ y_1' & y_2' & y_3' & \cdots & y_n' \\ y_1'' & y_2'' & y_3'' & \cdots & y_n'' \\ \vdots & \vdots & \vdots & \ddots & \vdots \\ y_1^{(n-2)} & y_2^{(n-2)} & y_3^{(n-2)} & \cdots & y_n^{(n-2)} \\ y_1^{(n)} & y_2^{(n)} & y_3^{(n)} & \cdots & y_n^{(n)} \end{vmatrix}.$$

To simplify this final determinant, we can observe that each $y_i(x)$ satisfies the original homogeneous equation, so

$$y_i^{(n)}(x) = -p_{n-1}(x)y^{(n-1)} - \cdots - p_2(x)y'' - p_1(x)y' - p_0(x)y.$$

We can then add $p_0(x)$ times the first row, $p_1(x)$ times the second row, etc., to the last row, giving us

$$W'(x) = \begin{vmatrix} y_1 & y_2 & y_3 & \cdots & y_n \\ y_1' & y_2' & y_3' & \cdots & y_n' \\ y_1'' & y_2'' & y_3'' & \cdots & y_n'' \\ \vdots & \vdots & \vdots & \ddots & \vdots \\ y_1^{(n-2)} & y_2^{(n-2)} & y_3^{(n-2)} & \cdots & y_n^{(n-2)} \\ -p_{n-1}y_1^{(n-1)} & -p_{n-1}y_2^{(n-1)} & -p_{n-1}y_3^{(n-1)} & \cdots & -p_{n-1}y_n^{(n-1)} \end{vmatrix}.$$

Finally, if we factor out $-p_{n-1}(x)$ from the last row, we reproduce the determinant for $W(x)$. Thus,

$$W'(x) = -p_{n-1}(x)W(x),$$

so the identity is established. ∎

Abel's identity can be used to show the relationship between the solutions to a homogeneous linear equation.

Example 5.13

Find the exact value of the Wronskian $W(\mathrm{Ai}(x), \mathrm{Bi}(x))$.

SOLUTION: Both $\mathrm{Ai}(x)$ and $\mathrm{Bi}(x)$ are solutions to the equation $y'' - xy = 0$. Note that for this equation, $p_1(x) = 0$, so Abel's theorem states that $W(\mathrm{Ai}(x), \mathrm{Bi}(x)) = Ke^{\int 0\,dx} = K$. So the Wronskian is a constant, but we need to determine which constant. We will use the asymptotic approximations of $\mathrm{Ai}(x)$and $\mathrm{Bi}(x)$ near ∞, derived from equation 5.4:

$$\mathrm{Ai}(x) \sim \frac{1}{2\sqrt{\pi}} x^{-1/4} e^{-2x^{3/2}/3}, \quad \text{and} \quad \mathrm{Bi}(x) \sim \frac{1}{\sqrt{\pi}} x^{-1/4} e^{2x^{3/2}/3} \quad \text{as } x \to \infty \tag{5.19}$$

Then $\mathrm{Ai}'(x) \sim -x^{1/4}e^{-2x^{3/2}/3}/(2\sqrt{\pi})$, and $\mathrm{Bi}'(x) \sim x^{1/4}e^{-2x^{3/2}/3}/\sqrt{\pi}$, so

$$\mathrm{Ai}(x)\mathrm{Bi}'(x) - \mathrm{Bi}(x)\mathrm{Ai}'(x) \sim \frac{1}{2\pi} - \frac{-1}{2\pi} \sim \frac{1}{\pi}.$$

Since we know that $W(\mathrm{Ai}(x), \mathrm{Bi}(x))$ will be an exact constant, we have that

$$\mathrm{Ai}(x)\mathrm{Bi}'(x) - \mathrm{Bi}(x)\mathrm{Ai}'(x) = \frac{1}{\pi}. \tag{5.20}$$

∎

Variation of parameters can be used to express the exact answer to an inhomogeneous equation in terms of some integrals. From this stage, we can perform local analysis on the solution.

Example 5.14

Find an integral representation of a solution which satisfies

$$y'' - xy = -\frac{1}{x^2} \quad \text{and} \quad \lim_{x\to\infty} y(x) = 0.$$

SOLUTION: We begin by finding the general solution to the equation, and then perform asymptotic analysis on the integrals to insure that the condition $\lim_{x\to\infty} y(x) = 0$ is satisfied.

The solution to the homogeneous equation $y'' - xy = 0$ can be expressed in terms of Airy functions, $y = c_1\mathrm{Ai}(x) + c_2\mathrm{Bi}(x)$. We see from equation 5.20 that the Wronskian of these two solutions is $1/\pi$. Thus, applying equation 4.10, we have a particular solution to be

$$y_p = -\mathrm{Ai}(x)\int \frac{-\pi\mathrm{Bi}(x)}{x^2}\,dx + \mathrm{Bi}(x)\int \frac{-\pi\mathrm{Ai}(x)}{x^2}\,dx.$$

In order to specify one particular solution, we need to use definite integrals. This is accomplished by integrating from a to x, where a is some convenient point where the integrand is defined. In this case, we cannot use $a = 0$, so we will use $a = 1$. Thus we can express the general solution as

$$y_p = \pi \mathrm{Ai}(x) \int_1^x \frac{\mathrm{Bi}(t)}{t^2}\,dt - \pi \mathrm{Bi}(x) \int_1^x \frac{\mathrm{Ai}(t)}{t^2}\,dt + c_1 \mathrm{Ai}(x) + c_2 \mathrm{Bi}(x). \quad (5.21)$$

The condition $\lim_{x\to\infty} y(x) = 0$ should allow us to compute c_2. Let us do the first order approximation of these integrals as $x \to \infty$, using integration by parts.

$$\begin{aligned}
\int_1^x \frac{\mathrm{Bi}(t)}{t^2}\,dt &\sim \frac{1}{\sqrt{\pi}} \int_1^x t^{-9/4} e^{2t^{3/2}/3}\,dt \\
&= \frac{1}{\sqrt{\pi}} t^{-11/4} e^{2t^{3/2}/3} \Big|_1^x - \frac{1}{\sqrt{\pi}} \int_1^x \frac{-11}{4} t^{-15/4} e^{2t^{3/2}/3}\,dt \\
&\sim \frac{1}{\sqrt{\pi}} x^{-11/4} e^{2x^{3/2}/3} \qquad \text{as } x \to \infty.
\end{aligned}$$

Thus, the first term of equation 5.21 is asymptotic to $x^{-3}/2$ as $x \to \infty$. So this term goes to 0 as $x \to \infty$.

The second integral in equation 5.21 is easier to approximate to first order.

$$-\pi \mathrm{Bi}(x) \int_1^x \frac{\mathrm{Ai}(t)}{t^2}\,dt \sim -\pi \mathrm{Bi}(x) \int_1^\infty \frac{\mathrm{Ai}(t)}{t^2}\,dt.$$

In order for $y(x) \to 0$ as $x \to \infty$, we need for the first order of this term to cancel with $c_2 \mathrm{Bi}(x)$. Thus,

$$c_2 = \pi \int_1^\infty \frac{\mathrm{Ai}(t)}{t^2}\,dt.$$

The expression for y_p can now be simplified to the form

$$y_p = \pi \mathrm{Ai}(x) \int_1^x \frac{\mathrm{Bi}(t)}{t^2}\,dt + \pi \mathrm{Bi}(x) \int_x^\infty \frac{\mathrm{Ai}(t)}{t^2}\,dt + c_1 \mathrm{Ai}(x).$$

The second integral can now be analyzed via integration by parts.

$$\begin{aligned}
\int_x^\infty \frac{\mathrm{Ai}(t)}{t^2}\,dt &\sim \frac{1}{2\sqrt{\pi}} \int_x^\infty t^{-9/4} e^{-2t^{3/2}/3}\,dt \\
&= -\frac{1}{2\sqrt{\pi}} t^{-11/4} e^{-2t^{3/2}/3} \Big|_x^\infty - \frac{1}{2\sqrt{\pi}} \int_x^\infty \frac{11}{4} t^{-15/4} e^{-2t^{3/2}/3}\,dt \\
&\sim \frac{1}{2\sqrt{\pi}} x^{-11/4} e^{-2x^{3/2}/3} \qquad \text{as } x \to \infty.
\end{aligned}$$

So we find that $y_p \sim x^{-3}$ regardless of the value of c_1, which agrees with the result of example 5.11. The value of c_1 cannot be determined from the condition $\lim_{x\to\infty} y(x) = 0$, but if we set $c_1 = 0$, we get the solution whose graph was used in figure 5.6. ∎

Although the method used in example 5.11 produced the leading order much more efficiently, there is an advantage of having the exact solution expressed in terms of integrals. This is because an integral solution contains *global analysis* information about the function, whereas even the full asymptotic series contains only *local* information. There are some problems that require knowledge of the global behavior in order to solve, which the asymptotic series does not give.

For example, it is clear that the y_p in figure 5.5 is not the same as the y_p in figure 5.6, even though both solve the same differential equation. The difference between the two functions would solve the homogeneous equation, so this difference can be written as $k_1\mathrm{Ai}(x) + k_2\mathrm{Bi}(x)$ for some k_1 and k_2. Consider the problem of determining the values of k_1 and k_2. Since we only have the local analysis of one function at $x = 0$, and the other at $x = \infty$, there is no way to glean this information from the two asymptotic series. However, the integral representation contains global information, so we can use this to link the behavior at $x = 0$ with the behavior at $x = \infty$. Problem 20 outlines how we can find the values of k_1 and k_2.

Example 5.15

Express the first two terms of the series both at 0 and ∞ to the boundary value problem

$$xy'' + (x-1)y' - y = \frac{-1}{x+1}, \qquad \lim_{x\to 0^+} y(x) = 0,\ \lim_{x\to\infty} y(x) \text{ is bounded}.$$

Utilize the fact that $y_1 = e^{-x}$ and $y_2 = x - 1$ solve the homogeneous equation.

SOLUTION: Since this is a boundary value problem, we need global information about the solutions to the equation in order to utilize the boundary conditions. This can be done using variation of parameters, which would give the general solution in terms of several integrals. Then we can do the asymptotics on the integrals.

Since we have two exact solutions to the homogeneous equations, we can compute the Wronskian $W(e^{-x}, x-1) = e^{-x} - (x-1)(-e^{-x}) = xe^{-x}$. Before we can use equation 5.17, we need to determine $g(x)$. Currently, the equation is not expressed in standard form, so we divide by the coefficient in front of the highest derivative to produce

$$y'' + \frac{x-1}{x}y' - \frac{1}{x}y = \frac{-1}{x(x+1)}.$$

Thus, $g(x) = -1/(x^2 + x)$. Equation 5.17 now gives us

$$y_p = e^{-x}\int \frac{(x-1)e^x}{x^2(x+1)}\,dx - (x-1)\int \frac{1}{x^2(x+1)}\,dx.$$

One of these integrals can be calculated via partial fractions, but the other must be approximated asymptotically. Since we need to incorporate the boundary conditions, we will use a definite integral in the general solution.

$$y = e^{-x}\int_1^x \frac{(t-1)e^t}{t^2(t+1)}\,dt - (x-1)\left(\ln(x+1) - \ln(x) - \frac{1}{x}\right) + c_1 e^{-x} + c_2(x-1).$$

We can find the behavior of the integral as $x \to \infty$ using integration by parts.

$$\int_1^x \frac{(t-1)e^t}{t^2(t+1)}\,dt = \frac{t-1}{t^2(t+1)}e^t\Big|_1^x - \int_1^x \frac{2+2t-2t^2}{t^3(t+1)^2}e^t\,dt,$$

so the first term is asymptotic to $1/x^2$ as $x \to \infty$. The second term is a straightforward asymptotic analysis

$$\ln(x+1) - \ln(x) = \ln(1+1/x) \sim 1/x - 1/(2x^2) + 1/(3x^3) \qquad \text{as } x \to \infty,$$

so

$$-(x-1)\left(\ln(x+1) - \ln(x) - \frac{1}{x}\right) \sim -(x-1)\left(-\frac{1}{2x^2} + \frac{1}{3x^3}\right)$$
$$\sim \frac{1}{2x} - \frac{5}{6x^2} + \cdots.$$

Thus, we have the asymptotic analysis of the general solution as $x \to \infty$,

$$y \sim c_1 e^{-x} + c_2(x-1) + \frac{1}{2x} + \frac{1}{6x^2} + \cdots.$$

In order for $y(x)$ to remain bounded as $x \to \infty$, we need to have $c_2 = 0$. However, the c_1 term will be subdominant to the series, so we need to use the condition at $x = 0$ to solve for c_1.

As x approaches 0, the integral becomes divergent because of the singularity of the integrand at $x = 0$. We can expand the integrand into a Frobenius series

$$\frac{(t-1)e^t}{t^2(t+1)} \sim -\frac{1}{t^2} + \frac{1}{t} - \frac{1}{2} + \frac{5t}{6} + \cdots.$$

Only the first two terms have a problem near $t = 0$, so we can separate these two terms from the remainder of the integral.

$$e^{-x}\int_1^x \frac{(t-1)e^t}{t^2(t+1)}\,dt = -e^{-x}\int_x^1 \frac{(t-1)e^t}{t^2(t+1)} + \frac{1}{t^2} - \frac{1}{t}\,dt - e^{-x}\int_x^1 -\frac{1}{t^2} + \frac{1}{t}\,dt.$$

The first integrand is now an analytic function, and the second integral can easily be computed. Thus,

$$e^{-x}\int_1^x \frac{(t-1)e^t}{t^2(t+1)}\,dt = -e^{-x}\int_0^1 \frac{(t-1)e^t + 1 - t^2}{t^2(t+1)}\,dt$$
$$+ e^{-x}\int_0^x \frac{(t-1)e^t + 1 - t^2}{t^2(t+1)}\,dt + e^{-x}\left(\ln x + \frac{1}{x} - 1\right).$$

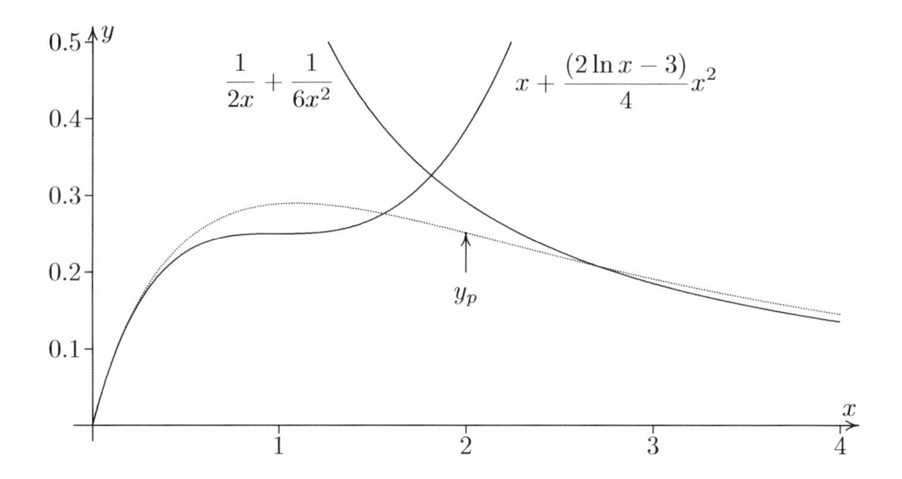

FIGURE 5.8: This shows the particular solution to the equation $xy'' + (x-1)y' - y = -1/(x+1)$, along with the first two terms of the asymptotic series at both 0 and ∞.

One of these terms is a constant times e^{-x}, which we can absorb into the homogeneous solution c_1e^{-x}. So the general solution can be written as

$$y = e^{-x}\int_0^x \frac{(t-1)e^t + 1 - t^2}{t^2(t+1)}\,dt + e^{-x}\left(\ln x + \frac{1}{x} - 1\right)$$
$$+ c_1e^{-x} + (1-x)\left(\ln(x+1) - \ln x - \frac{1}{x}\right).$$

The series expansion of this expression as $x \to 0$ is

$$y \sim c_1 - 1 + (2-c_1)x + \frac{2\ln x - 5 + 2c_1}{4}x^2 + \cdots.$$

In order for $\lim_{x\to 0} y(x) = 0$, we must have $c_1 = 1$, and the behavior at $x = 0$ is

$$y_p \sim x + (2\ln x - 3)x^2/4 + \cdots.$$

The solution is shown in figure 5.8. ∎

Problems for §5.3

For problems **1** through **10**: Find the first three terms of the asymptotic series as $x \to 0$ of a solution to the inhomogeneous equation which is not asymptotic to a solution to the associated homogeneous equation.

1 $y'' - xy = 1/x^3$

2 $y'' - y = 1/x^4$

3 $y'' + y = 1/x^3$

4 $xy'' - y = 1/x^3$

5 $xy'' + y = 1/x^2$

6 $y'' - xy' + y = 1/x^3$

7 $x^2y'' + y' - y = 1/x^2$

8 $x^3y'' - y = 1/x^3$

9 $xy'' + y' - y = 1/x^2$

10 $x^3y'' + xy' - y = 1/x^3$

For problems **11** through **18**: Find the first three terms of the asymptotic series as $x \to \infty$ of a solution to the inhomogeneous equation for which $\lim_{x\to\infty} y_p(x) = 0$.

11 $y'' - y = 1/x$

12 $y'' + y = 1/x^2$

13 $xy'' - y = 1/x$

14 $xy'' + 2y = 1/x^2$

15 $y'' + y' - xy = 1/x$

16 $y'' - xy' + x^2y = 1$

17 $xy'' + xy' - y = 1/x$

18 $xy'' - 2xy' + y = 1/x^2$

19 Use variation of parameters to find the exact solution to example 5.12 in terms of integrals, using the fact that $y_1 = 1/x$ solves the homogeneous equation.

Hint: To find a second solution y_2 to the homogeneous equation, use Abel's identity to find a first order linear equation for y_2.

20 Let $y_{p,0}$ be the solution to $y'' - xy = -1/x^2$ whose asymptotic series at 0 is $y_{p,0} \sim \ln x + (6\ln x - 5)x^3/36 + \cdots$, and let $y_{p,\infty}$ be the solution given by

$$y_{p,\infty} = \pi \mathrm{Ai}(x) \int_1^x \frac{\mathrm{Bi}(t)}{t^2}\,dt + \pi \mathrm{Bi}(x) \int_x^\infty \frac{\mathrm{Ai}(t)}{t^2}\,dt.$$

Find the constants k_1 and k_2 such that $y_{p,\infty} - y_{p,0} = k_1\mathrm{Ai}(x) + k_2\mathrm{Bi}(x)$.

Hint: Equations 5.5 and 5.6 give the local behavior of the Airy functions at $x = 0$. One can also write

$$\int_x^\infty \frac{\mathrm{Ai}(t)}{t^2}\,dt = \int_1^\infty \frac{\mathrm{Ai}(t)}{t^2}\,dx + \int_0^1 \frac{\mathrm{Ai}(t) - c_1 - c_2 t}{t^2}\,dt - \int_0^x \frac{\mathrm{Ai}(t) - c_1 - c_2 t}{t^2}\,dt + \int_x^1 \frac{c_1 + c_2 t}{t^2}\,dt,$$

$$\int_1^x \frac{\mathrm{Bi}(t)}{t^2}\,dt = -\int_0^1 \frac{\mathrm{Bi}(t) - d_1 - d_2 t}{t^2}\,dt + \int_0^x \frac{\mathrm{Bi}(t) - d_1 - d_2 t}{t^2}\,dt - \int_x^1 \frac{d_1 + d_2 t}{t^2}\,dt,$$

where the constants c_1, c_2, d_1, d_2 are defined by $\mathrm{Ai}(x) \sim c_1 + c_2 x$ and $\mathrm{Bi}(x) \sim d_1 + d_2 x$ as $x \to 0$.

For problems **21** through **26**: Use Abel's identity (equation 5.18) to determine the Wronskian of a fundamental set of solutions, up to a multiplicative constant.

21 $y'' + xy' - y = 0$

22 $xy'' - y' + 2y = 0$

23 $(1 + x^2)y'' + 2xy' - 6y = 0$

24 $y''' + x^2y' + e^x y = 0$

25 $x^3y''' + 2x^2y'' + 3xy' + 4y = 0$

26 $x^2y^{(4)} - 2y''' - y = 0$

27 First use Abel's identity (equation 5.18) to determine the Wronskian of $J_0(x)$ and $Y_0(x)$, up to a multiplicative constant, knowing that these both solve $xy'' + y' + xy = 0$. Then use the leading behaviors as $x \to \infty$ to determine this constant.

28 First use Abel's identity (equation 5.18) to determine the Wronskian of $D_{-1/2}(x)$ and $D_{-1/2}(-x)$, up to a multiplicative constant, knowing that these both solve $y'' - x^2y/4 = 0$. Then use the leading behaviors as $x \to \infty$ to determine this constant.

29 Use variation of parameters to find an integral representation of the solution to $y'' - y = 1/(x + 1)$, with $y(0) = 0$ and $\lim_{x\to\infty}(x) = 0$. Then find the first two terms of the asymptotic series of this solution both at $x = 0$ and $x = \infty$. Note that e^x and e^{-x} solve the homogeneous equation.

5.4 Local Analysis for Non-linear Equations

Very rarely can non-linear differential equations be solved exactly, so there is a high demand for approximation methods. The asymptotic series for the solutions can be very unpredictable, so it is important to develop techniques for finding terms in these series. The method of dominant balance will again be the primary tool, but the existence of special solutions allow for several possible leading behaviors to the same equation.

Example 5.16

Find the first two terms for the asymptotic series as $x \to \infty$ for a real solution to the equation

$$y' = \sqrt{y^2 - x^2}.$$

SOLUTION: It is actually more convenient if we square both sides of the equation, giving us $(y')^2 = y^2 - x^2$. However, we can only consider solutions for which $y' \geq 0$. There are 3 cases to consider.

Case 1) $y^2 \ll x^2$ as $x \to \infty$.

The equation becomes $(y')^2 \sim -x^2$, which would contradict the assumption that the solution is real.

Case 2) $(y')^2 \ll y^2$ as $x \to \infty$.

This leads to $y^2 \sim x^2$, or $y \sim \pm x$. Since $y' \geq 0$, we consider the case $y \sim x$. Then $(y')^2 \sim 1 \ll y^2$, so the assumption is consistent. To verify this leading behavior, we let $y = x + g(x)$, so the equation becomes

$$1 + 2g' + (g')^2 = x^2 + 2xg(x) + g(x)^2 - x^2.$$

The x^2 terms cancel, and since $g(x) \ll x$, we see that $g' \ll 1$ and $g(x)^2 \ll xg(x)$. So $1 \sim 2xg(x)$, or $g(x) \sim 1/(2x)$. Thus, the first two terms of the asymptotic series for a solution is

$$y \sim x + \frac{1}{2x} + \cdots.$$

Unfortunately, this is only one solution, and we expect the general solution to involve an arbitrary constant.

Case 3) $x^2 \ll y^2$ as $x \to \infty$.

This leads to $y' \sim |y|$, which is satisfied for either $y \sim Ke^x$ or $y \sim -Ke^{-x}$, where $K > 0$. The latter solution fails to satisfy $x^2 \ll y^2$, but the solution $y \sim Ke^x$ is consistent, and includes the arbitrary constant that we are looking for. To verify this is the correct leading behavior, we let $y(x) = Ke^x + g(x)$, so

$$K^2e^{2x} + 2Ke^x g' + (g')^2 = K^2e^{2x} + 2Ke^x g + g^2 - x^2.$$

The e^{2x} terms cancel, and clearly $(g')^2 \ll e^x g'$ and $g^2 \ll e^x g$, since we can assume both g and g' are smaller than e^x. This leaves us with

$$g'(x) - g(x) \sim -\frac{x^2}{2Ke^x}.$$

This first order equation can be solved to produce

$$g(x) \sim \frac{2x^2 + 2x + 1}{8Ke^x}.$$

This is indeed smaller than the first term, so we have the first two terms of the asymptotic series as $x \to \infty$,

$$y \sim Ke^x + \frac{2x^2 + 2x + 1}{8Ke^x} + \cdots.$$

Figure 5.9 shows the solutions to several equations, along with their asymptotic approximations. Note that the region between $y = -x$ and $y = x$ is a "dead zone," since any solution crossing into this region becomes complex. ◻

In this last example, the singularities can be explained by the discontinuities in the differential equation. However, the solutions may exhibit spontaneous singularities, which are difficult to predict.

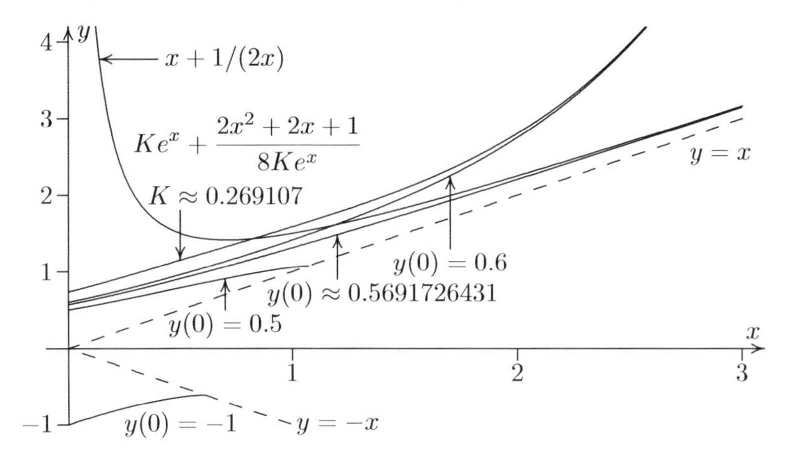

FIGURE 5.9: This shows several solutions to the equation $y' = \sqrt{y^2 - x^2}$. There is a special solution with $y(0) \approx 0.5691726431$ in which the solution behaves like $x + 1/(2x)$ as $x \to \infty$. For initial conditions below this value, the solution becomes complex when it crosses $y = \pm x$. For initial conditions above this value, the solution behaves like an exponential function as $x \to \infty$.

Example 5.17

Investigate the behavior of the real solutions to the equation $y' = y^2 + x^2$.

SOLUTION: Note that if we assume that y' is the smallest term as $x \to \infty$, we get $y = \pm ix$, which is complex. Assuming that y^2 is small as $x \to \infty$ leads to $y \sim x^3/3$, so this assumption is inconsistent. However, assuming that $x^2 \ll y^2$ as $x \to \infty$ produces the equation $y' = y^2$, whose exact solution is $y = -1/(x-c)$. Note that this solution has a spontaneous singularity at $x = c$, so we might expect the behavior of the solutions to the original equation to have spontaneous singularities.

Instead of analyzing the solution as $x \to \infty$, let us first analyze the function near a spontaneous singularity. That is, we will assume that the solution behaves like $-1/(x - c)$ as $x \to c$, where c is the location of the singularity. We can peel off this behavior by letting $y = -1/(x - c) + g(x)$ to produce

$$g' = \frac{-2g}{x - c} + g^2 + x^2.$$

Since $g \ll 1/(x - c)$, we have that $g^2 \ll g/(x - c)$, so we know that the g^2 term is small. It is unclear which other terms are small, but fortunately we have a first order linear equation, which we can solve. So

$$g(x) \sim \frac{\int x^2(x-c)^2\,dx}{(x-c)^2}.$$

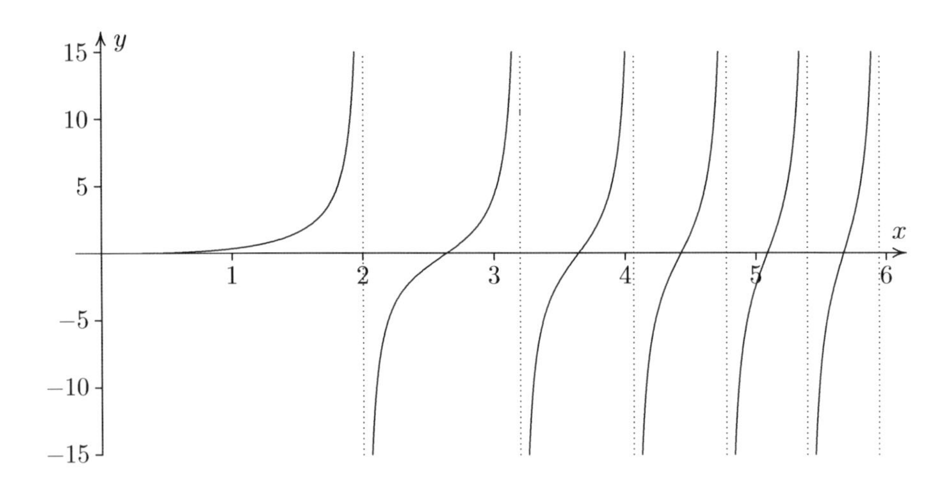

FIGURE 5.10: A plot of the solution for $y' = y^2 + x^2$ with $y(0) = 0$. Note that the poles get closer together as x increases.

When x is close to c, $x^2 \sim c^2$, so

$$\frac{\int x^2(x-c)^2\,dx}{(x-c)^2} \sim \frac{c^2\int (x-c)^2\,dx}{(x-c)^2} \sim \frac{c^2(x-c)}{3}.$$

Note that the arbitrary constant must be 0, so that g is indeed smaller than $1/(x-c)^2$ as $x \to c$. We now have the first two terms of the series

$$y \sim \frac{-1}{x-c} + \frac{c^2(x-c)}{3} + \cdots.$$

Apparently, the solution has the form of a Frobenius series

$$y = \sum_{n=0}^{\infty} a_n(x-c)^{n-1}$$

with $a_0 = -1$, $a_1 = 0$, and $a_2 = c^2/3$.

Although we have looked at the vicinity of one singularity, with some simple logic we can see that in fact, there will be an infinite number of these singularities in any given solution. By the nature of the solution, y' is always non-negative, so the solution will always be increasing. Thus, beyond each pole, the solution must increase to the point where it creates another pole. Figure 5.10 shows a plot of the solution with $y(0) = 0$.

The reason why we had trouble with the dominant balance method as $x \to \infty$ is that usually y^2 is large in comparison to x^2, but for regions of the graph close to a root, x^2 is large in comparison to y^2. In order to understand the distribution of the spontaneous singularities, we can make the substitution $y = xu$. This causes y^2+x^2 to separate into $x^2u^2+x^2 = x^2(u^2+1)$. Unfortunately, this substitution will not make the equation separable, since we also have to consider that $y' = u + xu'$. The new equation becomes

$$u' = x(u^2+1) - \frac{u}{x}. \tag{5.22}$$

Regardless of whether u is large or small, u/x will be small in comparison with $x(u^2+1)$ as $x \to \infty$. In fact, since $(u-1)^2 \geq 0$, one can see that $u^2+1 \geq 2u$ for all real u. Thus, we have the asymptotic equation $u' \sim x(u^2+1)$, whose solution is $u = \tan(x^2/2 + C)$. Because the zeros and poles will not be coincident with the true solution, we will apply the technique from subsection 2.4.4, and write

$$y = x\tan(w(x)), \qquad \text{where} \quad w(x) \sim \frac{x^2}{2} + C \quad \text{as } x \to \infty.$$

We can find further corrections to this series by finding the differential equation for which $w(x)$ solves. Substituting $u(x) = \tan(w(x))$ into equation 5.22 to obtain, after using some trig identities,

$$w' = x - \frac{\sin(2w)}{2x}. \tag{5.23}$$

It is now routine to find the first several terms for the asymptotic series of $w(x)$. See problem 1. □

In this example, there was a convergent Frobenius series for the solution centered at the spontaneous singularity, so the singularities were true poles. Not always will the spontaneous singularity be a pole, but we can easily test whether $y \sim A(x-c)^{-k}$ is a consistent leading behavior for the solution, where c is the location of the singularity.

Example 5.18

Find the nature of the spontaneous singularities of the solutions to the second order equation

$$y'' = y^2 + x^2.$$

SOLUTION: Let us see if a leading term of $y \sim A(x-c)^{-k}$ is consistent. Then $y'' \sim k(k+1)A(x-c)^{-k-2}$, so we get the equation $k(k+1)A(x-c)^{-k-2} \sim A^2(x-c)^{-2k}$. Since the exponents must be the same, we get $k = 2$, but then setting the coefficients equal produces $6A = A^2$, so $A = 6$. Thus, we have a consistent behavior of $y \sim 6(x-c)^{-2}$, which seems to indicate that the solution has a second order pole.

If the spontaneous singularity were indeed a second order pole, there would be a Frobenius series solution of the form

$$y = \sum_{n=0}^{\infty} a_n(x-c)^{n-2} = a_0(x-c)^{-2} + a_1(x-c)^{-1} + a_2 + a_3(x-c) + a_4(x-c)^2 + a_5(x-c)^3 + a_6(x-c)^4 + \cdots.$$

Then

$$y'' = \sum_{n=0}^{\infty}(n-2)(n-3)a_n(x-c)^{n-4} = 6a_0(x-c)^{-4} + 2a_1(x-c)^{-3} + 2a_4 + 6a_5(x-c) + 12a_6(x-c)^2 + \cdots.$$

Computing the series for y^2 is a bit trickier. Using the distributive law, we get

$$y^2 = a_0^2(x-c)^{-4} + 2a_0a_1(x-c)^{-3} + (2a_0a_2 + a_1^2)(x-c)^{-2} + (2a_0a_3 + 2a_1a_2)(x-c)^{-1} + (2a_0a_4 + 2a_1a_3 + a_2^2) + (2a_0a_5 + 2a_1a_4 + 2a_2a_3)(x-c) + (2a_0a_6 + 2a_1a_5 + 2a_2a_4 + a_3^2)(x-c)^2 + \cdots.$$

Finally we can write the x^2 term in powers of $(x-c)$.

$$x^2 = (c + (x-c))^2 = c^2 + 2c(x-c) + (x-c)^2.$$

We can now equate the coefficients for each power of $(x-c)$ in the equation $y'' = y^2 + x^2$.

$(x-c)^{-4}$: $6a_0 = a_0^2$, so $a_0 = 6$.
$(x-c)^{-3}$: $2a_1 = 2a_0a_1$, so $a_1 = 0$.
$(x-c)^{-2}$: $0 = 2a_0a_2 + a_1^2$, so $a_2 = 0$.
$(x-c)^{-1}$: $0 = 2a_0a_3 + 2a_1a_2$, so $a_3 = 0$.
$(x-c)^{0}$: $2a_4 = 2a_0a_4 + 2a_1a_3 + a_2^2 + c^2$, so $a_4 = -c^2/10$.
$(x-c)^{1}$: $6a_5 = 2a_0a_5 + 2a_1a_4 + 2a_2a_3 + 2c$, so $a_5 = -c/3$.
$(x-c)^{2}$: $12a_6 = 2a_0a_6 + 2a_1a_5 + 2a_2a_4 + a_3^2 + 1$, which simplifies to $12a_6 = 12a_6 + 1$.

This last equation has no solution. This is similar to the situation we encountered with Frobenius solutions when the indicial roots differed by an integer. In section 4.4, an inconsistent equation for one of the coefficients indicated that two solutions were interfering with each other. The secret was to introduce terms that were multiplied by a logarithm. Thus, we should replace the term $a_6(x-c)^4$ with $a_6(x-c)^4 + b_6\ln(x-c)(x-c)^4$. Then y'' will have the terms $12a_6(x-c)^2 + 7b_6(x-c)^2 + 12b_6\ln(x-c)(x-c)^2$, and y^2 will have the terms $2a_0a_6(x-c)^2 + 2a_0b_6\ln(x-c)(x-c)^2$. So equating the coefficients of similar terms, we have

$\ln(x-c)(x-c)^2$: $12b_6 = 2a_0b_6$, which is always true, since $a_0 = 6$.

$(x-c)^2$: $12a_6 + 7b_6 = 2a_0a_6 + 1$, so $b_6 = 1/7$.

We cannot determine a_6 from these equations. In fact, this becomes the second arbitrary constant in our second order equation solution (the first arbitrary constant being c). The presence of a logarithm term indicates that the spontaneous singularity is *not* a second order pole, but rather a more complicated singularity. We do, however, have many terms of the asymptotic series

$$y \sim 6(x-c)^{-2} - \frac{c^2(x-c)^2}{10} - \frac{c(x-c)^3}{3} + \ln(x-c)\frac{(x-c)^4}{7} + a_6(x-c)^4 + \cdots.$$

□

When there are only two terms in the equation, the method of dominant balance may seem pointless. However, the technique of searching for a solution of the form $y \sim A(x-c)^{-k}$ can prove very useful.

Example 5.19

Find the first three terms of the leading behavior as $x \to \infty$ for the solutions of the equation

$$y^2y'' = 6.$$

SOLUTION: If $y \sim A(x-c)^{-k}$, then $y'' \sim Ak(k+1)(x-c)^{-k-2}$, so we have $A^3k(k+1)(x-c)^{-3k-2} \sim 6$. Thus, $k = -2/3$ and $A = -3$, giving us $y \sim -3(x-c)^{2/3}$. However, this is not the complete solution, since it only involves one arbitrary constant, and this is a second order equation. There is no hope of a second constant appearing later on in the asymptotic series, because $y = -3(x-c)^{2/3}$ is an *exact* solution. In fact, this is a special solution, for the general solution has a totally different behavior.

In order to determine another possible leading behavior, we need to consider the possibility that the leading term of y goes to zero as we take two derivatives. That is, the leading behavior of y might be Cx. If we assume that $y = Cx + g(x)$, where $g \ll x$ as $x \to \infty$, then $y'' = g''$, so the equation becomes $C^2x^2g'' \sim 6$. Solving for g, we have $g \sim -6\ln x/C^2 + C_2$. Thus, we have consistency if the first three terms are $y \sim Cx - 6\ln x/C^2 + C_2$ as $x \to \infty$. This would give us a solution with two arbitrary constants, so we would have the behavior of the general solution.

To verify that these are indeed the first three terms, let us find the forth term. Letting $y = Cx - 6\ln x/C^2 + C_2 + h(x)$, we get the equation

$$\left(Cx - \frac{6\ln x}{C^2} + C_2 + h(x)\right)^2\left(\frac{6}{C^2x^2} + h''\right) = 6.$$

Multiplying this out, we get

$$6 - \frac{18\ln x}{C^3x} + O\left(\frac{\ln(x)^2}{x^2}\right) + C^2x^2h'' + O(h''x\ln x) = 6.$$

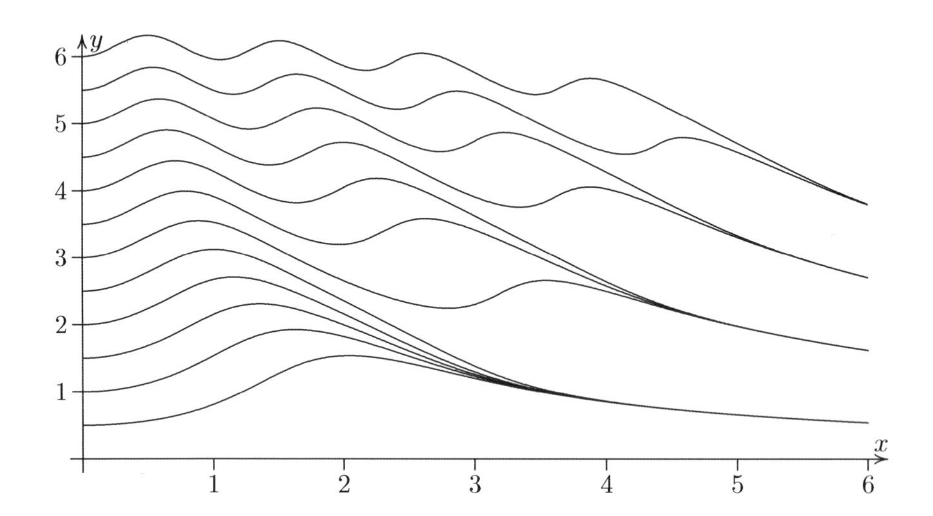

FIGURE 5.11: A plot of the solutions to $y' = \sin(xy)$ for the initial conditions $y(0) = 0.5, 1.0, 1.5, \ldots, 6.0$. Note that the solutions bunch together as $x \to \infty$. See problem 2 for an explanation.

The 6's cancel, and in order to get dominant balance, we need

$$h'' \sim \frac{18 \ln x}{C^5 x^3}.$$

Integrating twice, we have

$$h(x) \sim \frac{9 \ln(x)}{C^5 x} + \frac{27}{2C^5 x}.$$

Since this is indeed smaller than a constant, we have verified that the first terms of the asymptotic series of the solutions are

$$y \sim Cx - \frac{6 \ln x}{C^2} + C_2 + \frac{9 \ln(x)}{C^5 x} + \frac{27}{2C^5 x} + \cdots. \qquad \square$$

This section has given only a small sample of the type of behaviors that can be exhibited by non-linear differential equations. There are, in fact, examples for which the solutions appear to be chaotic, having *no* asymptotic behavior as $x \to \infty$. One example is the damped and driven Duffing equation,

$$y'' + \delta y' + \alpha y + \beta y^3 = \gamma \cos(\omega x).$$

Although we can analyze this equation if $\gamma = 0$ (see examples 9.2 and 9.8), there is no analysis possible if both $\delta \neq 0$ and $\gamma \neq 0$ [12].

Problems for §5.4

1 Find the first three terms for the asymptotic series for $w(x)$ in equation 5.23.

2 Find the possible leading behaviors as $x \to \infty$ for the equation $y' = \sin(xy)$. See figure 5.11 for the graphs of the solution. What are the second order corrections to these leading behaviors?

For problems **3** through **13**: Use the method of dominant balance to find the first two terms of the asymptotic series as $x \to \infty$ of all real solutions to the non-linear equation.

3 $y' = \sqrt{x^2 + y^2}$
4 $y' = \sqrt{y^2 - x}$
5 $y' = \sqrt[3]{y^3 - x^3}$
6 $y' = \sqrt{x^2 - y^2}$
7 $y'' = \sqrt{y^2 - x^2}$
8 $y' = \sqrt{y} - \sqrt{x}$
9 $y' = (1/y) + x$
10 $y' = 1/(y + x)$
11 $y' = y - x\sqrt{y}$
12 $y' = \sqrt{xy} - 1$
13 $y' = \sqrt{xy} - y$

For problems **14** through **19**: Find the first three terms of the asymptotic series for a solution to the non-linear differential equation, centered at a spontaneous singularity $x = c$.

14 $y' = y^2 - x^2$
15 $y' = y^2 + e^x$
16 $y' = y^2 + y$
17 $y' = y^2 - y + x^2$
18 $y' = y^3 + x^3$
19 $y'' = y^2 + y$

For problems **20** through **27**: Find the asymptotic series for a solution to the non-linear differential equation, centered at a spontaneous singularity $x = c$. Continue the series until a second arbitrary constant appears. Does the series contain logarithm terms, or is the nature of the singularity a true pole?

20 $y'' = y^2 + e^{-x}$
21 $y'' = 6y^2 + x$
22 $y'' = 6y^2 + y$
23 $y'' = 6y^2 + y + x$
24 $y'' = 6y^2 + y + x^2$
25 $y'' = 2y^3 + xy$
26 $y'' = 2y^3 + xy + \pi$
27 $y'' = 2y^3 + x^2y$

For problems **28** through **33**: For the following non-linear differential equations, first find a special solution of the form $y = A(x - c)^k$. Then find the asymptotic series as $x \to \infty$ for the general solution, up to the first five non-zero terms, which contains as many arbitrary constants as the order of the equation.

28 $y^3y'' = -4$
29 $y^3y''' = 60$
30 $y^3y'y'' = 450$
31 $y^2y''y''' = -300$
32 $y^2y'' = 6$
33 $yy'y''' = -288$

Chapter 6

Difference Equations

Whereas differential equations relate a function to its derivatives, a difference equation will be a relationship on a *sequence* of numbers. We have already seen difference equations in sections 4.3 and 4.4, when we found a recursion relationship for the coefficients of a series. This recursion formula is an example of a difference equation on the sequence of coefficients. Sometimes we were able to solve the recursion relationship, which allowed us to express the nth coefficient as a function of n, and hence can express the solution to the differential equation as an infinite sum. But there are also times where the recursion relationship cannot be solved in closed form. Nonetheless, we will be able to use asymptotics on this recursion relationship to extract important information from the sequence, such as the radius of convergence. This chapter will focus on the rich topic of difference equations, their uses, and how to find the asymptotic approximations to the solutions.

6.1 Classification of Difference Equations

Difference equations can be expressed in two different ways. The most common way is to express a relationship between consecutive terms of the sequence, such as a_n, a_{n+1}, and a_{n+2}. If we solve the equation for the term with the largest subscript, we get a recursion formula of the form

$$a_{n+k} = F(n, a_n, a_{n+1}, \ldots, a_{n+k-1}). \tag{6.1}$$

For example, we might have the equation

$$a_{n+2} = (n+4)a_{n+1} - (n+1)a_n. \tag{6.2}$$

If we are given the first k terms of the sequence, we can use the recursion formula to determine all of the remaining terms. This recursion formula is called a *difference equation of order k*. For example, equation 6.2 is a second order difference equation, and if we start with the initial conditions $a_0 = 0$ and $a_1 = 1$, we can plug in $n = 0, 1, 2, \cdots$ to determine the sequence

$$\{0, 1, 4, 18, 96, 600, 4320, \ldots\}.$$

However, simply finding the first few terms of the sequence is not considered a solution. A *solution* to the difference equation is a formula for the nth term of a sequence which satisfies the recursion relation. For example, $a_n = n \cdot n!$ would be a solution to equation 6.2, since we can verify that it satisfies the difference equation.

$$\begin{aligned} a_{n+2} &= (n+2)\cdot(n+2)! = [(n+2)(n+2)](n+1)n! \\ &= [(n+4)(n+1) - n](n+1)n! \\ &= (n+4)(n+1)(n+1)! - (n+1)n\cdot n! = (n+4)a_{n+1} - (n+1)a_n. \end{aligned}$$

A difference equation can also be expressed in terms of the *difference operator,* hence the name. The difference of a sequence a_n is given by

$$\Delta a_n = a_{n+1} - a_n.$$

For example, for the sequence $a_n = n^2$, $\Delta a_n = (n+1)^2 - n^2 = (n^2+2n+1) - n^2 = 2n+1$. The difference operator is the discrete analog of a derivative, since it measures the change of a quantity that varies over discrete time intervals (such as interest in a savings account.) We can iterate the difference operator, producing

$$\begin{aligned} \Delta^2 a_n &= \Delta a_{n+1} - \Delta a_n \\ &= (a_{n+2} - a_{n+1}) - (a_{n+1} - a_n) = a_{n+2} - 2a_{n+1} + a_n, \\ \Delta^3 a_n &= a_{n+3} - 3a_{n+2} + 3a_{n+1} - a_n, \\ \Delta^4 a_n &= a_{n+4} - 4a_{n+3} + 6a_{n+2} - 4a_{n+1} + a_n. \end{aligned}$$

It is fairly easy to prove using induction that the patten of binomial coefficients continues (See problem 1).

$$\Delta^k a_n = \sum_{i=0}^{k} (-1)^{k-i} \binom{k}{i} a_{n+i}, \qquad \text{where } \binom{k}{i} = \frac{k!}{i!(k-i)!}. \tag{6.3}$$

Every difference equation of the form $a_{n+k} = F(n, a_n, a_{n+1}, \ldots, a_{n+k-1})$ can be converted into an equation involving the difference operator. It is easy to show that

$$\begin{aligned} a_{n+1} &= \Delta a_n + a_n, \\ a_{n+2} &= \Delta^2 a_n + 2\Delta a_n + a_n, \\ a_{n+3} &= \Delta^3 a_n + 3\Delta^2 a_n + 3\Delta a_n + a_n. \end{aligned}$$

In fact, we can use a similar induction argument (see problem 2) to show that

$$a_{n+k} = \sum_{i=0}^{k} \binom{k}{i} \Delta^i a_n. \tag{6.4}$$

Example 6.1
Convert the *Fibonacci* difference equation

$$a_n = a_{n-1} + a_{n-2}$$

into a form involving only differences.
SOLUTION: Although this equation expresses a relationship between three consecutive terms, it is *not* in the standard form of equation 6.1, since it involves terms previous to a_n. This is easy to fix by shifting the index, that is, replacing every n with $n + 2$ in the equation. The new equation is then

$$a_{n+2} = a_{(n+2)-1} + a_{(n+2)-2} = a_{n+1} + a_n.$$

Using equation 6.4, we express the equation as

$$(\Delta^2 a_n + 2\Delta a_n + a_n) = (\Delta a_n + a_n) + a_n.$$

Simplifying, and solving for the highest difference, we get

$$\Delta^2 a_n = a_n - \Delta a_n.$$

∎

Many of the properties of differential equations can also be applied to difference equations. Having the difference equation in terms of the difference operator allows us to determine the *corresponding differential equation,* found by replacing the discrete variable n with a continuous variable x, so the sequence a_n is replaced with the function $y(x)$, Δa_n is replaced with $y'(x)$, $\Delta^2 a_n$ becomes $y''(x)$, etc. So the corresponding differential equation to the difference equation in example 6.1 is

$$y'' = y - y'.$$

Difference equations are classified the way that their corresponding differential equations are classified. That is, a *linear* difference equation is one of the form

$$a_{n+k} + p_{k-1}(n)a_{n+k-1} + p_{k-2}(n)a_{n+k-2} + \cdots + p_1(n)a_{n+1} + p_0(n)a_n = g(n). \tag{6.5}$$

A non-linear difference equation is one that cannot be expressed in the form of equation 6.5. A linear equation is *homogeneous* if $g(n) = 0$, but *inhomogeneous* if $g(n) \neq 0$. As with linear differential equations, the general solution can be created using a handful of particular solutions.

THEOREM 6.1
For a kth order linear difference equation in the form of equation 6.5, the general solution will be in the form

$$a_n = c_1 f_1(n) + c_2 f_2(n) + \cdots + c_k f_k(n) + f_p(n),$$

where $f_p(n)$ is a particular solution to the difference equation, and the $f_i(n)$ are solutions to the associated homogeneous equation. In the case of a homogeneous equation, $f_p(n)$ will be 0.

PROOF: We let $f_p(n)$ be the sequence which solves the equation with $a_0 = a_1 = \cdots a_{k-1} = 0$. Likewise, we can let $h_i(n)$ be the sequence which solves the equation with $a_{i-1} = 1$, and all other a_j with $0 \le j \le k-1$ being 0. Finally, we can let $f_i(n) = h_i(n) - f_p(n)$, which will be a solution to the associated homogeneous equation. By the linearity of the equation, the solution with initial conditions $a_0 = c_1$, $a_1 = c_2, \ldots, a_{k-1} = c_k$ would be given by $a_n = c_1 f_1(n) + c_2 f_2(n) + \cdots + c_k f_k(n) + f_p(n)$. Thus, this is the general solution to the equation. If $g(n) = 0$, then the sequence $a_n = 0$ would solve the equation, so $f_p(x)$ would be 0. ∎

6.1.1 Anti-differences

The main tool for solving differential equations is the anti-derivative, or integral, so naturally the primary tool that we will be using for solving difference equations is the anti-difference, or the sum of the sequence,

$$\Delta^{-1} a_n = \sum_{k=0}^{n-1} a_k + C.$$

Note that an arbitrary constant is added, as in the integral. Because of this constant, it does not matter what our starting point is in the sum. Often, we will start the sum at $k = 1$ because a_k is undefined at $k = 0$. The notation Δ^{-1} is referred to as the *indefinite sum*. It is easy to check that the difference of the indefinite sum reproduces a_n,

$$\Delta(\Delta^{-1} a_n) = \Delta^{-1} a_{n+1} - \Delta^{-1} a_n = \left(\sum_{k=0}^{n} a_k + C \right) - \left(\sum_{k=0}^{n-1} a_k + C \right) = a_n.$$

A *definite sum* is written with a definite starting point, so there will not be an arbitrary constant added. One can form a table of definite sums, which can be used to sum any polynomial using the linear properties of the sum.

$$\sum_{k=0}^{n-1} 1 = n,$$
$$\sum_{k=0}^{n-1} k = \frac{n(n-1)}{2},$$
$$\sum_{k=0}^{n-1} k^2 = \frac{n(n-1)(2n-1)}{6},$$

$$\sum_{k=0}^{n-1} k^3 = \frac{n^2(n-1)^2}{4}.$$

This table can be extended using the following property of binomial coefficients (See problem 10):

$$\sum_{k=0}^{n-1} \binom{k}{m} = \binom{n}{m+1} \qquad \text{for all } m \geq 0. \tag{6.6}$$

Example 6.2

Use equation 6.6 to find $\sum_{k=0}^{n-1} k^4$.

SOLUTION: For a fixed non-negative integer m, $\binom{k}{m}$ is an mth degree polynomial in k. Thus, we ought to be able to find constants A, B, C, D, and E such that

$$k^4 = A\binom{k}{4} + B\binom{k}{3} + C\binom{k}{2} + D\binom{k}{1} + E\binom{k}{0}.$$

Expanding the binomial coefficients, we get

$$k^4 = A\frac{k(k-1)(k-2)(k-3)}{24} + B\frac{k(k-1)(k-2)}{6} + C\frac{k(k-1)}{2} + Dk + E.$$

If we plug in $k = 0$ into both sides of the equation, all but one term will go to 0, so we have $E = 0$. Likewise, plugging in $k = 1$ causes most of the terms to go to zero, giving us $D = 1$. Plugging in $k = 2$, we obtain $16 = C + 2D$, so $C = 14$. Continuing this way, we can find $B = 36$ and $A = 24$. Thus, we can rewrite k^4 as

$$k^4 = 24\binom{k}{4} + 36\binom{k}{3} + 14\binom{k}{2} + \binom{k}{1}.$$

Using equation 6.6, we have

$$\begin{aligned}
\sum_{k=0}^{n-1} k^4 &= 24\binom{n}{5} + 36\binom{n}{4} + 14\binom{n}{3} + \binom{n}{2} \\
&= 24\frac{n(n-1)(n-2)(n-3)(n-4)}{120} + 36\frac{n(n-1)(n-2)(n-3)}{24} \\
&\qquad + 14\frac{n(n-1)(n-2)}{6} + \frac{n(n-1)}{2} \\
&= \frac{n(n-1)(2n-1)(3n^2-3n-1)}{30}.
\end{aligned}$$

□

We can also find the anti-difference for geometric sequences, valid if $r \neq 1$.

$$\sum_{k=0}^{n-1} r^k = \frac{r^n - 1}{r - 1},$$
$$\sum_{k=0}^{n-1} kr^k = \frac{r^n(nr - n - r) + r}{(r-1)^2}.$$

This table of anti-differences will be an invaluable tool for solving difference equations.

Example 6.3

Solve the inhomogeneous difference equation

$$a_{n+2} - 2a_{n+1} + a_n = n.$$

SOLUTION: Converting the recursion relation to difference equation form, we have the equation

$$\Delta^2 a_n = n.$$

Thus, we must take the anti-difference twice. From the above table, we find that

$$\Delta^{-1} n = \frac{n(n-1)}{2} + C_1 = \frac{n^2}{2} - \frac{n}{2} + C_1.$$

We now take the anti-difference of this quantity, using the property that the anti-difference is a linear operator.

$$\begin{aligned}
\Delta^{-1}\left(\frac{n^2}{2} - \frac{n}{2} + C_1\right) &= \frac{1}{2}(\Delta^{-1}n^2) - \frac{1}{2}(\Delta^{-1}n) + C_1\Delta^{-1}1 \\
&= \frac{n(n-1)(2n-1)}{12} - \frac{n(n-1)}{4} + C_1 n + C_2 \\
&= \frac{n(n-1)(n-2)}{6} + C_1 n + C_2.
\end{aligned}$$

□

Occasionally, the sum for the sequence a_n will telescope. This happens whenever there is a function $g(x)$ for which $a_n = g(n+1) - g(n)$. In which case,

$$\sum_{k=0}^{n-1} a_k = (g(1) - g(0)) + (g(2) - g(1)) + (g(3) - g(2)) + \cdots + (g(n) - g(n-1)).$$

All but two of the terms will cancel, leaving

$$\sum_{k=0}^{n-1} a_k = g(n) - g(0).$$

Example 6.4
Find the anti-difference of the sequence $a_n = 1/(n^2 + 3n + 2)$.
SOLUTION: Using partial fraction decomposition, we find that

$$\frac{1}{n^2 + 3n + 2} = \frac{1}{n+1} - \frac{1}{n+2}.$$

If the right hand side is to be $g(n+1) - g(n)$, then $g(n)$ must be $-1/(n+1)$. This means that

$$\sum_{k=0}^{n-1} \frac{1}{k^2 + 3k + 2} = g(n) - g(0) = \frac{-1}{n+1} - (-1) = 1 - \frac{1}{n+1}.$$

Since we will add an arbitrary constant, the 1 would be absorbed by this constant. Thus,

$$\Delta^{-1} \frac{1}{n^2 + 3n + 2} = C - \frac{1}{n+1}.$$

□

6.1.2 Regular and Irregular Singular Points

There is only one direction that we can do the analysis for a difference equation, and that is the behavior of the solutions as $n \to \infty$. We classify the point at ∞ for a linear homogeneous difference equation as either an ordinary point, a regular singular point, or an irregular singular point by classifying the point $x = \infty$ for the corresponding differential equation. That is, for a difference equation of the form

$$\Delta^k a_n + q_{k-1}(n)\Delta^{k-1} a_n + \cdots q_2(n)\Delta^2 a_n + q_1(n)\Delta a_n + q_0(n) a_n = 0,$$

we first form the corresponding differential equation

$$y^{(k)}(x) + q_{k-1}(x) y^{(k-1)}(x) + \cdots q_2(x) y''(x) + q_1(x) y'(x) + q_0(x) y(x) = 0.$$

Then we make the transformation $x = 1/t$ to move the point at infinity to 0. Recall that this transformation causes the replacements

$$\begin{aligned} x &\Rightarrow 1/t, \\ y'(x) &\Rightarrow -t^2 y'(t), \\ y''(x) &\Rightarrow t^4 y''(t) + 2t^3 y'(t), \\ y'''(x) &\Rightarrow -t^6 y'''(t) - 6t^5 y''(t) - 6t^4 y'(t), \\ y^{(4)}(x) &\Rightarrow t^8 y^{(4)}(t) + 12t^7 y'''(t) + 36t^6 y''(t) + 24t^5 y'(t), \ldots. \end{aligned}$$

After simplifying, we will have an equation of the form

$$y^{(k)}(t) + p_{k-1}(t) y^{(k-1)}(t) + \cdots + p_2(t) y''(t) + p_1(t) y'(t) + p_0(t) y(t) = 0.$$

Then the classification of the original difference equation depends on the classification of the point $t = 0$ of this differential equation. So if $p_0(t)$, $p_1(t), \ldots$,

$p_{k-1}(t)$ are analytic at $t = 0$, then the point at infinity of the difference equation is an ordinary point. If at least one of these functions is not analytic, but $tp_{k-1}(t)$, $t^2 p_{k-2}(t), \ldots, t^{k-1}p_1(t)$, and $t^k p_0(t)$ all have removable singularities at $t = 0$, then the point at infinity of the difference equation is a regular singular point. Finally, if neither of these conditions are satisfied, then the point at ∞ for the difference equation is an irregular singular point.

Example 6.5
Classify the point at infinity for the difference equation

$$n^4 a_{n+2} - 2n^4 a_{n+1} + (n^2 - 1)^2 a_n = 0.$$

SOLUTION: First, we must convert the equation into the difference notation. Replacing a_{n+1} with $\Delta a_n + a_n$, and a_{n+2} with $\Delta^2 a_n + 2\Delta a_n + a_n$ yields

$$n^4(\Delta^2 a_n + 2\Delta a_n + a_n) - 2n^4(\Delta a_n + a_n) + (n^4 - 2n^2 + 1)a_n = 0, \qquad \text{or}$$

$$n^4 \Delta^2 a_n + (1 - 2n^2)a_n = 0.$$

Thus, the corresponding differential equation is $x^4 y'' + (1 - 2x^2)y = 0$. We must classify the point at infinity for this equation, so we replace x with $1/t$, $y'(x)$ with $-t^2 y'(t)$, and $y''(x)$ with $t^4 y''(t) + 2t^3 y'(t)$, to get

$$y''(t) + \frac{2}{t} y'(t) + \left(1 - \frac{2}{t^2}\right) y(t) = 0.$$

Since $p_1(t) = 2/t$ and $p_0(t) = 1 - (2/t^2)$ are not analytic at $t = 0$, this is not an ordinary point. However, $tp_1(t) = 2$ and $t^2 p_0(t) = t^2 - 2$ are analytic at $t = 0$, so this is a regular singular point. Thus the point at infinity for the difference equation is a regular singular point. □

Understanding the nature of the point at infinity is important, since it determines the nature of the solution as $n \to \infty$. If the point at infinity is an ordinary point, then all of the solutions can be expressed in terms of a convergent Taylor series:

$$a_n = \sum_{k=0}^{\infty} A_k n^{-k} = A_0 + \frac{A_1}{n} + \frac{A_2}{n^2} + \frac{A_3}{n^3} + \cdots.$$

That is, for sufficiently large n, this series will converge to a solution of the difference equation. For an mth order equation, A_0 through A_{m-1} will be the arbitrary constants, and all of the remaining terms will depend on these constants.

For difference equations with a regular singular point at infinity, there will be at least one solution in the form of a convergent Frobenius series:

$$a_n = \sum_{k=0}^{\infty} A_k n^{r-k} = n^r \left(A_0 + \frac{A_1}{n} + \frac{A_2}{n^2} + \frac{A_3}{n^3} + \cdots\right).$$

Other solutions will either be in this form, or possibly multiplied by factors of $\ln n$. Thus, the nature of the series for all solutions is determined by the leading behavior, and there is an easy way to determine the leading behaviors for a difference equation with a regular singular point at infinity.

THEOREM 6.2

If $f(x)$ is a solution to a linear homogeneous differential equation with an ordinary point or regular singular point at infinity, then the corresponding difference equation will have a solution with the same leading behavior, replacing x with n. In fact, if $f(x)$ is an algebraic solution to *any* differential equation defined for all $x > 0$, then a solution to the corresponding difference equation will have the same leading behavior as $n \to \infty$.

See [4] for the proof. The key is to notice that if $y(n)$ is an algebraic function, then $y'(n) \sim \Delta y(n)$. Thus, the leading behavior of the differential equation will be consistent for the difference equation as well.

Example 6.6

Find the possible leading behaviors for the linear difference equation

$$n^4 a_{n+2} - 2n^4 a_{n+1} + (n^2 - 1)^2 a_n = 0.$$

SOLUTION: We have already seen in example 6.5 that the corresponding differential equation is $x^4 y'' + (1 - 2x^2)y = 0$, which has a regular singular point at infinity. Thus, the solutions to the differential equation will behave like x^k for some k. Substituting $y \sim x^k$ and $y'' \sim k(k-1)x^{k-2}$ produces an equation for k: $k(k-1) - 2 = 0$. The two solutions for this quadratic equation are $k = -1$ and $k = 2$. Thus, $y \sim 1/x$ and $y \sim x^2$ are two possible behaviors to the differential equation as $x \to \infty$, so by theorem 6.2 the two possible behaviors to the difference equation are $a_n \sim 1/n$ and $a_n \sim n^2$. □

Note that theorem 6.2 can apply to non-linear equations as well.

Example 6.7

Find the leading behavior for the non-linear difference equation

$$a_{n+1} = a_n + \frac{2}{n a_n}.$$

SOLUTION: The corresponding differential equation becomes

$$y' = \frac{2}{xy},$$

which is separable. Separating the variables and integrating, we obtain

$$\frac{y^2}{2} = 2 \ln x + C,$$

so the leading behavior of the solution is $y \sim 2\sqrt{\ln x}$. Since this grows as an algebraic function, we can apply theorem 6.2, so the leading behavior to the difference equation is also $a_n \sim 2\sqrt{\ln n}$.

There is a trick for finding the next terms of the asymptotic behavior. Since the leading behavior of a_n involves a square root, we should consider the sequence $b_n = a_n^2$. To find the difference equation for b_n, we square both sides of the original equation.

$$a_{n+1}^2 = a_n^2 + \frac{4}{n} + \frac{4}{n^2 a_n^2}, \qquad \text{so} \qquad b_{n+1} = b_n + \frac{4}{n} + \frac{4}{n^2 b_n}.$$

If we let $b_n = 4\ln(n) + c_n$, then c_n satisfies

$$4\ln\left(1 + \frac{1}{n}\right) + c_{n+1} = c_n + \frac{4}{n} + \frac{4}{n^2(4\ln n + c_n)}.$$

Clearly the c_n in the denominator is small, and we can expand $\ln(1 + 1/n)$ in a Maclaurin series. This produces

$$\Delta c_n \sim \frac{2}{n^2} + \frac{1}{n^2 \ln n} \sim \frac{2}{n^2}.$$

So $c_n \sim -2/n + C$, and we have $a_n \sim \sqrt{4\ln n + C - 2/n + \cdots}$ as $n \to \infty$. □

If the behavior of the solution to the corresponding differential equation is exponential, then the leading behavior of the solution to the difference equation will not be the same. However, there is a way to determine the leading behavior of the difference equation solution from the differential equation solution, but it is complicated. The interested reader can refer to [20].

Problems for §6.1

1 Use induction to prove equation 6.3.

2 Use induction to prove equation 6.4.

For problems **3** through **9**: Find the corresponding differential equation for each of these difference equations, and determine whether the point at infinity is ordinary, a regular singular point, or an irregular singular point.

3 $a_n = a_{n-1} + (a_{n-1}/n^2)$

4 $a_n = 2a_{n-1} - 2a_{n-2}$

5 $a_{n+2} = 2a_{n+1} - a_n - (a_n/n^2)$

6 $a_{n+2} = 2a_{n+1} + (a_n/n^3)$

7 $a_{n+2} = a_{n+1} - (a_n/n^2)$

8 $a_{n+2} = 2a_{n+1} - a_n + (2a_n/n^4)$

9 $n^4 a_{n+2} = 2n^4 a_{n+1} - 2n^3 a_{n+1} - n^4 a_n + 2n^3 a_n - a_n$

10 Use induction to prove equation 6.6.

11 Use the method of example 6.2 to find

$$\sum_{k=0}^{n-1} k^5.$$

For problems **12** through **21**: Compute the following sums, using the formulas from the text, or by finding a telescoping sum.

12 $\sum_{k=0}^{n-1} k^2 + 3k - 4$

13 $\sum_{k=0}^{n-1} (k+1)^4$

14 $\sum_{k=0}^{n-1} 3^k$

15 $\sum_{k=0}^{n-1} k(-1)^k$

16 $\sum_{k=0}^{n-1} 2^k(k+1)$

17 $\sum_{k=0}^{n-1} \frac{1}{4k^2 - 1}$

18 $\sum_{k=0}^{n-1} \frac{1}{k^2 + 4k + 3}$

19 $\sum_{k=0}^{n-1} \ln\left(\frac{k+2}{k+1}\right)$

20 $\sum_{k=0}^{n-1} \cos(k)$

21 $\sum_{k=0}^{n-1} \sec(k)\sec(k+1)$

For problems **22** through **31**: Use theorem 6.2 to find the possible leading behaviors of the solutions to the difference equation.

22 $a_{n+1} = a_n + (1/n)$

23 $a_{n+1} = a_n + (2a_n/n)$

24 $a_{n+1} = a_n - (1/n^2)$

25 $a_{n+1} = a_n + (1/a_n)$

26 $a_{n+1} = a_n + n/a_n$

27 $a_{n+1} = a_n + 6n^2/a_n$

28 $a_{n+1} = a_n - (n/a_n^2)$

29 $n^2 a_{n+2} - 2n^2 a_{n+1} + (n^2 - 6)a_n = 0$

30 $n^2 a_{n+2} = (2n^2 - n)a_{n+1} + (n + 1 - n^2)a_n$

31 $n^2 a_{n+2} = (2n^2 + n)a_{n+1} - (n^2 + n + 1)a_n$

6.2 First Order Linear Equations

Just as all first order linear differential equations can be solved in terms of integrals, first order linear difference equations can be solved in terms of anti-differences. We will begin by studying the homogeneous equations, which have the form

$$a_{n+1} = p(n)a_n.$$

It helps to study the pattern of the first few terms.

$$a_1 = a_0 p(0), \quad a_2 = a_0 p(0)p(1), \quad a_3 = a_0 p(0)p(1)p(2), \ldots.$$

We can express the pattern as a lemma.

LEMMA 6.1
If $a_{n+1} = p(n)a_n$, then

$$a_n = a_0 \prod_{k=0}^{n-1} p(k). \tag{6.7}$$

PROOF: We will proceed using induction. It is clearly true for $n = 1$, and if we assume that it is true for the previous case, we have

$$a_n = p(n-1)a_{n-1} = p(n-1)a_0 \prod_{k=0}^{n-2} p(k) = a_0 \prod_{k=0}^{n-1} p(k).$$

Hence, equation 6.7 holds for all n. ∎

Computing the product of $p(k)$ is easy if $p(k)$ is a rational function. In principle, any rational function can be factored with linear terms in the numerator and denominator. Thus, we only have to determine

$$\prod_{k=0}^{n-1} (k+a) = a(a+1)(a+2)\cdots(a+n-1).$$

If we multiply this expression by $\Gamma(a)$, then we have already seen how the gamma function has a "zipper" property that forms $\Gamma(a+n)$. Thus, we have

$$\boxed{\prod_{k=0}^{n-1} (k+a) = \frac{\Gamma(n+a)}{\Gamma(a)}, \qquad \text{provided} \quad a \neq 0, -1, -2, -3, \ldots.} \tag{6.8}$$

Example 6.8
Compute

$$\prod_{k=0}^{n-1} \frac{k^2+4k+4}{k^2+4k+3}.$$

SOLUTION: Factoring the numerator and denominator, we have

$$\frac{k^2+4k+4}{k^2+4k+3} = \frac{(k+2)^2}{(k+1)(k+3)}.$$

The product can be computed factor by factor, so

$$\prod_{k=0}^{n-1} \frac{(k+2)^2}{(k+1)(k+3)} = \frac{\Gamma(n+2)^2}{\Gamma(2)^2} \frac{\Gamma(1)}{\Gamma(n+1)} \frac{\Gamma(3)}{\Gamma(n+3)}.$$

This now simplifies to

$$\frac{2(n+1)^2\Gamma(n+1)^2}{\Gamma(n+1)^2(n+1)(n+2)} = \frac{2n+2}{n+2}.$$ ∎

By using equation 6.8 along with the obvious product

$$\prod_{k=0}^{n-1} r = r^n,$$

we can take the product of virtually any rational function, provided that none of the factors in the product evaluates to 0 or ∞.

Example 6.9

Compute $\displaystyle\prod_{k=0}^{n-1} \frac{k^2+1}{4k^2-1}$.

SOLUTION: We can factor this rational function by using complex numbers

$$\frac{k^2+1}{4k^2-1} = \frac{(k+i)(k-i)}{4\left(k+\frac{1}{2}\right)\left(k-\frac{1}{2}\right)}.$$

So we can apply equation 6.8 to each factor

$$\prod_{k=0}^{n-1} \frac{(k+i)(k-i)}{4\left(k+\frac{1}{2}\right)\left(k-\frac{1}{2}\right)} = \frac{\Gamma(n+i)}{\Gamma(i)} \frac{\Gamma(n-i)}{\Gamma(-i)} \frac{1}{4^n} \frac{\Gamma\left(\frac{1}{2}\right)}{\Gamma\left(n+\frac{1}{2}\right)} \frac{\Gamma\left(-\frac{1}{2}\right)}{\Gamma\left(n-\frac{1}{2}\right)}.$$

This can be simplified somewhat. Obviously, $\Gamma(1/2) = \sqrt{\pi}$, and $\Gamma(-1/2) = -2\sqrt{\pi}$, but we can also simplify $\Gamma(i)\Gamma(-i) = i\Gamma(i)\Gamma(1-i) = i\pi/\sin(\pi i) = \pi/\sinh(\pi)$. So we have

$$\prod_{k=0}^{n-1} \frac{(k+i)(k-i)}{4\left(k+\frac{1}{2}\right)\left(k-\frac{1}{2}\right)} = \frac{-2\sinh(\pi)\Gamma(n+i)\Gamma(n-i)}{4^n\Gamma\left(n+\frac{1}{2}\right)\Gamma\left(n-\frac{1}{2}\right)}.$$

□

6.2.1 Solving General First Order Linear Equations

We are now ready to consider the general first order linear difference equation $a_{n+1} = p(n)a_n + g(n)$. Once again, we will examine the first several terms to find a pattern, and then use induction to prove that the pattern continues.

$$\begin{aligned}
a_1 &= p(0)a_0 + g(0), \\
a_2 &= p(0)p(1)a_0 + p(1)g(0) + g(1), \\
a_3 &= p(0)p(1)p(2)a_0 + p(1)p(2)g(0) + p(2)g(1) + g(2), \\
a_4 &= p(0)p(1)p(2)p(3)a_0 + p(1)p(2)p(3)g(0) + p(2)p(3)g(1) + p(3)g(2) + g(3).
\end{aligned}$$

Clearly, the term involving a_0 is $a_0\prod_{i=0}^{n-1} p(i)$, but the pattern for the other terms is unclear. Let us factor out this product from all of the terms.

$$a_1 = p(0)\left(a_0 + \frac{g(0)}{p(0)}\right),$$

$$a_2 = p(0)p(1)\left(a_0 + \frac{g(0)}{p(0)} + \frac{g(1)}{p(0)p(1)}\right),$$
$$a_3 = p(0)p(1)p(2)\left(a_0 + \frac{g(0)}{p(0)} + \frac{g(1)}{p(0)p(1)} + \frac{g(2)}{p(0)p(1)p(2)}\right).$$

The pattern is becoming clear, so the final step is to prove that the pattern will continue.

THEOREM 6.3
The solution to the first order linear difference equation,

$$a_{n+1} = p(n)a_n + g(n)$$

is given by

$$\boxed{a_n = \left[a_0 + \sum_{k=0}^{n-1}\left(g(k)\Big/\prod_{i=0}^{k} p(i)\right)\right]\cdot\prod_{i=0}^{n-1} p(i).} \tag{6.9}$$

PROOF: Clearly, the formula works for $n = 1$. Let us assume that the formula works for the previous case, that is,

$$a_{n-1} = \left[a_0 + \sum_{k=0}^{n-2}\left(g(k)\Big/\prod_{i=0}^{k} p(i)\right)\right]\cdot\prod_{i=0}^{n-2} p(i).$$

Multiplying both sides by $p(n-1)$, we get

$$a_{n-1}p(n-1) = \left[a_0 + \sum_{k=0}^{n-2}\left(g(k)\Big/\prod_{i=0}^{k} p(i)\right)\right]\cdot\prod_{i=0}^{n-1} p(i).$$

Adding $g(n-1)$ to the right side amounts to adding $g(n-1)/\left(\prod_{i=0}^{n-1} p(i)\right)$ inside the brackets. Thus,

$$a_{n-1}p(n-1) + g(n-1) = \left[a_0 + \sum_{k=0}^{n-1}\left(g(k)\Big/\prod_{i=0}^{k} p(i)\right)\right]\cdot\prod_{i=0}^{n-1} p(i).$$

Since $a_n = a_{n-1}p(n-1) + g(n-1)$, we have proven that the formula is true for a_n, hence the pattern continues for all n. ∎

Note the similarity between equation 6.9 and equation 4.22, especially if we consider $\mu(n)$ to be the "summing factor"

$$\mu(n) = \prod_{k=0}^{n} p(k)^{-1}.$$

Then equation 6.9 can be expressed as

$$a_n = \frac{a_0 + \sum_{k=0}^{n-1} \mu(k)g(k)}{\mu(n-1)}, \qquad \text{where } \mu(n) = \prod_{k=0}^{n} p(k)^{-1}. \tag{6.10}$$

Example 6.10

Solve the first order linear difference equation

$$a_{n+1} = (n+1)a_n/(n+2) + n^2, \qquad a_0 = 1.$$

SOLUTION: Since $p(n) = (n+1)/(n+2)$, the first task is to compute the summing factor

$$\mu(n) = \prod_{k=0}^{n} \frac{k+2}{k+1} = \frac{\Gamma(n+3)}{\Gamma(2)} \frac{\Gamma(1)}{\Gamma(n+2)} = n+2.$$

Using $g(n) = n^2$, equation 6.10 gives us

$$\begin{aligned} a_n &= \frac{1 + \sum_{k=0}^{n-1}(k^3 + 2k^2)}{n+1} = \frac{1 + n(n-1)(3n^2 + 5n - 4)/12}{n+1} \\ &= \frac{3n^3 - n^2 - 8n + 12}{12}. \end{aligned}$$

□

In many applications, $p(n)$ and $g(n)$ are constants. In this case, the solution becomes an easier formula.

$$\text{If } a_{n+1} = ra_n + b, \quad \text{then } a_n = \frac{b}{1-r} + r^n \left(a_0 - \frac{b}{1-r}\right). \tag{6.11}$$

Example 6.11

A home mortgage for \$100,000 has an interest rate of 0.5% per month. Determine the monthly payments that would pay off the home in 30 years.

SOLUTION: We begin by setting up a difference equation for the amount owed on the house. If a_n represents the money owned on the house, then Δa_n would represent the change in the debt. Since 0.5% interest is added to the loan each month, and a payment of an unknown quantity p will be subtracted from the loan, we have

$$\Delta a_n = 0.005a_n - p.$$

Expressing this as a recursion relationship, we have

$$a_{n+1} = 1.005a_n - p,$$

where $a_0 = 100,000$, the original amount of the loan. Equation 6.11 gives the amount of the loan after n months to be

$$a_n = 200p + (1.005)^n(100,000 - 200p).$$

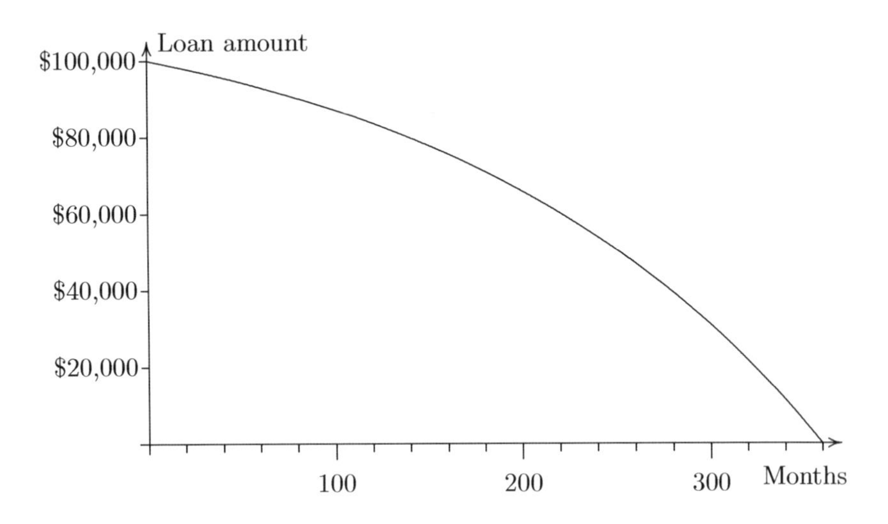

FIGURE 6.1: This shows the plot of a_n in example 6.11, showing the amount of the loan as a function of time. Note that after half of the payments are made, less that 30% of the loan has been paid off.

Finally, the loan is to be paid off in 30 years, or 360 months, so $a_{360} = 0$. This gives the equation

$$0 = 200p + (1.005)^{360}(100,000 - 200p) \approx 602257.52 - 1004.52p$$

which we can solve for p, yielding $p \approx \$599.55$. Figure 6.1 shows the amount of loan after n months. □

Many times, evaluating equation 6.9 will involve summing a rational function of n. Sometimes, the rational function can be expressed as $g(n+1)-g(n)$, as in example 6.4, which will allow the sum to telescope. Usually, though, we are not this fortunate. We need a general technique that can be used to find the anti-difference of *any* rational function.

We can, in principle, use partial fraction decomposition to express any rational function of n as a sum of terms of the form $A/(n + a)^k$, where k is a positive integer. Note that a may be a complex number. Thus, if we can find an anti-difference to a term in this form, we can find the anti-difference to any rational function.

6.2.2 The Digamma Function

At this point we introduce the *digamma* function, which is the logarithmic derivative of the gamma function.

$$\psi(z) = \frac{d}{dz}\ln(\Gamma(z)) = \frac{\Gamma'(z)}{\Gamma(z)}.$$

Let us determine the recursion relation that this function satisfies. Since $\Gamma(z+1) = z\Gamma(z)$, we can take the derivative of both sides to obtain

$$\Gamma'(z+1) = \Gamma(z) + z\Gamma'(z).$$

Dividing both sides by $\Gamma(z+1) = z\Gamma(z)$, we get

$$\frac{\Gamma'(z+1)}{\Gamma(z+1)} = \frac{1}{z} + \frac{\Gamma'(z)}{\Gamma(z)}.$$

Hence,

$$\boxed{\psi(z+1) = \psi(z) + \frac{1}{z} \qquad \text{if } z \neq 0, -1, -2, -3, \ldots.} \tag{6.12}$$

Hence, $\psi(n)$ becomes an anti-difference of $1/n$. In fact,

$$\Delta^{-1}\frac{1}{n+a} = \psi(n+a) + C.$$

However, in order to determine the arbitrary constant, we need to determine some special values of $\psi(n)$. In problem 23 from section 2.3, we found that $\Gamma'(1) = -\gamma$, where γ is the Euler-Mascheroni constant introduced in equation 2.1. This means that $\psi(1) = -\gamma$, and using the recursion formula, we have for positive integer n,

$$\psi(n) = -\gamma + \sum_{k=1}^{n-1}\frac{1}{k} = -\gamma + 1 + \frac{1}{2} + \frac{1}{3} + \cdots + \frac{1}{n-1}. \tag{6.13}$$

The asymptotic behavior of $\psi(n)$ is fairly easy to determine. Since $a_n = \psi(n)$ satisfies the difference equation $\Delta a_n = 1/n$, the corresponding differential equation is $y'(x) = 1/x$. The solution, $y = \ln(x) + C$, is algebraic, so by theorem 6.2, $\psi(n) \sim \ln n + C$ as $n \to \infty$. In fact, comparing equation 6.13 to the definition of γ,

$$\gamma = \lim_{n\to\infty}\left(1 + \frac{1}{2} + \frac{1}{3} + \cdots + \frac{1}{n} - \ln(n)\right),$$

we see that C must be zero, that is, $\lim_{n\to\infty}\psi(n) - \ln(n) = 0$. This is evident in figure 6.2. Because of the similarities between $\psi(x)$ and $\ln(x)$, the digamma function is often referred to as the *discrete logarithm*.

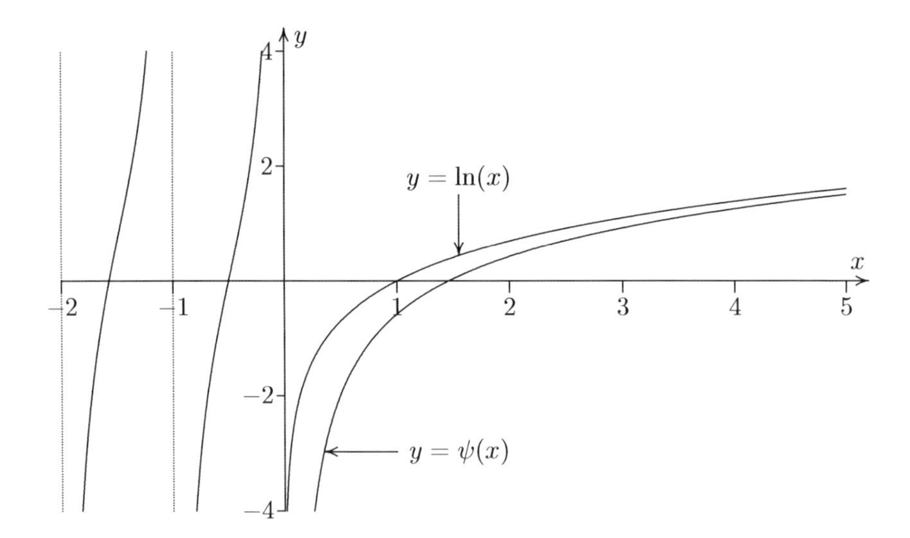

FIGURE 6.2: This shows the graph of the digamma function, $\psi(x)$, which is asymptotic to $\ln(x)$ as $x \to \infty$. There are first order poles at $x = 0, -1, -2, \ldots$.

If we need to compute the anti-difference of $1/(n+a)^2$, we can take the derivative of equation 6.12 to obtain $\psi'(z+1) = \psi'(z) - 1/z^2$. Hence,

$$\Delta^{-1}\frac{1}{(n+a)^2} = -\psi'(n+a) + C.$$

Higher powers of $(n+a)$ in the denominator can be handled by taking further derivatives of equation 6.12.

Example 6.12
Find a formula for the sum

$$\sum_{k=0}^{n-1} \frac{2k^2 - 5}{2k^3 + 3k^2 - 1}.$$

SOLUTION: Factoring the cubic $2k^3 + 3k^2 - 1$ produces $(k+1)^2(2k-1)$, so a straightforward partial fractions decomposition produces

$$\frac{2k^2 - 5}{2k^3 + 3k^2 - 1} = \frac{1}{(k+1)^2} + \frac{2}{k+1} - \frac{1}{k - \frac{1}{2}}.$$

Thus,

$$\sum_{k=0}^{n-1} \frac{1}{(k+1)^2} + \frac{2}{k+1} - \frac{1}{k - \frac{1}{2}} = -\psi'(n+1) + 2\psi(n+1) - \psi\left(n - \frac{1}{2}\right) + C.$$

To determine the value of C, we note that when $n = 0$, the sum will be 0, since there are no terms. Thus, $C = \psi'(1) - 2\psi(1) + \psi(-1/2)$, and we have the sum

$$\sum_{k=0}^{n-1} \frac{2k^2 - 5}{2k^3 + 3k^2 - 1} = -\psi'(n+1) + 2\psi(n+1) - \psi\left(n - \tfrac{1}{2}\right) + \psi'(1) + 2\gamma + \psi\left(\tfrac{1}{2}\right) + 2.$$

Here, we used the fact that $\psi(1/2) = -2 + \psi(-1/2)$. The exact values of $\psi(1/2)$ and $\psi'(1)$ can be found in problems 20 and 21. ∎

Since we now have a technique for summing rational functions, we can solve many first order difference equations in terms of the digamma function.

Example 6.13
Find the solution to the difference equation $(n+1)a_{n+1} = (n+2)a_n + n + 3$.
SOLUTION: By rewriting the equation as

$$a_{n+1} = \frac{n+2}{n+1} a_n + \frac{n+3}{n+1},$$

we find that $p(n) = (n+2)/(n+1)$, and $g(n) = (n+3)/(n+1)$. The summing factor is then

$$\mu(n) = \prod_{k=0}^{n} \frac{k+1}{k+2} = \frac{1}{n+2}.$$

Thus, we must find the anti-difference

$$\begin{aligned} \sum_{k=0}^{n-1} \mu(k) g(k) &= \sum_{k=0}^{n-1} \frac{k+3}{(k+1)(k+2)} \\ &= \sum_{k=0}^{n-1} \frac{2}{k+1} - \frac{1}{k+2} = 2\psi(n+1) - \psi(n+2) + C. \end{aligned}$$

This last expression simplifies, since $\psi(n+2) = \psi(n+1) + 1/(n+1)$. Also, we can compute C by noting that the sum must be 0 when $n = 0$, so $C = 1 + \gamma$. Thus, we have

$$\begin{aligned} a_n &= (n+1)\left(a_0 + \psi(n+1) - \frac{1}{n+1} + \gamma + 1\right) \\ &= a_0(n+1) + n + (\psi(n+1) + \gamma)(n+1). \end{aligned}$$

∎

Problems for §6.2

For problems **1** through **10**: Solve the following difference equations, expressing the answer in terms of the gamma function or digamma function.

1 $a_{n+1} = \dfrac{n+1}{2n+1}\, a_n$

2 $a_{n+1} = \dfrac{2n-1}{n+2}\, a_n$

3 $a_{n+1} = \dfrac{n^2+1}{n^2+2n+2}\, a_n$

4 $a_{n+1} = \dfrac{n^2-n+1}{n^2+n+1}\, a_n$

5 $a_{n+1} = \dfrac{n+1}{n+2}\, a_n + n$

6 $a_{n+1} = \dfrac{n+1}{n+3}\, a_n + 1$

7 $a_{n+1} = \dfrac{n+2}{n+1}\, a_n + n$

8 $a_{n+1} = \dfrac{n+3}{n+1}\, a_n + 1$

9 $a_{n+1} = \dfrac{n+1}{n+2}\, a_n + 3^n$

10 $a_{n+1} = \dfrac{n+2}{n+3}\, a_n + 2^n$

11 A person opens an IRA with \$2000 at age 25, and adds \$100 a month. The interest rate of the IRA is fixed at 1% per month. If the person plans to retire at age 65, what will be the value of the IRA?

12 A person at an age of 35 wants to retire at the age of 65. If he opens an IRA with \$1000 with an interest rate of 1% per month, how much will he have to add each month so that he would have a million dollars to retire on?

13 A \$20,000 car loan is to be paid in 5 years. If the monthly interest is 1%, find the monthly payment for the car.

14 Suppose that, in example 6.11, the monthly payments for the mortgage were not constant, but instead gradually decreased until the loan was paid off in 30 years. That is, if the first monthly payment is p dollars, then the payment after n months would be $(360-n)p/360$. Thus, the difference equation becomes

$$a_{n+1} = 1.005a_n - (360-n)p/360.$$

Solve this equation, and determine the value of the first monthly payment p such that $a_{360} = 0$.

15 A person has won a court settlement of \$12,000, to be given out as \$100 a month for the next 10 years, the first payment coming in one month. However, the person would prefer to have a lump sum of money today. This means he must know the *present value* of his annuity, assuming a discount rate of 0.5% per month. The present value is defined to be the amount of money that must be invested at the discount rate in order to make the payments which he is to receive. Determine the present value of this annuity.

16 An oil rig is currently producing a profit of \$60,000 per month. However, the oil field is running dry, and the amount of profit is expected to decrease by \$1000 each month, so that 5 years from now, the rig will cease being profitable. That is, the profit after n months will be $1000(60-n)$. Find the present value of the oil rig, assuming a discount rate of 0.5% per month. See problem 15 for the definition of present value.

17 Let z be a complex number other than $0, -1, -2 \ldots$. If we define

$$H(z) = \sum_{n=0}^{\infty} \frac{z-1}{(n+1)(n+z)},$$

show that $H(z)$ satisfies the identity $H(z+1) - H(z) = 1/z$. Note that this equation is also satisfied by the digamma function.

18 Show that if we define

$$H(z) = \sum_{n=0}^{\infty} \frac{z-1}{(n+1)(n+z)},$$

then $H(z) \sim \ln(z)$ as $z \to \infty$. Also show that the next correction will be some constant.

Hint: Approximate the sum with an integral.

19 Prove that for all complex numbers other than $0, -1, -2 \ldots$, we can express $\psi(z)$ in terms of an infinite sum:

$$\psi(z) = -\gamma + \sum_{n=0}^{\infty} \frac{z-1}{(n+1)(n+z)}.$$

See problems 17 and 18.

Hint: The only periodic functions that have a limit as $z \to \infty$ are constant.

20 Use problem 19 to find the exact value of $\psi(1/2)$.

Hint: Relate the sum to the alternating harmonic series, $1 - 1/2 + 1/3 + 1/4 - 1/5 + \cdots = \ln(2)$.

21 Use problem 19 to find the exact value of $\psi'(1)$.

Hint: Take the derivative of the sum with respect to z, and then plug in $z = 1$. Relate the resulting sum to the zeta function.

For problems **22** through **30**: Compute the following sums of rational functions in terms of the digamma function.

22 $\displaystyle\sum_{k=0}^{n-1} \frac{2k+3}{k^2+3k+2}$

23 $\displaystyle\sum_{k=0}^{n-1} \frac{2k}{k^2+4k+3}$

24 $\displaystyle\sum_{k=0}^{n-1} \frac{k}{k^2+2k+1}$

25 $\displaystyle\sum_{k=0}^{n-1} \frac{2k^2+8k}{2k^3+3k^2-1}$

26 $\displaystyle\sum_{k=0}^{n-1} \frac{2}{k^2+1}$

27 $\displaystyle\sum_{k=0}^{n-1} \frac{2k}{k^2+1}$

28 $\displaystyle\sum_{k=0}^{\infty} \frac{1}{2k^2+3k+1}$

29 $\displaystyle\sum_{k=0}^{\infty} \frac{9k}{2k^3+3k^2-1}$

30 $\displaystyle\sum_{k=0}^{\infty} \frac{2}{k^2+4}$

For problems **31** through **36**: Solve the following difference equations. Occasionally, the answer must be expressed in terms of a sum.

31 $a_{n+1} = (n+1)a_n + n, \qquad a_0 = 1$
32 $a_{n+1} = na_n + (n+1), \qquad a_1 = 0$
33 $(n+1)a_{n+1} = (n+2)a_n + n, \qquad a_0 = 2$
34 $(n+1)a_{n+1} = (n+3)a_n + n, \qquad a_0 = -1$
35 $(n+1)a_{n+1} = a_n + 2^n, \qquad a_0 = 0$
36 $(n+1)a_{n+1} = 2a_n + 1, \qquad a_0 = 0$

6.3 Analysis of Linear Difference Equations

We now turn our attention to finding the approximate solution to a linear homogeneous difference equation as $n \to \infty$. The cases where the point at infinity is an ordinary point or a regular singular point will be deferred until subsection 6.3.2 and section 6.5. For equations with an irregular singular point at infinity, we expect the solution to grow exponentially. Hence, the technique used in section 5.1 for differential equations, namely substituting $y = e^{S(x)}$, can be modified for difference equations. If we let $a_n = e^{S_n}$, then $a_{n+1} = e^{S_{n+1}}$, but we will eventually have to divide the equation by e^{S_n}. Hence, we can rewrite

$$a_{n+1} = e^{S_{n+1}} = e^{S_n} e^{S_{n+1}-S_n} = e^{S_n} e^{\Delta S_n}.$$

Higher order equations can also be expressed in terms of ΔS_n.

$$\boxed{\text{If } a_n = e^{S_n}, \text{ then } a_{n+1} = e^{S_n} e^{\Delta S_n},\ a_{n+2} = e^{S_n} e^{\Delta S_n} e^{\Delta S_{n+1}}, \text{ etc.}} \qquad (6.14)$$

This actually will reduce the order of the equation by 1, but often it produces a complicated non-linear equation, so the substitution is mainly used for finding the asymptotic approximation to the solutions.

Example 6.14
The gamma function satisfies the difference equation

$$a_{n+1} = na_n, \qquad a_1 = 1.$$

Find the leading behavior for this equation.
SOLUTION: Note that the corresponding differential equation is $y' = (x-1)y$, whose solution is an exponential function, $y = Ce^{x^2/2-x}$. Hence, we cannot use theorem 6.2 directly to find the leading behavior of the gamma function. However, by using the substitution in equation 6.14, we get

$$e^{\Delta S_n} = n, \qquad \text{or } \Delta S_n = \ln n.$$

The corresponding differential equation for this new difference equation is $y' = \ln x$, which has an algebraic solution, $y = x \ln x - x + C$. Thus, $S_n \sim n \ln n$ as $n \to \infty$. However, as in the case of differential equations, we need to find more terms in the asymptotic series for S_n in order to find the leading behavior of e^{S_n}. As with the corresponding situation with differential equations, we need to find enough terms so that we have a term which is going to 0 as $n \to \infty$.

By letting $S_n = n \ln n + g_n$, we find the new difference equation for g_n is

$$\Delta g_n = \ln n - (n+1)\ln(n+1) + n \ln n.$$

This simplifies using the asymptotic expansion of $\ln(n+1)$.

$$\ln(n+1) = \ln(n) + \ln\left(1 + \frac{1}{n}\right) \sim \ln(n) + \frac{1}{n} - \frac{1}{2n^2} + \frac{1}{3n^3} - \frac{1}{4n^4} + \cdots. \quad (6.15)$$

Now all of the logarithm terms cancel in computing Δg_n.

$$\Delta g_n \sim -1 - \frac{1}{2n} + \frac{1}{6n^2} - \frac{1}{12n^3} + \cdots.$$

So $g_n \sim -n$, and letting $g_n = -n + h_n$ causes $\Delta h_n \sim -1/(2n)$, so $h_n \sim -\ln(n)/2$. We still haven't found a term that is going to zero as $n \to \infty$, so we substitute $g_n = -n - \ln(n)/2 + h_n$ to get

$$\Delta h_n \sim \frac{\ln(n+1)}{2} - \frac{\ln(n)}{2} - \frac{1}{2n} + \frac{1}{6n^2} - \frac{1}{12n^3} + \cdots \sim -\frac{1}{12n^2} + \frac{1}{12n^3} + \cdots.$$

Here, we used equation 6.15 to eliminate the logarithms from the series. The corresponding differential equation is now $y' \sim -1/(12x^2)$, which is satisfied by $y \sim 1/(12x) + C$. Thus, $h_n \sim C + 1/(12n)$, and we finally have a term that is going to zero as $n \to \infty$.

Putting all of the terms together, we have $S_n \sim n\ln(n) - n - \ln(n)/2 + C + 1/(12n)$, and so

$$a_n = e^{S_n} \sim K n^n e^{-n} n^{-1/2} \left(1 + \frac{1}{12n}\right).$$

□

It is impossible to determine the constant K from the difference equation, since this constant is arbitrary in the solution of $a_{n+1} = n a_n$. To find the value of K corresponding to the solution for which $a_1 = 1$ requires *global analysis*, that is, global information about the difference equation, not just the local analysis near infinity. However, in example 2.12, we determined Stirling's formula, $n! \sim n^n e^{-n}\sqrt{2\pi n}$. Since $\Gamma(n) = n!/n$, we can compare this formula to the result of example 6.14, and determine that $K = \sqrt{2\pi}$. However, example 6.14 gives us another term of the series,

$$\Gamma(n) \sim \sqrt{2\pi} n^n e^{-n} n^{-1/2} \left(1 + \frac{1}{12n}\right). \quad (6.16)$$

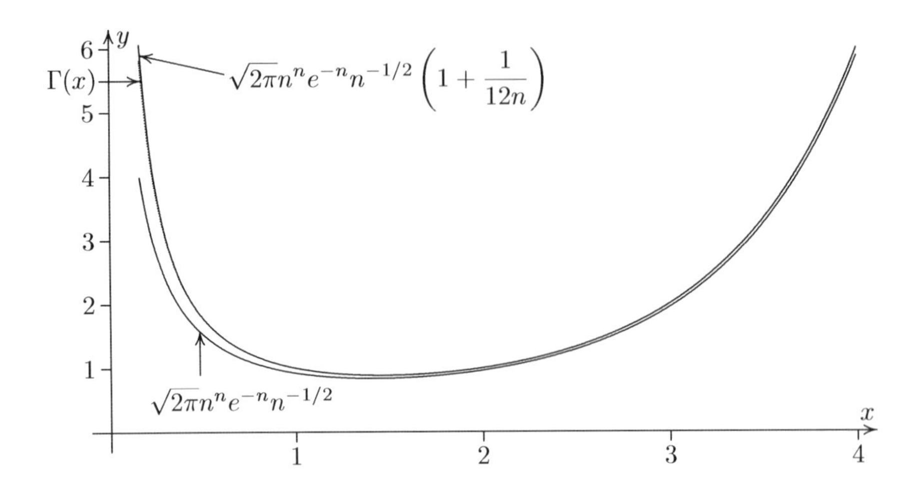

FIGURE 6.3: Plot of the gamma function for $0.16 \le x \le 4$, along with the leading term of the asymptotic approximation as $x \to \infty$, and the first order correction to the leading behavior. Note that the latter is almost indistinguishable from the gamma function, except for the far left portion of the graph.

The two term approximation is extremely accurate, even for small values of n. See figure 6.3.

For this example, we had to deal with the series for the logarithm, which got messy. But now that we have the asymptotic behavior of the gamma function, we can use the gamma function to describe the behavior in other difference equations. Because of the recursion relationship of the gamma function, expressing the leading behavior in terms of the gamma function makes the leading behavior particularly easy to peel off.

Example 6.15

Find the possible leading behaviors for the solutions to the difference equation

$$a_{n+2} = 4na_{n+1} - 3n^2 a_n.$$

SOLUTION: Letting $a_n = e^{S_n}$, we get

$$e^{\Delta S_n} e^{\Delta S_{n+1}} = 4ne^{\Delta S_n} - 3n^2.$$

Although a_n may grow exponentially, we can be fairly certain that $S_n = \ln(a_n)$ will be algebraic. Hence, it is reasonable to assume that $\Delta S_{n+1} \sim \Delta S_n$. So to find the leading order of S_n, it will usually be true that $e^{\Delta S_{n+k}} \sim e^{\Delta S_n}$ for all k, resulting in a *polynomial* equation for $e^{\Delta S_n}$. This is akin to assuming that $S''(x) \ll (S'(x))^2$ for differential equations.

If we assume that $e^{\Delta S_{n+1}} \sim e^{\Delta S_n}$, the equation becomes

$$(e^{\Delta S_n})^2 \sim 4ne^{\Delta S_n} - 3n^2.$$

We can solve this for $e^{\Delta S_n}$, to obtain

$$e^{\Delta S_n} \sim \frac{4n \pm \sqrt{16n^2 - 12n^2}}{2} = \begin{cases} 3n & \text{or} \\ n. \end{cases}$$

Thus, $\Delta S_n \sim \ln n$ or $\Delta S_n \sim \ln(3n) = \ln n + \ln 3$, so in both cases, $S_n \sim n \ln n$. We can now check that our assumption was valid. Since $\Delta S_n \sim \ln n$, $e^{\Delta S_n} \sim n$, so indeed $e^{\Delta S_n} \sim e^{\Delta S_{n+1}}$.

We still need to find more terms for the series in S_n to find the leading behavior of a_n. Rather than peeling off a logarithm term from S_n, let us peel off a factor from a_n that has the behavior of $e^{n \ln n} = n^n$. Of course, letting $a_n = n^n b_n$ will produce an unwieldy equation, but we can accomplish the same thing by letting $a_n = \Gamma(n) b_n$. The difference equation for b_n will in fact by fairly simple.

$$\Gamma(n+2)b_{n+2} = 4n\Gamma(n+1)b_{n+1} - 3n^2\Gamma(n)b_n.$$

But $\Gamma(n+1) = n\Gamma(n)$, and $\Gamma(n+2) = (n+1)n\Gamma(n)$. So the equation simplifies to

$$(n+1)b_{n+2} = 4nb_{n+1} - 3nb_n.$$

Now we find the leading behavior of b_n, by letting $b_n = e^{T_n}$, producing the equation

$$(n+1)e^{\Delta T_n}e^{\Delta T_{n+1}} = 4ne^{\Delta T_n} - 3n.$$

By assuming that $e^{\Delta T_n} \sim e^{\Delta T_{n+1}}$, we find that

$$e^{\Delta T_n} \sim \frac{4n \pm \sqrt{16n^2 - 12n(n+1)}}{2(n+1)} \sim \begin{cases} 3 - 9/(2n) + O(n^{-2}) & \text{or} \\ 1 + 1/(2n) + O(n^{-2}). \end{cases}$$

We can keep two terms in this equation, since the first term is a constant, so $e^{\Delta T_n} = e^{\Delta T_{n+1}}$ for this term. Hence the series for $e^{\Delta T_n}$ and for $e^{\Delta T_{n+1}}$ will agree for the two terms given.

Solving for b_n, we have two possibilities. If $e^{\Delta T_n} \sim 3 - 9/(2n) + O(n^{-2})$, then $\Delta T_n \sim \ln(3) - 3/(2n) + O(n^{-2})$, so $T_n \sim n \ln(3) - 3\ln(n)/2 + C_1 + O(n^{-1})$, so $b_n = e^{T_n} \sim K_1 3^n n^{-3/2}$. On the other hand, if $e^{\Delta T_n} \sim 1 + 1/(2n) + O(n^{-2})$, then $\Delta T_n \sim 1/(2n) + O(n^{-2})$, so $T_n \sim \ln(n)/2 + C_2 + O(n^{-1})$, giving $b_n = e^{T_n} \sim K_2\sqrt{n}$. In both cases, the next term in the exponent goes to 0, so we have the full leading behavior. Since $b_n \sim K_1 3^n n^{-3/2} + K_2\sqrt{n}$, and $a_n = \Gamma(n)b_n$, we have

$$a_n \sim K_1 n^n e^{-n} 3^n n^{-2} + K_2 n^n e^{-n}. \qquad \square$$

6.3.1 Full Stirling Series

Since the gamma function will be used to express the asymptotic series of other solutions to difference equations, we need to know the complete asymptotic behavior of this important function. The strategy will be to peel off the leading behavior of the series from the difference equation $a_{n+1} = na_n$. That is, we will assume that

$$a_n = n^n e^{-n} n^{-1/2} b_n,$$

and find the series expansion for b_n. However, this substitution creates some complications:

$$(n+1)^{n+1} e^{-n-1} (n+1)^{-1/2} b_{n+1} = n\, n^n e^{-n} n^{-1/2} b_n.$$

To simplify this, we can multiply both sides by e^n, and group factors with the same base together.

$$(n+1)^{n+1/2} e^{-1} b_{n+1} = n^{n+1/2} b_n.$$

Finally, we can utilize the fact that $(n+1)/n = 1 + (1/n)$.

$$b_{n+1} = Q(n) b_n, \qquad \text{where } Q(n) = e(1 + 1/n)^{-n-1/2}.$$

To find the series expansion for $Q(n)$ as $n \to \infty$, we first take the logarithm of both sides.

$$\begin{aligned}
\ln(Q(n)) &= 1 - \left(n + \frac{1}{2}\right) \ln\left(1 + \frac{1}{n}\right) \\
&= 1 - \left(n + \frac{1}{2}\right)\left(\frac{1}{n} - \frac{1}{2n^2} + \frac{1}{3n^3} - \frac{1}{4n^4} + \frac{1}{5n^5} - \frac{1}{6n^6} + \frac{1}{7n^7} - \cdots\right) \\
&= -\frac{1}{12n^2} + \frac{1}{12n^3} - \frac{3}{40n^4} + \frac{1}{15n^5} - \frac{5}{84n^6} + \cdots.
\end{aligned}$$

We can now take the exponential of this to find the series for $Q(n)$.

$$\begin{aligned}
Q(n) &= 1 + \left(-\frac{1}{12n^2} + \frac{1}{12n^3} - \frac{3}{40n^4} + \frac{1}{15n^5} - \frac{5}{84n^6} + \cdots\right) \\
&\quad + \frac{1}{2}\left(-\frac{1}{12n^2} + \frac{1}{12n^3} - \frac{3}{40n^4} + \frac{1}{15n^5} - \frac{5}{84n^6} + \cdots\right)^2 \\
&\quad + \frac{1}{6}\left(-\frac{1}{12n^2} + \frac{1}{12n^3} - \frac{3}{40n^4} + \frac{1}{15n^5} - \frac{5}{84n^6} + \cdots\right)^3 + \cdots \\
&= 1 - \frac{1}{12n^2} + \frac{1}{12n^3} - \frac{103}{1440n^4} + \frac{43}{720n^5} - \frac{18107}{362880n^6} + \cdots.
\end{aligned}$$

Thus, the equation for b_n satisfies

$$b_{n+1} = b_n \left(1 - \frac{1}{12n^2} + \frac{1}{12n^3} - \frac{103}{1440n^4} + \frac{43}{720n^5} - \frac{18107}{362880n^6} + \cdots\right). \tag{6.17}$$

We have already established that $b_n \sim K(1 + 1/(12n) + \cdots)$. The plan for finding the full asymptotic series for b_n is to assume that it is in the form of a Taylor series in powers of $1/n$. That is, we assume

$$b_n \sim A_0 + \frac{A_1}{n} + \frac{A_2}{n^2} + \frac{A_3}{n^3} + \frac{A_4}{n^4} + \frac{A_5}{n^5} + \cdots. \tag{6.18}$$

But there is now another complication. We need to find the power series for

$$b_{n+1} \sim A_0 + \frac{A_1}{n+1} + \frac{A_2}{(n+1)^2} + \frac{A_3}{(n+1)^3} + \frac{A_4}{(n+1)^4} + \frac{A_5}{(n+1)^5} + \cdots,$$

which will require the binomial series, equation 1.31. Using this equation, we have

$$\frac{1}{(n+1)^k} = \frac{1}{n^k}\left(1 + \frac{1}{n}\right)^{-k} = \frac{1}{n^k} - \frac{k}{n^{k+1}} + \frac{k(k+1)}{2n^{k+2}} - \frac{k(k+1)(k+2)}{6n^{k+3}} + \cdots.$$

We can express the series in terms of binomial coefficients.

$$\frac{1}{(n+1)^k} = \sum_{i=0}^{\infty}(-1)^i \binom{k+i-1}{i} \frac{1}{n^{k+i}}, \qquad \text{for } |n| > 1. \tag{6.19}$$

Thus, we can express b_{n+1} at a series

$$b_{n+1} \sim A_0 + \sum_{k=1}^{\infty} A_k \sum_{i=0}^{\infty}(-1)^i \binom{k+i-1}{i} \frac{1}{n^{k+i}}.$$

We can reorder the double sum so that like powers of n are combined. That is, we can let $m = i + k$, so $i = m - k$.

$$\boxed{b_{n+1} \sim A_0 + \sum_{m=1}^{\infty} \frac{1}{n^m} \sum_{i=0}^{m-1} (-1)^i \binom{m-1}{i} A_{m-i}.} \tag{6.20}$$

Expanding out this series, the first few terms are

$$b_{n+1} = A_0 + \frac{A_1}{n} + \frac{A_2 - A_1}{n^2} + \frac{A_3 - 2A_2 + A_1}{n^3} + \frac{A_4 - 3A_3 + 3A_2 - A_1}{n^4} + \cdots.$$

We now can equate the like powers of n in equation 6.17, to obtain

$$\begin{aligned} A_0 &= A_0 \\ A_1 &= A_1 \\ A_2 - A_1 &= A_2 - \frac{A_0}{12} \\ A_3 - 2A_2 + A_1 &= A_3 - \frac{A_1}{12} + \frac{A_0}{12} \end{aligned}$$

$$A_4 - 3A_3 + 3A_2 - A_1 = A_4 - \frac{A_2}{12} + \frac{A_1}{12} - \frac{103A_0}{1440}$$

$$A_5 - 4A_4 + 6A_3 - 4A_2 + A_1 = A_5 - \frac{A_3}{12} + \frac{A_2}{12} - \frac{103A_1}{1440} + \frac{43A_0}{720}$$

$$A_6 - 5A_5 + 10A_4 - 10A_3 + 5A_2 - A_1 = A_6 - \frac{A_4}{12} + \frac{A_3}{12} - \frac{103A_2}{1440} + \frac{43A_1}{720} - \frac{18107A_0}{362880}.$$

Notice that the A_k term will always cancel in the equation for n^{-k}, so we can solve that equation for A_{k-1} in terms of A_0. Thus, we have

$$A_1 = \frac{A_0}{12} \approx 0.083333A_0$$

$$A_2 = \frac{A_0}{288} \approx 0.003472A_0$$

$$A_3 = -\frac{139A_0}{51840} \approx -0.002681A_0$$

$$A_4 = -\frac{571A_0}{2488320} \approx -0.000229A_0$$

$$A_5 = \frac{163879A_0}{209018880} \approx 0.000784A_0.$$

We have already established that for the gamma function, $A_0 = \sqrt{2\pi}$. Thus, we have the full *Stirling series*:

$$\Gamma(n) \sim \sqrt{2\pi}n^n e^{-n} n^{-1/2} \times \left(1 + \frac{1}{12n} + \frac{1}{288n^2} - \frac{139}{51840n^3} - \frac{571}{2488320n^4} + \frac{163879}{209018880n^5} + \cdots\right).$$

The Stirling series is one of the oldest examples of an asymptotic series. Although the coefficients in the series start out getting smaller, they eventually turn around and grow without bounds. In fact, the series diverges for all n in its present form, which is typical for an asymptotic series. However, we have seen in chapter 3 that there are ways of converting a divergent asymptotic series into a form which converges. For example, the power series portion can be converted into a continuous fraction by first making the substitution $x = 1/n$. That is, we will find coefficients c_m such that

$$\Gamma(n) \sim \sqrt{2\pi}n^n e^{-n} n^{-1/2} \Big/ \left(1 + \frac{c_1}{n}\Big/\left(1 + \frac{c_2}{n}\Big/\left(1 + \frac{c_3}{n}\Big/\left(1 + \frac{c_4}{n}\Big/\left(1 + \cdots\right.\right.\right.\right.\right. \tag{6.21}$$

We actually find a pattern within the coefficients c_m:

$$c_1 = -\frac{1}{12}$$

TABLE 6.1: This table shows how the continued fraction approximates converge to $\Gamma(n)$, even for small values of n.

	$n = 1/2$	$n = 1$	$n = 3/2$	$n = 2$
$F_0(n)$	1.520346901	0.922137009	0.838956553	0.959502176
$F_1(n)$	1.824416281	1.005967646	0.888306938	1.001219662
$F_2(n)$	1.796773610	1.002322836	0.886896927	1.000332056
$F_3(n)$	1.882029427	0.989512341	0.885325375	0.999744532
$F_4(n)$	1.761140955	0.999582319	0.886165070	0.999981569
$F_5(n)$	1.785732285	1.000286466	0.886253057	1.000005334
$F_6(n)$	1.777681693	1.000113004	0.886236896	1.000001915
$F_7(n)$	1.791803963	1.000858823	0.885979409	0.999992177
$F_8(n)$	1.768899063	0.999954488	0.886224346	0.999999664
$F_9(n)$	1.778514890	1.000054001	0.886229065	1.000000204
$F_{10}(n)$	1.774639672	1.000020054	0.886227728	1.000000075
$\Gamma(n)$	1.772453851	1.0	0.886226925	1.0

$$\begin{aligned} c_2 &= \frac{1}{24} \\ -c_3 = c_4 &= \frac{293}{360} \\ -c_5 = c_6 &= \frac{4406147}{14422632} \\ -c_7 = c_8 &= \frac{805604978502319}{465939153254616} \\ -c_9 = c_{10} &= \frac{5224772492188649105544325380 7}{8981016262141825384927752132 0}. \end{aligned}$$

We find that the coefficients alternate in sign, and that $c_{2k} = -c_{2k-1}$ for $k > 1$. Because there are negative coefficients in the continued fraction, the Stirling series as given is not a Stieltjes series. Nonetheless, the continued fraction representation will converge to $\Gamma(n)$ whenever $\mathrm{Re}(n) > 0$. See table 6.1.

6.3.2 Taylor Series Solution

The technique that was used to form the Taylor series solution for the equation

$$b_{n+1} = e(1 + 1/n)^{-n-1/2} b_n$$

can be used to solve any equation of the form

$$b_{n+1} = b_n \left(1 + \frac{c_2}{n^2} + \frac{c_3}{n^3} + \frac{c_4}{n_4} + \cdots \right),$$

as long as there is not a $1/n$ term. (If a $1/n$ term is present, we can peel off a power of n to eliminate this term.) We can again use equation 6.20 to find the Taylor series for b_{n+1}, and align like powers of n.

Example 6.16
Find the first six terms of the Taylor series solution to the difference equation

$$b_{n+1} = \frac{1+n^2}{n^2} b_n.$$

SOLUTION: We assume b_n is of the form

$$b_n \sim A_0 + \frac{A_1}{n} + \frac{A_2}{n^2} + \frac{A_3}{n^3} + \frac{A_4}{n^4} + \frac{A_5}{n^5} + \cdots.$$

Then

$$\frac{1+n^2}{n^2} b_n = A_0 + \frac{A_1}{n} + \frac{A_2 + A_0}{n^2} + \frac{A_3 + A_1}{n^3} + \frac{A_4 + A_2}{n^4} + \frac{A_5 + A_3}{n^5} + \cdots.$$

By equating like powers of n with equation 6.20, we get

$$\begin{aligned} A_0 &= A_0 \\ A_1 &= A_1 \\ A_2 - A_1 &= A_2 + A_0 \\ A_3 - 2A_2 + A_1 &= A_3 + A_1 \\ A_4 - 3A_3 + 3A_2 - A_1 &= A_4 + A_2 \\ A_5 - 4A_4 + 6A_3 - 4A_2 + A_1 &= A_5 + A_3 \\ A_6 - 5A_5 + 10A_4 - 10A_3 + 5A_2 - A_1 &= A_6 + A_4. \end{aligned}$$

Thus, we have $A_1 = -A_0$, $A_2 = 0$, $A_3 = A_0/3$, $A_4 = A_0/6$, and $A_5 = -A_0/6$. Hence, we have

$$b_n = A_0 \left(1 - \frac{1}{n} + \frac{1}{3n^3} + \frac{1}{6n^4} - \frac{1}{6n^5}\right).$$

□

For higher order equations, we have to express b_{n+k}, for $k = 2, 3, \ldots$, as a Taylor series. Using the same logic that was used to generate equation 6.20 (see problem 14), we get

$$\boxed{b_{n+k} \sim A_0 + \sum_{m=1}^{\infty} \frac{1}{n^m} \sum_{i=0}^{m-1} (-k)^i \binom{m-1}{i} A_{m-i}.} \tag{6.22}$$

For example, when $k = 2$, we get

$$b_{n+2} = A_0 + \frac{A_1}{n} + \frac{A_2 - 2A_1}{n^2} + \frac{A_3 - 4A_2 + 4A_1}{n^3} + \frac{A_4 - 6A_3 + 12A_2 - 8A_1}{n^4} + \cdots$$

where the coefficients in each term are the same as for the expansion of $(x-2)^{m-1}$. Hence, we can use the same procedure to find the Taylor series solution to a higher order equation, provided that a solution has such a series.

Example 6.17

Find the Taylor series solutions, up to 6 terms, for the difference equation

$$a_{n+2} + a_{n+1} = 2a_n \left(\frac{1+n^2}{n^2} \right).$$

SOLUTION: We again assume that a_n is of the form

$$a_n \sim A_0 + \frac{A_1}{n} + \frac{A_2}{n^2} + \frac{A_3}{n^3} + \frac{A_4}{n^4} + \frac{A_5}{n^5} + \cdots,$$

and equate like powers of $1/n$.

$$\begin{aligned}
A_0 + A_0 &= 2A_0 \\
A_1 + A_1 &= 2A_1 \\
(A_2 - 2A_1) + (A_2 - A_1) &= 2(A_2 + A_0) \\
(A_3 - 4A_2 + 4A_1) + (A_3 - 2A_2 + A_1) &= 2(A_3 + A_1) \\
(A_4 - 6A_3 + 12A_2 - 8A_1) + (A_4 - 3A_3 + 3A_2 - A_1) &= 2(A_4 + A_2) \\
(A_5 - 8A_4 + 24A_3 - 32A_2 + 16A_1) \qquad & \\
+(A_5 - 4A_4 + 6A_3 - 4A_2 + A_1) &= 2(A_5 + A_3) \\
(A_6 - 10A_5 + 40A_4 - 80A_3 + 80A_2 - 32A_1) \qquad & \\
+(A_6 - 5A_5 + 10A_4 - 10A_3 + 5A_2 - A_1) &= 2(A_6 + A_4)
\end{aligned}$$

Thus, we have $A_1 = -2A_0/3$, $A_2 = -A_0/3$, $A_3 = 5A_0/27$, $A_4 = 79A_0/162$, and $A_5 = 11A_0/405$. So one solution to the difference equation has a Taylor series solution

$$a_n \sim 1 - \frac{2}{3n} - \frac{1}{3n^2} + \frac{5}{27n^3} + \frac{79}{162n^4} + \frac{11}{405n^5} + \cdots.$$

However, there should be two linearly independent solutions to a second order equation. To find the leading behavior of the second solution, let us make the substitution $a_n = e^{S_n}$. The new equation becomes

$$e^{\Delta S_{n+1}} e^{\Delta S_n} + e^{\Delta S_n} = 2 + \frac{2}{n^2}.$$

If we assume that $e^{\Delta S_{n+1}} \sim e^{\Delta S_n}$, we can solve for $e^{\Delta S_n}$ via the quadratic equation

$$e^{\Delta S_n} \sim \frac{-1 \pm \sqrt{1 + 8 + 8/n^2}}{2} \sim \begin{cases} 1 + 4/(3n^2) & \text{or} \\ -2 - 4/(3n^2). \end{cases}$$

The first case leads to $\Delta S_n \sim 4/(3n^2)$, so $S_n \sim -2/(3n) + C$, so $a_n \sim K(1 - 2/(3n))$, which is the solution that we already have. But the second case produces $\Delta S_n \sim \ln(-2) + 2/(3n^2)$, so $S_n \sim n\ln(-2) - 1/(3n) + C$. Then

$a_n = e^{S_n} \sim K(-2)^n(1 - 1/(3n))$. Notice that we temporarily had to use complex numbers in our computation, but the result came out real. We have found that the second possible leading behavior is $(-2)^n$, so we can peel off the leading behavior by letting $a_n = (-2)^n b_n$. The equation for b_n becomes

$$2b_{n+2} - b_{n+1} = b_n \left(\frac{1+n^2}{n^2}\right).$$

We assume that b_n is in the form

$$b_n \sim B_0 + \frac{B_1}{n} + \frac{B_2}{n^2} + \frac{B_3}{n^3} + \frac{B_4}{n^4} + \frac{B_5}{n^5} + \cdots,$$

then we can equate like powers of $1/n$.

$$\begin{aligned}
2B_0 - B_0 &= B_0 \\
2B_1 - B_1 &= B_1 \\
2(B_2 - 2B_1) - (B_2 - B_1) &= B_2 + B_0 \\
2(B_3 - 4B_2 + 4B_1) - (B_3 - 2B_2 + B_1) &= B_3 + B_1 \\
2(B_4 - 6B_3 + 12B_2 - 8B_1) - (B_4 - 3B_3 + 3B_2 - B_1) &= B_4 + B_2 \\
2(B_5 - 8B_4 + 24B_3 - 32B_2 + 16B_1) & \\
-(B_5 - 4B_4 + 6B_3 - 4B_2 + B_1) &= B_5 + B_3 \\
2(B_6 - 10B_5 + 40B_4 - 80B_3 + 80B_2 - 32B_1) & \\
-(B_6 - 5B_5 + 10B_4 - 10B_3 + 5B_2 - B_1) &= B_6 + B_4
\end{aligned}$$

This produces $B_1 = -B_0/3$, $B_2 = -B_0/3$, $B_3 = -5B_0/27$, $B_4 = 14B_0/81$, and $B_5 = 244B_0/405$. So the second linearly independent solution to the original equation is

$$a_n \sim (-2)^n \left(1 - \frac{1}{3n} - \frac{1}{3n^2} - \frac{5}{27n^3} + \frac{14}{81n^4} + \frac{244}{405n^5} + \cdots\right).$$

□

Problems for §6.3

For problems **1** through **10**: Find the possible leading behaviors for the following difference equations.

1 $a_{n+1} = a_n(2 + 1/n)$

2 $a_{n+1} = a_n(-1 + 1/n)$

3 $a_{n+1} = (2n+1)a_n$

4 $a_{n+1} = (n^2 + 2n + 3)a_n$

5 $na_{n+2} + (2n+1)a_n = 3na_{n+1}$

6 $(n+1)a_{n+2} - 4na_{n+1} + 3na_n = 0$

7 $(n+1)a_{n+2} - 5na_{n+1} + 6na_n = 0$

8 $na_{n+2} + (n+1)a_{n+1} = 2na_n$

9 $a_{n+2} - 4na_{n+1} + 3n^2a_n = 0$

10 $a_{n+2} - 5na_{n+1} + 6n^2a_n = 0$

11 There is another remarkable property of the continued fraction portion of equation 6.21. If we compute the N, N Padé approximate from the continued fraction, we find that for each rational function, the numerator differs from the denominator only in sign changes. In fact, it can be shown that $P_N^N(-n) = 1/P_N^N(n)$. Verify this property by computing $P_1^1(n)$, $P_2^2(n)$ and $P_3^3(n)$ for the continued fraction

$$1\Big/\Big(1 - \frac{1}{12n}\Big/\Big(1 + \frac{1}{24n}\Big/\Big(1 - \frac{293}{360n}\Big/\Big(1 + \frac{293}{360n}\Big/\Big(1 + \cdots.$$

12 Use equation 6.21 to compute $\Gamma(3/5)$ to eight places of accuracy.

Hint: Compute $\Gamma(10.6)$ using equation 6.21, and then use the properties of the gamma function to relate $\Gamma(10.6)$ to $\Gamma(0.6)$.

13 A computer programmer needed to write a program that computes $n!$ for all non-negative integer values of n. Rather than using a recursive routine, he used the fact that $n! = \Gamma(n+1)$, and used equation 6.16 to approximate $\Gamma(n+1)$. He then rounded this approximation to the nearest integer, since the factorial will always be an integer. Since the two term approximation is clearly accurate enough for $n = 1$, and only gets more accurate as $n \to \infty$, he figured this would work for all n. What was wrong with his logic?

14 Derive equation 6.22.

For problems **15** through **24**: The following difference equations have a solution in the form of a Taylor series. Find the first 4 non-zero terms of the Taylor series solution to the equation.

15 $n^2 a_{n+1} = (n^2 - 2)a_n$

16 $n^2 a_{n+1} = (n^2 + 2)a_n$

17 $n^3 a_{n+1} = (n^3 + 1)a_n$

18 $n^3 a_{n+1} = (n^3 + n + 1)a_n$

19 $a_{n+1} = a_n e^{1/n^2}$

20 $a_{n+2} - 3a_{n+1} + 2a_n(1 + 1/n^2) = 0$

21 $2a_{n+2} - 3a_{n+1} + a_n(1 + 1/n^2) = 0$

22 $a_{n+2} - 4a_{n+1} + 3a_n(1 - 1/n^2) = 0$

23 $3a_{n+2} - 4a_{n+1} + a_n(1 - 1/n^2) = 0$

24 $a_{n+2} - 2a_{n+1} + a_n(1 - 1/n^3) = 0$

For problems **25** through **30**: First find the possible leading behaviors of the following difference equations. Then, after peeling off each leading behavior, determine the first three terms of the Taylor series for the resulting equation. You may have to use the trick from example 6.15.

25 $a_{n+2} - 5a_{n+1} + 6a_n(1 + 1/n^2) = 0$

26 $a_{n+2} - a_{n+1} - 2a_n(1 + 1/n^2) = 0$

27 $a_{n+2} + a_{n+1} - 2a_n(1 + 3/n + 3/n^2) = 0$

28 $a_{n+2} - a_{n+1} - 2a_n(1 + 3/n + 6/n^2) = 0$

29 $a_{n+2} + na_{n+1} - 2(n^2 + 2n + 3)a_n = 0$

30 $a_{n+2} = (n^2 + 3n + 1)a_n$

6.4 The Euler-Maclaurin Formula

Because solving difference equations often involves finding the sum, or anti-difference, of a function of n,

$$\Delta^{-1} f(n) = \sum_{x=0}^{n-1} f(x),$$

it is useful to know how to find the asymptotic series for such expressions. If $f(n)$ is an algebraic function, then theorem 6.2 shows that

$$\sum_{x=0}^{n-1} f(x) \sim \int_0^n f(x)\,dx + C \qquad \text{as } n \to \infty. \tag{6.23}$$

This gives us the first order approximation for the anti-difference. The goal of this section is to find the full asymptotic expansion for $\Delta^{-1} f(n)$ in terms of $f(n)$. It turns out that there is an elegant formula that can be used for general functions.

We can think of the approximation in equation 6.23 as having the sum be a crude Riemann approximation for the integral, with $\Delta x = 1$. Figure 6.4 shows the relationship between the sum and the integral. Of course, there are better ways of approximating an integral that using a Riemann sum. For example, the trapezoidal rule gives a more accurate approximation to the integral than the Riemann sum.

The trapezoidal rule for approximating the integral, using $\Delta x = 1$, can be written in terms of the sum

$$\int_0^n f(x)\,dx \approx \sum_{x=0}^{n-1} f(x) - \frac{f(0)}{2} + \frac{f(n)}{2}.$$

The error in this approximation is considerably less than for the Riemann approximation. See figure 6.5. Solving for the sum, we obtain

$$\sum_{x=0}^{n-1} f(x) \sim \int_0^n f(x)\,dx - \frac{f(n)}{2} + C \qquad \text{as } n \to \infty.$$

Here, the $f(0)$ term was incorporated into the constant C.

To find the next term of the asymptotic series, we need to analyze the tiny area above each of the trapezoids in figure 6.5. The area A_k for the $(k+1)$-st segment can be represented by

$$A_k = \int_k^{k+1} f(x)\,dx - \frac{f(k) + f(k+1)}{2}.$$

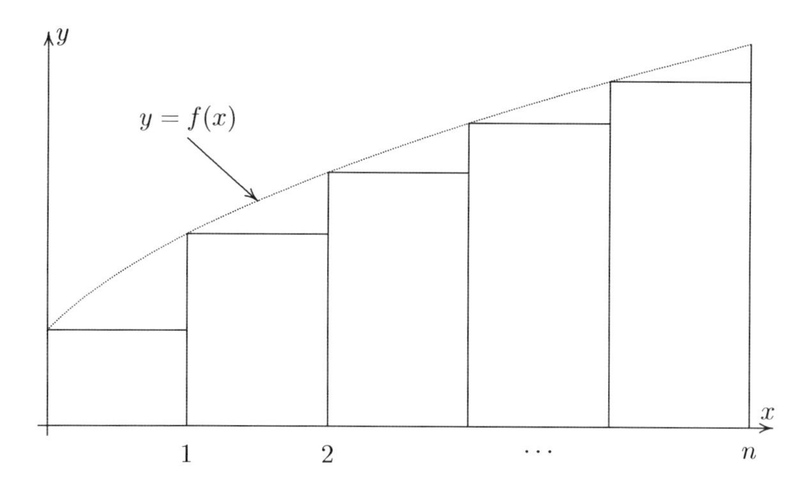

FIGURE 6.4: This graph compares the sum $\sum_{x=0}^{n-1} f(x)$, denoted by the area inside of the rectangles, with the area under the curve $\int_0^n f(x)\,dx$. Although the difference can increase as $n \to \infty$, the ratio of the two areas approaches 1.

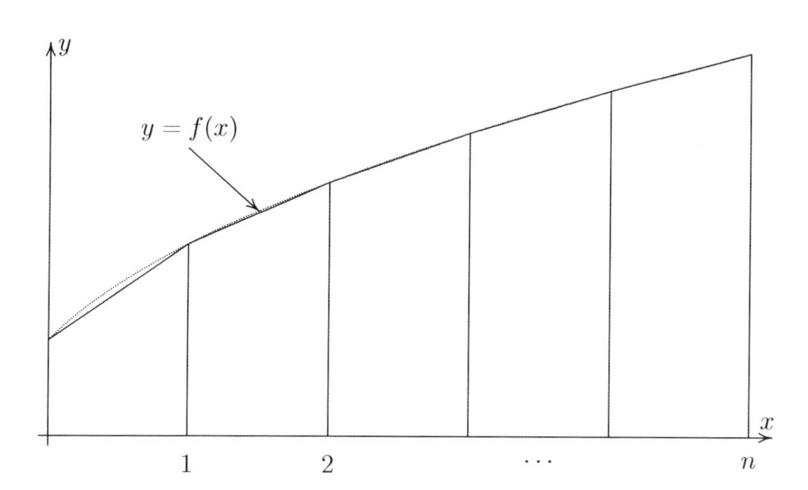

FIGURE 6.5: This graph compares the trapezoidal rule approximation, denoted by the area inside of the trapezoids, with the area under the curve $\int_0^n f(x)\,dx$. In places, the two are barely distinguishable. If the slope of the curve approaches a constant, then the total difference between the two areas will approach a constant as $n \to \infty$.

We can combine this into a single integral via integration by parts. Letting $u = f(x)$, $du = f'(x)\,dx$, $v = x - k - 1/2$, $dv = dx$, we have

$$\int_k^{k+1} f(x)\,dx = f(x)\left(x - k - \frac{1}{2}\right)\Big|_k^{k+1} - \int_k^{k+1} f'(x)\left(x - k - \frac{1}{2}\right)dx$$
$$= \frac{f(k+1)}{2} - \frac{-f(k)}{2} - \int_k^{k+1} f'(x)\left(x - k - \frac{1}{2}\right)dx.$$

Thus,

$$A_k = -\int_k^{k+1} f'(x)\left(x - k - \frac{1}{2}\right)dx.$$

The total error in the trapezoid rule is the sum of all of these areas. Thus,

$$\sum_{x=0}^{n-1} f(x) = \int_0^n f(x)\,dx - \frac{f(n)}{2} + \frac{f(0)}{2} + \sum_{k=0}^{n-1}\int_k^{k+1} f'(x)\left(x - k - \frac{1}{2}\right)dx. \tag{6.24}$$

We can integrate by parts again, with $u = f'(x)$, $du = f''(x)\,dx$, $dv = x - k - 1/2$, and $v = (x-k)^2/2 - (x-k)/2$, giving

$$A_k = -f'(x)\frac{(x-k)^2 - (x-k)}{2}\Big|_k^{k+1} + \int_k^{k+1} f''(x)\frac{(x-k)^2 - (x-k))}{2}\,dx$$
$$= \int_k^{k+1} f''(x)\frac{(x-k)^2 - (x-k)}{2}\,dx.$$

We are now ready to find the next term in the series, by estimating the sum of the A_k's. Since $(x-k) - (x-k)^2 \geq 0$ within the interval of integration, we have by the mean value theorem that

$$\int_k^{k+1} f''(x)\frac{(x-k) - (x-k)^2}{2}\,dx = f''(\xi_k)\int_k^{k+1} \frac{(x-k) - (x-k)^2}{2}\,dx$$

for some ξ_k between k and $k+1$. This last integral evaluates to $f''(\xi_k)/12$. This means that the total error in the trapezoid rule can be expressed by

$$\sum_{k=0}^{n-1} \frac{f''(\xi_k)}{12},$$

which is a Riemann sum of the integral $\int_0^n f''(x)/12\,dx = f'(n)/12 - f'(0)/12$. Thus,

$$\sum_{x=0}^{n-1} f(x) \sim \int_0^n f(x)\,dx - \frac{f(n)}{2} + \frac{f(0)}{2} + \frac{f'(n)}{12} - \frac{f'(0)}{12}.$$

Once again, we can incorporate the $f'(0)$ and $f(0)$ terms into the constant, and obtain

$$\sum_{x=0}^{n-1} f(x) \sim \int_0^n f(x)\,dx + C - \frac{f(n)}{2} + \frac{f'(n)}{12}.$$

This gives us four terms of the asymptotic series.

6.4.1 The Bernoulli Numbers

At this point, we need to introduce the *Bernoulli numbers*. These numbers are based on the Maclaurin series for the function

$$\frac{x}{e^x - 1} = 1 - \frac{x}{2} + \frac{x^2}{12} - \frac{x^4}{720} + \frac{x^6}{30240} - \frac{x^8}{1209600} + \frac{x^{10}}{47900160} - \frac{691x^{12}}{1307674368000} + \cdots \tag{6.25}$$

which has a removable singularity at $x = 0$, but a pole at $x = \pm 2\pi i$. Hence, the radius of convergence of this series is 2π. We define the *Bernoulli number* B_n to be $n!$ times the nth coefficient of this series. In other words,

$$B_n = \left. \frac{d^n}{dx^n} \frac{x}{e^x - 1} \right|_{x=0}.$$

So the Maclaurin series can be expressed in terms of the Bernoulli numbers.

$$\frac{x}{e^x - 1} = \sum_{n=0}^{\infty} \frac{B_n}{n!} x^n. \tag{6.26}$$

Comparing this definition to the series in equation 6.25, we see that the first several Bernoulli numbers are

$$B_0 = 1,\ B_1 = -\frac{1}{2},\ B_2 = \frac{1}{6},\ B_3 = 0,\ B_4 = -\frac{1}{30},\ B_5 = 0,\ \text{and } B_6 = \frac{1}{42}.$$

There are some patterns that we can observe from this series. We see that $B_{2n+1} = 0$ for $n \geq 1$. See problem 1 for an explanation. Also, the non-zero terms alternate (see problem 4). There is, in fact, an explicit formula for the nth Bernoulli number [8].

$$B_n = \sum_{k=0}^{n} \sum_{i=0}^{k} (-1)^i \binom{k}{i} \frac{i^n}{k+1}. \tag{6.27}$$

Here, one must interpret 0^0 as 1 to get the formula to work for $n = 0$. There is also a connection between the Bernoulli numbers and the Riemann zeta function

$$B_n = (-1)^{n+1} n \zeta(1 - n).$$

Since $\zeta(1)$ is undefined, one must consider a limit as $n \to 0$ to evaluate B_0. For $n > 0$, the formula has no issues.

The Bernoulli numbers can also be computed recursively using *Bernoulli polynomials,* which is an extension of the Bernoulli numbers. The Bernoulli polynomial $B_n(x)$ is defined by the series

$$\frac{te^{tx}}{e^t - 1} = \sum_{n=0}^{\infty} \frac{B_n(x)}{n!} t^n. \tag{6.28}$$

The first few Bernoulli polynomials are

$$B_0(x) = 1, \quad B_1(x) = x - \frac{1}{2}, \quad B_2(x) = x^2 - x + \frac{1}{6}, \quad B_3(x) = x^3 - \frac{3x^2}{2} + \frac{x}{2}.$$

Plugging in $x = 0$ into equation 6.28 shows that $B_n(0) = B_n$, so we can quickly calculate the Bernoulli numbers from the Bernoulli polynomials. But the Bernoulli polynomials satisfy a recursion relationship (see problems 6 through 8)

$$B_n'(x) = nB_{n-1}(x) \qquad \text{and} \qquad \int_0^1 B_n(x)\,dx = 0 \qquad \text{if } n \geq 1. \tag{6.29}$$

These two equations allow us to find the Bernoulli polynomial in terms of the previous one. For example, we can find $B_4(x)$ by integrating $4B_3(x)$,

$$\int 4x^3 - 6x^2 + 2x\,dx = x^4 - 2x^3 + x^2 + C.$$

This determines $B_4(x)$ up to a arbitrary constant. To determine C, we use the second property, and integrate again.

$$\int_0^1 x^4 - 2x^3 + x^2 + C\,dx = \frac{x^5}{5} - \frac{x^4}{2} + \frac{x^3}{3} + Cx\Big|_0^1 = \frac{1}{5} - \frac{1}{2} + \frac{1}{3} + C.$$

In order for this to be 0, C must be $-1/30$, so we have $B_4(x) = x^4 - 2x^3 + x^2 - 1/30$, and in particular, $B_4 = -1/30$. The graphs of the first several Bernoulli polynomials are given in figure 6.6.

Although the first few Bernoulli polynomials seem to be getting smaller in magnitude on the interval from 0 to 1, they actually grow as $n \to \infty$, since the Bernoulli numbers increase without bound. However, it is possible to put bounds on how large the Bernoulli polynomials can become [16].

$$|B_n(x)| \leq \frac{2\zeta(n)n!}{(2\pi)^n} \quad \text{for } 0 \leq x \leq 1, \qquad n > 1.$$

We can use the Bernoulli polynomials to prove the following important theorem, which gives us the full asymptotic series for the sum.

THEOREM 6.4

Let $f(x)$ be a function which possesses all derivatives, and which for all k,

$$f^{(k+1)}(x) \ll f^{(k)}(x) \text{ as } x \to \infty.$$

Then as $n \to \infty$,

$$\sum_{x=0}^{n-1} f(x) \sim \int_0^n f(x)\,dx + \sum_{k=1}^{\infty} \frac{B_k}{k!}(f^{(k-1)}(n) - f^{(k-1)}(0)). \tag{6.30}$$

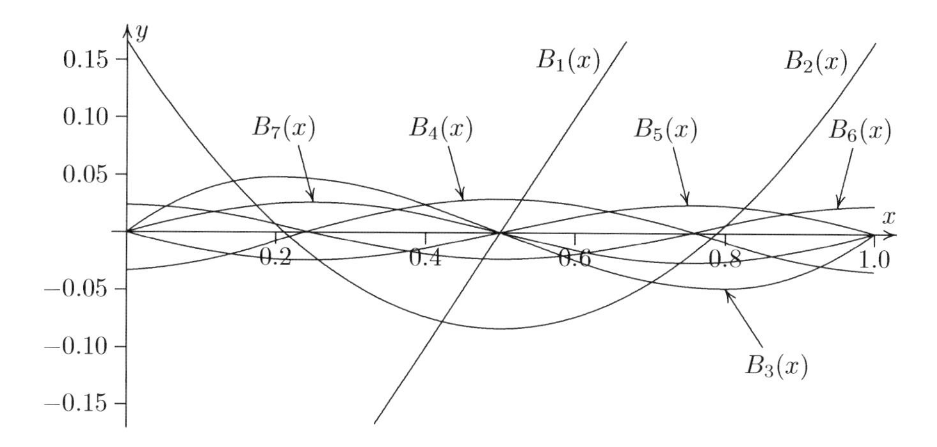

FIGURE 6.6: This shows the graph of the Bernoulli polynomials B_1 through B_7. Except for the case $m = 1$, we find that $B_m(0) = B_m(1)$. See problem 7 for the explanation.

Equation 6.30 is referred to as the *Euler-Maclaurin formula.*

PROOF: We will actually prove a stronger statement, for which we can use inductive reasoning.

$$\sum_{x=0}^{n-1} f(x) \sim \int_0^n f(x)\,dx + \sum_{k=1}^{m} \left(\frac{B_k}{k!}(f^{(k-1)}(n) - f^{(k-1)}(0)) \right) + R_m(n), \quad (6.31)$$

where $R_m(n)$ is the "remainder term"

$$R_m(x) = (-1)^{m-1} \sum_{k=0}^{n-1} \int_k^{k+1} f^{(m)}(x) \frac{B_m(x-k)}{m!}\,dx.$$

The case for $m = 1$ amounts to equation 6.24. Thus, we can assume that the statement is true for the previous m, with the remainder term begin

$$R_{m-1}(x) = (-1)^{m} \sum_{k=0}^{n-1} \int_k^{k+1} f^{(m-1)}(x) \frac{B_{m-1}(x-k)}{(m-1)!}\,dx.$$

We integrate the integral in the remainder term by parts, using $u = f^{(m-1)}(x)$, $du = f^{(m)}(x)\,dx$, $dv = B_{m-1}(x-k)/(m-1)! = B'_m(x-k)/m!$, so $v = B_m(x-k)/m!$. This gives us

$$\int_k^{k+1} f^{(m-1)}(x) \frac{B_{m-1}(x-k)}{(m-1)!}\,dx$$

$$= f^{(m-1)}(x)\frac{B_m(x-k)}{m!}\Big|_k^{k+1} - \int_k^{k+1} f^{(m)}(x)\frac{B_m(x-k)}{m!}\,dx$$
$$= \frac{f^{(m-1)}(k+1)B_m(1) - f^{(m-1)}(k)B_m(0)}{m!}$$
$$- \int_k^{k+1} f^{(m)}(x)\frac{B_m(x-k)}{m!}\,dx.$$

From problem 7, since $m > 1$, $B_m(1) = B_m(0) = B_m$, and so summing this expression from $k = 0$ to $n - 1$ causes many of the terms to telescope. This gives us

$$R_{m-1}(n) = (-1)^m B_m \frac{f^{(m-1)}(n) - (-1)^m f^{(m-1)}(0)}{m!}$$
$$-(-1)^m \sum_{k=0}^{n-1} \int_k^{k+1} f^{(m)}(x)\frac{B_m(x-k)}{m!}\,dx$$
$$= (-1)^m B_m \frac{f^{(m-1)}(n) - (-1)^m f^{(m-1)}(0)}{m!} + R_m(n).$$

If $m > 1$ is odd, then $B_m = 0$. If m is even, then $(-1)^m = 1$. So in either case, $(-1)^m B_m = B_m$, so we have produced the next term in the sum of equation 6.31. Hence, we have shown that

$$\sum_{x=0}^{n-1} f(x) \sim \int_0^n f(x)\,dx + \sum_{k=1}^{m}\left(\frac{B_k}{k!}(f^{(k-1)}(n) - f^{(k-1)}(0))\right) + R_m(n).$$

By induction, equation 6.31 has been proven for all $m \geq 1$.

Since $f^{(k+1)}(x) \ll f^{(k)}(x)$ as $x \to \infty$ for all k, each term in the sum is asymptotically smaller than the previous term as $n \to \infty$, so all we have left to prove is that the remainder term is asymptotically smaller than the final term of the sum. That is, we need to show that

$$\sum_{k=0}^{n-1} \int_k^{k+1} f^{(m)}(x)\frac{B_m(x-k)}{m!}\,dx \ll f^{(m-1)}(n) \qquad \text{as } n \to \infty.$$

If we let $M = 2\zeta(m)m!/(2\pi)^m$, then $M + B_m(x)$ will always be positive, so we can estimate the integral by the mean value theorem.

$$\int_k^{k+1} f^{(m)}(x)\frac{B_m(x-k)}{m!}\,dx$$
$$= \int_k^{k+1} f^{(m)}(x)\frac{B_m(x-k)+M}{m!}\,dx - \frac{M}{m!}\int_k^{k+1} f^{(m)}(x)\,dx$$
$$= f^{(m)}(\xi_k)\int_k^{k+1} \frac{B_m(x-k)+M}{m!}\,dx - \frac{M}{m!}\int_k^{k+1} f^{(m)}(x)\,dx$$
$$= \frac{M}{m!}(f^{(m)}(\xi_k) - f^{(m-1)}(k+1) + f^{(m-1)}(k))$$

for some ξ_k between k and $k+1$. Thus,

$$R_m(n) = \frac{M}{m!}\left(f^{(m-1)}(0) - f^{(m-1)}(n) + \sum_{k=0}^{n-1} f^{(m)}(\xi_k)\right).$$

The sum is now a Riemann sum for the integral $\int_0^n f^{(m)}(x)\,dx$, so this will be an approximation to the integral, $f^{(m-1)}(n) - f^{(m-1)}(0)$. Hence, the difference will be small in comparison to the integral, so

$$R_m(n) \ll 2\zeta(m)/(2\pi)^m (f^{(m-1)}(n) - f^{(m-1)}(0)).$$

Therefore, the remainder term is asymptotically smaller than the last term of the sum, and the theorem is proved. ∎

We often will combine all of the terms of equation 6.30 involving $f(0)$ and its derivatives into a single constant. Also, the starting point of the sum does not have to be 0, since changing the starting point only affects the formula by a constant. So we have

$$\boxed{\Delta^{-1} f(n) \sim \int f(n)\,dn + C + \sum_{k=1}^{\infty} \frac{B_k}{k!} f^{(k-1)}(n).} \tag{6.32}$$

This formula can now be used in cases where $f(0)$ is undefined. Here, it is understood that the constant may not be in its proper position in the asymptotic series.

Example 6.18

Find the full asymptotic series for the digamma function $\psi(n)$ as $n \to \infty$.

SOLUTION: From equation 6.13,

$$\psi(n) = -\gamma + \sum_{x=1}^{n-1} \frac{1}{x},$$

so we begin by finding the asymptotic series for $\sum_{x=1}^{n-1} 1/x$. Since

$$\int \frac{1}{n}\,dn = \ln(n) + C, \quad \text{and} \quad \frac{d^k}{dx^k}\frac{1}{n} = (-1)^k k!\, n^{-(k+1)},$$

we have by equation 6.32,

$$\sum_{x=1}^{n-1} \frac{1}{x} \sim \ln(n) + C + \sum_{k=1}^{\infty} \frac{B_k(-1)^{k-1}}{k} \frac{1}{n^k}$$
$$\sim \ln(n) + C - \frac{1}{2n} - \frac{1}{12n^2} + \frac{1}{120n^4} - \frac{1}{252n^6} + \cdots$$

To find the arbitrary constant, consider the limit as $n \to \infty$. The terms involving $1/n^k$ will go to zero, so

$$C = \lim_{n\to\infty} \left(\sum_{x=1}^{n-1} \frac{1}{x} \right) - \ln(n).$$

From equation 2.1, this C is the Euler-Mascheroni constant, γ. This means that for the asymptotic series for $\psi(n)$, the γ's will cancel out, and we have

$$\psi(n) \sim \ln(n) + \sum_{k=1}^{\infty} \frac{(-1)^{k-1} B_k}{kn^k} \tag{6.33}$$

$$\sim \ln(n) - \frac{1}{2n} - \frac{1}{12n^2} + \frac{1}{120n^4} - \frac{1}{252n^6} + \cdots. \qquad \square$$

6.4.2 Applications of the Euler-Maclaurin Formula

The Euler-Maclaurin formula has many different applications, both in finding asymptotic series and exact solutions to problems. For example, if we apply the formula to a case where $f(x)$ is a polynomial, eventually the derivatives will be 0, so the series will truncate to the exact answer. This gives us an alternative to using binomial coefficients to find the sum of the first n powers, as was done in example 6.2.

Example 6.19

Use equation 6.32 to find $\sum_{i=0}^{n-1} i^4$.

Solution: Using $f(x) = x^4$, we find $\int x^4\,dn = x^5/5 + C$, $f'(x) = 4x^3$, $f'''(x) = 24x$, and $f^{(5)}(x) = 0$. The even order derivatives are not needed, since the odd Bernoulli numbers, for $n > 1$, are 0. So we have

$$\sum_{i=0}^{n-1} i^4 = \frac{n^5}{5} + C - \frac{1}{2}n^4 + \frac{1}{6}\frac{1}{2!}4n^3 - \frac{1}{30}\frac{1}{4!}24n = \frac{n^5}{5} - \frac{n^4}{2} + \frac{n^3}{3} - \frac{n}{30} + C.$$

Since the sum is 0 when $n = 0$, we see that $C = 0$. This can be seen to be equivalent to the answer in example 6.2. $\square$

Another application of the Euler-Maclaurin formula is to speed up the convergence of a series that is known to converge. In this case, the constant becomes the true sum of the series, and the other terms are corrections to this sum. Solving for the constant, we have

$$C \approx \sum_{x=a}^{n-1} f(x) + \int_n^{\infty} f(x)\,dx - \sum_{k=1}^{m} \frac{B_k}{k!} f^{(k-1)}(n). \tag{6.34}$$

TABLE 6.2: This table shows how increasing m in equation 6.34 greatly speeds up the convergence of the series $\sum_{n=1}^{\infty} 1/n^2$. The true sum is $\pi^2/6 \approx 1.6449340668848$.

n	$m=0$	$m=1$	$m=2$	$m=4$	$m=6$
1	1.0000000	1.50000000	1.666666667	1.633333333	1.65714285714
2	1.5000000	1.62500000	1.645833333	1.644791667	1.64497767857
3	1.5833333	1.63888889	1.645061728	1.644924554	1.64493544103
4	1.6111111	1.64236111	1.644965278	1.644932726	1.64493417891
5	1.6236111	1.64361111	1.644944444	1.644933778	1.64493408254
6	1.6302778	1.64416667	1.644938272	1.644933985	1.64493406996
7	1.6342460	1.64445011	1.644936022	1.644934039	1.64493406764
8	1.6367970	1.64460955	1.644935073	1.644934056	1.64493406709
9	1.6385332	1.64470600	1.644934626	1.644934062	1.64493406693
10	1.6397678	1.64476773	1.644934398	1.644934064	1.64493406688

The endpoints of the integral were chosen so that, as $n \to \infty$, the integral would approach 0. Note that since the sum is assumed to converge, so will the integral.

Example 6.20
Estimate the sum

$$\sum_{x=1}^{\infty} \frac{1}{x^2}$$

using equation 6.34 with $n = 10$, and $m = 4$.
SOLUTION: With $m = 4$, equation 6.34 has the constant C approximated by

$$C \approx \sum_{x=1}^{n-1} \frac{1}{x^2} + \frac{1}{n} - \frac{-1}{2}\frac{1}{n^2} - \frac{1}{12}\frac{-2}{n^3} - \frac{-1}{720}\frac{-24}{n^5}.$$

Now we can plug in $n = 10$, to obtain

$$C \approx \sum_{x=1}^{9} \frac{1}{x^2} + \frac{1}{10} + \frac{1}{200} + \frac{1}{6000} - \frac{1}{3000000} \approx 1.64493406449.$$

The true value of the sum is 1.6449340668482264365, so we have 9 places of accuracy using only the first 9 terms of the series. Table 6.2 shows how increasing m can greatly improve the rate of convergence. In fact, Euler used this method to find the sum to 20 places in 1735. This convinced him that the sum was $\pi^2/6$, which he was able to prove in the same year [21]. □

The Euler-Maclaurin formula can provide an alternative method of finding the full asymptotic behavior of $\Gamma(n)$. This is because the logarithm of the gamma function satisfies the relation

$$\ln(\Gamma(n)) = \sum_{k=1}^{n-1} \ln(k).$$

Thus, we can substitute $f(x) = \ln(x)$ into equation 6.32, so that $f^{(k)}(x) = (-1)^{k+1}(k-1)!x^{-k}$ for $k \geq 1$. We obtain

$$\ln(\Gamma(n)) \sim \int_1^n \ln(x)\,dx + C - \frac{1}{2}\ln(n) + \sum_{k=2}^{\infty} \frac{B_k}{k!}\frac{(-1)^k(k-2)!}{n^{k-1}}.$$

$$\sim n\ln(n) - n + C - \frac{\ln(n)}{2} + \frac{1}{12n} - \frac{1}{360n^3} + \frac{1}{1260n^5} - \frac{1}{1680n^7} + \cdots.$$

If we exponentiate both sides, we find that

$$\Gamma(n) \sim n^n e^{-n} n^{-1/2} e^C \exp\left(\frac{1}{12n} - \frac{1}{360n^3} + \frac{1}{1260n^5} - \frac{1}{1680n^7} + \cdots\right).$$

We know that $\Gamma(n) \sim n^n e^{-n}\sqrt{2\pi/n}$, so e^C must be $\sqrt{2\pi}$. But this provides another way to determine the Sterling series, by exponentiating the series

$$\sum_{k=1}^{\infty} \frac{B_{2k}}{2k(2k-1)n^{2k-1}} \sim \frac{1}{12n} - \frac{1}{360n^3} + \frac{1}{1260n^5} - \frac{1}{1680n^7} + \frac{1}{1188n^9} - \cdots. \tag{6.35}$$

Here, we were able to replace every k with $2k$, since only the even terms are non-zero.

If we compute the continued fraction form for this series, we get a pleasant surprise.

$$\sum_{k=1}^{\infty} \frac{B_{2k}}{2k(2k-1)n^{2k-1}} \sim \frac{1}{12n}\Big/\Big(1+\frac{1}{30n^2}\Big/\Big(1+\frac{53}{210n^2}\Big/\Big(1+\frac{195}{371n^2}\Big/\Big(1+\cdots.$$

All of the coefficients in the continued fraction expansion turn out to be positive, so this is an S-fraction. This means that the series is related to a Stieltjes function, expressed by the convergent continued fraction representation

$$S(x) = \frac{1}{12}\Big/\Big(1+\frac{x}{30}\Big/\Big(1+\frac{53x}{210}\Big/\Big(1+\frac{195x}{371}\Big/\Big(1+\frac{22999x}{22737}\Big/\Big(1+\cdots. \tag{6.36}$$

The asymptotic series for this Stieltjes function is

$$S(x) \sim \sum_{k=1}^{\infty} \frac{B_{2k}}{2k(2k-1)} x^{k-1}. \tag{6.37}$$

This means that we can express the gamma function in terms of this Stieltjes function

$$\Gamma(n) = \sqrt{2\pi}n^{n-1/2}e^{-n}e^{S(1/n^2)/n}.$$

This Stieltjes series in equation 6.37 is important enough to determine its Borel sum of the series, since this would help prove that the continued fraction does indeed converge to the key part of the gamma function. We will utilize equation 6.26, from which we can determine that

$$\sum_{k=1}^{\infty} \frac{B_{2k}}{(2k)!} x^{2k} = \frac{x}{e^x - 1} - 1 + \frac{x}{2}, \tag{6.38}$$

so the plan is to show the relationship between the series for $S(x)$ and this series. We begin by defining

$$\phi(x) = \sum_{k=1}^{\infty} \frac{B_{2k}}{(2k)!2k(2k-1)} x^{k-1}.$$

Then

$$x\phi(x^2) = \sum_{k=1}^{\infty} \frac{B_{2k}}{(2k)!2k(2k-1)} x^{2k-1},$$

so taking the derivative eliminates the $2k - 1$ in the denominator.

$$[x\phi(x^2)]' = \sum_{k=1}^{\infty} \frac{B_{2k}}{(2k)!2k} x^{2k-2}.$$

Multiplying by x^2, and taking another derivative, will eliminate the $2k$ in the denominator.

$$[x^2[x\phi(x^2)]']' = \sum_{k=1}^{\infty} \frac{B_{2k}}{(2k)!} x^{2k-1}.$$

If we multiply by x, we get the series for $x/(e^x - 1) - 1 + x/2$, so we have that

$$[x^2[x\phi(x^2)]']' = \frac{1}{e^x - 1} - \frac{1}{x} + \frac{1}{2}.$$

We now unravel this by integrating. We choose the constant of integration so that the integral evaluates to 0 when $x = 0$.

$$x^2[x\phi(x^2)]' = \ln(e^x - 1) - \ln(x) - \frac{x}{2} = \ln\left(\frac{e^{x/2} - e^{-x/2}}{x}\right) = \ln(2\sinh(x/2)/x).$$

Now we must divide by x^2, and integrate again, producing

$$x\phi(x^2) = \int \frac{\ln(2\sinh(x/2)/x)}{x^2}\, dx.$$

This is a trickier integral, but we can make progress if we integrate by parts, letting $u = \ln(2\sinh(x/2)/x)$ and $dv = 1/x^2\, dx$, so that $v = -1/x$, and $du = (\coth(x/2)/2 - 1/x)\, dx$. Thus,

$$x\phi(x^2) = -\frac{\ln(2\sinh(x/2)/x)}{x} - \int \frac{1}{x^2} - \frac{\coth(x/2)}{2x}\,dx.$$

If we define the "hyperbolic cotangent integral function" to be

$$\text{Cthi}(x) = -\frac{1}{x} + \int_0^x \left(\frac{\coth(t)}{t} - \frac{1}{t^2}\right) dt, \tag{6.39}$$

which is designed so that $\text{Cthi}'(x) = \coth(x)/x$. Then we have

$$x\phi(x^2) = \frac{1}{x} + \frac{\text{Cthi}(x/2)}{2} - \frac{\ln(2\sinh(x/2)/x)}{x}.$$

Thus,

$$\phi(x) = \frac{1}{x} + \frac{\text{Cthi}(\sqrt{x}/2)}{2\sqrt{x}} - \frac{\ln(2\sinh(\sqrt{x}/2)/\sqrt{x})}{x}.$$

Then by equation 3.10, the Borel sum of the series is

$$\begin{aligned} S(x) &= \int_0^\infty e^{-t}\phi(xt^2)\,dt \\ &= \int_0^\infty e^{-t}\left(\frac{1}{xt^2} + \frac{\text{Cthi}(t\sqrt{x}/2)}{2t\sqrt{x}} - \frac{\ln(2\sinh(t\sqrt{x}/2)/(t\sqrt{x}))}{xt^2}\right) dt. \end{aligned}$$

Since the gamma function can be expressed in terms of this Stieltjes function, we have $\Gamma(x) =$

$$\sqrt{2\pi}x^{x-1/2}e^{-x}\exp\left[\int_0^\infty e^{-t}\left(\frac{x}{t^2} + \frac{\text{Cthi}\left(\frac{t}{2x}\right)}{2t} - \frac{x\ln\left(2\sinh\left(\frac{t}{2x}\right)/t\right)}{t^2}\right) dt\right].$$

Actually, this isn't a very useful formula for the gamma function, since we have much easier formulas that compute $\Gamma(n)$ more efficiently. But this does point out that we were able to find the Borel sum of the Sterling series using the Euler-Maclaurin formula. It also gives us another continued fraction representation for the gamma function,

$$\sqrt{2\pi}x^{x-1/2}e^{-x}\exp\left[\frac{1}{12x}\Big/\left(1+\frac{1}{30x^2}\Big/\left(1+\frac{53}{210x^2}\Big/\left(1+\frac{195}{371x^2}\Big/\left(1+\cdots\right.\right.\right.\right.\right]. \tag{6.40}$$

Even though equation 6.40 has the same region of convergence as equation 6.21, this seems to converge much faster [19].

Problems for §6.4

1 Show that $B_{2n+1} = 0$ for $n \geq 0$.
Hint: Show that

$$f(x) = \frac{x}{e^x - 1} + \frac{x}{2}$$

is an even function.

2 Show that

$$x \coth(x) = \sum_{n=0}^{\infty} \frac{2^{2n} B_{2n}}{(2n)!} x^{2n}.$$

Hint: Begin with equation 6.26, and use the fact that

$$\coth(x) = \frac{e^x + e^{-x}}{e^x - e^{-x}}.$$

3 Use the result of problem 2 to show that

$$\tan(x) = \sum_{n=1}^{\infty} \frac{(-1)^{n+1}(4^{2n} - 2^{2n}) B_{2n}}{(2n)!} x^{2n-1}.$$

Hint: Use the identity $\coth(2x) = (\coth(x) + \tanh(x))/2$ to find the Maclaurin series for $\tanh(x)$, and then use the fact that $\tan(x) = -i \tanh(ix)$.

4 Show that $(-1)^{n+1} B_{2n} \geq 0$ for all $n \geq 1$.
Hint: First prove by induction that all derivatives of $\tan(x)$ involve positive terms involving $\tan^n(x) \sec^m(x)$. Then utilize problem 3.

5 Use equation 6.27 to find B_8.

6 Show that for $n \geq 1$, $B_n'(x) = n B_{n-1}(x)$.
Hint: Take the derivative of equation 6.28 with respect to x.

7 Show that for $n > 1$, $B_n(0) = B_n(1)$.
Hint: Subtract the two series in equation 6.28 evaluated at $x = 1$ and $x = 0$.

8 Use the results of problems 6 and 7 to show that for $n \geq 1$,

$$\int_0^1 B_n(x)\, dx = 0.$$

9 Use equation 6.29 to prove inductively that $B_n(1 - x) = (-1)^n B_n(x)$.

10 Use equation 6.29 to prove inductively that $B_n(x + 1) - B_n(x) = n x^{n-1}$.

11 Use equation 6.29 to find $B_5(x)$.

12 The series for the digamma function in equation 6.33 diverges for all n, but we can create a continued fraction representation of this series that will converge for $\mathrm{Re}(n) > 0$. Find the continued fraction representation, up to F_4, of the asymptotic series for $\psi(n) - \ln(n)$.
Hint: Substitute $x = 1/n$ to put the asymptotic series into the standard form.

13 Because the series in equation 6.33 diverges for all n, one can ask what the Borel sum of the series is. Consider the series

$$\psi(1/x) - \ln(1/x) = -\frac{x}{2} - \frac{x^2}{12} + \frac{x^4}{120} - \frac{x^6}{252} + \cdots.$$

Show that the Borel sum of the series is

$$\int_0^\infty e^{-t} \ln\left(\frac{xt}{e^{xt} - 1}\right) dt.$$

Hint: Utilize equation 6.25.

14 Use the result of problem 13 to show that

$$\psi(z) = -\gamma - \int_0^\infty e^{-t} \ln(e^{t/z} - 1)\, dt.$$

This formula is valid whenever $\text{Re}(z) > 0$.

15 Use equation 6.32, to find $\sum_{i=0}^{n-1} i^5$.

16 Use equation 6.32, to find $\sum_{i=0}^{n-1} i^6$.

For problems **17** through **22**: Estimate the following sums using the Euler-Maclaurin formula (equation 6.34), using $n = 9$ and the given value for m.

17 $\sum_{x=1}^{\infty} \frac{1}{x^3}$, $m = 4$

18 $\sum_{x=1}^{\infty} \frac{1}{x^4}$, $m = 4$

19 $\sum_{x=1}^{\infty} \frac{1}{x\sqrt{x}}$, $m = 4$

20 $\sum_{x=0}^{\infty} \frac{1}{x^2 + 1}$, $m = 2$

21 $\sum_{x=0}^{\infty} e^{-x}$, $m = 4$

22 $\sum_{x=1}^{\infty} \frac{e^{-\sqrt{x}}}{\sqrt{x}}$, $m = 2$

23 Use the result of problem 2 to find the Frobenius series for the $\text{Cthi}(x)$ function, defined by equation 6.39.

24 Use the Euler-Maclaurin series, equation 6.32, to find the asymptotic series for

$$\sum_{x=1}^{n-1} \frac{1}{\sqrt{x}}$$

up to the $n^{-7/2}$ term, in terms of a constant.

25 Plug in $n = 9$ into the series in problem 24 to find the numerical approximation for

$$\zeta(1/2) = \lim_{n\to\infty}\left(\sum_{x=1}^{n-1}\frac{1}{\sqrt{x}} - 2\sqrt{n}\right).$$

26 Use the Euler-Maclaurin series to find the asymptotic series for

$$\sum_{x=1}^{n-1}\sqrt{x}$$

up to the $n^{-9/2}$ term, in terms of a constant.

27 Plug in $n = 9$ into the series in problem 26 to find the value of the constant to 6 decimal places.

6.5 Taylor-like and Frobenius-like Series Expansions

Although in subsection 6.3.1 we found a way to find the Taylor series solution for difference equations, the method was rather awkward. Each term in the series require more terms which involve binomial coefficients. Even though it was not difficult to compute the first few terms of the series, it is nearly impossible to find a pattern for *all* of the series' terms.

In this section, we will explore an alternative to the Taylor series, for which we will often be able to determine the pattern for all of the terms of the series. This will allow us to express the asymptotic series in closed form.

The key to this method is to find an expression that behaves like $1/n^k$ for large values of k, yet its discrete derivative is similar to the original expression. We found one such expression, resulting from equation 6.6:

$$\Delta\binom{n}{k} = \binom{n}{k-1}.$$

The problem is that $\binom{n}{k}$ is a kth degree polynomial in n, and so as $n \to \infty$, this behaves as Cn^k, not n^{-k}. Also, we later want to consider fractional powers of n, so let us consider using the gamma function instead of the factorial function. One such function would be

$$\frac{\Gamma(n)}{\Gamma(n+k)} = \frac{1}{n(n+1)(n+2)\cdots(n+k-1)}.$$

Clearly, this will behave like n^{-k} for large k. We then find that discrete derivative to this function is

$$\Delta\frac{\Gamma(n)}{\Gamma(n+k)} = \frac{\Gamma(n+1)}{\Gamma(n+k+1)} - \frac{\Gamma(n)}{\Gamma(n+k)}$$

$$= \frac{n\Gamma(n)}{\Gamma(n+k+1)} - \frac{(n+k)\Gamma(n)}{\Gamma(n+k+1)}$$
$$= \frac{-k\Gamma(n)}{\Gamma(n+k+1)}.$$

Note that this is analogous to the differential property $(n^{-k})' = -kn^{-(k+1)}$. Because of this property, the function $\Gamma(n)/\Gamma(n+k)$ becomes the discrete analog to n^{-k}. Our plan is to express a_n as a series involving these new functions.

$$a_n = \sum_{k=0}^{\infty} B_k \frac{\Gamma(n)}{\Gamma(n+k)} = B_0 + \frac{B_1}{n} + \frac{B_2}{n(n+1)} + \frac{B_3}{n(n+1)(n+2)} + \cdots.$$

Then the discrete derivatives of a_n can be expressed by a similar sum.

$$\Delta a_n = \sum_{k=1}^{\infty} -kB_k \frac{\Gamma(n)}{\Gamma(n+k+1)}, \quad \Delta^2 a_n = \sum_{k=1}^{\infty} k(k+1)B_k \frac{\Gamma(n)}{\Gamma(n+k+2)}, \text{ etc.} \tag{6.41}$$

The approach then becomes very similar to that of section 4.3. First of all, any polynomial times $\Gamma(n)/\Gamma(n+k)$ can be expressed as a linear combination of terms of the form $\Gamma(n)/\Gamma(n+k-\alpha)$, where the coefficients may depend on k, but not on n, and α is an integer. For example, to simplify the expression

$$n\frac{\Gamma(n)}{\Gamma(n+k-\alpha)},$$

we can rewrite

$$n = (n+k-\alpha-1) + (1-k+\alpha),$$

so that

$$n\frac{\Gamma(n)}{\Gamma(n+k-\alpha)} = \frac{(n+k-\alpha-1)\Gamma(n)}{(n+k-\alpha-1)\Gamma(n+k-\alpha-1)} + \frac{(1-k+\alpha)\Gamma(n)}{\Gamma(n+k-\alpha)}.$$

Thus,

$$n\frac{\Gamma(n)}{\Gamma(n+k-\alpha)} = \frac{\Gamma(n)}{\Gamma(n+k-\alpha-1)} + (1-k+\alpha)\frac{\Gamma(n)}{\Gamma(n+k-\alpha)}. \tag{6.42}$$

By repeated use of this formula, we can determine how any polynomial times $\Gamma(n)/\Gamma(n+k)$ will expand to a form where the coefficients do not depend on n. For example,

$$n^2\frac{\Gamma(n)}{\Gamma(n+k)} = n\left(\frac{\Gamma(n)}{\Gamma(n+k-1)} + (1-k)\frac{\Gamma(n)}{\Gamma(n+k)}\right)$$
$$= n\frac{\Gamma(n)}{\Gamma(n+k-1)} + (1-k)n\frac{\Gamma(n)}{\Gamma(n+k)}$$

$$= \frac{\Gamma(n)}{\Gamma(n+k-2)} + (2-k)\frac{\Gamma(n)}{\Gamma(n+k-1)}$$
$$+(1-k)\left(\frac{\Gamma(n)}{\Gamma(n+k-1)} + (1-k)\frac{\Gamma(n)}{\Gamma(n+k)}\right)$$
$$= \frac{\Gamma(n)}{\Gamma(n+k-2)} + \frac{(3-2k)\Gamma(n)}{\Gamma(n+k-1)} + \frac{(1-k)^2\Gamma(n)}{\Gamma(n+k)}.$$

So we obtain the formula

$$\frac{n^2\Gamma(n)}{\Gamma(n+k)} = \frac{\Gamma(n)}{\Gamma(n+k-2)} + \frac{(3-2k)\Gamma(n)}{\Gamma(n+k-1)} + \frac{(1-k)^2\Gamma(n)}{\Gamma(n+k)}. \tag{6.43}$$

Multiplying the equation by n again, and applying equation 6.42 to each term produces

$$\frac{n^3\Gamma(n)}{\Gamma(n+k)} = \frac{\Gamma(n)}{\Gamma(n+k-3)} + \frac{(6-3k)\Gamma(n)}{\Gamma(n+k-2)}$$
$$+ \frac{(3k^2-9k+7)\Gamma(n)}{\Gamma(n+k-1)} + \frac{(1-k)^3\Gamma(n)}{\Gamma(n+k)}, \tag{6.44}$$

and so on. Thus, we can eliminate the n variable from the coefficients. This gives us a strategy for obtaining a series solution to any linear homogeneous difference equation in which all of the coefficient functions are rational functions.

1. We can first multiply through by the common denominator, so that the coefficient functions are polynomials.
2. We replace a_n with the series $\sum_{k=0}^{\infty} B_k\Gamma(n)/\Gamma(n+k)$, and use equation 6.41 to replace Δa_n, $\Delta^2 a_n$, etc.
3. We must rewrite each series so that n only appears inside of a gamma function, using equations 6.42 through 6.44.
4. We are now ready to shift the index on the sums so that all sums are of the form $\Gamma(n)/\Gamma(n+k)$.
5. Finally, we can equate the coefficients to obtain a recursion formula for the B_k.
6. We can use the recursion formula to compute as many coefficients as desired.

Example 6.21

Find the Taylor-like series, up to the $\Gamma(n)/\Gamma(n+6)$ term, for the solution to the difference equation

$$b_{n+1} = \frac{1+n^2}{n^2}b_n.$$

SOLUTION: If we multiply by the denominator, we can express this equation in the form $n^2\Delta b_n = b_n$. We will assume that the solution is of the form

$$b_n = \sum_{k=0}^{\infty} B_k \frac{\Gamma(n)}{\Gamma(n+k)}$$

then the difference equation becomes

$$\sum_{k=0}^{\infty} -kn^2 B_k \frac{\Gamma(n)}{\Gamma(n+k+1)} = \sum_{k=0}^{\infty} B_k \frac{\Gamma(n)}{\Gamma(n+k)}.$$

Replacing k with $k+1$ in equation 6.43 gives us

$$\frac{n^2\Gamma(n)}{\Gamma(n+k+1)} = \frac{\Gamma(n)}{\Gamma(n+k-1)} + \frac{(1-2k)\Gamma(n)}{\Gamma(n+k)} + \frac{k^2\Gamma(n)}{\Gamma(n+k+1)}.$$

Thus, our difference equation can now be written as

$$\sum_{k=0}^{\infty} -kB_k \frac{\Gamma(n)}{\Gamma(n+k-1)} + \sum_{k=0}^{\infty} -k(1-2k)B_k \frac{\Gamma(n)}{\Gamma(n+k)} + \sum_{k=0}^{\infty} -k(k^2)B_k \frac{\Gamma(n)}{\Gamma(n+k+1)} = \sum_{k=0}^{\infty} B_k \frac{\Gamma(n)}{\Gamma(n+k)}.$$

In order to get all of the sums of the form $\Gamma(n)/\Gamma(n+k)$, we need to shift the index up on the first term, and down on the third term.

$$\sum_{k=-1}^{\infty} -(k+1)B_{k+1} \frac{\Gamma(n)}{\Gamma(n+k)} + \sum_{k=0}^{\infty} -k(1-2k)B_k \frac{\Gamma(n)}{\Gamma(n+k)} + \sum_{k=1}^{\infty} -(k-1)^3 B_{k-1} \frac{\Gamma(n)}{\Gamma(n+k)} = \sum_{k=0}^{\infty} B_k \frac{\Gamma(n)}{\Gamma(n+k)}.$$

This gives us the recursion formula for B_k:

$$-(k+1)B_{k+1} - k(1-2k)B_k - (k-1)^3 B_{k-1} = B_k, \qquad \text{for } k \geq 1.$$

Note that the recursion formula is also true for $k = 0$, if we return to the convention that $B_{-1} = 0$. This verifies that there is indeed a Taylor-like series for the solution. Thus, we have

$$B_1 = -B_0, \quad B_2 = 0, \quad B_3 = \frac{B_0}{3}, \quad B_4 = \frac{7B_0}{6}, \quad B_5 = \frac{9B_0}{2}, \ldots.$$

So the Taylor-like series can be written as

$$b_n = B_0\left(1 - \frac{1}{n} + \frac{1}{3n(n+1)(n+2)} + \frac{7}{6n(n+1)(n+2)(n+3)} + \frac{9}{2n(n+1)(n+2)(n+3)(n+4)} + \cdots\right).$$

□

The Taylor-like series has a clear advantage in that there will be a simple recursion formula for the coefficients B_k. One of the disadvantages is that we cannot form a continued fraction form of the series directly from the Taylor-like series. Also, the notation is a bit awkward. One solution is to introduce the notation

$$n^{[-k]} = \frac{\Gamma(n)}{\Gamma(n+k)}.$$

With this notation, the analogy between the difference operator and the derivative becomes even more pronounced.

$$\Delta n^{[-k]} = -kn^{[-k-1]}.$$

Equations 6.42 through 6.44 can be expressed in terms of this new notation.

$$nn^{[-k+\alpha]} = n^{[-k+\alpha+1]} + (1-k+\alpha)n^{[-k+\alpha]}, \tag{6.45}$$

$$n^2 n^{[-k]} = n^{[-k+2]} + (3-2k)n^{[-k+1]} + (k-1)^2 n^{[-k]}, \tag{6.46}$$

$$n^3 n^{[-k]} = n^{[-k+3]} + (6-3k)n^{[-k+2]} + (3k^2-9k+7)n^{[-k+1]} - (k-1)^3 n^{[-k]}. \tag{6.47}$$

Example 6.22

Find the Taylor-like series for the solution to the equation

$$n(n+2)a_{n+1} = (n+1)^2 a_n.$$

SOLUTION: We can rewrite this difference equation as $(n^2+2n)\Delta a_n = a_n$. Expressing a_n by a Taylor-like series,

$$a_n = \sum_k B_k n^{[-k]}, \qquad \Delta a_n = \sum_k -kB_k n^{[-k-1]}.$$

Letting $\alpha = -1$ in equations 6.45 and 6.46 gives us

$$nn^{[-k-1]} = n^{[-k]} - kn^{[-k-1]}, \quad n^2 n^{[-k-1]} = n^{[-k+1]} + (1-2k)n^{[-k]} + k^2 n^{[-k-1]}.$$

So the difference equation can be written as

$$\sum_k -kB_k(n^{[-k+1]} + (1-2k)n^{[-k]} + k^2 n^{[-k-1]} + 2n^{[-k]} - 2kn^{[-k-1]})$$
$$= \sum_k B_k n^{[-k]}.$$

Combining like terms, we get

$$\sum_k -kB_k n^{[-k+1]} + \sum_k (2k^2-3k-1)B_k n^{[-k]} + \sum_k (2k^2-k^3)B_k n^{[-k-1]} = 0.$$

Shifting the index so that all sums involve $n^{[-k]}$, we get

$$\sum_k -(k+1)B_{k+1}n^{[-k]} + \sum_k (2k^2 - 3k - 1)B_k n^{[-k]} + \sum_k (k-1)^2(3-k)B_{k-1}n^{[-k]}) = 0.$$

Thus, we have the recursion formula

$$-(k+1)B_{k+1} + (2k^2 - 3k - 1)B_k + (k-1)^2(3-k)B_{k-1} = 0.$$

Plugging in $k = 0, 1, 2, \ldots$ produces

$$B_1 = -B_0, \quad B_2 = B_0, \quad B_3 = 0, \quad B_4 = 0, \ldots.$$

Since we have two consecutive zeros, all further coefficients will be 0. So this series truncates, and we have the exact solution

$$a_n = B_0\left(1 - \frac{1}{n} + \frac{1}{n(n+1)}\right).$$

□

If the leading behavior of the difference equation is of the form n^α, we could peel off the leading behavior to obtain an equation that has a Taylor series solution, but this often becomes messy, especially if α is not an integer. A better alternative is to modify the Taylor-like series to form a Frobenius-like series

$$a_n = \sum_{k=0}^{\infty} B_k \frac{\Gamma(n)}{\Gamma(n+k-\alpha)} = \sum_k B_k n^{[-k+\alpha]}.$$

Then the difference operators will behave in a manner similar to derivatives of standard power functions.

$$\Delta a_n = \sum_k (-k+\alpha)B_k n^{[-k-1+\alpha]},$$

$$\Delta^2 a_n = \sum_k (-k+\alpha)(-k-1+\alpha)B_k n^{[-k-2+\alpha]}, \text{ etc.}$$

If the original difference equation had rational functions for the coefficients, there will be a simple recursion formula for the B_k.

Example 6.23

Find the Frobenius-like series for the solution to the difference equation

$$n\Delta a_n = a_n/2.$$

SOLUTION: The corresponding differential equation is $xy' = y/2$, which has the solution $y = C\sqrt{x}$. Thus, $\alpha = 1/2$, and so we want to consider a series of the form

$$a_n = \sum_{k=0}^{\infty} B_k n^{[-k+1/2]}.$$

Then

$$n\Delta a_n = \sum_k \left(\frac{1}{2} - k\right) n B_k n^{[-k-1/2]},$$

and using $\alpha = -1/2$ in equation 6.45 produces

$$\begin{aligned} n\Delta a_n &= \sum_k \left(\frac{1}{2} - k\right) B_k \left(n^{[-k+1/2]} + \left(\frac{1}{2} - k\right) n^{[-k-1/2]}\right) \\ &= \sum_k B_k n^{[-k+1/2]}/2. \end{aligned}$$

Grouping terms of like order, we find that

$$\sum_k -k B_k n^{[-k+1/2]} + \sum_k \left(\frac{1}{2} - k\right)^2 B_k n^{[-k-1/2]} = 0.$$

Shifting the index of the first sum up by 1, we obtain

$$\sum_k -(k+1) B_{k+1} n^{[-k-1/2]} + \sum_k \left(\frac{1}{2} - k\right)^2 B_k n^{[-k-1/2]} = 0.$$

Thus, we have the recursion formula

$$B_{k+1} = \frac{\left(k - \frac{1}{2}\right)^2}{k+1} B_k,$$

which has the solution

$$B_k = \frac{\left(\Gamma\left(k - \frac{1}{2}\right)\right)^2}{\left(\Gamma\left(-\frac{1}{2}\right)\right)^2 k!} B_0.$$

So we have the full Frobenius-like series solution to the difference equation

$$a_n = B_0 \sum_{k=0}^{\infty} \frac{\left(\Gamma\left(k - \frac{1}{2}\right)\right)^2}{4\pi k!} \frac{\Gamma(n)}{\Gamma\left(n + k - \frac{1}{2}\right)}.$$

□

In section 4.4, when we used Frobenius series to solve differential equations at regular singular points, we occasionally had a series with a logarithm, typically due to two series solutions conflicting with one another. This can also happen with difference equations, where the leading behavior will involve a logarithm. However, introducing the logarithm into a difference equation can get extremely complicated. A better approach would be to use the digamma function $\psi(n)$, which has the leading behavior of a logarithm, but possesses the property $\Delta\psi(n) = 1/n$. This in turn creates more identities for the digamma function.

$$\begin{aligned}
\Delta(\psi(n)n^{[-k]}) &= \frac{\psi(n+1)\Gamma(n+1)}{\Gamma(n+k+1)} - \frac{\psi(n)\Gamma(n)}{\Gamma(n+k)} \\
&= \frac{(\psi(n)+1/n)n\Gamma(n)}{\Gamma(n+k+1)} - \frac{(n+k)\psi(n)\Gamma(n)}{(n+k)\Gamma(n+k)} \\
&= \frac{n\psi(n)\Gamma(n)+\Gamma(n)-(n+k)\psi(n)\Gamma(n)}{\Gamma(n+k+1)} \\
&= \frac{\Gamma(n)}{\Gamma(n+k+1)} - k\frac{\psi(n)\Gamma(n)}{\Gamma(n+k+1)} = n^{[-k-1]} - k\psi(n)n^{[-k-1]}.
\end{aligned}$$

Notice the similarity of this rule to the derivative rule $(x^{-k}\ln x)' = x^{-k-1} - kx^{-k-1}\ln x$. We can take further differences of the digamma function, to produce

$$\Delta^2(\psi(n)n^{[-k]}) = (k^2+k)\psi(n)n^{[-k-2]} - (2k+1)n^{[-k-2]}. \tag{6.48}$$

Example 6.24

Find the Frobenius-like series for the solutions to the difference equation

$$4n^2a_{n+2} - 8n^2a_{n+1} + (4n^2+1)a_n = 0.$$

SOLUTION: This equation can be written as $4n^2\Delta^2a_n + a_n = 0$. The corresponding differential equation is $4x^2y'' + y = 0$, for which the solutions are $y = C_1\sqrt{x} + C_2\ln(x)\sqrt{x}$. Thus, we expect one solution to have a Frobenius-like series with $\alpha = 1/2$, but the other solution will have terms involving $\psi(n)$.

For the solution of the form

$$a_n = \sum_{k=0}^{\infty} A_k n^{[-k+1/2]},$$

we have

$$\Delta^2a_n = \sum_k(-k+1/2)(-k-1/2)A_kn^{[-k-3/2]} = \sum_k(k^2-1/4)A_kn^{[-k-3/2]}.$$

Thus,

$$4n^2\Delta^2a_n = \sum_k(4k^2-1)A_kn^2n^{[-k-3/2]}.$$

Replacing k with $k+3/2$ in equation 6.46 produces

$$n^2n^{[-k-3/2]} = n^{[-k+1/2]} - 2kn^{[-k-1/2]} + (k+1/2)^2n^{[-k-3/2]}.$$

So our difference equation becomes

$$\sum_k (4k^2-1)A_k n^{[-k+1/2]} - \sum_k (4k^2-1)2kA_k n^{[-k-1/2]} \\ + \sum_k (4k^2-1)A_k(k+1/2)^2 n^{[-k-3/2]} + \sum_k A_k n^{[-k+1/2]} = 0.$$

The first and the last sums can combine, and we shift the index on the second and third sums down by one and two, respectively.

$$\sum_k 4k^2 A_k n^{[-k+1/2]} - \sum_k (4k^2-8k+3)2(k-1)A_{k-1} n^{[-k+1/2]} \\ + \sum_k (4k^2-16k+15)A_{k-2}(k-3/2)^2 n^{[-k+1/2]} = 0.$$

This gives us the recursion formula

$$4k^2 A_k - 2(4k^2-8k+3)(k-1)A_{k-1} + (4k^2-16k+15)(k-3/2)^2 A_{n-2} = 0.$$

Plugging in $k = 1, 2, 3, \ldots$ produces

$$A_1 = 0, \quad A_2 = \frac{A_0}{64}, \quad A_3 = \frac{5A_0}{192}, \quad A_4 = \frac{1025A_0}{16384}, \quad A_5 = \frac{5005A_0}{24576}, \ldots.$$

So the first solution can be expressed as

$$a_n = A_0 \left(\frac{\Gamma(n)}{\Gamma\left(n-\frac{1}{2}\right)} + \frac{\Gamma(n)}{64\,\Gamma\left(n+\frac{3}{2}\right)} + \frac{5\,\Gamma(n)}{192\,\Gamma\left(n+\frac{5}{2}\right)} \right. \\ \left. + \frac{1025\,\Gamma(n)}{16384\,\Gamma\left(n+\frac{7}{2}\right)} + \frac{5005\,\Gamma(n)}{24576\,\Gamma\left(n+\frac{9}{2}\right)} + \cdots \right).$$

The second solution, b_n, will have the leading behavior of $\ln(x)\sqrt{x}$, so we will assume that there is a series of the form

$$b_n = \sum_{k=0}^{\infty} (A_k \psi(n) + B_k) n^{[-k+1/2]}.$$

By replacing k with $k-1/2$ in equation 6.48, we find that the second difference of this sum becomes

$$\Delta^2 b_n = \sum_k \left(A_k \left(\left(k^2 - \frac{1}{4} \right) \psi(n) - 2k \right) + B_k \left(k^2 - \frac{1}{4} \right) \right) n^{[-k-3/2]}.$$

Thus,

$$4n^2 \Delta^2 b_n = \sum_k \; (A_k((4k^2-1)\psi(n) - 8k) + B_k(4k^2-1)) \times \\ \left(n^{[-k+1/2]} - 2kn^{[-k-1/2]} + \left(k + \frac{1}{2} \right)^2 n^{[-k-3/2]} \right).$$

Expanding this, and shifting the index as we go, we get

$$4n^2\Delta^2 b_n = \sum_k \Big(\; (A_k((4k^2-1)\psi(n) - 8k) + B_k(4k^2-1))$$
$$-2(k-1)A_{k-1}((4(k-1)^2-1)\psi(n) - 8(k-1))$$
$$-2(k-1)B_{k-1}(4(k-1)^2-1)$$
$$+\left(k-\frac{3}{2}\right)^2 A_{k-2}((4(k-2)^2-1)\psi(n) - 8(k-2))$$
$$+\left(k-\frac{3}{2}\right)^2 B_{k-2}(4(k-2)^2-1)\Big) n^{[-k+1/2]}.$$

Now we can easily add b_n, and set this equal to zero. Both the terms involving $\psi(n)$ and those without $\psi(n)$ must cancel, so we have the set of recursion relations

$$4k^2 A_k - 2(k-1)(4k^2-8k+3)A_{k-1} + \left(k-\frac{3}{2}\right)^2 (4k^2-16k+15)A_{k-2} = 0,$$

$$-8kA_k + 4k^2 B_k + 16(k-1)^2 A_{k-1} - 2(k-1)(4k^2-8k+3)B_{k-1}$$
$$-8\left(k-\frac{3}{2}\right)^2 (k-2)A_{k-2} + \left(k-\frac{3}{2}\right)^2 (4k^2-16k+15)B_{k-2} = 0.$$

The first equation is the same as for the first solution. Also, we can add a constant multiple of the first solution to the second solution, so we can assume that $B_0 = 0$. We then find that

$$B_1 = 0, \quad B_2 = \frac{A_0}{64}, \quad B_3 = \frac{A_0}{64}, \quad B_4 = \frac{835A_0}{32768}, \quad B_5 = \frac{2915A_0}{49152}, \ldots.$$

Thus, we have found the second solution as

$$b_n = \psi(n)a_n + A_0\left(\frac{\Gamma(n)}{64\,\Gamma\left(n+\frac{3}{2}\right)} + \frac{\Gamma(n)}{64\,\Gamma\left(n+\frac{5}{2}\right)} + \frac{835\,\Gamma(n)}{32768\,\Gamma\left(n+\frac{7}{2}\right)} + \frac{2915\,\Gamma(n)}{49152\,\Gamma\left(n+\frac{9}{2}\right)} + \cdots \right).$$

□

We now have the tools needed for finding a full asymptotic analysis for a difference equation. We can first peel off the leading behaviors, and then create a Taylor-like or Frobenius-like series of the remainder.

Example 6.25

Find the full asymptotic series for the solutions to the equation

$$a_{n+2} = 4na_{n+1} - 3n^2 a_n.$$

SOLUTION: We saw in example 6.15 that the possible leading behaviors are

$$a_n \sim K_1 n^n e^{-n} 3^n n^{-2} + K_2 n^n e^{-n}.$$

We will now peel off the possible leading behaviors, and form a Frobenius series from the remaining equation. Rather than letting $a_n = n^n e^{-n} 3^n b_n$, we will use the gamma function, and let $a_n = \Gamma(n) 3^n b_n$. The new equation for b^n will still involve rational coefficient functions,

$$9(n+1)n\Gamma(n)3^n b_{n+2} = 4n(3n\Gamma(n)3^n b_{n+1}) - 3n^2\Gamma(n)3^n b_n,$$

which simplifies to

$$3(n+1)b_{n+2} - 4nb_{n+1} + nb_n = 0.$$

This converts to the equation

$$3(n+1)\Delta^2 b_n + 2(n+3)\Delta b_n + 3b_n = 0.$$

If we assume that b_n can be expressed by a Frobenius-like series,

$$b_n = \sum_k B_k n^{[-k+\alpha]}, \qquad \Delta b_n = \sum_k (\alpha - k) B_k n^{[-k+\alpha-1]},$$

$$\Delta^2 b^n = \sum_k (\alpha - k)(\alpha - k - 1) B_k n^{[-k+\alpha-2]},$$

then the equation becomes

$$\sum_k 3(n+1)(\alpha - k)(\alpha - k - 1)B_k n^{[-k+\alpha-2]} + 2(n+3)(\alpha - k)B_k n^{[-k+\alpha-1]} + 3B_k n^{[-k+\alpha]} = 0.$$

Applying equation 6.45 to this, we get

$$\begin{aligned}\sum_k \; & 3(\alpha - k)(\alpha - k - 1)B_k(n^{[-k+\alpha-1]} + (\alpha - k)n^{[-k+\alpha-2]})) \\ & +2(\alpha - k)B_k(n^{[-k+\alpha]} + (\alpha - k + 3)n^{[-k+\alpha-1]})) \\ & +3B_k n^{[-k+\alpha]} = 0.\end{aligned}$$

Combining like terms, and shifting the index so that all terms involve $n^{[-k+\alpha]}$ produces

$$\begin{aligned}\sum_k \; & 3(\alpha - k + 2)^2(\alpha - k + 1)B_{k-2} n^{[-k+\alpha]} \\ & + \sum_k (\alpha - k + 1)(5\alpha - 5k - 4)B_{k-1} n^{[-k+\alpha]} \\ & + \sum_k (2\alpha - 2k + 3)B_k n^{[-k+\alpha]}.\end{aligned}$$

Thus, the recursion formula is

$$3(\alpha-k+2)^2(\alpha-k+1)B_{k-2}+(\alpha-k+1)(5\alpha-5k-4)B_{k-1}+(2\alpha-2k+3)B_k=0.$$

The indicial polynomial is found by plugging in $k=0$, since only one term will involve a non-negative subscript. This produces $(2\alpha+3)B_0=0$, so $\alpha=-3/2$. There is only one solution since the point at ∞ is an irregular singular point to the difference equation. This means that the series may not converge, but we will still have a Frobenius-like asymptotic series. Plugging in $\alpha=-3/2$ simplifies the recursion formula

$$3\left(-k+\frac{1}{2}\right)^2\left(-k-\frac{1}{2}\right)B_{k-2}+\left(-k-\frac{1}{2}\right)\left(-5k-\frac{33}{2}\right)B_{k-1}-2kB_k=0.$$

This gives us

$$B_1=\frac{129}{8}B_0,\quad B_2=\frac{33645}{128}B_0,\quad B_3=\frac{4765215}{1024}B_0,\quad B_4=\frac{2952696915}{32768}B_0,$$

and so on. So we have the series for one of the two solutions

$$a_n\sim B_0\Gamma(n)3^n\left(\frac{\Gamma(n)}{\Gamma\left(n+\frac{3}{2}\right)}+\frac{129\,\Gamma(n)}{8\,\Gamma\left(n+\frac{5}{2}\right)}+\frac{33645\,\Gamma(n)}{128\,\Gamma\left(n+\frac{7}{2}\right)}\right.$$
$$\left.+\frac{4765215\,\Gamma(n)}{1024\,\Gamma\left(n+\frac{9}{2}\right)}+\frac{2952696915\,\Gamma(n)}{32768\,\Gamma\left(n+\frac{11}{2}\right)}+\cdots\right).$$

To find the other solution, which has a leading behavior of n^ne^{-n}, we substitute $a_n=\Gamma(n)d_n$ into the original equation, to obtain

$$(n+1)d_{n+2}-4nd_{n+1}+3nd_n=0,$$

which converts to the equation

$$(n+1)\Delta^2d_n-2(n-1)\Delta d_n+d_n=0.$$

We again assume that d_n can be expressed by the series

$$d_n=\sum_k D_kn^{[-k+\alpha]},\qquad \Delta d_n=\sum_k(\alpha-k)D_kn^{[-k+\alpha-1]},$$

$$\Delta^2d^n=\sum_k(\alpha-k)(\alpha-k-1)D_kn^{[-k+\alpha-2]},$$

so that the equation becomes

$$\sum_k(n+1)(\alpha-k)(\alpha-k-1)D_kn^{[-k+\alpha-2]}-2(n-1)(\alpha-k)D_kn^{[-k+\alpha-1]}$$
$$+D_kn^{[-k+\alpha]}=0.$$

Using equation 6.45, this produces

$$\sum_k (\alpha - k)(\alpha - k - 1)D_k(n^{[-k+\alpha-1]} + (\alpha - k)n^{[-k+\alpha-2]})$$
$$-2(\alpha - k)D_k(n^{[-k+\alpha]} + (\alpha - k - 1)n^{[-k+\alpha-1]})$$
$$+D_k n^{[-k+\alpha]} = 0.$$

Combining the terms, and shifting the index as before, we obtain

$$(\alpha - k + 2)^2(\alpha - k + 1)D_{k-2} - (\alpha - k + 1)(\alpha - k)D_{k-1} + (2k - 2\alpha + 1)D_k = 0.$$

Setting $k = 0$ gives us the indicial equation $(1 - 2\alpha)D_0 = 0$, so $\alpha = 1/2$. This simplifies the recursion formula to

$$\left(-k + \frac{5}{2}\right)^2 \left(-k + \frac{3}{2}\right) D_{k-2} - \left(-k + \frac{3}{2}\right)\left(-k + \frac{1}{2}\right) D_{k-1} + 2kD_k = 0.$$

This gives us

$$D_1 = -\frac{D_0}{8}, \quad D_2 = \frac{D_0}{128}, \quad D_3 = \frac{-3D_0}{1024}, \quad D_4 = \frac{75D_0}{32768}, \quad \text{etc.}$$

So the second solution to the equation has the asymptotic series

$$a_n \sim D_0\Gamma(n)\left(\frac{\Gamma(n)}{\Gamma\left(n - \frac{1}{2}\right)} - \frac{\Gamma(n)}{8\,\Gamma\left(n + \frac{1}{2}\right)} + \frac{\Gamma(n)}{128\,\Gamma\left(n + \frac{3}{2}\right)} - \frac{3\,\Gamma(n)}{1024\,\Gamma\left(n + \frac{5}{2}\right)} + \frac{75\,\Gamma(n)}{32768\,\Gamma\left(n + \frac{7}{2}\right)} + \cdots\right).$$

□

Problems for §6.5

1 Use equations 6.42 and 6.44 to express

$$\frac{n^4\Gamma(n)}{\Gamma(n + k)}$$

in a form where the n only appears inside of a gamma function.

2 Examples 6.16 and 6.21 solve the same problem in two different ways, both to the same number of terms. Plug in $n = 10$ to see which comes closest to the exact answer,

$$a_{10} = \frac{544792625\,\pi\text{csch}(\pi)}{164602368} \approx 0.90034806145672.$$

For consistency, we used $A_0 = B_0 = 1$.

For problems **3** through **12**: The following difference equations have a solution in the form of a Taylor series. Find the recursion formula for the Taylor-like series, and then find the first 4 non-zero terms of the Taylor-like series solution to the equation.

3 $n^2 a_{n+1} = (n^2 - 2)a_n$
4 $n^2 a_{n+1} = (n^2 + 2)a_n$
5 $n^3 a_{n+1} = (n^3 + 1)a_n$
6 $n^3 a_{n+1} = (n^3 + n + 1)a_n$
7 $n^3 a_{n+1} = (n^3 + n - 1)a_n$
8 $a_{n+2} - 3a_{n+1} + 2a_n(1 + 1/n^2) = 0$
9 $2a_{n+2} - 3a_{n+1} + a_n(1 + 1/n^2) = 0$
10 $a_{n+2} - 4a_{n+1} + 3a_n(1 - 1/n^2) = 0$
11 $3a_{n+2} - 4a_{n+1} + a_n(1 - 1/n^2) = 0$
12 $a_{n+2} - 2a_{n+1} + a_n(1 - 1/n^3) = 0$

For problems **13** through **22**: These difference equations have a solution for which there is a Frobenius-like series for this solution. Find the recursion formula for this series solution, and then find the first 4 non-zero terms of the series.

13 $(2n + 2)a_{n+1} = (2n + 1)a_n$
14 $2na_{n+1} = (2n + 1)a_n$
15 $(3n + 1)a_{n+1} = (3n + 2)a_n$
16 $(3n + 2)a_{n+1} = (3n + 1)a_n$
17 $(3n + 3)a_{n+1} = (3n + 1)a_n$
18 $a_{n+2} + na_{n+1} = (n + 1/2)a_n$
19 $a_{n+2} + na_{n+1} = (n - 1/2)a_n$
20 $a_{n+2} + (n - 2)a_{n+1} = (n - 1/2)a_n$
21 $a_{n+2} + (n - 2)a_{n+1} = (n - 3)a_n$
22 $a_{n+2} + (n - 1)a_{n+1} = (n - 2)a_n$

23 The equation $a_{n+2} + na_{n+1} = na_n$ for $n > 0$ has a Frobenius-like series, and the recursion formula can be solved exactly. Find the complete series in closed form.

24 The equation $4n^2 a_{n+2} - 8n^2 a_{n+1} + (4n^2 - 3)a_n = 0$ has a Frobenius-like series solution for $\alpha = -1/2$. Find the first 4 terms of this series.

25 The equation $4n^2 a_{n+2} - 8n^2 a_{n+1} + (4n^2 - 3)a_n = 0$ also has a Frobenius-like series solution for $\alpha = 3/2$, but there is interference from the solution found in problem 24. If we assume that the solution is of the form

$$a_n = \sum_{k=0}^{\infty} A_k n^{[-k+3/2]} + \sum_{k=0}^{\infty} B_k \psi(n) n^{[-k-1/2]},$$

determine the recursion formula for the A_k and B_k. Assuming $A_2 = 0$, find the first 5 non-zero terms of the series, two of which will involve $\psi(n)$.

26 The equation $n^2 a_{n+2} + (3n - 2n^2)a_{n+1} + (n^2 - 3n + 1)a_n = 0$ has a Frobenius-like solution for $\alpha = -1$, but the other solution is asymptotic to $n^{-1} \ln n$ as $n \to \infty$. If we assume that the solution is of the form

$$a_n = \sum_{k=0}^{\infty} A_k n^{[-k-1]} + \sum_{k=0}^{\infty} B_k \psi(n) n^{[-k-1]},$$

determine the recursion formula for the A_k and B_k. Find the first 4 non-zero terms of two solutions, one with $A_0 = 1$ and $B_0 = 0$, and one with $A_0 = 0$ and $B_0 = 1$.

For problems **27** through **32**: First find the possible leading behaviors of the following difference equations. Then for each leading behavior, peel off the exponential portion of the behavior to produce a new equation that has a Frobenius-like series solution, as was done in example 6.25. Find the recursion formula for the Frobenius-like series, and find the first three non-zero terms of the series, unless the series truncates.

27 $na_{n+2} + (2n+1)a_n = 3na_{n+1}$

28 $(n+1)a_{n+2} - 4na_{n+1} + 3na_n = 0$

29 $(n+1)a_{n+2} - 5na_{n+1} + 6na_n = 0$

30 $na_{n+2} + (n+1)a_{n+1} = 2na_n$

31 $a_{n+2} - 4na_{n+1} + 3n^2 a_n = 0$

32 $a_{n+2} - 5na_{n+1} + 6n^2 a_n = 0$

Chapter 7

Perturbation Theory

Perturbation theory is a collection of methods used to find the approximate solution to a problem for which the exact solution cannot be solved in closed form. The strategy is to introduce a parameter ϵ into the problem, so that the problem *is* solvable when $\epsilon = 0$, or as ϵ approaches 0. Although we have made the problem more complicated by adding an additional parameter, we can determine a solution to the original problem as a power series in ϵ.

$$y = f_0(x) + \epsilon f_1(x) + \epsilon^2 f_2(x) + \cdots .$$

Perturbation theory originated as a way to tackle the three body problem in celestial mechanics. Using Newtonian mechanics, the Earth would revolve about the sun in a perfect ellipse if it weren't for the moon and other planets. However, the moon causes the motion of the Earth to be *perturbed*, so that the Earth wiggles slightly as it goes around the sun. Other planets also affect the Earth's orbit by a tiny amount. Likewise, the moon's orbit around the Earth is perturbed by the presence of the sun. Using perturbation theory, we could accurately approximate the deviation from the elliptical orbit, and this variance was verified by observation. Since then, perturbation theory has been used for algebraic equations, differential equations, and quantum mechanics. The techniques can be used in a variety of problems in applied mathematics.

7.1 Introduction to Perturbation Theory

In order to introduce the technique of perturbation theory, let us consider solving polynomial equations.

Example 7.1
Estimate the roots of the equation

$$x^3 - 1.01x + .03 = 0.$$

SOLUTION: This is currently not a perturbation problem, since there is no small parameter. However, it is clear that the problem is close to another problem that can be solved fairly easily: $x^3 - x = 0$. Thus, we can consider the polynomial equation to be a special case of a *one-parameter family* of polynomial equations

$$x^3 - (1+\epsilon)x + 3\epsilon = 0.$$

Note that when $\epsilon = 0$, the equation becomes easy to solve, $x = \pm 1$ and $x = 0$. When $\epsilon = 0.01$, we reconstruct the original problem. This is the key feature needed to convert a problem to a perturbation problem. The parameter ϵ is added to the problem (if it isn't there already) so that when $\epsilon = 0$, the equation reduces to a much simpler problem. Another value of ϵ must reconstruct the original problem. Note that there may be more than one way to meet these criteria.

The plan is to assume that the solution can be expressed as a series in ϵ,

$$x = a_0 + a_1\epsilon + a_2\epsilon^2 + a_3\epsilon^3 + \cdots.$$

If we plug this series into the perturbation problem, and expand in powers of ϵ, we get

$$(a_0^3 - a_0) + (3 - a_0 - a_1 + 3a_0^2 a_1)\epsilon + (3a_0a_1^2 - a_1 - a_2 + 3a_0^2a_2)\epsilon^2 \\ + (a_1^3 - a_2 + 6a_0a_1a_2 - a_3 + 3a_0^2a_3)\epsilon^3 + \cdots.$$

Because ϵ is a variable, we can set each power of ϵ equal to zero separately. Thus, we have that

$$a_0^3 - a_0 = 0, \quad 3 - a_0 - a_1 + 3a_0^2a_1 = 0, \quad 3a_0a_1^2 - a_1 - a_2 + 3a_0^2a_2 = 0,$$

$$a_1^3 - a^2 + 6a_0a_1a_2 - a_3 + 3a_0^2a_3 = 0, \ \ldots.$$

The first equation is easy to solve for a_0, since we designed the problem to be simple when $\epsilon = 0$, otherwise known as the *unperturbed problem*. Once we have a_0 known, the next equation will solve for a_1 as a *linear* equation, hence it is much easier than the original cubic. In fact, solving each equation for the next coefficient will always involve solving a linear equation. Table 7.1 shows the three possible solutions. ▯

Example 7.2

Use perturbative methods up to order ϵ^3 to approximate the roots of the equation

$$x^3 - (1.01)x^2 + .04 = 0.$$

SOLUTION: This equation is very close to the equation $x^3 - x^2 = 0$, which can be solved easily, with $x = 1$ or $x = 0$. Thus, we can generalize the problem to a perturbation problem

$$x^3 - (1+\epsilon)x^2 + 4\epsilon = 0. \tag{7.1}$$

TABLE 7.1: Table of the possible a_n for which the series $\sum a_n \epsilon^n$ solves the perturbation problem $x^3 - (1+\epsilon)x + 3\epsilon = 0$. Note that the third order perturbation is accurate to 6 places when $\epsilon = 0.01$.

	First solution	Second solution	Third solution
a_0	-1	0	1
a_1	-2	3	-1
a_2	5	-3	-2
a_3	-23.5	30	-6.5
$\epsilon = .01$	-1.0195235	0.0297300	0.9897935
Exact root	-1.0195222	0.029728985	0.98979323

The original problem is reproduced by setting $\epsilon = 0.01$.

If we let $x = a_0 + a_1\epsilon + a_2\epsilon^2 + a_3\epsilon^3 + \cdots$, and expand in powers of ϵ, we find that

$$\begin{aligned} a_0^3 - a_0^2 &= 0, \\ 4 - a_0^2 - 2a_0a_1 + 3a_0^2a_1 &= 0, \\ 3a_0a_1^2 - 2a_0a_1 - a_1^2 - 2a_0a_2 + 3a_0^2a_2 &= 0, \\ a_1^3 - a_1^2 - 2a_0a_2 - 2a_1a_2 + 6a_0a_1a_2 - 2a_0a_3 + 3a_0^2a_3 &= 0. \end{aligned}$$

If we let $a_0 = 1$ to solve the first equation, then the other equations can be solved to produce $a_1 = -3$, $a_2 = -24$, and $a_3 = -300$. So one solution to the perturbation problem is $x = 1 - 3\epsilon - 24\epsilon^2 - 300\epsilon^3 + \cdots$, from which if we plug in $\epsilon = 0.01$ produces $x = 0.9673$. This is very close to the true solution $x \approx 0.96724498987$.

If we try letting $a_0 = 0$, the second equation causes a contradiction: $4 = 0$. Apparently, the perturbation series for the other solutions are not in the form $x = a_0 + a_1\epsilon + a_2\epsilon^2 + a_3\epsilon^3 + \cdots$. We can use dominant balance to determine the true form of the perturbation series. If the solution is close to 0, then $x^3 \ll x^2$ and $\epsilon x^2 \ll x^2$, so we find that $4\epsilon \sim x^2$, hence $x \sim \pm 2\sqrt{\epsilon}$. This would indicate that the perturbation series is of the form

$$a_1\epsilon^{1/2} + a_2\epsilon + a_3\epsilon^{3/2} + a_4\epsilon^2 + a_5 + \epsilon^{5/2} + \epsilon^3 + \cdots.$$

If we plug this into equation 7.1, we get

$$\begin{aligned} 0 = {} & (4 - a_1^2)\epsilon + (a_1^3 - 2a_1a_2)\epsilon^{3/2} + (3a_1^2a_2 - a_1^2 - a_2^2 - 2a)1a_3)\epsilon^2 + \\ & (3a_1a_2^2 - 2a_1a_2 + 3a_1^2a_3 - 2a_2a_3 - 2a_1a_4)\epsilon^{5/2} + \\ & (a_2^3 - a_2^2 - 2a_1a_3 + 6a_1a_2a_3 - a_3^2 + 3a_1^2a_4 - 2a_2a_4 - 2a_1a_5)\epsilon^3 + \\ & (3a_2^2a_3 - 2a_2a_3 + 3a_1a_3^2 - 2a_1a_4 \\ & \quad +6a_1a_2a_4 - 2a_3a_4 + 3a_1^2a_5 - 2a_2a_5 - 2a_1a_6)\epsilon^{7/2} + \cdots. \end{aligned}$$

If $a_1 = 2$, we find that $a_2 = 2$, $a_3 = 4$, $a_4 = 12$, $a_5 = 41$, and $a_6 = 150$. If, on the other hand, we assume that $a_1 = -2$, then $a_2 = 2$, $a_3 = -4$, $a_4 = 12$,

$a_5 = -41$, and $a_6 = 150$. So both perturbation solutions can be expressed as

$$x \sim \pm 2\epsilon^{1/2} + 2\epsilon \pm 4\epsilon^{3/2} + 12\epsilon^2 \pm 41\epsilon^{5/2} + 150\epsilon^3 + \cdots.$$

When $\epsilon = 0.01$, we get the approximate solutions 0.22576 and -0.18306. Compare these to the true solutions $x \approx 0.22585627428$ and $x \approx -0.18310126415$, we see that we obtained 4 places of accuracy. ☐

In the last example, we saw that the perturbation series are not always simple Taylor series in ϵ. Notice that there was a *fundamental difference* between the solution to the unperturbed problem (double roots) and the perturbed problem (3 unequal roots). Any time that there is a change of behavior as ϵ changes from 0 to a non-zero number, the problem is called a *singular perturbation problem.* If there is no fundamental change to the nature of the solutions as ϵ moves away from 0, the problem is a *regular perturbation problem.* The distinction is analogous to the singular points and ordinary points of subsection 4.1.4. For regular perturbation problems, the perturbation series will be a Taylor series in ϵ, and will converge for at least some non-zero values of ϵ. For singular perturbation problems, there may be fractional powers of ϵ involved, and the series may diverge for all non-zero ϵ, so that the methods of chapter 3 must be invoked.

We define the *zeroth-order solution* to be the first term of the perturbation series solution to the problem. Note that in singular perturbation problems, this is not always the same as the solution to the unperturbed problem. For example, the solution to the unperturbed problem from example 7.2 is $x = 1$ or $x = 0$, but the zeroth-order solution is $x \sim 1$ or $x \sim \pm 2\sqrt{\epsilon}$. In fact, the unperturbed problem may not even be solvable!

Example 7.3

Find the third order perturbation series solution for the perturbation problem

$$\epsilon x^3 - (1+\epsilon)x + 2 = 0.$$

SOLUTION: Note that when $\epsilon = 0$, we obtain the linear equation $-x + 2 = 0$, giving us one solution, $x = 2$. However, for non-zero ϵ, there will be 3 solutions, so this is a singular perturbation problem. The solution near 2 will have a Taylor series of the form $x \sim 2 + a_1\epsilon + a_2\epsilon^2 + a_3\epsilon^3 + \cdots$, and plugging this into the equation produces

$$(6 - a_1)\epsilon + (11a_1 - a_2)\epsilon^2 + (6a_1^2 + 11a_2 - a_3)\epsilon^3 + \cdots.$$

Thus, $a_1 = 6$, $a_2 = 66$, and $a_3 = 942$, so one solution to the perturbation problem is

$$x \sim 2 + 6\epsilon + 66\epsilon^2 + 942\epsilon^3 + \cdots.$$

The other two solutions must go the infinity as $\epsilon \to 0$. We can use dominant balance to find the zeroth order solutions. Since $2 \ll x$ and $\epsilon x \ll x$, we have $\epsilon x^3 \sim x$, so $x \sim \pm\epsilon^{-1/2}$.

If we assume that the perturbation series has the form

$$x \sim a_0\epsilon^{-1/2} + a_1 + a_2\epsilon^{1/2} + a_3\epsilon + \cdots,$$

we can plug this into the problem and expand in powers of ϵ:

$$0 = (a_0^3 - a_0)\epsilon^{-1/2} + (2 - a_1 + 3a_0^2a_1) + (3a_0a_1^2 - a_0 - a_2 + 3a_0^2a_2)\epsilon^{1/2} \\ + (a_1^3 - a_1 + 6a_0a_1a_2 - a_3 + 3a_0^2a_3)\epsilon + \cdots.$$

From this, we get $a_0 = \pm 1$, $a_1 = -1$, $a_2 = \mp 1$, and $a_3 = -3$, so we can find the other two perturbation series solutions

$$x \sim \pm\epsilon^{-1/2} - 1 \mp \epsilon^{1/2} - 3\epsilon + \cdots. \qquad \square$$

Even though this example was a singular perturbation problem, it can be made into a regular perturbation problem by using a *scale transformation*. Since dominant balance revealed that the roots behaved as $O(\epsilon^{-1/2})$, we can make the substitution $x = t\epsilon^{-1/2}$, so that t will be of order 1. This produces the new equation for t,

$$t^3 - (1 + \epsilon)t + 2\epsilon^{1/2} = 0.$$

Since this equation involves $\epsilon^{1/2}$, we can eliminate the fractional powers by substituting $\delta = \epsilon^{1/2}$. Then the new equation is then

$$t^3 - (1 + \delta^2)t + 2\delta = 0.$$

We now have a regular perturbation problem, which has the three solutions

$$t \sim 1 - \delta - \delta^2 - 3\delta^3,$$
$$t \sim -1 - \delta + \delta^2 - 3\delta^3, \text{ and}$$
$$t \sim 2\delta + 6\delta^3 + 66\delta^5 + 943\delta^7.$$

Notice that by converting back to the original variables, we reproduce the solutions of example 7.3, including the regular perturbation series solution.

Example 7.4

Use a scale transformation to convert the singular perturbation problem

$$\epsilon x^4 - x^2 + x + 2 = 0 \tag{7.2}$$

into a regular perturbation problem. Then find the third order perturbation series solutions for the roots.

SOLUTION: The unperturbed solution, $-x^2 + x + 2 = 0$, has only two solutions, so two of the solutions must approach ∞ as $\epsilon \to 0$. For these solutions, $x + 2 \ll x^2$, so we have $\epsilon x^4 \sim x^2$, or $x \sim \pm 1/\sqrt{\epsilon}$. Thus, we can make the scale transformation $t = \delta x$, where $\delta = \sqrt{\epsilon}$. This causes t to be of order 1. The new equation becomes

$$t^4 - t^2 + t\delta + 2\delta^2 = 0.$$

This new equation is now regular, and the unperturbed problem is $t^4 - t^2 = 0$, which has solutions 1, -1, and a double root at 0. If we let $t = a_0 + a_1\delta + a_2\delta^2 + a_3\delta^3 + \cdots$, and expand in powers of δ, we obtain

$$\begin{aligned} 0 &= (a_0^4 - a_0^2) + (a_0 - 2a_0a_1 + 4a_0^3a_1)\delta \\ &\quad + (2 + a_1 - a_1^2 + 6a_0^2a_1^2 - 2a_0a_2 + 4a_0^3a_2)\delta^2 \\ &\quad + (a_2 - 2a_1a_2 - 2a_0a_3 + 4a_0a_1^3 + 3a_0^2a_1a_2 + a_0^3a_3)\delta^3 + \cdots . \end{aligned}$$

When $a_0 = 1$, we have $a_1 = -1/2$, $a_2 = -11/8$, and $a_3 = -5/2$. For $a_0 = -1$, the results are slightly different: $a_1 = -1/2$, $a_2 = 11/8$, and $a_3 = -5/2$. So two of the series can be combined into

$$t \sim \pm 1 - \delta/2 \mp 11\delta^2/8 - 5\delta^3/2 + \cdots .$$

Converting to the original variables, we find

$$x \sim \pm\frac{1}{\sqrt{\epsilon}} - \frac{1}{2} \mp \frac{11\sqrt{\epsilon}}{8} - \frac{5\epsilon}{2} + \cdots .$$

The other two roots can be handled by using regular perturbation on the original equation. Letting $x = a_0 + a_1\epsilon + a_2\epsilon^2 + a_3\epsilon^3 + \cdots$ in equation 7.2 produces

$$\begin{aligned} 0 &= (2 + a_0 - a_0^2) + (a_0^4 + a_1 - 2a_0a_1)\epsilon + (4a_0^3a_1 - a_1^2 + a_2 - 2a_0a_2)\epsilon^2 \\ &\quad + (6a_0^2a_1^2 + 4a_0^3a_2 - 2a_1a_2 + a_3 - 2a_0a_3)\epsilon^3 + \cdots . \end{aligned}$$

When $a_0 = 2$, $a_1 = 16/3$, $a_2 = 1280/27$, and $a_3 = 137216/243$. However, when $a_0 = -1$, then $a_1 = -1/3$, $a_2 = -11/27$, and $a_3 = -164/243$. So the two remaining solutions are

$$\begin{aligned} x &\sim 2 + \frac{16\epsilon}{3} + \frac{1280\epsilon^2}{27} + \frac{137216\epsilon^3}{243} + \cdots, \text{ and} \\ x &\sim -1 - \frac{\epsilon}{3} - \frac{11\epsilon^2}{27} - \frac{164\epsilon^3}{243} + \cdots . \end{aligned}$$

□

For a regular perturbation problem, or even for a singular perturbation problem which can become regular via an appropriate scale transformation, the resulting series solution is guaranteed to have a finite radius of convergence. However, this radius may be extremely small, especially if the degree of the polynomial is large. A classical example is the Wilkinson polynomial [25].

TABLE 7.2: List of the first order perturbation solution to the Wilkinson polynomial. Even with ϵ as small as 10^{-9}, the perturbation approximation for the roots are off, since the true roots are complex.

First order perturbation solution	Perturbation at $\epsilon = 10^{-9}$	True roots of polynomial
$1 + 8.22064 \times 10^{-18}\epsilon$	1.0000000000	1.0000000000
$2 - 8.18896 \times 10^{-11}\epsilon$	2.0000000000	2.0000000000
$3 + 1.63382 \times 10^{-6}\epsilon$	3.0000000000	3.0000000000
$4 - 0.00218962\epsilon$	4.0000000000	4.0000000000
$5 + 0.60774197\epsilon$	5.0000000006	5.0000000006
$6 - 58.2484213\epsilon$	5.9999999418	5.9999999418
$7 + 2542.43552\epsilon$	7.0000025424	7.0000025425
$8 - 59695.5905\epsilon$	7.9999403044	7.9999403147
$9 + 839327.471\epsilon$	9.0008393275	9.0008410335
$10 - 7.59406 \times 10^{6}\epsilon$	9.9924059416	9.9925181240
$11 + 4.64446 \times 10^{7}\epsilon$	11.0464445706	11.0506225921
$12 - 1.98503 \times 10^{8}\epsilon$	11.8014968360	11.8329359874
$13 + 6.05559 \times 10^{8}\epsilon$	13.6055586238	13.3490180355 $\pm$
$14 - 1.33297 \times 10^{9}\epsilon$	12.6670315658	0.5327657500 i
$15 + 2.11907 \times 10^{9}\epsilon$	17.1190652221	15.4577907243 $\pm$
$16 - 2.40751 \times 10^{9}\epsilon$	13.5924860203	0.8993415262 i
$17 + 1.90440 \times 10^{9}\epsilon$	18.9044021453	17.6624344772 $\pm$
$18 - 9.95587 \times 10^{8}\epsilon$	17.0044133003	0.7042852369 i
$19 + 3.09013 \times 10^{8}\epsilon$	19.3090134607	19.2337033342
$20 - 4.30998 \times 10^{7}\epsilon$	19.9569001959	19.9509496543

Example 7.5

Determine the first order perturbation solution for the roots of the *Wilkinson polynomial*, the 20th degree polynomial given by

$$(x-1)(x-2)(x-3)(x-4)\cdots(x-19)(x-20)+\epsilon x^{19}=0.$$

SOLUTION: The unperturbed solutions are clearly the integers from 1 to 20. If we let $x = a_0 + a_1\epsilon$, and keep only terms of order ϵ, we get

$$f(a_0)+\epsilon a_0^{19}+\epsilon a_1 f'(a_0)=0, \tag{7.3}$$

where $f(x) = (x-1)(x-2)(x-3)\cdots(x-20)$. By letting a_0 range from 1 to 20, we can compute the corresponding value of a_1, giving us the first order correction. The results are given in table 7.2. The table reveals a surprising result: for some of the roots, the value of a_1 is astronomically large! As a result, even with ϵ as small as 10^{-9} causes a huge change in the location of the roots. The perturbation approximation indicates the roots at 13 and 14 will exchange places. Actually, though, as ϵ increases, the roots will collide into a double root, and then become a pair of complex roots. Figure 7.1 shows

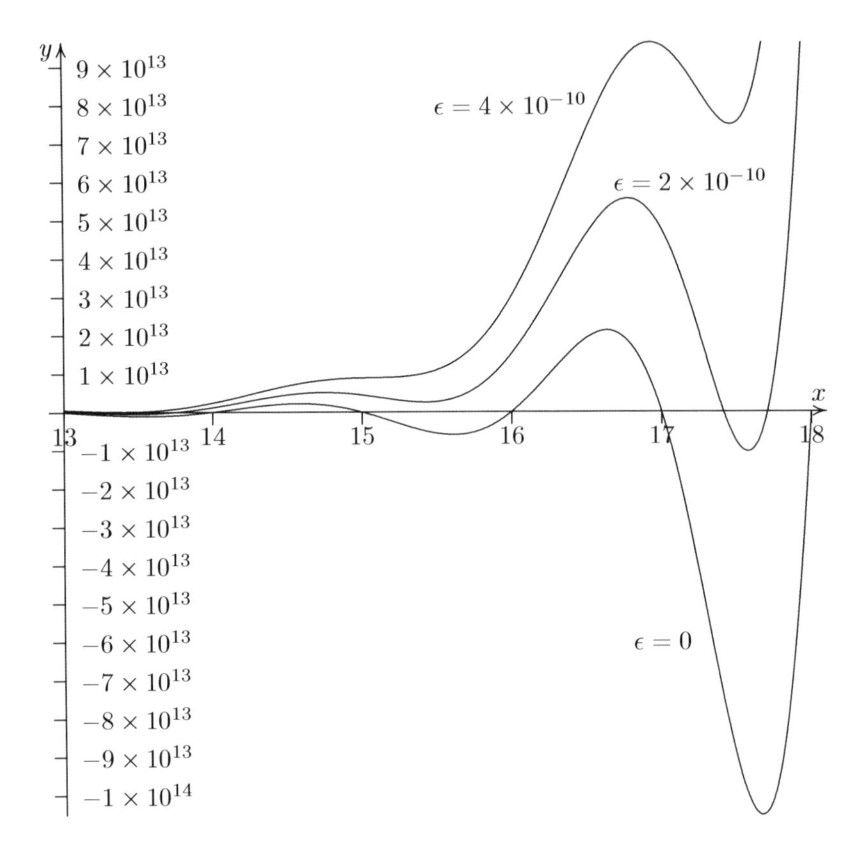

FIGURE 7.1: This shows the plot of the Wilkinson polynomial $(x-1)(x-2)\cdots(x-20)+\epsilon x^{19}$ in the critical range from 13 to 18. When $\epsilon = 4\times 10^{-10}$, the polynomial is totally above the x-axis in this range, so that the 6 roots have become complex.

the progression of the polynomial as ϵ slowly increases. By $\epsilon = 4 \times 10^{-10}$, six of the real roots have become complex roots. This example is used in numerical analysis to show the problems associated with finding the roots of a high degree polynomial with a computer. If the coefficients are stored as machine precision numbers (usually 10-12 digits of accuracy), there still could be massive errors in the computed roots. □

Problems for §7.1

For the regular perturbation polynomial problems **1** through **10**: Determine the second order solutions (up to ϵ^2) for the roots.

1 $x^2 + \epsilon x - 1 = 0$
2 $x^2 + (1+\epsilon)x - (2-\epsilon) = 0$
3 $(1+\epsilon)x^2 - 4x + 3 = 0$
4 $(1+\epsilon)x^2 - 3x + (2+\epsilon) = 0$
5 $x^3 - 3x^2 + (2+\epsilon)x + \epsilon = 0$
6 $x^3 + \epsilon x^2 - (4+\epsilon)x + \epsilon = 0$
7 $x^3 + \epsilon x^2 - 7x + (6+\epsilon) = 0$
8 $x^3 + (1+\epsilon)x^2 - (4+\epsilon)x - 4 = 0$
9 $x^4 - 5x^2 + \epsilon x + 4 = 0$
10 $x^4 - 6x^3 + 11x^2 - 6x + \epsilon = 0$

For the singular perturbation problems **11** through **26**: Determine the first three terms of the perturbation series for each of the roots. Note that many of these singular perturbation problems can be transformed into regular perturbation problems via a scale transformation.

11 $x^3 + (1+2\epsilon)x^2 - \epsilon = 0$
12 $x^3 - 2x^2 + (1+\epsilon)x - 2\epsilon = 0$
13 $x^3 + (2+\epsilon)x^2 + (1+\epsilon)x + e = 0$
14 $x^3 - 2x^2 + \epsilon x + 2\epsilon = 0$
15 $x^4 - 2x^3 + (1+\epsilon)x^2 + 2\epsilon x - 4\epsilon = 0$
16 $x^4 + (2+\epsilon)x^3 + x^2 + 2\epsilon x - \epsilon = 0$
17 $x^3 - 3x^2 + (3-\epsilon)x - 1 = 0$
18 $x^3 - (3+2\epsilon)x^2 + 3x + (2\epsilon - 1) = 0$
19 $\epsilon x^3 - x^2 + 1 = 0$
20 $\epsilon x^3 - x + 1 = 0$
21 $\epsilon x^3 - x + 2 + \epsilon = 0$
22 $\epsilon^2 x^3 - \epsilon x^2 - 2x + 2 = 0$
23 $\epsilon^2 x^3 - 3\epsilon x^2 + 2x - 2 = 0$
24 $\epsilon x^4 - x^2 + 3x - 2 = 0$
25 $\epsilon x^4 - x^2 - x + 2 = 0$
26 $\epsilon x^4 - x^2 + 2x - 1 = 0$

27 Show that for the Wilkinson polynomial, example 7.5, the first order approximation for the root at k, where $1 \le k \le 20$, is given by

$$k + \frac{\epsilon(-1)^{k+1}k^{19}}{(20-k)!(k-1)!}.$$

Hint: In order to compute $f'(k)$ in equation 7.3, use the product rule. Which term of the product rule will *not* go to zero when evaluated at $x = k$?

28 Using the first order estimates given in table 7.2, determine the value of ϵ for which the roots at 15 and 16 will merge into a single double root. Also determine the value of ϵ for which the roots at 16 and 17 will merge together. From this information, estimate the radius of convergence for the perturbation series for the root at 16.

29 Consider the general regular perturbation problem

$$f(x) + \epsilon g(x) = 0$$

where $f(x)$ and $g(x)$ are polynomials, and r is a root of $f(x)$, but not a multiple root, then determine the second order perturbation series for the root r.

Hint: Consider the Taylor series for $f(x)$ and $g(x)$ centered at r.

7.2 Regular Perturbation for Differential Equations

Perturbation theory is particularly useful in analyzing differential equations. A small parameter ϵ can be added to the equation so that when $\epsilon = 0$, the differential equation is simplified to the point of being almost trivial to solve. We can then find the perturbation series solution to the perturbed equation, and finally replace $\epsilon = 1$ to reproduce the original equation.

Example 7.6

Consider the second order linear homogeneous equation in which the y' term is missing, along with initial conditions

$$y'' + q(x)y = 0 \qquad y(0) = 0, \quad y'(0) = 1. \tag{7.4}$$

An equation of this form is called a *Schrödinger equation*, not to be confused with the partial differential equation of quantum physics. In fact, any second order linear homogeneous equation can be converted to a Schrödinger equation (see problem 1). There is no closed form solution for this problem, except for special cases of $q(x)$. Nonetheless, we can find a perturbation series solution to this problem.

SOLUTION: The first task is to introduce an ϵ into the problem so that the unperturbed problem is solvable. Clearly the difficulty lies in the $q(x)$ term, so we add the ϵ to this term.

$$y'' + \epsilon q(x)y = 0 \qquad y(0) = 0, \quad y'(0) = 1.$$

Now when $\epsilon = 0$, the equation becomes $y'' = 0$, which is trivial to solve. The solution that satisfies the initial conditions is $y = x$.

The next step is to assume that the solution has the form of a perturbation series

$$y(x) = y_0(x) + \epsilon y_1(x) + \epsilon^2 y_2(x) + \epsilon^2 y_3(x) + \cdots.$$

For the initial conditions, we let $y_0(x)$ satisfy the original initial conditions, $y_0(0) = 0$ and $y_0'(0) = 1$, but for all of the other terms in the series, we set

$y_n(0) = 0$, $y'_n(0) = 0$ for $n \geq 1$ so that these terms will not affect the initial conditions of $y(x)$.

If we plug the perturbation series into the differential equation and aligning powers of epsilon, we get, for $n \geq 1$,

$$y''_n(x) + y_{n-1}(x)q(x) = 0, \qquad y_n(0) = y'_n(0) = 0.$$

Thus, each term of the perturbation series can be obtained with two integrations.

$$y_n = \int_0^x \int_0^t -y_{n-1}(u)q(u)\, du\, dt. \tag{7.5}$$

Notice that we have replaced an unsolvable differential equation 7.4 with a sequence of very easy equations.

The final step is to replace ϵ with 1, and sum the series

$$y(x) = y_0(x) + y_1(x) + y_2(x) + \cdots.$$

The natural question is whether or not this series converges. It is fairly easy to show that, for a fixed value of x, this series will converge as long as $q(x)$ is defined from 0 to x (see problem 2). The rate of convergence could depend on the value of x, so this is referred to as *point-wise convergence*. This is very different than the Taylor series solutions found in section 4.3, where the radius of convergence for the Taylor series was limited by the closest singularity of $q(x)$. ∎

We can consider special cases of example 7.6 by specifying $q(x)$.

Example 7.7

Compute the first four terms of the perturbation series for the equation

$$y'' + \epsilon e^{-x/2} y = 0, \qquad y(0) = 0, \quad y'(0) = 1.$$

SOLUTION: This is example 7.6 with $q(x) = e^{-x/2}$. We have that $y_0 = x$, and each successive term is found by multiplying by $-e^{x/2}$, and integrating twice.

$$\int_0^x -te^{-t/2}\, dt = (4 + 2x)e^{-x/2} - 4$$

so

$$y_1 = \int_0^x (4 + 2t)e^{-t/2} - 4\, dt = 16 - 4x - (16 + 4x)e^{-x/2}.$$

We multiply by $-e^{-x/2}$ and perform two more integrations to find y_2.

$$\int_0^x (4t - 16)e^{-t/2} + (16 + 4t)e^{-t}\, dt = 4 + (16 - 8x)e^{-x/2} - (20 + 4x)e^{-x}$$

so

$$y_2 = \int_0^x 4+(16-8t)e^{-t/2}-(20+4t)e^{-t}\,dt = 4x-24+16xe^{-x/2}+(24+4x)e^{-x}.$$

Finally, to find y_3, we integrate

$$\int_0^x (24-4t)e^{-t/2} - 16te^{-t} - (24+4t)e^{-3t/2}\,dt$$
$$= -\frac{16}{9} + (8x-32)e^{-x/2} + (16x+16)e^{-x} + \frac{160+24x}{9}e^{-3x/2}$$

so

$$y_3 = \int_0^x -\frac{16}{9} + (8t-32)e^{-t/2} + (16t+16)e^{-t} + \frac{160+24t}{9}e^{-3t/2}\,dt$$
$$= \frac{352-48x}{27} + (32-16x)e^{-x/2} - (32+16x)e^{-x} - \frac{352+48x}{27}e^{-3x/2}.$$

Hence, the perturbation solution is

$$\begin{aligned} y &= x + \epsilon(16-4x-(16+4x)e^{-x/2}) \\ &\quad + \epsilon^2(4x-24+16xe^{-x/2}+(24+4x)e^{-x}) \\ &\quad + \epsilon^3\left(\frac{352-48x}{27} + (32-16x)e^{-x/2} - (32+16x)e^{-x} - \frac{352+48x}{27}e^{-3x/2}\right) \\ &\quad + \cdots. \end{aligned}$$

The plot of the approximations to the solution when $\epsilon = 1$ is given in figure 7.2. Notice that the perturbation solution converges much better than the Taylor series solution. □

So far, most of the analysis that we have done with differential equations has been *local analysis*, meaning we have focused on the behavior of the solutions near one point. This analysis helps us in solving initial value problems, but is useless in solving boundary value problems, since boundary value problems require the knowledge of the behavior at *two* different points. Boundary value problems requires *global analysis*, and perturbation theory provides us with the means of performing global analysis. The strategy will be the same as for initial value problems. We will let the zeroth order solution satisfy the boundary values, and each correction will be 0 at the boundaries, so that the sum of the series will correctly satisfy the boundary conditions.

Example 7.8

Use perturbation theory to approximate the solution to the boundary value problem

$$y'' - xy = 0, \qquad y(0) = 0, \quad y(2) = 2.$$

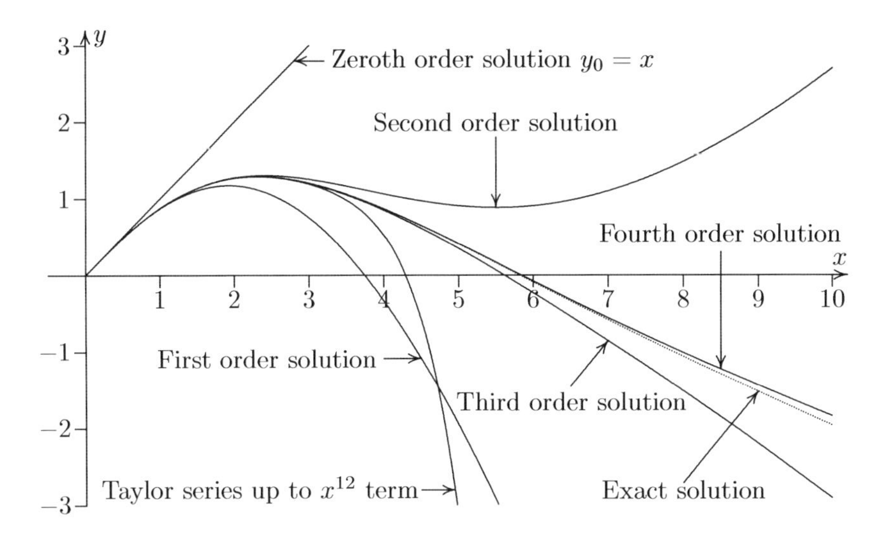

FIGURE 7.2: This shows the successive perturbation approximations for the initial value problem $y'' + e^{-x/2}y = 0$, $y(0) = 0$, $y'(0) = 1$. Note that even the third order perturbation does a better job at approximating the solution than the 12th order Taylor series solution.

SOLUTION: The equation is the Airy equation, and in fact the exact solution can be given by

$$y = \frac{2\sqrt{3}\text{Ai}(x) - 2\text{Bi}(x)}{\sqrt{3}\text{Ai}(2) - \text{Bi}(2)}.$$

But we can make this into a perturbation problem by adding an ϵ, as we did in example 7.6, to get

$$y'' - \epsilon xy = 0, \qquad y(0) = 0, \quad y(2) = 2.$$

Solving the unperturbed problem $y'' = 0$ with the boundary conditions produces $y_0 = x$, and the other orders are obtained by solving

$$y_n'' = xy_{n-1}, \qquad y_n(0) = y_n(2) = 0.$$

So for y_1, we integrate x^2 twice, to produce $x^4/12 + Cx + D$, and the boundary conditions give us $C = -2/3$, $D = 0$. Repeating this procedure, we obtain

$$y_1 = \frac{x^4 - 8x}{12}, \qquad y_2 = \frac{x^7 - 28x^4 + 160x}{504},$$

$$y_3 = \frac{x^{10} - 60x^7 + 1200x^4 - 6272x}{45360},$$

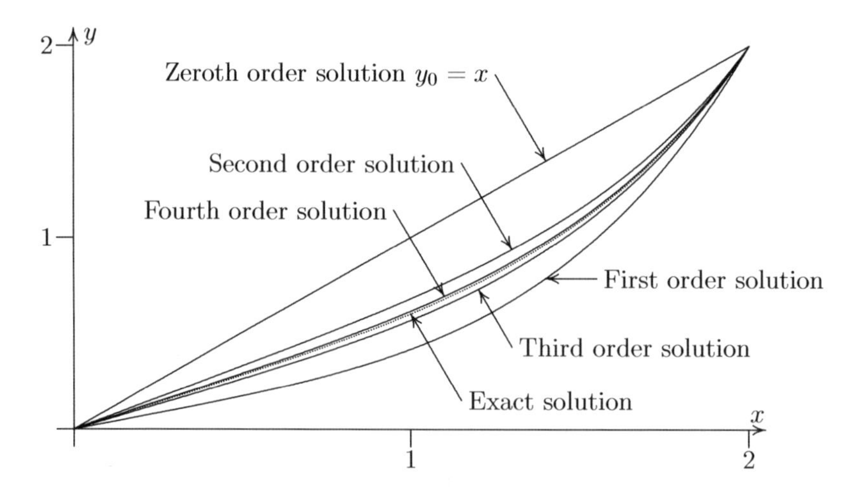

FIGURE 7.3: This shows the successive perturbation approximations for the boundary value problem $y'' = xy$, $y(0) = 0$, $y(2) = 2$. Although local analysis is unable to solve this problem, perturbation theory gives a rapidly converging sequence of functions.

$$y_4 = \frac{7x^{13} - 728x^{10} + 31200x^7 - 570752x^4 + 2913280x}{49533120}, \ldots.$$

Thus, the perturbation series can be given by

$$y = x + \epsilon\frac{x^4 - 8x}{12} + \epsilon^2\frac{x^7 - 28x^4 + 160x}{504} + \epsilon^3\frac{x^{10} - 60x^7 + 1200x^4 - 6272x}{45360}$$
$$+ \epsilon^4\frac{7x^{13} - 728x^{10} + 31200x^7 - 570752x^4 + 2913280x}{49533120} + \cdots.$$

When ϵ is set to 1, the series converges to the solution to the boundary value problem. See figure 7.3. ▯

Since we can study global behavior of a differential equation using perturbation theory, we can also apply perturbation theory to an *eigenvalue problem.* The key difference is that we need two perturbation series, one for the function $y(x)$, and one for the corresponding eigenvalue E. Also, it is vital that the unperturbed eigenvalue problem be completely solvable, which limits the choice as to where we can introduce the perturbation parameter.

Example 7.9

Use perturbation theory to analyze the Sturm-Liouville eigenvalue problem

$$y'' - xy + Ey = 0 \qquad y(0) = y(\pi) = 0.$$

SOLUTION: There are very few eigenvalue problems that have an exact solution, but one of these was found in example 4.8, $y'' + Ey = 0$. Hence, we will introduce the parameter ϵ so that the unperturbed equation becomes this known example.

$$y'' - \epsilon xy + Ey = 0 \qquad y(0) = y(\pi) = 0.$$

The unperturbed solution is $E_0 = n^2$, $y_0 = \sin(nx)$. Let us assume that the true solution can be expressed as a perturbation series

$$y = y_0 + \epsilon y_1 + \epsilon^2 y_2 + \epsilon^3 y_3 + \cdots, \qquad E = E_0 + \epsilon E_1 + \epsilon^2 E_2 + \epsilon^3 E_3 + \cdots.$$

The only difficult aspect of plugging these series into the equation is the term Ey, which will require us to multiply the two series together.

$$\begin{aligned} Ey = {} & E_0 y_0 + \epsilon(E_0 y_1 + E_1 y_0) + \epsilon^2(E_0 y_2 + E_1 y_1 + E_2 y_0) \\ & + \epsilon^3(E_0 y_3 + E_1 y_2 + E_2 y_1 + E_0 y_3) + \cdots. \end{aligned}$$

When we equate the orders of epsilon, we get the sequence of equations

$$\begin{aligned} y_0'' + E_0 y_0 &= 0, \\ y_1'' + E_0 y_1 &= xy_0 - E_1 y_0, \\ y_2'' + E_0 y_2 &= xy_1 - E_1 y_1 - E_2 y_0, \\ y_3'' + E_0 y_3 &= xy_2 - E_1 y_2 - E_2 y_1 - E_3 y_0, \ldots. \end{aligned}$$

Each of these inhomogeneous equations has the same associated homogeneous equation, which is that of the unperturbed problem. Thus, these equations can be solved via variation of parameters, equation 4.10. For example, for the first eigenfunction $n = 1$, we let $y_0 = \sin(x)$ and $E_0 = 1$, so the general solution to the equation for y_1 is

$$y_1 = c_1 \cos x + c_2 \sin x + \frac{(2E_1 x - x^2)\cos x + x \sin x}{4}.$$

For the initial conditions, we need $y_1(0) = y_1(\pi) = 0$. This determines that $c_1 = 0$, and that $E_1 = \pi/2$, but cannot determine c_2. This is because the eigenfunctions are only determined up to a multiplicative constant. So we need to add a "normalizing condition" in order to fix the scaling of the eigenfunction. For example, we can insist that $y(a) = y_0(a)$ for some a in which $y_0 \neq 0$, so that $y_k(a) = 0$ for all $k \geq 1$. For the first eigenfunction $y_0 = \sin(x)$, we can choose $a = \pi/2$. The additional information $y_1(\pi/2) = 0$ gives us $c_2 = -\pi/8$, so we have

$$E_1 = \pi/2, \qquad y_1 = \frac{(\pi - x)x\cos x}{4} + \frac{(2x - \pi)\sin x}{8}.$$

For the next order, we have to solve

$$y_2'' + y_2 = \left(x - \frac{\pi}{2}\right)\left(\frac{(\pi - x)x\cos x}{4} + \frac{(2x - \pi)\sin x}{8}\right) - E_2 \sin x.$$

Using variation of parameters again, we have the general solution as

$$y_2 = c_1 \sin x + c_2 \cos x + \frac{15\pi x^2 - 10x^3 + (15 - 6\pi^2 + 48E_2)x}{96} \cos x$$
$$+ \frac{2\pi x^3 - x^4 + (5 - \pi^2)x^2 - 5\pi x}{32} \sin x.$$

Forcing $y_2(0) = y_2(\pi) = 0$ gives $c_1 = 0$ and $E_2 = (\pi^2 - 15)/48$. Finally, the normalizing condition $y_2(\pi/2) = 0$ gives $c_2 = (\pi^4 + 20\pi^2)/512$. This produces the second order correction $y_2 =$

$$(\pi - 2x)\frac{80x(x - \pi)\cos x + 3(\pi^3 + 2\pi^2 x - 40x + 8x^3 + 20\pi - 12\pi x^2)\sin x}{1536}.$$

We can do a similar analysis on the other eigenfunctions. For example, for $n = 2$, we let $y_0 = \sin(2x)$ and $E_0 = 4$. The solution of the equation for y_1 is now

$$y_1 = c_1 \cos(2x) + c_2 \sin(2x) + \frac{(4E_1 x - 2x^2)\cos(2x) + x\sin(2x)}{16}.$$

Once again, the initial conditions $y(0) = y(\pi) = 0$ determines $c_1 = 0$ and $E_1 = \pi/2$. However, we cannot use $a = \pi/2$ for the normalizing point, since $y_0(\pi/2) = 0$. So for this eigenfunction, we can choose $a = \pi/4$, and insist that $y_1(\pi/4) = 0$. This gives us $c_2 = -\pi/64$, so we have

$$y_1 = \frac{8x(\pi - x)\cos(2x) + (4x - \pi)\sin(2x)}{64}.$$

□

Regular perturbation can also be applied to non-linear differential equation problems, especially if the unperturbed problem becomes linear.

Example 7.10

Consider the *projectile motion*, or path of an object that is thrown at an angle α near the earth's surface, taking into consideration the *air resistance*. One popular model for air resistance is for the magnitude to be proportional to the velocity squared, but in the direction opposing the motion. Thus, if $x(t)$ and $y(t)$ denote the coordinates at time t, then the air resistance can be expressed by the vector

$$-k\sqrt{x'(t)^2 + y'(t)^2}\langle x'(t), y'(t)\rangle,$$

where k is the coefficient of resistance. Thus, the equations of motion are given by

$$mx''(t) = -k\sqrt{x'(t)^2 + y'(t)^2}x'(t),$$
$$my''(t) = -mg - k\sqrt{x'(t)^2 + y'(t)^2}y'(t),$$

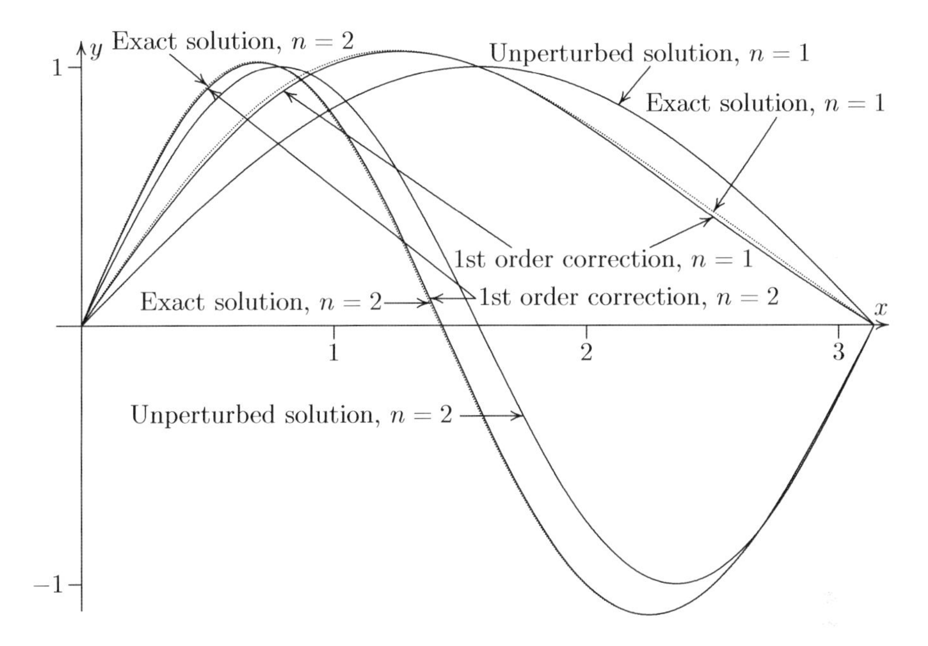

FIGURE 7.4: This shows the zeroth and first order perturbation solutions to the eigenvalue problem $y'' - \epsilon xy + y = 0$, $y(0) = y(\pi) = 0$. Note that the first order solutions are nearly indistinguishable from the exact solution, even with $\epsilon = 1$. The graphs of higher order perturbation solutions truly are indistinguishable.

where m is the mass of the object, and g is the constant gravity. These equations are highly non-linear, and cannot be solved in closed form. However, if we suppose that $\epsilon = k/m$ is small, we can make this into a perturbation problem. Find the first order corrections to the problem.

SOLUTION: If we let v_0 denote the initial velocity, we have the initial conditions as $x(0) = 0$, $x'(0) = v_0 \cos(\alpha)$, $y(0) = 0$, and $y'(0) = v_0 \sin(\alpha)$. The unperturbed problem is routine to solve:

$$x_0(t) = v_0 t \cos(\alpha) \qquad y_0(t) = v_0 t \sin(\alpha) - \frac{gt^2}{2}.$$

If we assume that $x(t) = x_0(t) + \epsilon x_1(t) + \epsilon^2 x_2(t) + \cdots$ and $y(t) = y_0(t) + \epsilon y_1(t) + \epsilon^2 y_2(t) + \cdots$, we find that the equation for x_1 and y_1 are straightforward.

$$x_1''(t) = -x_0' \sqrt{x_0'^2 + y_0'^2}, \qquad y_1''(t) = -y_0' \sqrt{x_0'^2 + y_0'^2}.$$

So the first order correction can be computed by integrations, using the initial conditions $x_1(0) = x_1'(0) = y_1(0) = y_1'(0) = 0$.

By computing $\sqrt{x_0'^2 + y_0'^2} = \sqrt{g^2t^2 + v_0^2 - 2gtv_0 \sin(\alpha)}$, we can integrate once to find

$$x_1'(t) = \frac{v_0 \cos(\alpha)}{2g} \Bigg(v_0 \sin(\alpha)(v(t) - v_0) - gtv(t) + v_0^2 \cos^2(\alpha) \ln\left(\frac{v_0 \sin(\alpha) - gt + v(t)}{v_0(\sin(\alpha) + 1)} \right) \Bigg)$$

where we let $v(t)$ represent $\sqrt{g^2t^2 + v_0^2 - 2gtv_0 \sin(\alpha)}$. We can integrate this again to get

$$\begin{aligned} x_1(t) = \frac{v_0 \cos(\alpha)}{6g^2} \Bigg(& v_0^2(3\cos^2(\alpha) - 1)(v(t) - v_0) - 3gtv_0^2 \sin(\alpha) \\ & + 2gtv_0 \sin(\alpha) v(t) - g^2t^2 v(t) \\ & + 3v_0^2 \cos^2(\alpha)(gt - v_0 \sin(\alpha)) \ln\left(\frac{v_0 \sin(\alpha) - gt + v(t)}{v_0(\sin(\alpha) + 1)} \right) \Bigg). \end{aligned}$$

Solving for $y_1(t)$ is easier. Integrating

$$y_1''(t) = -(v_0 \sin(\alpha) - gt) \sqrt{g^2t^2 + v_0^2 - 2gtv_0 \sin(\alpha)}$$

produces

$$y_1'(t) = \frac{v(t)^3 - v_0^3}{3g}.$$

Integrating once again, we obtain

$$\begin{aligned} y_1(t) = \frac{1}{24g^2} \Bigg(& v_0^3 \sin(\alpha)(3\cos^2(\alpha) + 2)(v_0 - v(t)) - 8v_0^3 gt \\ & + v_0^2(\sin^2(\alpha) + 5) gtv(t) - 6v_0 \sin(\alpha) g^2t^2 v(t) + 2g^3t^3 v(t) \\ & - 3v_0^4 \cos^4(\alpha) \ln\left(\frac{v_0 \sin(\alpha) - gt + v(t)}{v_0(\sin(\alpha) + 1)} \right) \Bigg). \end{aligned}$$

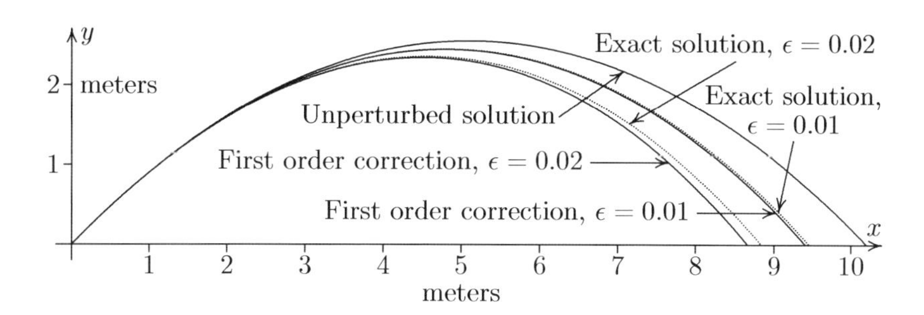

FIGURE 7.5: This shows how the perturbation approximation can be used to estimate the projectile's path if the amount of air resistance is low. The air resistance is assumed to have a magnitude of kv^2, where $\epsilon = k/m$ is small. In this scenario, a projectile is thrown at an angle of $\pi/4$ at 10 m/s, with $g = 9.8$ m/s^2. When $\epsilon = 0.01$, the first order correction is almost indistinguishable from the exact solution.

Figure 7.5 shows the relationship between the first order perturbation solution and the exact solution. Even with $\epsilon = 0.01$, the air resistance can make a large difference in the path of the projectile, and the first order perturbation accurately estimates this change. The difference between the exact solution and the first order perturbation solutions is of order ϵ^2, so doubling the size of ϵ will give us about 4 times as much error. ∎

Problems for §7.2

1 Show that the general second order linear homogeneous equation

$$y'' + p(x)y' + q(x)y = 0$$

can be made into a Schrödinger equation with the proper substitution. That is, the y' term can be eliminated.

Hint: Substitute $y = u(x)f(x)$, and determine $u(x)$ so that the equation has no $f'(x)$ term.

2 Show that if $|q(x)| \le M$ for $0 \le x \le x_0$, then by induction,

$$|y_n(x)| \le \frac{M^n x^{2n+1}}{(2n+1)!}$$

for $0 \le x \le x_0$, where $y_0 = x$ and y_n is defined by equation 7.5. From this, show that the perturbation series from example 7.6 converges at x_0 for all ϵ.

3 The Airy equation is given by $y'' - xy = 0$. Use example 7.6, with $q(x) = -x$, to find the first four terms of the perturbation series, using the same initial conditions. How does this compare to the Taylor series solution, done in example 4.18?

4 The parabolic cylinder equation, introduced in example 5.2, is given by $y'' + (v + 1/2 - x^2/4)y = 0$. Use example 7.6, with $q(x) = (2v + 1)/2 - x^2/4$, to find the first three terms of the perturbation series, using the same initial conditions.

For problems **5** through **14**: Determine the second order perturbation solutions (up to y_2) for the following initial value problems.

5 $y' + \epsilon e^x y = 0, \quad y(0) = 1$

6 $y' + \epsilon \sin(x)y = 0, \quad y(0) = 2$

7 $y' + \epsilon xy = e^x, \quad y(0) = 1$

8 $y' - \epsilon e^x y = 1, \quad y(0) = 0$

9 $y' - \epsilon xy = 1/x, \quad y(1) = 0$

10 $y'' - \epsilon e^x y = 0, \quad y(0) = 1, y'(0) = 0$

11 $y'' - \epsilon e^x y = 0, \quad y(0) = 0, y'(0) = 1$

12 $y'' + \epsilon y' + y = 0, \quad y(0) = 1, y'(0) = 0$

13 $y'' + \epsilon xy = x, \quad y(0) = y'(0) = 0$

14 $y'' - \epsilon xy = e^x, \quad y(0) = y'(0) = 1$

For problems **15** through **20**: Determine the second order perturbation solutions (up to y_2) for the following boundary value problems.

15 $y'' - \epsilon xy = 0, \quad y(0) = 0, y(1) = 1$

16 $y'' - \epsilon xy = 0, \quad y'(0) = 1, y(1) = 1$

17 $y'' + \epsilon x^2 y = 0, \quad y(0) = 0, y'(1) = 1$

18 $y'' + \epsilon y' + y = 0, \quad y(0) = 0, y(\pi/2) = 1$

19 $y'' + \epsilon xy = x^2, \quad y(0) = y(1) = 0$

20 $y'' + \epsilon y' + y = 2\sin(x), \quad y(0) = y(\pi/2) = 0$

21 Show that in example 7.9, $E_1 = \pi/2$ for all of the eigenfunction solutions. Determine the first order correction y_1 for the nth eigenfunction as a function of n. Normalize the solution so that $y(\pi/(2n)) = 1$.

22 The eigenvalue problem in example 7.9 is actually solvable in terms of the Airy functions. Find the exact solution to the first eigenfunction, in terms of $E_1 \approx 2.4659002963$.

23 Consider the eigenvalue problem $y'' + \epsilon y' + Ey = 0$, $y(0) = y(\pi) = 0$. Use perturbation theory to find the second order corrections for the first eigenfunction. Normalize the solution so that $y(\pi/2) = 1$.

24 Consider the eigenvalue problem $y'' + \epsilon y' + Ey = 0$, $y(0) = y(\pi) = 0$. Use perturbation theory to find the second order corrections for the second eigenfunction. Normalize the solution so that $y(\pi/4) = 1$.

25 Consider the eigenvalue problem $y'' - 2\epsilon xy' + Ey = 0$, $y(0) = y(\pi) = 0$. Use perturbation theory to find the first order corrections for the first eigenfunction. Normalize the solution so that $y(\pi/2) = 1$.

26 Consider the eigenvalue problem $y'' - 2\epsilon xy' + Ey = 0$, $y(0) = y(\pi) = 0$. Use perturbation theory to find the first order corrections for the second eigenfunction. Normalize the solution so that $y(\pi/4) = 1$.

27 In example 7.10, the *height* of the projectile is the maximum elevation of the object. Determine to first order in ϵ the affect of the air resistance on the height of the projectile.
Hint: Determine *when* the projectile is at its highest position in the unperturbed problem, and plug this value of time into the first order correction.

28 In example 7.10, the *range* of the projectile is the distance that it traveled, assuming that the ground is level. Determine to first order in ϵ the affect of the air resistance on the range of the projectile.
Hint: Determine when the projectile lands in the unperturbed problem, and plug this value of time into the first order correction. The y coordinate will not be 0, but we can use the fact that, to leading order, the angle of landing is also α.

29 An alternative model for air resistance is for the force to be proportional to the velocity. This is a particularly good model if the object is aeronautically designed, like a football. The new set of differential equations are much simpler than in example 7.10:

$$mx''(t) = -kx'(t), \qquad my''(t) = -mg - ky'(t).$$

If we again let $\epsilon = k/m$ be a small parameter, we can use perturbation theory to approximate the solution. Find the second order perturbation solution.

30 Find the exact solution for the set of equations given in problem 29.

7.3 Singular Perturbation for Differential Equations

We have seen how regular perturbation theory can be used to approximate the solutions to differential equations, including boundary value problems. However, the perturbation series tend to converge like Taylor series, and may only be useful for small epsilon. The situation is analogous to finding the Taylor series solution to a differential equation at an ordinary point, or regular singular point. Even though the asymptotic series at an irregular singular point may diverge, through the techniques of chapter 3 we can obtain better accuracy with a series expanded at an irregular singular point than an ordinary point. In the same way, once we develop the techniques for singular perturbation, they can prove to be more accurate than regular perturbation theory.

Let us consider the general second order boundary value problem.

$$y''(x) + p(x)y'(x) + q(x)y(x) = 0, \qquad y(x_0) = y_0, \quad y(x_1) = y_1.$$

We want to add an ϵ to the equation so that the zeroth order equation becomes trivial to solve. This can be accomplished by adding the ϵ to the highest derivative term.

$$\epsilon y''(x) + p(x)y'(x) + q(x)y(x) = 0, \qquad y(x_0) = y_0, \quad y(x_1) = y_1. \tag{7.6}$$

When $\epsilon = 0$ the equation becomes first order, and is in fact separable. The drawback is that we lose one of the solutions, since the solution to a first order equation only contains one arbitrary constant, and we still have two boundary conditions to satisfy. This is very similar to example 7.3, in which two of the polynomial roots approached ∞ as ϵ approaches 0. Likewise, as $\epsilon \to 0$, the solutions to equation 7.6 will approach a discontinuous function.

The best way to demonstrate what is happening is with an illustrative example.

Example 7.11

Consider the boundary value problem

$$\epsilon y'' - y = 0, \qquad y(0) = 0, \quad y(1) = 1.$$

Analyze the solution as $\epsilon \to 0$, particular near the endpoint $x = 1$.

SOLUTION: The equation is simple enough to solve in terms of ϵ. It is routine to check that $e^{x/\sqrt{\epsilon}}$ and $e^{-x/\sqrt{\epsilon}}$ satisfy the differential equations, so the general solution is $C_1 e^{x/\sqrt{\epsilon}} + C_2 e^{-x/\sqrt{\epsilon}}$. The boundary conditions allow us to solve for C_1 and C_2, yielding

$$y = \frac{e^{(x+1)/\sqrt{\epsilon}} - e^{(1-x)/\sqrt{\epsilon}}}{e^{2/\sqrt{\epsilon}} - 1}.$$

The graph of the solution for several values of ϵ are shown in figure 7.6. The graph reveals a region in which the solution varies rapidly as epsilon becomes small. This region is called the *boundary layer*.

Note that as ϵ approaches 0, the solution changes rapidly near $x = 1$. In order to understand the behavior of this portion of the function, we introduce a scale transformation. If we let $x = 1 - t\sqrt{\epsilon}$, then as t ranges from 0 to 1, x will range over this boundary layer. We let Y denote the function y in the new variable, so that $y(x) = Y(1 - t\sqrt{\epsilon})$. This transformation converts the complicated expression $e^{(1-x)/\sqrt{\epsilon}}$ to just e^t. The differential equation is easily converted to the new variable, since $y''(x) = Y''(t)/\epsilon$. So in terms of the new variable, we have

$$Y''(t) - Y(t) = 0, \qquad Y(0) = 1, \qquad Y(1/\sqrt{\epsilon}) = 0.$$

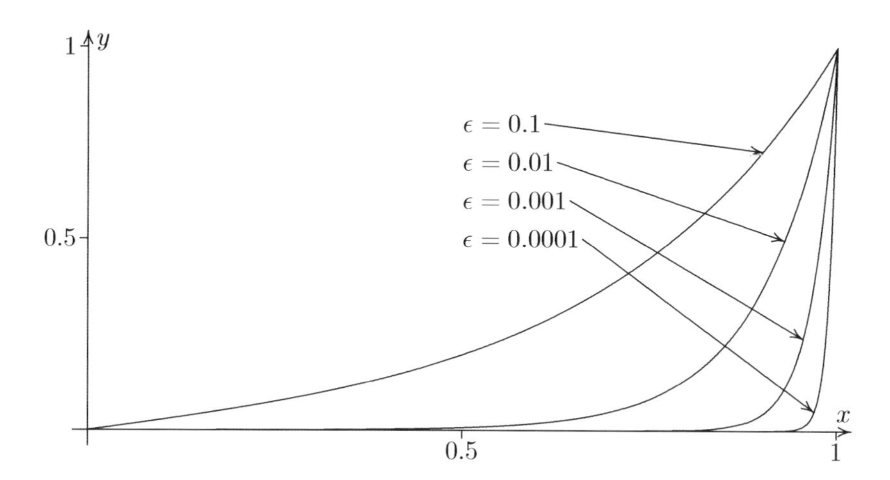

FIGURE 7.6: The exact solutions to the equation $\epsilon y'' - y = 0$, $y(0) = 0$, $y(1) = 1$ are shown for various ϵ. As $\epsilon \to 0$ the solution becomes rapidly varying in the region near 1. This region is called the *boundary layer*.

As $\epsilon \to 0$, this becomes the non-perturbation problem $Y'' = Y$, $Y(0) = 1$, $Y(\infty) = 0$, which has the unique solution $Y \sim e^{-t}$. Converting back to x, we have

$$y \sim e^{(x-1)/\sqrt{\epsilon}}. \tag{7.7}$$

The asymptotics were a bit tricky because there are *two* limits happening simultaneously. We are considering both x approaching 1, and ϵ approaching 0. This is why we had to consider the *relative* sizes of the two quantities ϵ and $1 - x$. ∎

The perturbation series which is valid on a region for which the solution is changing rapidly is called the *inner solution*, y_{in}. So the asymptotic solution found in equation 7.7 is the inner solution to the perturbation problem. The perturbation series which is valid on a region where the solution is not changing rapidly is called the *outer solution*, y_{out}. The outer solution in example 7.11 is just $y = 0$. Other problems have more interesting outer solutions.

Example 7.12

Find the first three terms of both the inner and outer solutions of the perturbation problem

$$\epsilon y'' + 2y' + y = 0, \qquad y(0) = 0, \quad y(1) = 1. \tag{7.8}$$

SOLUTION: We begin with the outer solution, which is found by assuming that ϵ is small. If we set $\epsilon = 0$, the problem becomes a first order separable equation, $2y' + y = 0$, which is easily solved using the methods of

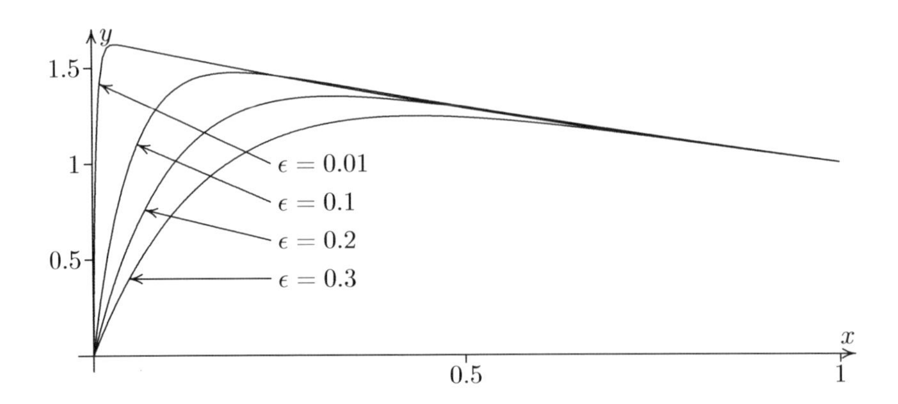

FIGURE 7.7: This shows the exact solutions to the equation $\epsilon y'' + 2y' + y = 0$ for various ϵ. We can clearly see that this time the boundary layer is in the region near 0. In fact, one can see that the location of the maximum is approximately proportional to ϵ. We say that the boundary layer has a *thickness* of order ϵ.

subsection 4.2.2. The general solution is $y_{\text{out}} \sim Ce^{-x/2}$. Clearly, both initial conditions cannot be satisfied simultaneously, so there will be a boundary layer on one side, but which side? In this case the equation can be solved in closed form (see problem 2), so we can observe the graphs of the solutions for various values of ϵ, shown in figure 7.7. The graph clearly indicates that the boundary layer will be near $x = 0$, so for the outer solution, we will use the endpoint $y(1) = 1$ to solve for C. Thus produces the zeroth order outer solution, $y_0 = e^{(1-x)/2}$.

To find higher orders of the outer solution, we assume that the outer solution is of the form

$$y_{\text{out}} \sim y_0 + \epsilon y_1 + \epsilon^2 y_2 + \cdots,$$

and substitute this expression into equation 7.8. After combining like powers of ϵ, we obtain the equations

$$\begin{aligned} 2y_0' + y_0 &= 0, \\ 2y_1' + y_1 &= -y_0'', \\ 2y_2' + y_2 &= -y_1'', \ldots . \end{aligned}$$

The first equation we already solved, and the other equations are first order inhomogeneous equations. Using the initial condition $y_i(1) = 0$ for $i \geq 1$ allows us to uniquely find each y_i:

$$y_1 = (1-x)e^{(1-x)/2}/8, y_2 = (x^2 - 10x + 9)e^{(1-x)/2}/128, \ldots .$$

Thus, the first three terms of the outer perturbation series are

$$y_{\text{out}} \sim e^{(1-x)/2} + \epsilon \frac{(1-x)e^{(1-x)/2}}{8} + \epsilon^2 \frac{(x^2 - 10x + 9)e^{(1-x)/2}}{128} + \cdots.$$

To find the inner solution, we will use a scale transformation. Figure 7.7 indicates that the width of the boundary layer is proportional to ϵ, so let us consider the substitution $x = \epsilon t$, producing the equation

$$Y''(t) + 2Y'(t) + \epsilon Y(t) = 0, \qquad Y(0) = 0, \quad Y(1/\epsilon) = 1. \tag{7.9}$$

This is now a regular perturbation problem, which confirms that the width of the boundary layer is indeed of order ϵ. We cannot use the boundary condition $Y(1/\epsilon) = 1$, since $1/\epsilon$ will be outside of the boundary layer. The general solution to the unperturbed problem, $Y_0'' + 2Y_0' = 0$, is $Y_0 = C_1 + C_2 e^{-2t}$. The initial condition $Y(0) = 0$ allows us to eliminate one of the constants, $Y_0 = C_1(1 - e^{-2t})$. If we assume that the perturbation solution is of the form

$$y_{\text{in}}(t) \sim Y_0 + \epsilon Y_1 + \epsilon^2 Y_2 + \cdots,$$

we can substitute this into equation 7.9 to produce the sequence of equations

$$\begin{aligned} Y_0'' + 2Y_0' &= 0, \\ Y_1'' + 2Y_1' &= -Y_0, \\ Y_2'' + 2Y_2' &= -Y_1, \ldots. \end{aligned}$$

Each of these equations can be thought of as a first order homogeneous equation in Y', so they can easily be solved. We have one boundary condition, $Y_i(0) = 0$, but there will still be an arbitrary constant added for each equation. The solutions to these equations are

$$\begin{aligned} Y_0 &= C_1(1 - e^{-2t}), \\ Y_1 &= -C_1 t(1 + e^{-2t})/2 + C_2(1 - e^{-2t}), \\ Y_2 &= C_1(t^2 - t - t^2 e^{-2t} - te^{-2t})/8 - C_2 t(1 + e^{-2t})/2 + C_3(1 - e^{-2t}), \ldots. \end{aligned}$$

Thus, the perturbation series for the inner solution is

$$\begin{aligned} Y_{\text{in}}(t) \sim{}& C_1(1 - e^{-2t}) + \epsilon\left(C_2(1 - e^{-2t}) - C_1 t \frac{1 + e^{-2t}}{2}\right) \\ &+ \epsilon^2\left(C_1 \frac{t^2 - t - t^2 e^{-2t} - te^{-2t}}{8} - C_2 t \frac{1 + e^{-2t}}{2} + C_3(1 - e^{-2t})\right) + \cdots. \end{aligned}$$

It seems strange that there are an infinite number of constants in the perturbation series, but we can factor this series into the form

$$\begin{aligned} y_{\text{in}}(t) \sim{}& (C_1 + \epsilon C_2 + \epsilon^2 C_3 + \cdots) \times \\ &\left((1 - e^{-2t}) - \epsilon t \frac{1 + e^{-2t}}{2} + \epsilon^2 \frac{t^2 - t - t^2 e^{-2t} - te^{-2t}}{8} + \cdots\right). \end{aligned}$$

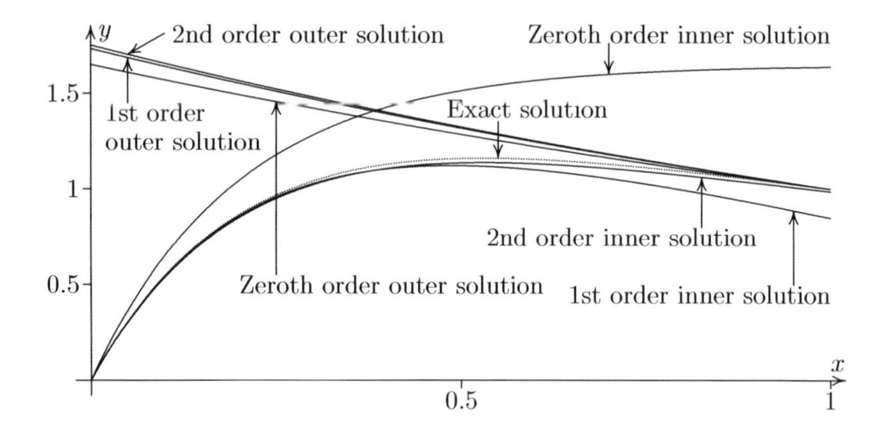

FIGURE 7.8: This shows the successive approximations to both the outer and inner solutions for the equation $\epsilon y'' + 2y' + y = 0$ with $\epsilon = 0.4$. Note that for this value of ϵ, the inner solution is a better approximation than the outer solution. At first it seems odd that the second order outer solution is worse than the first order, but recall that the perturbation series is a divergent series.

Thus, the inner solution can be expressed as

$$y_{\text{in}}(t) \sim K_\epsilon \left((1 - e^{-2t}) - \epsilon t \frac{1 + e^{-2t}}{2} + \epsilon^2 \frac{t^2 - t - t^2 e^{-2t} - te^{-2t}}{8} + \cdots \right),$$

where K_ϵ depends on ϵ. This allows us to express the inner solution in terms of a single constant. Converting back to the x variable, we get $y_{\text{in}}(x) \sim$

$$K_\epsilon \left((1 - e^{-2x/\epsilon}) - x \frac{1 + e^{-2x/\epsilon}}{2} + \frac{x^2 - \epsilon x - x^2 e^{-2x/\epsilon} - \epsilon x e^{-2x/\epsilon}}{8} + \cdots \right).$$ ☐

There is a way to determine the constant K_ϵ by comparing the outer and inner solutions. We will study this more in section 7.4, but we can determine K_ϵ to leading order using some simple logic. The leading order of in inner solution is $K_\epsilon(1 - e^{-2t})$, which approaches K_ϵ as $t \to \infty$. But as t goes to infinity, we leave the boundary layer. We see in figure 7.7 that the boundary layer ends around the point of the local maximum of the solution. But as $\epsilon \to 0$, this maximum approaches the limit of the outer solution, $\lim_{x\to 0} y_{\text{out}} = \sqrt{e}$. Thus, to leading order, $K_\epsilon \sim \sqrt{e}$.

Example 7.12 raises some important questions. How do we know whether there will be a boundary layer at one of the endpoints, which endpoint will the boundary layer be, and how thick will the boundary layer be? The solution is to rescale the problem within the boundary layer in such a way so that the problem will be converted to a regular perturbation problem. This involves the method of dominant balance, to determine which boundary layer thickness produces a consistent result. The consistent thickness of the boundary layer is called the *singular distinguished limit.*

To find the singular distinguished limit, we assume that the boundary layer at one of the endpoints is δ, where δ is a small quantity that will be related to ϵ. For a left-hand endpoint at x_1, we make the substitution $x = x_1 + \delta t$. This causes the substitutions

$$\boxed{x \Rightarrow x_1 + \delta t, \qquad y'(x) \Rightarrow y'(t)/\delta, \qquad y''(x) \Rightarrow y''(t)/\delta^2, \text{ etc.}} \tag{7.10}$$

After this substitution, we are ready to apply the method of dominant balance, which assumes two of the terms are of the same size as $\epsilon \to 0$, and the others are either the same size or smaller. We assume that t and $y(t)$, along with the derivatives, are of order 1 after the rescaling. The goal is to find a combination for which the zeroth order non-constant solution approaches a finite value as $t \to \infty$. This prevents the original solution from going to ∞ at a finite distance away from x_1 in the original variable x. Usually, letting $\delta = O(1)$ will be consistent, and this case is used in finding the outer solution. But the goal is to find *another* consistent combination, indicating the presence of a boundary layer. This will give us the singular distinguished limit.

Example 7.13

Find the singular distinguished limit for the two endpoints of the perturbation problem

$$\epsilon y'' + 2xy' + xy = 0, \qquad y(0) = 0, \quad y(1) = e^{-1/2}. \tag{7.11}$$

Assume that $\epsilon > 0$.

SOLUTION: For the left-hand endpoint $x_1 = 0$, we substitute $x = \delta t$, where δ is yet to be determined. The new equation is

$$\frac{\epsilon Y''(t)}{\delta^2} + \frac{2\delta t Y'(t)}{\delta} + \delta t Y(t) = 0.$$

There are three cases to consider for the left-hand endpoint.

Case 1) $2tY'(t) \sim -\delta t Y(t)$.

This can only happen if $\delta = O(1)$, which indicates that this combination finds a possible outer solution at this endpoint. Indeed, the remaining term would be small compared to the two terms we kept. If we replace δ with 1, the solution to the differential equation is $Y = Ce^{-t/2}$. The initial condition $Y(0) = 0$ shows that $C = 0$, so we have the possible outer solution to be $y = 0$.

Case 2) $\epsilon Y''(t)/\delta^2 \sim -\delta t Y(t)$.

This would indicate that $\delta = O(\epsilon^{1/3})$, but then the remaining term would be of order 1, which is larger than the terms we kept. So this combination is inconsistent.

Case 3) $\epsilon Y''(t)/\delta^2 \sim -2tY'(t)$.

For this to occur, we need to have $\delta = O(\sqrt{\epsilon})$. The remaining term is smaller than the terms we kept, so this could be consistent. However, we still have to test whether the leading order solutions approach a constant as $t \to \infty$.

If we make the substitution $\delta = \sqrt{\epsilon}$, we get the equation

$$Y''(t) + 2tY'(t) + \sqrt{\epsilon} t y(t) = 0.$$

This is a regular perturbation problem in $\delta = \sqrt{\epsilon}$. The unperturbed problem, $Y'' + 2tY' = 0$, is a separable equation for Y', so we can solve this to obtain $Y' = c_1 e^{-t^2}$. Integrating, we obtain

$$Y(t) = c_2 + c_1 \int_0^t e^{-s^2}\, ds.$$

Indeed, this does approach a finite value, namely $c_2 + c_1\sqrt{\pi}/2$. Thus, we have found a singular distinguished limit for the endpoint at $x = 0$. The initial condition $y(0) = 0$ forces $c_2 = 0$. So a possible boundary layer first order solution is $Y(t) = c_1\sqrt{\pi}\,\mathrm{erf}(t)/2$, where $\mathrm{erf}(t)$ is defined by equation 2.8.

Note that this does not prove that a boundary layer exists at $x = 0$. In fact, there may not be any boundary layer at all, for this depends on the endpoints of the problem. We need to examine the other endpoint before we can assess the boundary layers.

For the right hand endpoint, we make a similar substitution as equation 7.10.

$$\boxed{x \Rightarrow x_2 - \delta t, \qquad y'(x) \Rightarrow -Y'(t)/\delta, \qquad y''(x) \Rightarrow Y''(t)/\delta^2, \text{ etc.}} \tag{7.12}$$

So the new equation becomes

$$\frac{\epsilon Y''(t)}{\delta^2} - \frac{2(1-\delta t)Y'(t)}{\delta} + (1-\delta t)Y(t) = 0.$$

Case 1) $\delta = O(1)$.

This would give rise to the outer solution, $2Y'(t) = Y(t)$. The solution is $Y(t) = Ce^{t/2}$. The endpoint $t = 0$ corresponds to $x = 1$, so $Y(0) = e^{-1/2}$, hence $C = e^{-1/2}$. Translating to the original variables, the outer solution is $y(x) = e^{-x/2}$.

For the other cases, we must have $\delta \ll 1$. This forces $\delta t \ll 1$, and the $\epsilon Y''(t)/\delta^2$ term must be kept. Thus, there are only two other cases to consider.

Case 2) $\epsilon Y''(t)/\delta^2 \sim -Y(t)$.

This would indicate that $\delta = O(\epsilon^{1/2})$, but then we would have $2Y'/\delta$ being larger than the terms that we kept, so this case is inconsistent.

Case 3) $\epsilon Y''(t)/\delta^2 \sim 2Y'/\delta$.

This final case produces $\delta = O(\epsilon)$, and the remaining terms would indeed be smaller than the terms we kept. So the final test is to see if the solutions remain bounded as $t \to \infty$.

Replacing δ with ϵ, the equation becomes

$$Y'' + 2(\epsilon t - 1)Y' + \epsilon(1 - \epsilon t)Y = 0.$$

The unperturbed equation is $Y'' - 2Y' = 0$, whose solutions are $Y = c_1 + c_2 e^{2t}$. In order for this to approach a constant as $t \to \infty$, we must have $c_2 = 0$, so the solution is a constant, $y = c_1$. But if the function is a constant within the boundary layer, then the boundary layer vanishes. So this case is also inconsistent.

The only case that is consistent for the endpoint at $x = 1$ is the outer solution, $y(x) = e^{-x/2}$. Since this is not equal to 0 when $x = 0$, there must be a boundary layer near $x = 0$, which we found was possible, with a solution of $y(x) = c_1\sqrt{\pi}\,\mathrm{erf}(x/\sqrt{\epsilon})/2$. The boundary layer has a width of order $\sqrt{\epsilon}$. □

Once again, we can use some simple logic to determine the arbitrary constant in the inner solution. The inner solution, $y_{\text{in}}(x) \sim c_1\sqrt{\pi}\,\mathrm{erf}(x/\sqrt{\epsilon})/2$, approaches $c_1\sqrt{\pi}/2$ as $x \to \infty$, so this is the value of the function as we leave the boundary layer, at least to first order in ϵ. The outer solution $y_{\text{out}}(x) = e^{-x/2}$ approaches 1 as x approaches the boundary layer. These two limits must be the same, so $c_1 \sim 2/\sqrt{\pi}$.

There is a way to combine the inner and outer solutions to form a single function. If we let y_c be the common limit between the inner and outer solutions, we can let

$$\boxed{y_{\text{comp}}(x) = y_{\text{in}}(x) + y_{\text{out}}(x) - y_c,} \tag{7.13}$$

called the *uniformly valid composite approximation*. Notice that within the boundary layer, $y_{\text{out}}(x) \sim y_c$, so $y_{\text{comp}}(x) \sim y_{\text{in}}(x)$. Outside the boundary layer, $y_{\text{in}}(x) \sim y_c$, so $y_{\text{comp}}(x) \sim y_{\text{out}}(x)$. Hence, equation 7.13 constructs a function that is asymptotic to both the inner and outer solutions in the appropriate region, similar to the way that the two point Padé of subsection 3.6.1 consolidated two different series valid in different intervals. The composite approximation often gives a very impressive amount of accuracy, even for just the leading order. In example 7.11, the composite approximation would be $y_{\text{comp}} \sim \mathrm{erf}(x/\sqrt{\epsilon}) + e^{-x/2} - 1$, which is shown in figure 7.9.

Example 7.14

Find the leading order composite approximation to the singular perturbation problem

$$\epsilon y'' - 2xy' + xy = 0, \qquad y(0) = 1, \quad y(1) = 1. \tag{7.14}$$

SOLUTION: The outer solution is essentially the unperturbed problem, found by setting $\epsilon = 0$. The solution to this equation, $2y' = y$, is $y_{\text{out}} = C_1 e^{x/2}$. However, this solution cannot satisfy both of the boundary conditions, so there must be a boundary layer somewhere in the problem. Since the outer

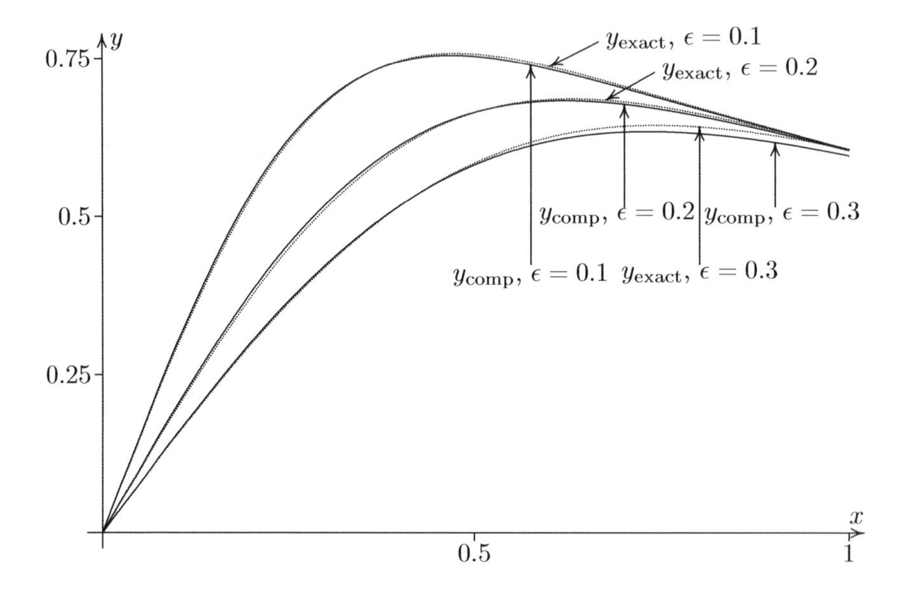

FIGURE 7.9: This compares the leading order composite approximation to the equation $\epsilon y'' + 2xy' + xy = 0$, $y(0) = 0$, $y(1) = e^{-1/2}$, to the exact solution for $\epsilon = 0.1$, 0.2, and 0.3. The composite approximations are drawn with solid lines, while the exact solutions use dotted lines. As ϵ gets smaller, the composite approximation becomes an excellent approximation to the exact value.

equation has no singular points on the interior of the interval, the boundary layer must appear near one of the endpoints, $x = 0$ or $x = 1$.

Let us suppose that the boundary layer is near $x = 0$. We substitute $x = \delta t$ to determine the new equation

$$\frac{\epsilon Y''(t)}{\delta^2} - 2tY'(t) + \delta t Y(t) = 0.$$

If there is a boundary layer near $x = 0$, then $\delta \ll 1$, so $\delta t Y \ll 2tY'$. This yields the relation $\epsilon Y''/\delta^2 \sim 2tY'$, which would indicate that the thickness of the boundary layer is $\delta = O(\sqrt{\epsilon})$. This leads to the equation $Y'' = 2tY'$. However, the solution to this equation, $Y = C_1 + C_2 \int_0^t e^{s^2}\,ds$, approaches ∞ as $t \to \infty$ unless $C_2 = 0$. But then the solution is a constant, so the boundary layer near $x = 0$ would disappear. Hence, the assumption that the boundary layer was at the endpoint $x = 0$ is inconsistent, so the boundary layer must occur near $x = 1$.

To determine the boundary layer near $x = 1$, we substitute $x = 1 - \delta t$ into equation 7.14 to produce

$$\frac{\epsilon Y''(t)}{\delta^2} + \frac{2(1 - \delta t)Y'(t)}{\delta} + (1 - \delta t)Y(t) = 0.$$

Since there must be a boundary layer here, we can determine that $t\delta \ll 1$ and $Y'(t)/\delta \gg Y(t)$, which will simplify this equation considerably. The resulting asymptotic relation is

$$\frac{\epsilon Y''}{\delta^2} \sim -\frac{2Y'}{\delta}.$$

This indicates that $\delta = O(\epsilon)$, and the resulting equation is $Y'' + 2Y' = 0$. The solution, $C_2 + C_3 e^{-2t}$, does indeed approach a constant as $t \to \infty$, which is consistent with the boundary layer.

We now will use the boundary conditions $y(0) = y(1) = 1$ to determine the arbitrary constants. Since $x = 0$ is in the outer solution $y_{\text{out}} = C_1 e^{x/2}$, we can determine that $C_1 = 1$. When $x = 1$, we are in the boundary layer, so this value corresponds to $t = 0$. Hence, $Y(0) = 1$, so $C_2 + C_3 = 1$, but this is not enough information to determine C_2 and C_3. To gain one more equation that will determine the arbitrary constants, we utilize the fact that the inner solution as $t \to \infty$ should match the outer solution as $x \to 1$, which has the limit $\sqrt{e}$. Hence, we find that $C_2 = \sqrt{e}$, and $C_3 = 1 - \sqrt{e}$.

We can now put all of the information together to form the leading order composite approximation to the solution. For the inner solution, we had $x = 1 - \epsilon t$, so $t = (1 - x)/\epsilon$. So the inner solution can be expressed by $y_{\text{in}} = \sqrt{e} + (1 - \sqrt{e})e^{2(x-1)/\epsilon}$. If we add the outer and inner solutions together, and subtract the common limit $\sqrt{e}$, we obtain

$$y_{\text{comp}}(x) = e^{x/2} - (\sqrt{e} - 1)e^{2(x-1)/\epsilon}.$$

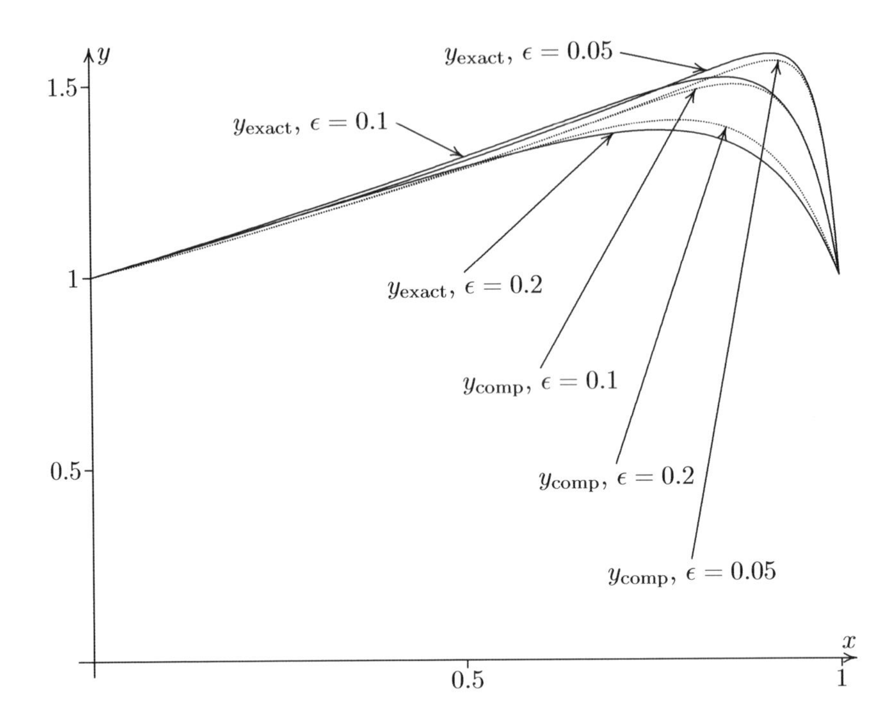

FIGURE 7.10: This compares the leading order composite approximation to the equation $\epsilon y'' - 2xy' + xy = 0$, $y(0) = 0$, $y(1) = 1$, to the exact solution for $\epsilon = 0.05$, 0.1, and 0.2. The composite approximations are drawn with dotted lines, while the exact solutions use solid lines. Notice that a small change in the equation shifts the boundary layer to the other side of the interval, and changes the width of the boundary layer.

Figure 7.10 compares the composite approximation to the exact solution for various values of ϵ. Although the approximations are not as close as in figure 7.9, it is clear that as $\epsilon \to 0$, the approximations become better. $\square$

In the last two examples, we did some analysis just to locate where the boundary layer will be. However, in most cases there is an easy way to determine the endpoint at which the boundary layer could appear.

PROPOSITION 7.1

Consider the second order linear perturbation problem as $\epsilon \to 0^+$

$$\epsilon y''(x) + p(x)y'(x) + q(x)y(x) = 0, \qquad y(x_1) = y_1, \quad y(x_2) = y_2 \tag{7.15}$$

where $p(x)$ and $q(x)$ are continuous on the interval $x_1 \leq x \leq x_2$, and $p(x) \neq 0$ on this interval. If $p(x) > 0$ on the interval, then if a boundary layer is necessary, it will have a thickness of order ϵ near the endpoint x_1. On the other hand, if $p(x) < 0$ on the interval, then the boundary layer, if necessary, will have a thickness of order ϵ near the endpoint x_2.

PROOF: For the outer solution, we consider the unperturbed problem

$$p(x)y'(x) + q(x)y(x) = 0.$$

Since $p(x) \neq 0$ in the interval, we can solve this first order separable equation to obtain

$$y_{\text{out}} = C_1 \exp\left(-\int \frac{q(x)}{p(x)}\,dx\right).$$

It is possible for both of the boundary conditions to be satisfied with the single arbitrary constant, but this is unlikely. Generally, we will only be able to satisfy one of the boundary conditions, so a boundary layer will develop somewhere in the problem. Notice that the outer solution remains bounded as x approaches the endpoints.

First suppose that the boundary layer occurs near $x = x_1$. By substituting $x = x_1 + \delta t$, we find the new equation is

$$\frac{\epsilon Y''(t)}{\delta^2} + p(x_1 + \delta t)\frac{Y'(t)}{\delta} + q(x_1 + \delta t)Y(t) = 0.$$

Since we are assuming that there is a boundary layer near $x = x_1$, $\delta \ll 1$, so we see that the term $q(x_1 + \delta t)y(t)$ is small. Also, we can replace $p(x_1 + \delta t)$ with $p(x_1)$, since we know that this is non-zero. The remaining equation is only consistent if $\delta = O(\epsilon)$, which produces the equation

$$Y''(t) + p(x_1)Y'(t) = 0.$$

The solution to this equation is $Y(t) = C_2 + C_3 e^{-p(x_1)t}$. Since this must remain bounded as $t \to \infty$ to match the outer solution, we must have $p(x_1) > 0$ to obtain a non-trivial solution. So if $p(x_1) < 0$, there cannot be a boundary layer at $x = x_1$.

To determine if there is a boundary layer occurring near $x = x_2$, we substitute $x = x_2 - \delta t$ to produce the equation

$$\frac{\epsilon Y''(t)}{\delta^2} - p(x_2 - \delta t)\frac{Y'(t)}{\delta} + q(x_2 - \delta t)Y(t) = 0.$$

Again, we can assume that $\delta \ll 1$, so $q(x_2 - \delta t)Y(t) \ll p(x_2 - \delta t)Y'(t)/\delta$. We can also replace $p(x_2 - \delta t)$ with $p(x_2)$, to obtain the equation

$$Y''(t) - p(x_2)Y'(t) = 0,$$

where $\delta = O(\epsilon)$. This time, the solution is $Y(t) = C_2 + C_3 e^{p(x_2)t}$, so we must have $p(x_2) < 0$ in order to have a non-trivial solution remain bounded as $t \to \infty$. So if $p(x_2) > 0$, there cannot be a boundary layer at $x = x_2$.

In summary, if $p(x) > 0$ on the interval $x_1 \le x \le x_2$, then the only possible location of a boundary layer is at $x = x_1$, with thickness of order ϵ. Likewise, if $p(x) < 0$ on this interval, then the only possible location for a boundary layer is an $x = x_2$, again with a thickness of order ϵ. ▯

Example 7.15

Find the leading order composite approximation to the perturbation problem

$$\epsilon y'' - (x+1)y' + 3y = 0, \qquad y(0) = 1, \quad y(1) = 0.$$

SOLUTION: Because $p(x) = -(x+1)$ is negative on the interval $0 \le x \le 1$, we know from proposition 7.1 that if a boundary layer exists, it will be near the point $x = 1$. The outer solution is obtained by setting $\epsilon = 0$, and solving the equation $-(x+1) + 3y = 0$. The solution to this separable equation is $C(1+x)^3$, and since $x = 0$ is in the outer solution, we can use the initial condition $y(0) = 1$ to find that $C = 1$. Since the outer solution $y_{\text{out}} = (1+x)^3$ does not satisfy the initial condition $y(1) = 0$, there will be a boundary layer near $x = 1$.

For the inner solution, we make the substitution $x = 1 - t\epsilon$, giving us

$$\frac{Y''(t)}{\epsilon} + (2 - t\epsilon)\frac{Y'(t)}{\epsilon} + 3Y(t) = 0.$$

If we multiply by ϵ, we get

$$Y''(t) + (2 - t\epsilon)Y'(t) + 3\epsilon Y(t) = 0.$$

The unperturbed equation, found by setting $\epsilon = 0$, is then $Y'' + 2Y' = 0$. This is a separable equation for $Y'(t)$, so it can easily be solved to produce $Y(t) = C_1 + C_2 e^{-2t}$. As $t \to \infty$, this must match the outer solution at $x = 1$, so we have $C_1 = 8$. Finally, we have that $Y(t) = 0$ when $x = 1$ and $t = 0$, so $C_2 = -8$.

Converting to the original variable, we have

$$y_{\text{in}} = 8 - 8e^{-2(1-x)/\epsilon}.$$

Since the constant in the overlap region is 8, we can compute the composite approximation to be

$$y_{\text{comp}} = (1+x)^3 - 8e^{-2(1-x)/\epsilon}.$$

The graph of the composite approximation is shown in figure 7.11. ▯

Note that neither example 7.13 nor example 7.14 can utilize proposition 7.1, since $p(0) = 0$ in both of these examples. Indeed, the thickness of the boundary layer in example 7.13 was of order $\sqrt{\epsilon}$ instead of ϵ. Nonetheless, this proposition can be applied to many of the homework problems.

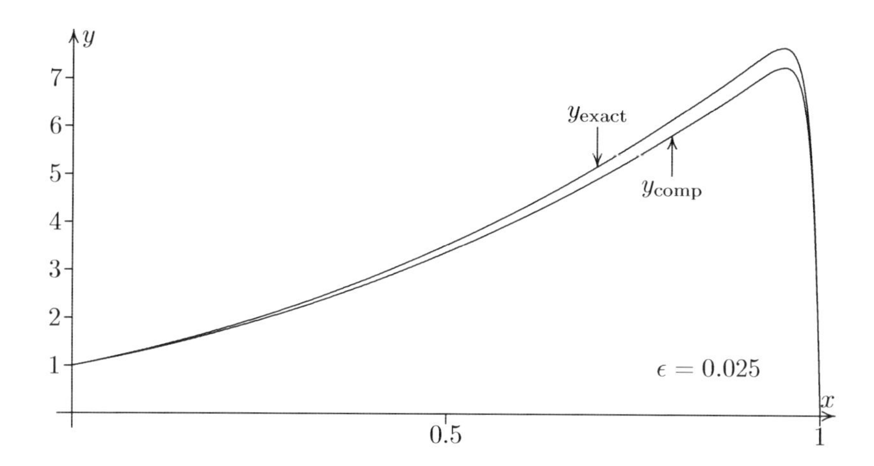

FIGURE 7.11: This compares the leading order composite approximation to the equation $\epsilon y'' - (x+1)y' + 3y = 0$, $y(0) = 1$, $y(1) = 0$, to the exact solution for $\epsilon = 0.025$.

Problems for §7.3

1 Show that in example 7.11, the substitution $\delta = (1-x)/\sqrt{\epsilon}$ converts the perturbation problem to a regular perturbation in δ. What is the new perturbation problem?

2 Show that both

$$e^{(-1+\sqrt{1-\epsilon})x/\epsilon} \quad \text{and} \quad e^{(-1-\sqrt{1-\epsilon})x/\epsilon}$$

satisfy equation 7.8. From these solutions, find a solution that also satisfies the boundary conditions.

For problems **3** through **8**: Determine the third order perturbation solution (up to ϵ^2) for the outer solution valid near $x = 1$ for the following perturbation problems,

3 $\epsilon y'' + y' + y = 0,\quad y(0) = 1,\quad y(1) = 2$

4 $\epsilon y'' + 2y' - 2y = 0,\quad y(0) = 2,\quad y(1) = 1$

5 $\epsilon y'' + (x+1)y' - 2y = 0,\quad y(0) = 0,\quad y(1) = 4$

6 $\epsilon y'' + (x+1)y' + y = 0,\quad y(0) = 2,\quad y(1) = 1/2$

7 $\epsilon y'' + y' + 2xy = 0,\quad y(0) = 1,\quad y(1) = 1$

8 $\epsilon y'' + y' - 2xy = 0,\quad y(0) = 2,\quad y(1) = 1$

For problems **9** through **14**: Determine the third order perturbation solution (up to ϵ^2) for the inner solution valid near $x = 0$ for the following perturbation problems. The answer will be expressed in terms of an arbitrary constant, which depends on ϵ. You can leave the answer as a function of t.

9 $\epsilon y'' + y' + y = 0,\quad y(0) = 1,\ y(1) = 2$
10 $\epsilon y'' + 2y' - 2y = 0,\quad y(0) = 2,\ y(1) = 1$
11 $\epsilon y'' + (x+1)y' - 2y = 0,\quad y(0) = 0,\ y(1) = 4$
12 $\epsilon y'' + (x+1)y' + y = 0,\quad y(0) = 2,\ y(1) = 1/2$
13 $\epsilon y'' + y' + 2xy = 0,\quad y(0) = 1,\ y(1) = 1$
14 $\epsilon y'' + y' - 2xy = 0,\quad y(0) = 2,\ y(1) = 1$

For problems **15** through **22**: The following equations are defined on the interval $0 \le x \le 1$. Determine the possible location for a boundary layer, find the singular distinguished limit, and determine the leading order solution for the boundary layer, in terms of two arbitrary constants.

15 $\epsilon y'' - 2y' + xy = 0$
16 $\epsilon y'' + 2xy' + xy = 0$
17 $\epsilon y'' + 3x^2y' + x^2y = 0$
18 $\epsilon y'' - 3x^2y' - 2x^2y = 0$
19 $\epsilon y'' + (x-1)y' + (x^2-1)y = 0$
20 $\epsilon y'' + (1-x)y' + (x-1)y = 0$
21 $\epsilon y'' + (x-x^2)y' + (x^2-x)y = 0$
22 $\epsilon y'' + (x^2-x)y' + (x^3-x^2)y = 0$

23 Use the result of proposition 7.1 to find the leading order composite approximation to the general perturbation problem

$$\epsilon y''(x) + p(x)y'(x) + q(x)y(x) = 0, \qquad y(x_1) = y_1, \quad y(x_2) = y_2$$

provided that $p(x)$ and $q(x)$ are continuous, and $p(x) > 0$ on the interval $x_1 \le x \le x_2$.

24 Use the result of proposition 7.1 to find the leading order composite approximation to the general perturbation problem

$$\epsilon y''(x) + p(x)y'(x) + q(x)y(x) = 0, \qquad y(x_1) = y_1, \quad y(x_2) = y_2$$

provided that $p(x)$ and $q(x)$ are continuous, and $p(x) < 0$ on the interval $x_1 \le x \le x_2$.

For problems **25** through **30**: Find the leading order composite approximation for the following perturbation problems.

25 $\epsilon y'' - 2y' + y = 0,\quad y(0) = 2,\ y(2) = 1$
26 $\epsilon y'' + (x+1)y' + y = 0,\quad y(0) = 4,\ y(2) = 1$
27 $\epsilon y'' - xy' + 3y = 0,\quad y(1) = 2,\ y(2) = 1$
28 $\epsilon y'' + 2xy' - xy = 0,\quad y(0) = 0,\ y(1) = 1$
29 $\epsilon y'' + (x-1)y' + (1-x^2)y = 0,\quad y(0) = 1,\ y(1) = 0$
30 $\epsilon y'' + 3x^2y' - 3x^2y = 0,\quad y(0) = 0,\ y(1) = 1$

7.4 Asymptotic Matching

In the last section, we saw a way to combine the inner and outer solutions into a composite approximation to the solution, where the boundary layer occurred near the point x_0. In fact, we were able to use the outer solution in order to find the arbitrary constant in the inner solution. The key to this method is that we set

$$\lim_{x \to x_0} y_{out}(x) = \lim_{t \to \infty} y_{in}(t). \tag{7.16}$$

This is at first baffling, since it says that the outer solution, when extended to values for which it is no longer valid, must equal the inner solution extended beyond the range of the inner solution. We will now justify this procedure, using the concept of *overlapping regions of validity*.

To introduce this concept, let us reexamine example 7.12. For the inner solution, we assumed that x was small, but just how small must x be for the inner solution $y_{in} \sim C_1 + C_2 e^{-2x/\epsilon}$ to be a valid approximation? In solving for the inner solution, we assumed that x was of order ϵ, so that $t = x/\epsilon$ was of order 1. But in problem 2 of the last section shows, we have the exact solutions to this equation as

$$e^{(-1+\sqrt{1-\epsilon})x/\epsilon} \quad \text{and} \quad e^{(-1-\sqrt{1-\epsilon})x/\epsilon}.$$

By forming a Taylor series of the square roots, we have

$$\frac{(-1+\sqrt{1-\epsilon})}{\epsilon} \sim -\frac{1}{2} - \frac{\epsilon}{8} + \cdots, \qquad \frac{(-1-\sqrt{1-\epsilon})}{\epsilon} \sim -\frac{2}{\epsilon} + \frac{1}{2} + \frac{\epsilon}{8} + \cdots.$$

Thus, the exact solutions can be expressed by the series

$$y = C_1 e^{-x/2 - \epsilon x/8 + \cdots} + C_2 e^{-2x/\epsilon + x/2 + \epsilon x/8 + \cdots}.$$

So the inner solution is in fact valid as long as $x \ll 1$, so that $e^{\pm x/2}$ will approach 1. For example, if we set $x = \sqrt{\epsilon}$, and then consider the limit as $\epsilon \to 0$, the inner solution will still be valid.

We can use the same argument for the outer solution, $y_{out} = Ce^{-x/2}$. As long as x/ϵ goes to infinity, the solution $e^{(-1-\sqrt{1-\epsilon})x/\epsilon}$ will approach 0, and the other solution $e^{(-1+\sqrt{1-\epsilon})x/\epsilon}$ will, by the above Taylor series, approach $e^{-x/2}$. Thus, the outer solution is valid as long as $x \gg \epsilon$. So if we set $x = \sqrt{\epsilon}$, and considered the limit as $\epsilon \to 0$, the outer solution will also be valid.

In general, we will always be able to extend the validity of the inner solution beyond the width of the boundary layer. This can be done via *extension theorems*, such as the following.

THEOREM 7.1

Suppose that, in terms of a local variable t, an inner solution approximation for a function $y_\epsilon(x)$ has the first few terms of the perturbation series given by

$$y_\epsilon(x) \sim Y_0(t) + \epsilon Y_1(t) + \epsilon^2 Y_2(t) + \cdots + \epsilon^{m-1} Y_{m-1}(t), \tag{7.17}$$

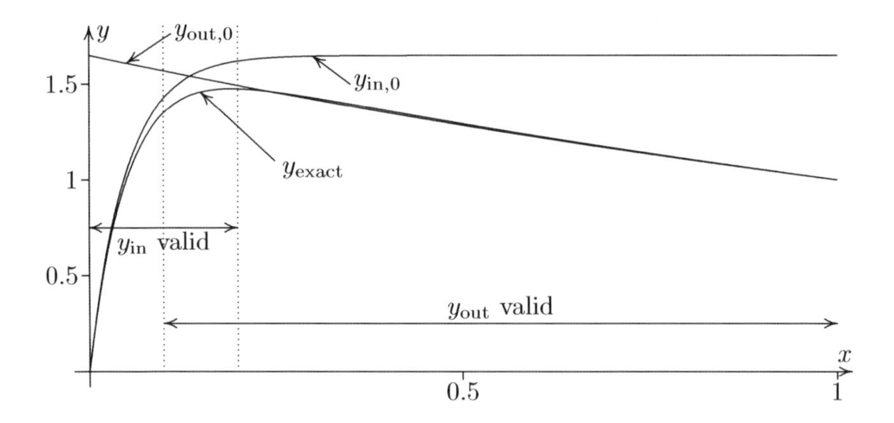

FIGURE 7.12: This shows the zeroth order inner and outer solution, along with the exact solution, for the equation $\epsilon y'' + 2y' + y = 0$, $y(0) = 1$, $y(1) = 1$ for $\epsilon = 0.1$. The outer solution is valid, up to order ϵ, when $x > \epsilon$. However, the inner solution is valid for larger values of x, even up to $\sqrt{\epsilon}$. As a result, there is an overlapping region for which *both* y_{in} and y_{out} are valid. We can use this overlapping region to determine the arbitrary constants of the inner solution.

which is valid for all $0 \le t \le A$, where A is an arbitrary positive number, independent of ϵ. Then there exists a function $\omega(\epsilon) \gg 1$ as $\epsilon \to 0$ such that the local expansion equation 7.17 remains valid in the extended domain

$$0 \le t \le A\omega(\epsilon).$$

See [13] for the proof. Essentially, this theorem states that a finite number of terms of the inner perturbation series will be valid *beyond* the width of the boundary layer. There are several comments that need to be made at this point. First of all, the function $\omega(\epsilon)$ depends on m, the number of terms kept in equation 7.17. Hence, the interval may not always be extended for the entire infinite sum, but only a truncation of the series. Nonetheless, we will apply this theorem on a term by term basis, so this result is sufficient for our purposes.

Also, we have seen many perturbation series for which the terms are not always integral powers of epsilon. There are some notoriously tricky examples in which terms involve

$$\epsilon^p \ln(1/\epsilon)^q, \qquad p \ge 0 \text{ and } q \in \mathbb{R}.$$

Theorem 7.1 can be extended to cover these cases. As long as we only consider a finite number of terms of the inner perturbation series, we can find a function $\omega(\epsilon) \gg 1$ such that truncated series is also valid for $0 \le t \le A\omega(\epsilon)$, where A is arbitrary.

There is also an extension theorem for the outer solution. The outer perturbation series is defined to be valid when x is fixed (not changing with ϵ). However, if we only consider a finite number of terms of the outer solution, we can find a function $\zeta(\epsilon)$, with $\zeta(0)$ being the location of the boundary layer, for which truncated series will be valid for $x \sim \zeta(\epsilon)$. In example 7.12, we could pick $\zeta(\epsilon) = \sqrt{\epsilon}$, or any other function for which $\zeta(0) = 0$ and $\zeta(\epsilon) \gg \epsilon$ as $\epsilon \to 0$.

The extension theorems are almost never used as an explicit analysis tool for a specific problem. The main use of these theorems is to explain why we usually can, by extending the region of validity of both the inner and outer solutions, find a region of overlap, as suggested in figure 7.12. We make the assumption that we can find a region for which both the inner and outer solutions are valid. This assumption is called the *overlap hypothesis*.

If this overlap exists, we can introduce an *intermediate variable* which will be of order 1 in a range that is within the overlapping region. In example 7.17, we found that when $x = O(\sqrt{\epsilon})$, then x was in the overlapping region between the inner and outer solutions. Thus, we can let $x = s\sqrt{\epsilon}$, and assume that s is of order 1. For the outer solution, we have

$$y_{\text{out}}(s) = Ce^{-s\sqrt{\epsilon}/2} \sim C \qquad \text{as } \epsilon \to 0$$

and for the inner solution, we have

$$y_{\text{in}}(s) = C_1 + C_2 e^{-2s/\sqrt{\epsilon}} \sim C_1 \qquad \text{as } \epsilon \to 0.$$

Because we are in the overlapping region, these two must agree, so $C = C_1$. We can now use the initial conditions to find C, C_1, and C_2.

This argument puts the method of asymptotic matching on a firm foundation. In fact, it allows us to match higher order terms of the outer and inner perturbation series, using a methodology developed by Kaplun [13]. We cannot determine in advance how the intermediate variable should depend on ϵ, so we will let $x = s\eta$, where $\eta = \epsilon^d$ for some d yet to be determined. We will demonstrate this methodology on the problem introduced in example 7.12.

Example 7.16

Use asymptotic matching on the results of example 7.12,

$$y_{\text{out}}(x) \sim e^{(1-x)/2} + \epsilon\frac{(1-x)e^{(1-x)/2}}{8} + \epsilon^2\frac{(x^2 - 10x + 9)e^{(1-x)/2}}{128} + \cdots$$

and

$$y_{\text{in}}(t) \sim K_\epsilon\left((1 - e^{-2t}) - t\frac{1 + e^{-2t}}{2}\epsilon + \frac{t^2 - t - t^2e^{-2t} - te^{-2t}}{8}\epsilon^2 + \cdots\right),$$

where $K_\epsilon = K_0 + K_1\epsilon + K_2\epsilon^2 + \cdots$, to determine the values of K_0, K_1, and K_2.

SOLUTION:
If we make the substitution $x = s\eta$ into the outer solution, and expand the Taylor series in powers of η, we obtain

$$y_{\text{out}}(s) \sim \sqrt{e}\Bigg(1 - \frac{s\eta}{2} + \frac{s^2\eta^2}{8} + \cdots + \frac{\epsilon}{8}(1 - s\eta)\left(1 - \frac{s\eta}{2} + \cdots\right) + \frac{\epsilon^2}{128}(9 - 10s\eta + s^2\eta^2)\left(1 - \frac{s\eta}{2} + \cdots\right) + \cdots\Bigg).$$

On the other hand, we can make the substitution $t = s\eta/\epsilon$ into the inner solution, we get

$$y_{\text{in}}(s) \sim (K_0 + K_1\epsilon + K_2\epsilon^2)\Bigg((1 - e^{-2s\eta/\epsilon}) - \frac{s\eta}{2}(1 + e^{-2s\eta/\epsilon}) + \frac{s^2\eta^2 - s\eta\epsilon - s^2\eta^2 e^{-2s\eta/\epsilon} - s\eta\epsilon e^{-2s\eta/\epsilon}}{8} + \cdots\Bigg).$$

If we assume that $\eta \gg \epsilon\ln(1/\epsilon)$, then $e^{-2s\eta/\epsilon} \ll \epsilon^N$ for all N, so this term would be subdominant in the series for y_{in}. Hence, we have

$$y_{\text{in}}(s) \sim (K_0 + K_1\epsilon + K_2\epsilon^2)\left(1 - \frac{s\eta}{2} + \frac{s^2\eta^2 - s\eta\epsilon}{8} + \cdots\right) \tag{7.18}$$
$$= K_0 - \frac{K_0 s\eta}{2} + K_1\epsilon + \frac{K_0 s^2\eta^2}{8} - \frac{(K_0 + 4K_1)s\eta\epsilon}{8} + K_2\epsilon^2 + \cdots.$$

Assuming the overlap hypothesis, there will be a range of orders of η for which $y_{\text{out}}(s)$ and $y_{\text{in}}(s)$ will be the same function. Hence, the terms of the form $C\eta^\alpha\epsilon^\beta$ for any given α and β must match. This gives us more than enough information to determine K_0, K_1, and K_2. Comparing the highest order terms, we see that $K_0 = \sqrt{e}$. This value of K_0 also causes the terms of order η and η^2 to match. The ϵ and $\eta\epsilon$ terms reveal that K_1 must be $\sqrt{e}/8$. Finally, we can deduce using the ϵ^2 terms that $K_2 = 9\sqrt{e}/128$. ☐

There are several observations to make from this example. First of all, the size of the overlapping region decreases as we increase the order of the match. To illustrate, we will use the notation $y_{\text{out},n}(x)$ and $y_{\text{in},n}(t)$ to denote the outer and inner solutions expanded to order ϵ^n in their respective variables. That is,

$$y_{\text{out},1}(x) = e^{(1-x)/2} + \epsilon\frac{(1-x)}{8}e^{(1-x)/2},$$
$$y_{\text{in},1}(t) = K_0(1 - e^{-2t}) - \epsilon K_0 t\frac{1 + e^{-2t}}{2} + \epsilon K_1(1 - e^{-2t}).$$

If we want $y_{\text{out},1} \sim y_{\text{in},1}$ after the substitutions $x = s\eta$ and $t = s\eta/\epsilon$, we find that $y_{\text{out},1}$ has terms of order η^2, but $y_{\text{out},1}$ does not. This means that $\eta^2 \ll \epsilon$, so we must have $\epsilon\ln(1/\epsilon) \ll \eta \ll \sqrt{\epsilon}$ in order for $y_{\text{out},1} \sim y_{\text{in},1}$. See problem 1.

Likewise, for the second order match, $y_{\text{out},2}$ will have terms of order η^3, but $y_{\text{in},2}$ will not. Yet we used the ϵ^2 term to determine K_2. So $\eta^3 \ll \epsilon^2$, so the match would only be valid if $\epsilon \ln(1/\epsilon) \ll \eta \ll \epsilon^{2/3}$. In fact, to get a match in order $y_{\text{out},n} \sim y_{\text{in},n}$, the range of the overlap will be $\epsilon \ln(1/\epsilon) \ll x \ll \epsilon^{n/(n+1)}$. Here we see the extension theorems in action. As long as we consider only a *finite* number of terms of the matching series, there will be a range for the order of η for which the series will match. There is no guarantee that there is a range that will work for all terms of the series at the same time, but we only deal with a finite number of terms at a time to determine the unknown constants in the inner solution.

We can now use the matching function to stitch together the inner and outer solutions, to form a *uniformly valid composite approximation*,

$$y_{\text{comp}}(x) = y_{\text{out}}(x) + y_{\text{in}}(t) - y_{\text{match}}(s).$$

Here, y_{match} represents the common expansion of y_{out} and y_{in} in the overlapping region.

We will let $y_{\text{comp},n}$ denote the nth order composite approximation. We have already seen how to compute the leading order of the composite approximation, $y_{\text{comp},0}(x)$, by using the leading orders of the inner and outer solutions. So $y_{\text{comp},0}(x) = e^{(1-x)/2} - e^{1/2}e^{-2x/\epsilon}$. To calculate $y_{\text{comp},n}$, we compute both the inner and outer solution, keeping terms which are no smaller than ϵ^n. For example, to compute $y_{\text{comp},1}$, we find the outer and inner solutions up to order ϵ, assuming for the outer solution that $x \gg \epsilon \ln(1/\epsilon)$, and assume that $x \ll \sqrt{\epsilon}$, or $t \ll \epsilon^{-1/2}$, for the inner solution. With these assumption, we find that

$$y_{\text{out},1}(x) = e^{(1-x)/2} + \epsilon\frac{(1-x)e^{(1-x)/2}}{8},$$

and

$$y_{\text{in},1}(x) = \sqrt{e}(1 - e^{-2x/\epsilon}) + \frac{\sqrt{e}}{8}\epsilon(1 - e^{-2x/\epsilon}) - \sqrt{e}x\frac{1 + e^{-2x/\epsilon}}{2}.$$

We let $y_{\text{match},1}$ be the expansion of either of these two series to order ϵ, assuming that x is in the overlapping region $\epsilon \ln(1/\epsilon) \ll x \ll \sqrt{\epsilon}$.

$$y_{\text{match},1}x = \sqrt{e}\left(1 - x/2 + \frac{\epsilon}{8}\right)$$

Putting the pieces together, we have

$$y_{\text{comp},1}(x) = e^{(1-x)/2} + \frac{(1-x)e^{(1-x)/2}}{8}\epsilon - e^{1/2}e^{-2x/\epsilon}\left(1 + \frac{x}{2} + \frac{\epsilon}{8}\right).$$

To determine how accurate this composite approximation is, observe that by the way we computed this composite function,

$$y_{\text{exact}} = y_{\text{out},1} + o(\epsilon) \quad \text{for} \quad x \gg \epsilon \ln(1/\epsilon),$$

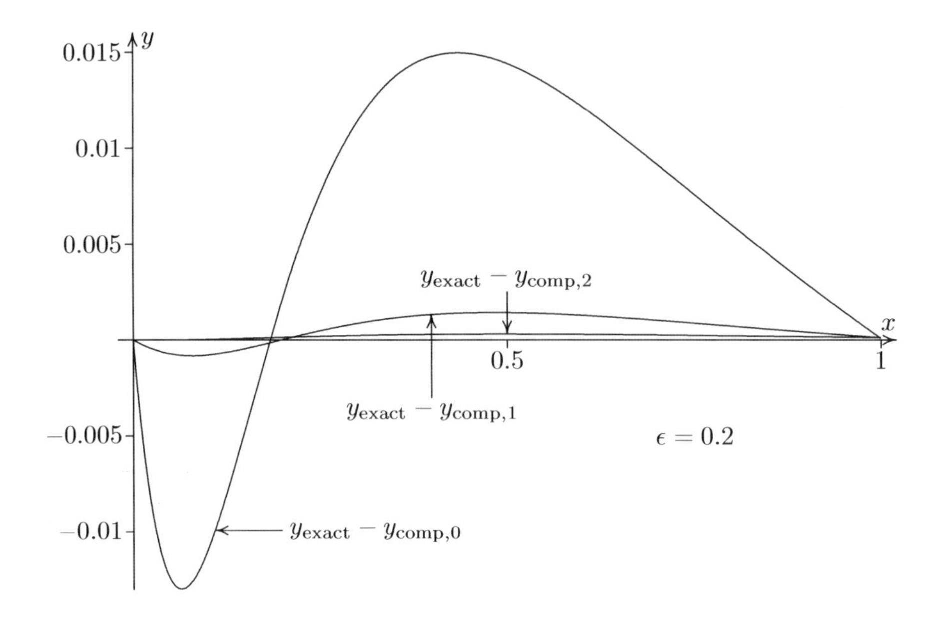

FIGURE 7.13: This shows the absolute error of the composite approximations compared to the true solution to the equation $\epsilon y'' + 2y' + y = 0$, with $y(0) = 0$ and $y(1) = 1$, using the value $\epsilon = 0.2$. For this value of ϵ, the composite approximation is clearly getting closer to the true solution as we increase the order. However, this is a singular perturbation problem, so we expect the perturbation series to be divergent. As a result, increasing the order sufficiently will eventually result in a decrease in precision.

$$y_{\text{exact}} = y_{\text{in},1} + o(\epsilon) \quad \text{for} \quad x \ll \sqrt{\epsilon},$$

and

$$y_{\text{exact}} = y_{\text{match},1} + o(\epsilon) \quad \text{for the overlap region.}$$

So we see that $y_{\text{exact}} = y_{\text{comp},1} + o(\epsilon)$ for all x in the interval. In general, we have $y_{\text{exact}} = y_{\text{comp},n} + o(\epsilon^n)$.

We can follow this same procedure to compute.

$$y_{\text{comp},2}(x) = e^{(1-x)/2}\left(1 + \frac{(1-x)}{8}\epsilon + \frac{(x^2 - 10x + 9)}{128}\epsilon^2\right) \qquad (7.19)$$
$$- e^{1/2}e^{-2x/\epsilon}\left(1 + \frac{\epsilon}{8} + \frac{9\epsilon^2}{128} + \frac{x}{2} + \frac{3x\epsilon}{16} + \frac{x^2}{8}\right).$$

Figures 7.13, 7.14, and 7.15 compare the uniform approximations to the exact solution for various ϵ. When $\epsilon = 0.2$, the composite approximation seems to be approaching the exact solution in figure 7.13. However, we are

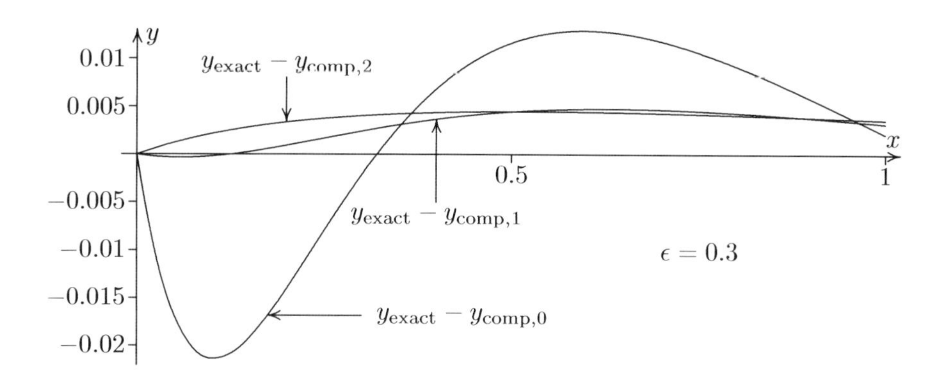

FIGURE 7.14: Same as figure 7.13, but with $\epsilon = 0.3$. Here, the first order and second order corrections give about the same amount of accuracy.

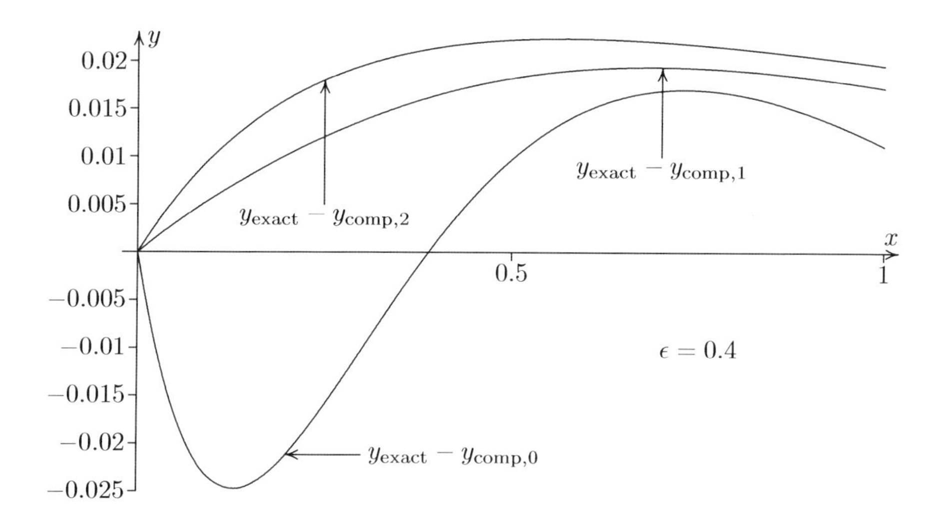

FIGURE 7.15: Same as figure 7.13, but with $\epsilon = 0.4$. For this value of ϵ, the approximations are actually getting worse as the order increases. This is because of the divergent nature of the perturbation series.

dealing with a singular perturbation problem, which will often produce divergent series for all ϵ. We saw in chapter 3 that when ϵ is sufficiently small, the series will begin to approach the correct function, but the higher orders will eventually pull away from the exact solution. We see this happening in figures 7.14 and 7.15.

When the boundary layer is not of width ϵ, it may take more terms in the inner solution to form a match.

Example 7.17

Find $y_{\text{comp},1}$ and $y_{\text{comp},2}$ for the perturbation problem

$$\epsilon y'' + x^2 y' - y = 0, \qquad y(0) = 2, \quad y(1) = 1.$$

SOLUTION: For the outer solution, we solve the unperturbed equation $x^2y' - y = 0$ to obtain $y = Ce^{-1/x}$. Note that as $x \to 0$, this solution will always approach 0. Thus, the outer solution cannot possibly satisfy the condition $y(0) = 2$, so there must be a boundary layer at $x = 0$. However, the other boundary condition, $y(1) = 1$, will be in the outer region. Applying this condition gives us $C = e$, so $y_0 = e^{1-1/x}$.

We can find the perturbation series for the outer solution by assuming it is of the form $y_{\text{out}} = y_0 + y_1\epsilon + y_2\epsilon^2 + \cdots$. The equations that must be solved are

$$\begin{aligned} x^2 y_0' - y_0 &= 0, & y(1) &= 1; \\ y_0'' + x^2 y_1' - y_1 &= 0, & y(1) &= 0; \\ y_1'' + x^2 y_2' - y_2 &= 0, & y(1) &= 0; \ldots. \end{aligned}$$

These equations turn out to be fairly routine to solve, producing

$$y_1 = e^{1-1/x}\frac{3x^5 - 5x + 2}{10x^5},$$

and

$$y_2 = e^{1-1/x}\frac{5617x^{10} - 1890x^6 + 756x^5 - 18000x^3 + 17325x^2 - 4060x + 252}{12600x^{10}}.$$

So the outer solution is

$$y_{\text{out}} \sim e^{1-1/x}\left(1 + \epsilon\frac{3x^5 - 5x + 2}{10x^5}\right.$$
$$\left. + \epsilon^2\frac{5617x^{10} - 1890x^6 + 756x^5 - 18000x^3 + 17325x^2 - 4060x + 252}{12600x^{10}} + \cdots\right).$$

For the inner solutions, we find the singular distinguished limit by substituting $x = \delta t$. In the new equation,

$$\frac{\epsilon Y''(t)}{\delta^2} + \delta t^2 Y'(t) - Y(t) = 0,$$

we find that $\delta t^2 Y'(t) \ll Y(t)$, so we must have $\delta = O(\sqrt{\epsilon})$ to have a balance. The new equation, $Y'' + \delta t^2 Y' - Y = 0$, has the unperturbed solution of $Y_0 = C_1 e^t + C_2 e^{-t}$. But if $C_1 \neq 0$, then there is no way that this could match the outer solution, since the function would blow up as $t \to \infty$. So $Y_0 = C_2 e^{-t}$, and since $Y(0) = 2$, we have $C_2 - 2$.

For higher order terms in the inner solution, we need to solve

$$\begin{aligned} y_0'' - y_0 &= 0, & y(0) &= 2; \\ y_1'' - y_1 &= -t^2 y_0', & y(0) &= 0; \\ y_2'' - y_2 &= -t^2 y_1', & y(0) &= 0; \ldots. \end{aligned}$$

These can be solved using variation of parameters (equation 4.10). In each case, we find that the solution involves two terms of the form $C_1 e^t + C_2 e^{-t}$. If $C_1 \neq 0$ for any of these, the inner solution would blow up as $t \to \infty$, so again we must have $C_1 = 0$. We can use the boundary condition at $Y(0)$ to determine C_2. This gives us the solutions

$$Y_1 = -te^{-t} \frac{2t^2 + 3t + 3}{6},$$

and

$$Y_2 = te^{-t} \frac{20t^5 + 24t^4 + 15t^3 - 30t^2 - 45t - 45}{720}.$$

Hence we have $y_{\text{in}} \sim$

$$e^{-t} \left(2 - \sqrt{\epsilon} \frac{2t^3 + 3t^2 + 3t}{6} + \epsilon \frac{20t^6 + 24t^5 + 15t^4 - 30t^3 - 45t^2 - 45t}{720} + \cdots \right).$$

We finally must substitute $t = x/\sqrt{\epsilon}$ to express this in terms of the original variable.

Notice that as $t \to \infty$, all of the terms in the inner solution are exponentially small. Also, all terms in the outer solution as $x \to 0$ become exponentially small. So surprisingly, there is no interaction between the inner and outer solutions. That is, $y_{\text{match},n} = 0$ for all n.

As we construct the uniformly valid composite approximation, we get another surprise. In order to have both the inner and outer solutions accurate to order ϵ, we need to include all three terms of the inner solution. Thus we have

$$\begin{aligned} y_{\text{comp},1} &= e^{1-1/x} \left(1 + \epsilon \frac{3x^5 - 5x + 2}{10x^5} \right) \\ &\quad + e^{-x/\sqrt{\epsilon}} \left(2 - \frac{2x^3 + 3x^2\sqrt{\epsilon} + 3x\epsilon}{6\epsilon} \right. \\ &\quad \left. + \frac{20x^6 + 24x^5\sqrt{\epsilon} + 15x^4\epsilon - 30x^3\epsilon^{3/2} - 45x^2\epsilon^2 - 45x\epsilon^{5/2}}{720\epsilon^2} \right). \end{aligned}$$

This means that to find $y_{\text{comp},2}$, we need two more terms of the inner series. Computing these, we get

$$
\begin{aligned}
Y_3 = \frac{te^{-t}}{25920} & (-40t^8 + 36t^7 + 414t^6 + 1719t^5 + 4995t^4 + 12285t^3 \\
& + 24300t^2 + 36450t + 36450) \quad \text{and} \\
Y_4 = \frac{te^{-t}}{43545600} & (2800t^{11} - 13440t^{10} - 84504t^9 - 312480t^8 - 847665t^7 \\
& - 1867860t^6 - 3059910t^5 - 3056130t^4 + 14175t^3 \\
& + 10234350t^2 + 15351525t + 15351525).
\end{aligned}
$$

Putting all of the pieces together, we have

$$
\begin{aligned}
y_{\text{comp},2} &= e^{1-1/x}\left(1 + \epsilon\frac{3x^5 - 5x + 2}{10x^5}\right. \\
&\quad \left. + \epsilon^2\frac{5617x^{10} - 1890x^6 + 756x^5 - 18000x^3 + 17325x^2 - 4060x + 252}{12600x^{10}}\right) \\
&\quad + e^{-x/\sqrt{\epsilon}}\left(2 - \frac{2x^3 + 3x^2\sqrt{\epsilon} + 3x\epsilon}{6\epsilon}\right. \\
&\quad + \frac{20x^6 + 24x^5\sqrt{\epsilon} + 15x^4\epsilon - 30x^3\epsilon^{3/2} - 45x^2\epsilon^2 - 45x\epsilon^{5/2}}{720\epsilon^2} \\
&\quad + \frac{-40x^9 + 36x^8\sqrt{\epsilon} + 414x^7\epsilon + 1719x^6\epsilon^{3/2} + 4995x^5\epsilon^2 + 12285x^4\epsilon^{5/2}}{25920\epsilon^3} \\
&\qquad + \frac{24300x^3\epsilon^3 + 36450x^2\epsilon^{7/2} + 36450x\epsilon^4}{25920\epsilon^3} \\
&\quad + \frac{2800x^{12} - 13440x^{11}\sqrt{\epsilon} - 84504x^{10}\epsilon - 312480x^9\epsilon^{3/2} - 847665x^8\epsilon^2}{43545600\epsilon^4} \\
&\qquad + \frac{-1867860x^7\epsilon^{5/2} - 3059910x^6\epsilon^3 - 3056130x^5\epsilon^{7/2} + 14175x^4\epsilon^4}{43545600\epsilon^4} \\
&\qquad + \frac{10234350x^3\epsilon^{9/2} + 15351525x^2\epsilon^5 + 15351525x\epsilon^{11/2}}{43545600\epsilon^4}.
\end{aligned}
$$

This example illustrates how the perturbation solutions can become very unwieldy even for relatively small orders. The graphs of the solutions are shown in figure 7.16. □

7.4.1 Van Dyke Method

Although Kaplun's method using an intermediate variable is rigorous, another method is popular because it eliminates the need for the intermediate variable. The Van Dyke method was originally criticized for being nonrigorous, but this is because Van Dyke's original presentation of the method

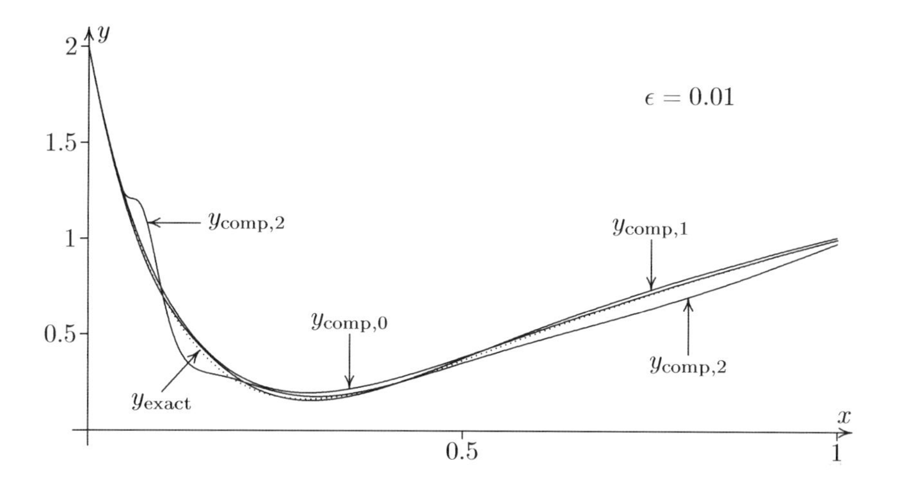

FIGURE 7.16: This compares the composite approximations to the equation $\epsilon y'' + x^2 y' - y = 0$, with $y(0) = 2$, $y(1) = 1$, and $\epsilon = 0.01$. The exact solution is shown with a dotted line. Note that although $y_{\text{comp},1}$ is a slightly better approximation than $y_{\text{comp},0}$, $y_{\text{comp},2}$ is actually much worse. It requires an even smaller value of ϵ before $y_{\text{comp},2}$ is more accurate than $y_{\text{comp},1}$.

TABLE 7.3: This table shows the maximum error, $\max_x\{|y_{\text{exact}}(x) - y_{\text{comp},n}x|\}$ for various values of ϵ. The higher level perturbation solutions only become more accurate for smaller values of ϵ.

ϵ	$n = 0$	$n = 1$	$n = 2$
0.1	0.176164	0.505133	23.1832
0.05	0.119226	0.223098	5.78034
0.02	0.0601087	0.069615	0.912867
0.01	0.0352904	0.0267276	0.222163
0.005	0.0355317	0.0101733	0.0526082
0.002	0.0297535	0.00290082	0.00735681
0.001	0.0232081	0.00118709	0.00156737
0.0005	0.0170863	0.000680302	0.000316288
0.0002	0.0109912	0.000188753	0.0000340831
0.0001	0.0078059	0.0000442819	0.00000596898

in [26] had a flaw. The corrected method is just as valid as the Kaplun's intermediate variable method. Unlike Kaplun's method, Van Dyke's method does not assume that there is an overlapping region between the inner and outer solution. Rather, Van Dyke's method starts by assuming the existence of a uniformly valid approximation $y_{\text{comp},n}$ which will match the inner and outer solution in their respective regions of validity, and which has the structure consistent with a perturbation series [7].

Van Dyke's method is based on the following simple rule:

$$\boxed{\begin{array}{c}\text{The inner expansion of } y_{\text{out},m} \text{ to order } \epsilon^n = \\ \text{the outer expansion of } y_{\text{in},n} \text{ to order } \epsilon^m.\end{array}} \tag{7.20}$$

This rule requires a bit of explanation. To find the inner expansion of an expression to order ϵ^n, we convert the expression to the inner variable by the substitution $x = t\delta$, where δ is the width of the boundary layer, and then re-expand this expression, throwing out terms that are smaller than order ϵ^n. Likewise, the outer expansion of an expression to order ϵ^n is found by substituting to the outer variable, re-expanding and keeping only terms of size ϵ^m or larger. To illustrate this idea, let us demonstrate using the same problem as example 7.16, using $m = n = 2$.

The two term outer expansion is

$$y_{\text{out},2}(x) = e^{(1-x)/2} + \epsilon\frac{(1-x)e^{(1-x)/2}}{8} + \epsilon^2\frac{(x^2-10x+9)e^{(1-x)/2}}{128}.$$

Convert to the inner variable by the substitution $x = \epsilon t$:

$$y_{\text{out},2}(t) = e^{(1-t\epsilon)/2} + \epsilon\frac{(1-t\epsilon)e^{(1-t\epsilon)/2}}{8} + \epsilon^2\frac{(t^2\epsilon^2-10t\epsilon+9)e^{(1-t\epsilon)/2}}{128}.$$

Re-expanding to order ϵ^2, and treating t to be of order 1, to give us

$$\sqrt{e}\left(1 - \frac{t\epsilon}{2} + \frac{t^2\epsilon^2}{8} + \frac{\epsilon}{8} - \frac{3\epsilon^2 t}{16} + \frac{9\epsilon^2}{128}\right). \tag{7.21}$$

Now we perform a similar operation to the inner solution. The two term inner expansion is

$$y_{\text{in},2}(t) = K_0(1-e^{-2t}) + \epsilon\left(K_1(1-e^{-2t}) - K_0\frac{t}{2}(1+e^{-2t})\right)$$
$$+ \epsilon^2\left(K_2(1-e^{-2t}) - K_1\frac{t}{2}(1+e^{-2t}) + K_0\frac{t^2-t-t^2e^{-2t}-te^{-2t}}{8}\right).$$

Convert to the outer variable by the substitution $t = x/\epsilon$:

$$y_{\text{in},2}(x) = K_0(1-e^{-2x/\epsilon}) + \epsilon\left(K_1(1-e^{-2x/\epsilon}) - K_0\frac{x}{2\epsilon}(1+e^{-2x/\epsilon})\right)$$

$$+ \epsilon^2 \left(K_2(1 - e^{-2x/\epsilon}) - K_1 \frac{x}{2\epsilon}(1 + e^{-2x/\epsilon}) + K_0 \frac{x^2 - x\epsilon - x^2 e^{-2e/\epsilon} - x\epsilon e^{-2x/\epsilon}}{8\epsilon^2} \right).$$

Since we now consider x to be of order 1, the terms involving $e^{-2x/\epsilon}$ will be subdominant. Re-expanding the remaining terms to order ϵ^2 produces

$$K_0 + \epsilon K_1 - \frac{K_0 x}{2} + \epsilon^2 K_2 - \epsilon \frac{K_1 x}{2} + K_0 \frac{x^2 - x\epsilon}{8}. \tag{7.22}$$

Van Dyke's matching principle (equation 7.20) now states that equation 7.21 is equal to equation 7.22. However, these equations are not in the same variable, so we have to replace $t \to x/\epsilon$ in equation 7.21 first. (We could also have replaced $x \to \epsilon t$ in equation 7.22.) So by Van Dyke's matching rule,

$$\sqrt{e}\left(1 - \frac{x}{2} + \frac{x^2}{8} + \frac{\epsilon}{8} - \frac{3x\epsilon}{16} + \frac{9\epsilon^2}{128}\right)$$
$$= K_0 + \epsilon K_1 - \frac{K_0 x}{2} + \epsilon^2 K_2 - \epsilon \frac{K_1 x}{2} + K_0 \frac{x^2 - x\epsilon}{8}.$$

This equation is satisfied provided that $K_0 = \sqrt{e}$, $K_1 = \sqrt{e}/8$, and $K_2 = 9\sqrt{e}/128$. Thus, we have found the asymptotic matching without introducing an intermediate variable.

Example 7.18

Use Van Dyke's matching principle to compute $y_{\text{comp},0}$, $y_{\text{comp},1}$, and $y_{\text{comp},2}$ to the problem

$$\epsilon y'' - (x+1)y' + 2y = 0, \qquad y(0) = 1, \quad y(1) = 0. \tag{7.23}$$

SOLUTION: We need to compute the inner and outer solutions to order ϵ^2. In this case, we can use proposition 7.1, since $p(x) = -(1+x) < 0$ on the interval $0 \le x \le 1$. Thus, there is a boundary layer of order ϵ near the endpoint $x = 1$. The unperturbed problem, $-(x+1)y' + 2y = 0$, has a general solution of $y_0 = C(x+1)^2$, and the initial condition $y(0) = 1$ determines that $C = 1$. Thus, we have $y_{\text{out},0} = (x+1)^2$. To find the outer perturbation series, we solve the series of equations

$$\begin{aligned} -(x+1)y_0' + 2y_0 &= 0, \quad y_0(0) = 1, \\ y_0'' - (x+1)y_1' + 2y_1 &= 0, \quad y_1(0) = 0, \\ y_1'' - (x+1)y_2' + 2y_2 &= 0, \quad y_2(0) = 0, \ldots . \end{aligned}$$

Since $y_0'' = 2$, we solve $-(x+1)y_1' + 2y_1 = -2$, which has a general solution of $C(x+1)^2 - 1$. The initial condition $y_1(0) = 0$ gives us $y_1 = x^2 + 2x$. In fact, $y_1'' = 2$ as well, so by the same logic, $y_2 = x^2 + 2x$. Because this pattern repeats, we can actually produce the full perturbation series for the outer solution.

$$y_{\text{out}} \sim (x+1)^2 + \epsilon(x^2+2x) + \epsilon^2(x^2+2x) + \epsilon^3(x^2+2x) + \cdots. \tag{7.24}$$

For the inner solution near $x = 1$, we make the substitution $x = 1 - \epsilon t$. This produces the new equation

$$Y''(t) + (2 - \epsilon t)Y'(t) + 2\epsilon Y(t) = 0.$$

This time, the unperturbed equation is $Y_0''(t) + 2Y_0'(t) = 0$, which has the solution $Y_0(t) = C_1 + C_2 e^{-2t}$. When $t = 0$, $x = 1$, so $Y(0) = 0$. Hence, $C_2 = -C_1$. To determine the value of C_1, we apply Van Dyke's method using $m = n = 0$. Thus, we substitute $x = 1 - \epsilon t$ to produce

$$y_{\text{out}}(t) \sim (2-\epsilon t)^2 + \epsilon(1-\epsilon t)(3-\epsilon) + \epsilon^2(1-\epsilon)(3-\epsilon) + \cdots.$$

If we re-expand this in powers of ϵ, we obtain

$$y_{\text{out}}(t) = 4 + \epsilon(3-4t) + \epsilon^2(3-4t+t^2) + \epsilon^3(3-4t+t^2) + \cdots. \tag{7.25}$$

Now we convert the inner solution that we have so far into the outer variable with the substitution $t = (1-x)/\epsilon$, we have

$$y_{\text{in}} \sim C_1(1 - e^{-2(1-x)/\epsilon}).$$

The leading order, C_1 must match the leading order in equation 7.25, so $C_1 = 4$. This produces

$$y_{\text{comp},0} = (x+1)^2 - 4e^{-2(1-x)/\epsilon}. \tag{7.26}$$

To find higher orders of the inner solution, we solve the system of equations

$$\begin{aligned} Y_1''(t) + 2Y_1'(t) &= tY_0'(t) - 2Y_0(t), \quad Y_1(0) = 0 \\ Y_2''(t) + 2Y_2'(t) &= tY_1'(t) - 2Y_1(t), \quad Y_2(0) = 0. \end{aligned}$$

These equations are difficult, but straightforward. We find that

$$Y_1(t) = C_3(1 - e^{-2t}) - 4t - (6t + 2t^2)e^{-2t}.$$

Hence,

$$y_{\text{in},1}(t) = 4(1 - e^{-2t}) + \epsilon(C_3(1-e^{-2t}) - 4t - (6t+2t^2)e^{-2t}).$$

We can substitute $t = (1-x)/\epsilon$, and eliminate all terms containing $e^{2(1-x)\epsilon}$, since these terms would be subdominant.

$$y_{\text{in},1}(x) \sim 4 + \epsilon\left(C_3 - 4\frac{1-x}{\epsilon}\right).$$

No terms in this expression are smaller than $O(\epsilon)$, so when we expand to this order, and substitute back in terms of the variable t, this becomes

$$y_{\text{in},1}(t) = 4 + \epsilon(C_3 - 4t).$$

Equating this to the first two terms of equation 7.25, we see that $C_3 = 3$. Thus, $Y_1 = 3 - 4t - (3 + 6t + 2t^2)e^{-2t}$, and

$$\begin{aligned} y_{\text{comp},1} &= (x+1)^2 + \epsilon(x^2 + 2x) \\ &\quad - e^{-2(1-x)/\epsilon}\left(4 + \epsilon\left(3 + 6\frac{1-x}{\epsilon} + 2\frac{(1-x)^2}{\epsilon^2}\right)\right) \\ &= (x+1)^2 + \epsilon(x^2+2x) - e^{-2(1-x)/\epsilon}\left(10 - 6x + 3\epsilon + 2\frac{(1-x)^2}{\epsilon}\right). \end{aligned}$$

Finally, we can solve for $Y_2(t)$ to obtain

$$Y_2(t) = C_4(1 - e^{-2t}) + t^2 - 4t - e^{-2t}\frac{21t + 15t^2 + 6t^3 + t^4}{2}.$$

When we convert to the outer variable, we can eliminate all terms containing e^{-2t}, since these will become subdominant. So

$$y_{\text{in},2}(x) \sim 4 + \epsilon\left(3 - 4\frac{1-x}{\epsilon}\right) + \epsilon^2\left(C_4 + \frac{(1-x)^2}{\epsilon^2} - 4\frac{1-x}{\epsilon}\right).$$

When we re-expand in powers of ϵ, no terms will be of order less than ϵ^2, so again, no terms are removed. Substituting back in terms of the variable t, we have

$$y_{\text{in},2}(t) = 4 + \epsilon(3 - 4t) + \epsilon^2(C_4 + t^2 - 4t).$$

Equating this with the first two terms of equation 7.25, we see that $C_4 = 3$. So

$$Y_2(t) = 3 - 4t + t^2 - e^{-2t}\frac{6 + 21t + 15t^2 + 6t^3 + t^4}{2}.$$

This gives us the final result

$$\begin{aligned} y_{\text{comp},2} &= (x+1)^2 + \epsilon(x^2+2x) + \epsilon^2(x^2+2x) \\ &\quad - e^{-2(1-x)/\epsilon}\left(4 + \epsilon(3 + 6t + 2t^2) + \epsilon^2\frac{6 + 21t + 15t^2 + 6t^3 + t^4}{2}\right) \\ &= 1 + (1 + \epsilon + \epsilon^2)(x^2 + 2x) - \frac{e^{-2(1-x)/\epsilon}}{2\epsilon^2}((1-x)^4 \\ &\quad + \epsilon(10 - 6x)(1-x)^2 + 35\epsilon^2 + 27\epsilon^3 + 6\epsilon^4 - 42\epsilon^2 x - 21\epsilon^3 x + 15\epsilon^2 x^2). \end{aligned}$$

$\square$

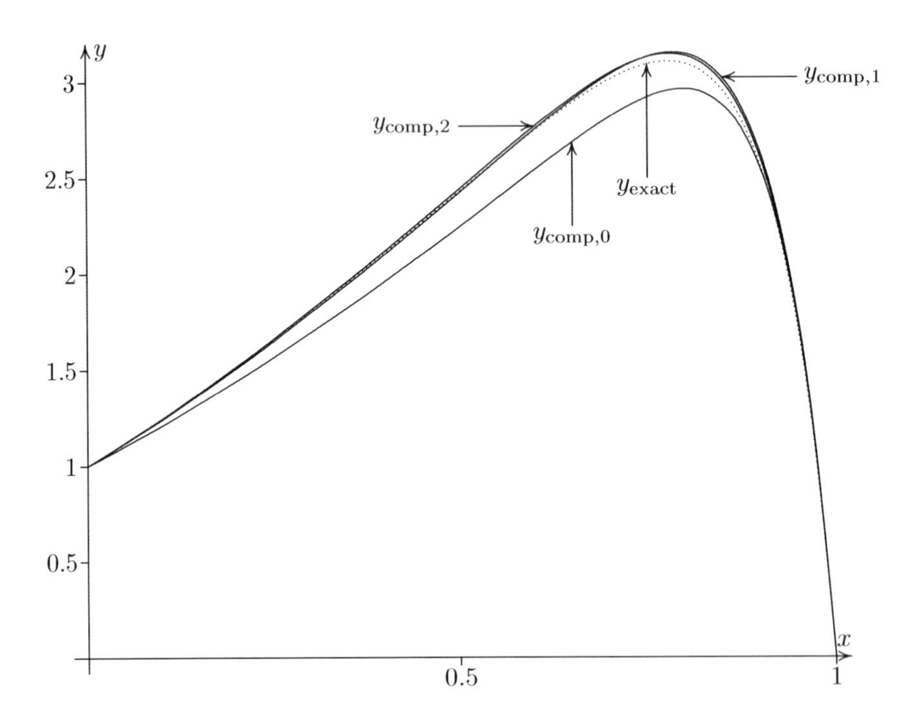

FIGURE 7.17: This compares the composite approximations to the equation $\epsilon y'' - (x+1)y' + 2y = 0$, with $y(0) = 1$, $y(1) = 0$, and $\epsilon = 0.15$. The exact solution is shown with a dotted line. Even for this fairly large value for ϵ, the first and second order composite approximations are barely distinguishable from the exact solution.

If we define $(f(x))_{\text{out},m}$ as the expansion of $f(x)$ converted to the outer variable, kept to order ϵ^m, and $(f(x))_{\text{in},n}$ as the inner expansion of $f(x)$ to order ϵ^n, we can express Van Dyke's rule very succinctly by the equation

$$\boxed{(y_{\text{out},m})_{\text{in},n} = (y_{\text{in},n})_{\text{out},m}.} \tag{7.27}$$

We can also use this notation to define the uniformly valid composite approximation using different orders for the inner and outer solutions. For $m \geq 0$ and $n \geq 0$, we can define

$$\begin{aligned} y_{\text{comp},m,n} &= y_{\text{out},m} + y_{\text{in},n} - (y_{\text{out},m})_{\text{in},n} \\ &= y_{\text{out},m} + y_{\text{in},n} - (y_{\text{in},n})_{\text{out},m}. \end{aligned}$$

Note that when $m = n$, this reproduces the uniform approximation $y_{\text{comp},m}$. However, this formula allows us to consider using a different order of approximation for the inner and outer solutions. This is particularly useful if the terms of the perturbation series for the outer solutions is much easier to find than for the inner solution, or vice versa.

Example 7.19

The results of example 7.18 to find $y_{\text{comp},4,2}$ for the equation

$$\epsilon y'' - (x+1)y' + 2y = 0, \qquad y(0) = 1, \quad y(1) = 0.$$

SOLUTION: Here is an example where the terms of the outer perturbation series were much easier to find than the inner. We already have the outer solution

$$y_{\text{out}} = (1+x)^2 + \epsilon(x^2+2x) + \epsilon^2(x^2+2x) + \epsilon^3(x^2+2x) + \epsilon^4(x^2+2x) + \cdots.$$

For the inner solution, we have

$$\begin{aligned} Y_0 &= 4(1 - e^{-2t}), \qquad Y_1 = (3 - 4t) - (3 + 6t + 2t^2)e^{-2t}, \\ Y_2(t) &= 3 - 4t + t^2 - e^{-2t}\frac{6 + 21t + 15t^2 + 6t^3 + t^4}{2}. \end{aligned}$$

We have already computed the constants required for the matching, so we only have to compute *either* $(y_{\text{out},4})_{\text{in},2}$ or $(y_{\text{in},2})_{\text{out},4}$. The easiest is $(y_{\text{out},4})_{\text{in},2}$. We substitute $x = 1 - \epsilon t$ into

$$y_{\text{out},4}(x) = (1+x)^2 + (\epsilon + \epsilon^2 + \epsilon^3 + \epsilon^4)(x^2 + 2x)$$

to produce

$$y_{\text{out},4}(t) = (2 - \epsilon t)^2 + (\epsilon + \epsilon^2 + \epsilon^3 + \epsilon^4)(3 - 4\epsilon t + \epsilon^2 t^2).$$

Expanding this, and keeping only terms of order up to ϵ^2 yields

$$(y_{\text{out},4})_{\text{in},2} = 4 + \epsilon(3 - 4t) + \epsilon^2(3 - 4t + t^2).$$

So we have $y_{\text{comp},4,2} = y_{\text{out},4} + y_{\text{in},2} - (y_{\text{out},4})_{\text{in},2} =$

$$(x+1)^2 + \epsilon(x^2+2x) + \epsilon^2(x^2+2x) + \epsilon^3(x^2+2x) + \epsilon^4(x^2+2x)$$
$$- \frac{e^{-2(1-x)/\epsilon}}{2\epsilon^2}\Bigg((1-x)^4 + \epsilon(10-6x)(1-x)^2 + 35\epsilon^2 + 27\epsilon^3$$
$$+6\epsilon^4 - 42\epsilon^2 x - 21\epsilon^3 x + 15\epsilon^2 x^2\Bigg).$$

The graph of this function is virtually indistinguishable from $y_{\text{comp},2}$ in figure 7.17. So it turns out not to be a significant advantage to include more terms in the outer solution than in the inner solution. □

Example 7.20

Consider the perturbation problem introduced in example 7.13,

$$\epsilon y'' + 2xy' + xy = 0, \qquad y(0) = 0, \quad y(1) = e^{-1/2}.$$

Recall that the boundary layer near $x = 0$ had thickness of order $\sqrt{\epsilon}$. Hence, we expect the inner perturbation series to involve powers of $\epsilon^{1/2}$, as in example 7.17. Determine the composite approximation $y_{\text{comp},1,1/2}$.

SOLUTION: Note that we will need two terms of both the inner and outer perturbation series. For the outer solution, we solve the unperturbed problem

$$2xy_0' + xy_0 = 0, \qquad y(1) = e^{-1/2},$$

to obtain $y_0 = e^{-x/2}$. For the next term of the perturbation series, we solve

$$y_0'' + 2xy_1' + xy_1 = 0, \qquad y(1) = 0.$$

Since $y_0'' = e^{-x/2}/4$, we can solve this to obtain

$$y_1 = -\frac{e^{-x/2}\ln(x)}{8}.$$

Thus, we have the first two terms of the outer solution being

$$y_{\text{out},1}(x) = e^{-x/2} - \epsilon\frac{e^{-x/2}\ln(x)}{8}.$$

For the inner solution, we substitute $x = \sqrt{\epsilon}t$ into the original equation to produce

$$Y'' + 2tY' + \sqrt{\epsilon}tY = 0. \tag{7.28}$$

We will assume that this has a perturbation series of the form

$$y_{\text{in}}(t) = Y_0(t) + \epsilon Y_1(t) + \epsilon^2 Y_2(t) + \cdots.$$

As we saw in example 7.13, the solution to the unperturbed problem which satisfies $Y(0) = 0$, and which matches the zeroth order outer solution, is

$$Y_0(t) = \text{erf}(t).$$

In order to find the next term in the inner series, we must solve

$$Y_1'' + 2tY_1' + tY_0 = 0, \qquad Y(0) = 0.$$

If we let $s = Y_1'$, then v satisfies the first order equation

$$s' + 2ts = -t\,\text{erf}(t).$$

This can be solved using equation 4.22, which gives us

$$s = e^{-t^2}\left(-\int te^{t^2}\text{erf}(t)\,dt + C_1\right).$$

Although this looks like an impossible integral, it actually is quite easy using integrating by parts. Let $u = \text{erf}(t)$, so that $du = 2e^{-t^2}/\sqrt{\pi}\,dt$, and $dv = te^{t^2}\,dt$, which we can integrate to produce $v = e^{t^2}/2$. Since vdu will be a constant, the new integral will be trivial, so we have

$$s = \frac{te^{-t^2}}{\sqrt{\pi}} - \frac{\text{erf}t}{2} + C_1 e^{-t^2}.$$

Yet we have to integrate this one more time to produce Y_1. The only tricky term is integrating $\text{erf}(t)$, but this can also be done via integrating by parts.

$$\int \text{erf}(t)\,dt = t\,\text{erf}(t) - \int \frac{2}{\sqrt{\pi}}te^{-t^2}\,dt = t\,\text{erf}(t) + \frac{1}{\sqrt{\pi}}e^{-t^2} + C.$$

Thus, we have

$$Y_1 = -\frac{e^{-t^2}}{2\sqrt{\pi}} - \frac{t\,\text{erf}(t)}{2} - \frac{e^{-t^2}}{2\sqrt{\pi}} + C_1\frac{\sqrt{\pi}\,\text{erf}(t)}{2} + C_2.$$

The initial condition $Y_1(0) = 0$ indicates that $C_2 = 1/\sqrt{\pi}$. So we have

$$y_{\text{in},1/2}(t) = \text{erf}(t) + \sqrt{\epsilon}\left(\frac{1 - e^{-t^2}}{\sqrt{\pi}} - \frac{t\,\text{erf}(t)}{2} + C_1\frac{\sqrt{\pi}\,\text{erf}(t)}{2}\right).$$

In order to determine C_1, we must match $(y_{\text{out},1})_{\text{in},1/2}$ with $(y_{\text{in},1/2})_{\text{out},1}$. We begin by replacing $x = t\sqrt{\epsilon}$ into the outer solution, to produce

$$y_{\text{out},1}(t) = e^{-t\sqrt{\epsilon}/2} - \epsilon\frac{e^{-t\sqrt{\epsilon}/2}\ln(t\sqrt{\epsilon})}{8}.$$

If we expand this as an asymptotic series in ϵ, we get

$$1 - \sqrt{\epsilon}\frac{t}{2} - \frac{\epsilon \ln(\epsilon)}{16} + \epsilon\frac{t^2 - \ln(t)}{8} + \cdots. \tag{7.29}$$

Since we only need to keep the terms up to order $\sqrt{\epsilon}$, we have

$$(y_{\text{out},1})_{\text{in},1/2}(t) = 1 - \sqrt{\epsilon}\frac{t}{2}. \tag{7.30}$$

Next, we replace $t = x/\sqrt{\epsilon}$ in the inner solution, to obtain $y_{\text{in},1/2}(x) =$

$$\text{erf}(x/\sqrt{\epsilon}) + \sqrt{\epsilon}\left(\frac{1 - e^{-x^2/\epsilon}}{\sqrt{\pi}} - \frac{x}{2\sqrt{\epsilon}}\text{erf}(x/\sqrt{\epsilon}) + C_1\frac{\sqrt{\pi}}{2}\text{erf}(x/\sqrt{\epsilon})\right).$$

We now keep only terms of order ϵ or larger. As complicated as this looks, it is actually quite easy, since from equation 2.8, $\text{erf}(x/\sqrt{\epsilon}) \sim 1+$ subdominant terms. Thus, we have

$$(y_{\text{in},1/2})_{\text{out},1}(x) = 1 - \frac{x}{2} + \sqrt{\epsilon}\left(\frac{1}{\sqrt{\pi}} + C_1\frac{\sqrt{\pi}}{2}\right). \tag{7.31}$$

In order to get equations 7.30 and 7.31 to match, we need to have $C_1 = -2/\pi$. We can then assemble the pieces to produce

$$\begin{aligned} y_{\text{comp},1,1/2} &= y_{\text{out},1} + y_{\text{in},1/2} - (y_{\text{in},1/2})_{\text{out},1} \\ &= e^{-x/2}\left(1 - \frac{\epsilon \ln x}{8}\right) + \left(1 - \frac{x}{2}\right)(\text{erf}(x/\sqrt{\epsilon}) - 1) \\ &\quad + \frac{\sqrt{\epsilon}}{\sqrt{\pi}}(1 - e^{-x^2/\epsilon} - \text{erf}(x/\sqrt{\epsilon})). \end{aligned} \tag{7.32}$$

The graphs of the composite approximation for various ϵ is shown in figure 7.18. □

The graphs of the composite approximations in figure 7.18 reveal something disconcerting. The function $y_{\text{comp},1,1/2}$ is undefined at $x = 0$, caused by the logarithm term in the outer solution, even though the exact solution is continuous at 0. The problem is not in the Van Dyke's method of matching the inner and outer solution. In fact, in the overlapping region, $\epsilon \ln x = O(\epsilon \ln(\epsilon))$, which is smaller than order $\sqrt{\epsilon}$. The term $\epsilon \ln(x)$ only becomes than larger than $\sqrt{\epsilon}$ if $x \ll e^{-1/\sqrt{\epsilon}}$, which goes to zero extremely rapidly as $\epsilon \to 0$. When $\epsilon = 0.01$, the graphs of the exact solution and the composite are indistinguishable.

The appearance of the asymptote lies in the fact that equation 7.32 is not a *uniform* approximation, in that we are not expanding the inner and outer solution to the same order of epsilon. If we keep both inner and outer to order ϵ, then the $\epsilon \ln(x)$ behavior should appear in the matching region, and hence this behavior would cancel out as we form the composite approximation. We will explore this further in the next subsection, but first we must understand how to handle different types of terms in the perturbation series.

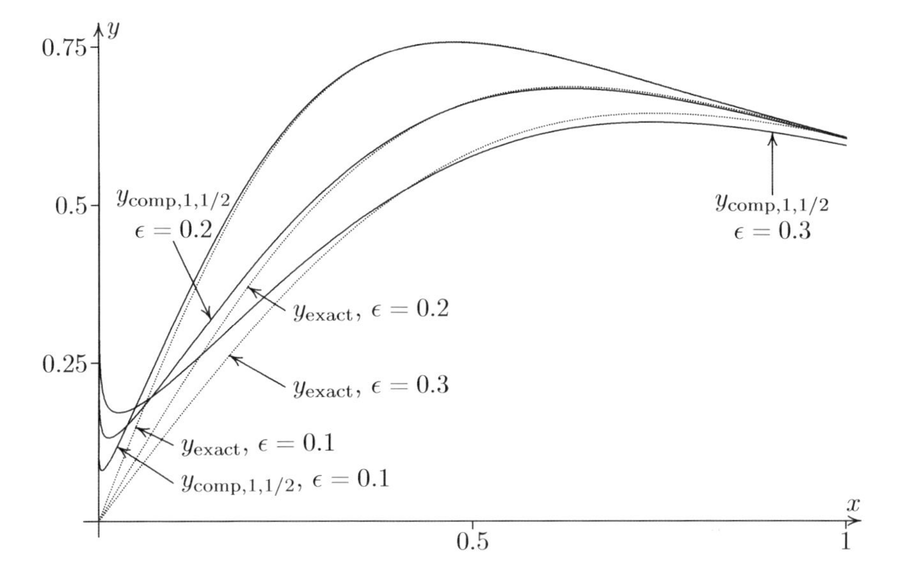

FIGURE 7.18: This compares the uneven composite approximations $y_{\text{comp},1,1/2}$ to the equation $\epsilon y'' + 2xy' + xy = 0$, with $y(0) = 0$ and $y(1) = e^{-1/2}$. The exact solutions are shown with a dotted line. Comparing these to the zeroth order uniform approximations, shown in figure 7.9, we see that the approximations are much closer near $x = 1$, especially for small ϵ. But the approximations have an asymptote at $x = 0$.

7.4.2 Dealing with Logarithmic Terms

Whenever we have logarithmic terms in the perturbation series, such as $\epsilon \ln(\epsilon)$ terms, we must modify the rules for Van Dyke's method slightly. When we refer to the expansion of the inner or outer solution up to order ϵ^n, we must also include terms of order $\epsilon^n \ln(\epsilon)^q$ for any q, even if q is negative. This assumes that there are only a finite number of such terms for each n. If one of the series has an infinite number of terms whose ratios are all of the order of a logarithm, as in equation 1.40, then Van Dyke's method cannot be applied. However, such series are very atypical, and usually the series will have only a finite number of logarithmic terms for each value of n. All such terms must be grouped together in expanding the outer or inner solution to order ϵ^n. Van Dyke phrased this concept by his famous warning, "Don't cut between logarithms."

Example 7.21
The non-linear initial value problem

$$y'' = \frac{\epsilon y y'}{x^2}, \qquad y(0) = 1, \quad y(1) = 0$$

is related to the problem for heat conduction outside a sphere with a small non-linear heat source (see problem 21). Determine the first order composite approximation, $y_{\text{comp},1}$, for this equation.
SOLUTION: This is a non-linear singular perturbation problem, which we have not dealt with before. However, such problems are solved in the same fashion as linear equations. If we assume that $y = y_0 + \epsilon y_1 + \epsilon^2 y_2 + \cdots$, then

$$yy' = y_0 y_0' + \epsilon(y_0 y_1' + y_1 y_0') + \epsilon^2 (y_0 y_2' + y_1 y_1' + y_2 y_0') + \cdots.$$

The ϵ is not on the highest derivative, so at first this looks like a regular perturbation problem, except that there will apparently be a problem near $x = 0$. If we try to proceed as if this were a regular perturbation problem, we would have the sequence of equations

$$\begin{aligned} y_0'' &= 0 \qquad y_0(0) = 1, \quad y_0(1) = 0; \\ y_1'' &= \frac{y_0 y_0'}{x^2}, \qquad y_1(0) = 0, \quad y_1(1) = 0; \\ y_2'' &= \frac{y_0 y_1' + y_1 y_0'}{x^2}, \qquad y_2(0) = 0, \quad y_2(1) = 0; \ldots . \end{aligned}$$

The first equation can be solved trivially to produce $y_0 = 1 - x$, but the general solution to the equation for y_1 is $x \ln(x) + \ln(x) + C_1 x + C_2$. This expression is undefined at $x = 0$, so we cannot impose both initial conditions. Hence, there must be a boundary layer near $x = 0$. We can, however, impose the condition $y_1(1) = 0$ to find that $C_2 = -C_1$.

For the inner solution near 0, we substitute $x = t\delta$ to find the singular distinguished limit. The new equation is

$$\frac{y''(t)}{\delta^2} = \frac{\epsilon y(t)y'(t)}{\delta^3 t^2}.$$

For these two terms to be of the same order, we have $\delta = O(\epsilon)$. However, if we let $\delta = \epsilon$, the equation becomes

$$Y''(t) = \frac{Y(t)Y'(t)}{t^2}.$$

There is no longer a small parameter in this new equation, so the inner equation is as formidable as the original equation. However, we know that the leading term of the outer solution was able to satisfy the inner boundary condition, so we know that the inner solution will have the series

$$y_{\text{in}} \sim 1 + \epsilon Y_1(t) + \epsilon^2 Y_2(t) + \cdots.$$

Thus, Y_1 will satisfy $Y_1'' = Y_1'/x^2$. This is easy to solve for Y_1', giving $Y_1' = C_3 e^{-1/t}$. Hence, we can integrate this to produce

$$Y_1 = C_3 \int_0^t e^{-1/u}\, du + C_4.$$

Note that this improper integral converges, since $e^{-1/u}$ approaches 0 as $u \to 0$. We can also impose the condition $Y_1(0) = 0$, which forces $C_4 = 0$. We can also make the substitution $s = 1/u$ to convert this integral into the form

$$Y_1 = C_3 \int_{1/t}^{\infty} \frac{e^{-s}}{s^2}\, ds.$$

Integration by parts, we can simplify this to

$$Y_1 = C_3 \left(te^{-1/t} - \int_{1/t}^{\infty} \frac{e^{-s}}{s}\, ds \right).$$

This integral we have seen before. Using the results of example 2.4, we have that

$$\int_x^{\infty} \frac{e^{-s}}{s}\, ds \sim -\ln x - \gamma + x - x^2/4 + x^3/18 + \cdots \qquad \text{as } x \to 0. \tag{7.33}$$

Thus, if we start with the inner solution expanded to order ϵ,

$$y_{\text{in},1}(t) = 1 + C_3 \epsilon t e^{-1/t} - C_3 \epsilon \int_{1/t}^{\infty} \frac{e^{-s}}{s}\, ds,$$

and in turn convert to the variable $x = t/\epsilon$, we obtain

$$y_{\text{in},1}(x) = 1 + C_3 x e^{-\epsilon/x} - C_3\epsilon \int_{\epsilon/x}^{\infty} \frac{e^{-s}}{s}\, ds.$$

Expanding this for small ϵ, we have

$$y_{\text{in},1}(x) \sim 1 + C_3 x \left(1 - \frac{\epsilon}{x} + \frac{\epsilon^2}{2x^2} + \cdots\right) - C_3\epsilon\left(-\ln(\epsilon/x) - \gamma + \frac{\epsilon}{x} + \cdots\right).$$

Keeping only terms of order ϵ we have

$$(y_{\text{in},1})_{\text{out},1}(x) = 1 + C_3 x - C_3\epsilon + C_3\epsilon\ln(\epsilon) - C_3\epsilon\ln(x) + C_3\epsilon\gamma. \tag{7.34}$$

We need to compare this to $(y_{\text{out},1})_{\text{in},1}$, so we form the outer solution up to order ϵ,

$$y_{\text{out},1}(x) = 1 - x + \epsilon\left(x\ln(x) + \ln(x) + C_1 x - C_1\right),$$

and replace x with ϵt, keeping only terms of order ϵ.

$$(y_{\text{out},1}(x))_{\text{in},1}(t) = 1 - t\epsilon + \epsilon\ln(t) + \epsilon\ln(\epsilon) - C_1\epsilon.$$

Converting back to the x variable, we get

$$(y_{\text{out},1}(x))_{\text{in},1}(x) = 1 - x + \epsilon\ln(x) - C_1\epsilon. \tag{7.35}$$

Unfortunately, no set of constants allows this equation to agree with equation 7.34, because C_3 must be -1 for the x terms to agree, only then there is a $\epsilon\ln(\epsilon)$ term in equation 7.34 that does not appear in $(y_{\text{out},1}(x))_{\text{in},1}(x)$. This indicates a presence of an additional $\epsilon\ln(\epsilon)$ term in one of the two series, but which one? If we added such a term to the outer solution,

$$y_{\text{out}}(x) \sim y_0 + \epsilon\ln\epsilon\,\overline{y}_1 + \epsilon y_1 + \cdots,$$

then y_0 and y_1 would solve the same equations as before, but $\overline{y}_1$ would satisfy $\overline{y}_1'' = 0$, $\overline{y}(1) = 0$. So $\overline{y}_1 = C_5(1-x)$, which would contribute an extra $C_5\epsilon\ln(\epsilon)$ term to equation 7.34. This would be sufficient to match equations 7.34 and 7.35, using $C_3 = -1$, $C_5 = -1$, and $C_1 = \gamma - 1$. The uniform composite approximation is given by

$$\begin{aligned}
y_{\text{comp},1} &= y_{\text{out},1} + y_{\text{in},1} - (y_{\text{in},1})_{\text{out},1}(x)\\
&= 1 - x + \epsilon\ln(\epsilon)(x-1) + \epsilon(x\ln(x) + \ln(x) + (\gamma-1)(x-1))\\
&\quad +1 - xe^{-\epsilon/x} + \epsilon\int_{\epsilon/x}^{\infty}\frac{e^{-s}}{s}\,ds\\
&\quad -(1 - x + \epsilon(1-\gamma) - \epsilon\ln(\epsilon) + \epsilon\ln(x))\\
&= 1 + x(\epsilon\ln(\epsilon) + \epsilon\ln x + \epsilon\gamma - \epsilon) - xe^{-\epsilon/x} + \epsilon\int_{\epsilon/x}^{\infty}\frac{e^{-s}}{s}\,ds.
\end{aligned}$$

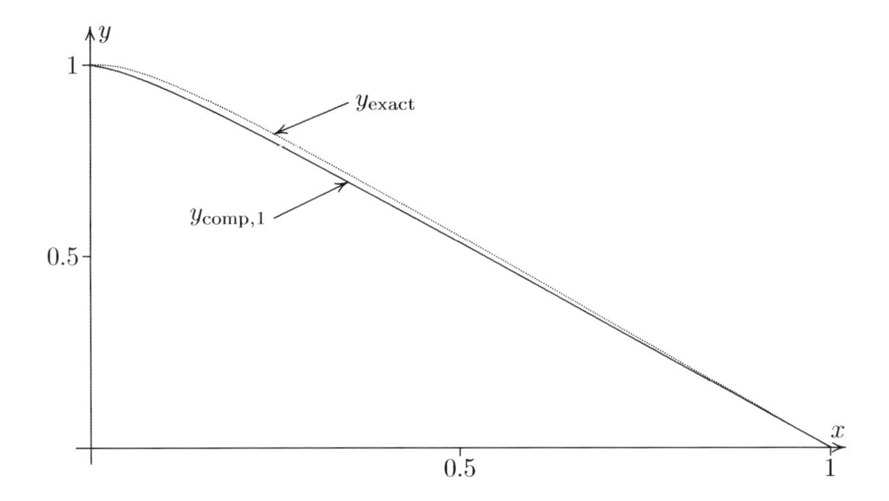

FIGURE 7.19: This compares the uniform composite approximations $y_{\text{comp},1}$ to the equation $y'' = \epsilon y y'/x^2$, with $y(0) = 1$ and $y(1) = 0$, using $\epsilon = 0.05$.

The graph for $\epsilon = 0.05$ is shown in figure 7.19, along with the exact solution. For smaller ϵ, the two graphs would be nearly indistinguishable.

Had we added the $\epsilon \ln(\epsilon)$ to the inner solution,

$$y_{\text{in}} \sim Y_0 + \epsilon \ln(\epsilon)\overline{Y}_1 + \epsilon Y_1 + \cdots,$$

then $\overline{Y}_1$ would satisfy the same equation that Y_1 satisfied. This would not have helped the situation, since it would introduce $\epsilon(\ln(\epsilon))^2$ terms. ▯

This example shows that logarithmic terms can appear spontaneously in a perturbation series, especially for non-linear equations. Their presence is revealed only as we asymptotically match the inner and outer solutions. A term that is forced by the asymptotic matching is sometimes referred to as a *switchback*. Switchbacks are often the result of one of the series involving a complicated integral, for which the series as $x \to 0$ is a standard Taylor series, but the series as $x \to \infty$ involves a logarithm. The strategy is to begin by assuming that the perturbation series is standard, and consider adjusting the series should the inner and outer solutions fail to match.

Example 7.22

The uneven composite approximation $y_{\text{comp},1,1/2}$ for the perturbation problem

$$\epsilon y'' + 2xy' + xy = 0, \qquad y(0) = 0, \quad y(1) = e^{-1/2}$$

was shown in example 7.20 to be discontinuous at $x = 0$. It was hinted that the uniform approximation $y_{\text{comp},1}$ would fix this problem. Verify that this is indeed the case.

SOLUTION: The complication lies in the fact that the next term in the inner solution, Y_2, is difficult to express in closed form. We must solve the linear equation

$$Y_2'' + 2tY_2' = \frac{t}{\sqrt{\pi}}\left(e^{-t^2} + \operatorname{erf}(t) - 1 + \frac{t\sqrt{\pi}}{2}\operatorname{erf}(t)\right).$$

This is still a first order linear equation in Y_2', so we can solve this in terms of an integral

$$Y_2' = C_1 e^{-t^2} + \frac{e^{-t^2}}{\sqrt{\pi}}\int\left(t + te^{t^2}\operatorname{erf}(t) - te^{t^2} + \frac{t^2 e^{t^2}\sqrt{\pi}}{2}\operatorname{erf}(t)\right)dt.$$

We have already seen how to integrate the $te^{t^2}\operatorname{erf}(t)$ term, and we can make progress on the $t^2 e^{t^2}\operatorname{erf}(t)$, using integration by parts with $u = t$, $dv = te^{t^2}\operatorname{erf}(t)$. The result is

$$Y_2'(t) = C_1 e^{-t^2} - \frac{e^{-t^2}}{4}\int_0^t e^{s^2}\operatorname{erf}(s)\,ds$$
$$+ \frac{1}{\sqrt{\pi}}\left(\frac{t^2}{2}e^{-t^2} - \frac{1}{2} - \frac{t}{\sqrt{\pi}}e^{-t^2} + \frac{1}{2}\operatorname{erf}(t) - \frac{t^2}{4}e^{-t^2} + \frac{t\sqrt{\pi}}{4}\operatorname{erf}(t)\right).$$

But we still need to integrate this expression to find Y_2, The answer can be expressed in terms of a double integral.

$$Y_2(t) = C_1\operatorname{erf}(t) + C_2 + \left(\frac{e^{-t^2}}{\pi} + \frac{t}{2\sqrt{\pi}}(\operatorname{erf}(t) - 1) + \frac{t^2}{8}\operatorname{erf}(t)\right)$$
$$- \frac{1}{4}\int_0^t e^{-u^2}\int_0^u e^{s^2}\operatorname{erf}(u)\,ds\,du.$$

We can use the initial condition $Y_2(0) = 0$ to determine that $C_2 = -1/\pi$, but we must match this with the outer solution to determine C_1. But how do we find $(y_{\text{in},1})_{\text{out},1}$ for an expression this complicated? In particular, we must find the asymptotic expansion of the double integral.

The key is to notice that $\operatorname{erf}(t) \sim 1$ plus *subdominant terms* for large t. Thus,

$$e^{-u^2}\int_0^u e^{s^2}\operatorname{erf}(u)\,ds \sim e^{-u^2}\int_0^u e^{s^2}\,ds,$$

where the difference will be subdominant to the series. We can use integration by parts to determine the asymptotics of the integral, as was done in problem 8 in section 2.2. We find that

$$e^{-u^2}\int_0^u e^{s^2}\,ds \sim \frac{1}{2u} + \frac{1}{4u^3} + \frac{3}{8u^5} + \cdots \qquad \text{as } u \to \infty.$$

Integrating, we have

$$\int_0^t e^{-u^2} \int_0^u e^{s^2}\, ds \sim \frac{\ln(t)}{2} + C_3 - \frac{1}{8t^2} - \frac{3}{32t^4} - \cdots$$

plus subdominant terms. We still have to determine the value of C_3, and the subdominant terms will affect this constant. *Mathematica* can determine that

$$\lim_{t\to\infty} \left(\int_0^t e^{-u^2} \int_0^u e^{s^2} \operatorname{erf}(s)\, ds\, du - \frac{\ln t}{2} \right) = \frac{\gamma}{4},$$

so we have

$$\int_0^t e^{-u^2} \int_0^u e^{s^2} \operatorname{erf}(s)\, ds\, du \sim \frac{\ln t}{2} + \frac{\gamma}{4} - \frac{1}{8t^2} - \frac{3}{32t^4} + \cdots.$$

So we can now determine the asymptotic series for $y_{\text{in},1}(x)$ to be

$$1 - \frac{x}{2} + \epsilon \left(C_1 - \frac{1}{\pi} + \frac{x^2}{8\epsilon} - \frac{\ln(x/\sqrt{\epsilon})}{8} - \frac{\gamma}{16} + \frac{\epsilon}{32x^2} + \cdots \right).$$

Keeping only terms of order ϵ or larger, and rewriting $\ln(x/\sqrt{\epsilon}) = \ln(x) - \ln(\epsilon)/2$, we have

$$(y_{\text{in},1})_{\text{out},1} = 1 - \frac{x}{2} + \frac{x^2}{8} + \epsilon C_1 - \frac{\epsilon}{\pi} - \frac{\epsilon}{8}\ln(x) + \frac{\epsilon \ln \epsilon}{16} - \frac{\epsilon\gamma}{16}.$$

To find $(y_{\text{out},1})_{\text{in},1}$, we replace x with $t\sqrt{\epsilon}$ and find the series up to order ϵ.

$$e^{-t\sqrt{\epsilon}/2} + \frac{\epsilon}{8} e^{-t\sqrt{\epsilon}/2} \ln(t\sqrt{\epsilon}) \sim 1 - \frac{t\sqrt{\epsilon}}{2} + \frac{t^2\epsilon}{8} - \frac{\epsilon \ln(t)}{8} + \frac{\epsilon \ln(\epsilon)}{16} + o(\epsilon).$$

Converting back to the variable x, we have

$$(y_{\text{out},1})_{\text{in},1}(x) = 1 - \frac{x}{2} + \frac{x^2}{8} - \frac{\epsilon \ln(x)}{8}.$$

We almost have a match if $C_1 = 1/\pi + \gamma/16$, but $(y_{\text{in},1})_{\text{out},1}$ has a $\epsilon \ln(\epsilon)$ term not in $(y_{\text{out},1})_{\text{in},1}$. Either the outer or the inner perturbation series must have a term of order $\epsilon \ln(\epsilon)$. Adding $\overline{y}_1 \epsilon \ln(\epsilon)$ to the outer solution forces $\overline{y}_1$ to satisfy $2x\overline{y}_1' + x\overline{y}_1 = 0$, $\overline{y}_1(1) = 0$, so that $\overline{y}_1 = 0$. Hence, the extra term must be in the inner perturbation series:

$$y_{\text{in}} \sim Y_0 + \sqrt{\epsilon} Y_1 + \epsilon \ln(\epsilon) \overline{Y}_2 + \epsilon Y_2 + \epsilon^{3/2} \ln(\epsilon) \overline{Y}_3 + \epsilon^{3/2} Y_3 \cdots. \tag{7.36}$$

When we plug the series of equation 7.36 into equation 7.28, and we find that

$$\overline{Y}_2'' + 2t\overline{Y}_2' = 0, \qquad \overline{Y}_2(0) = 0,$$

whereas the equations for Y_0, Y_1, and Y_2 remain the same as before. The solution for $\overline{Y}_2(t)$ is $C_3\text{erf}(t)$, so we fold this term into the inner solution that we found before.

$$(y_{\text{in},1})_{\text{out},1} = 1 - \frac{x}{2} + \frac{x^2}{8} + \epsilon C_1 - \frac{\epsilon}{\pi} - \frac{\epsilon}{8}\ln(x) + \frac{\epsilon\ln\epsilon}{16} - \frac{\gamma\epsilon}{16} + C_3\epsilon\ln\epsilon.$$

Now it is possible to apply Van Dyke's rule, and determine that $C_1 = 1/\pi + \gamma/16$ and $C_3 = -1/16$. The composite approximation becomes

$$\begin{aligned}
y_{\text{comp},1} &= y_{\text{out},1} + y_{\text{in},1} - (y_{\text{out},1})_{\text{in},1} \\
&= e^{-x/2}\left(1 - \frac{\epsilon\ln x}{8}\right) \\
&+ \text{erf}(x/\sqrt{\epsilon}) + \sqrt{\epsilon}\left(\frac{1-e^{-x^2/\epsilon}}{\sqrt{\pi}} - \left(\frac{1}{\sqrt{\pi}} + \frac{x}{2\sqrt{\epsilon}}\right)\text{erf}(x/\sqrt{\epsilon})\right) \\
&- \frac{\epsilon\ln\epsilon}{16}\text{erf}(x/\sqrt{\epsilon}) \\
&+ \epsilon\left(\frac{e^{-x^2/\epsilon} + \text{erf}(x/\sqrt{\epsilon}) - 1}{\pi} + \frac{x}{2\sqrt{\epsilon\pi}}(\text{erf}(x/\sqrt{\epsilon}) - 1) + \frac{x^2}{8\epsilon}\text{erf}(x/\sqrt{\epsilon})\right) \\
&+ \frac{\gamma\epsilon\,\text{erf}(x/\sqrt{\epsilon})}{16} - \frac{\epsilon}{4}\int_0^{x/\sqrt{\epsilon}} e^{-u^2}\int_0^u e^{s^2}\text{erf}(s)\,ds\,du \\
&- \left(1 - \frac{x}{2} + \frac{x^2}{8} - \frac{\epsilon}{8}\ln(x)\right).
\end{aligned}$$

This time, the uniform composite approximation does not have an asymptote at $x = 0$, since the logarithm terms cancel to first order. In fact, $y_{\text{comp},1}$ provides a much better approximation than any of the other approximations that we have computed. See figure 7.20. When $\epsilon = 0.1$, the graph is virtually indistinguishable from the exact solution. □

7.4.3 Multiple Boundary Layers

It is possible for boundary layers to occur at *both* endpoints, or even at an interior point. We always begin by finding the outer solution, and determine where it cannot possibly match an endpoint, or becomes discontinuous.

Example 7.23

Find the leading order composite approximation for the perturbation problem

$$\epsilon y'' - x^2 y' - 4y = 0, \qquad y(0) = 1, \quad y(1) = 1. \tag{7.37}$$

SOLUTION: The unperturbed problem is $-x^2y' - 4y = 0$, which has the solution $y = C_1 e^{4/x}$. Since there is no value of C_1 for which $y(0) = 1$, there

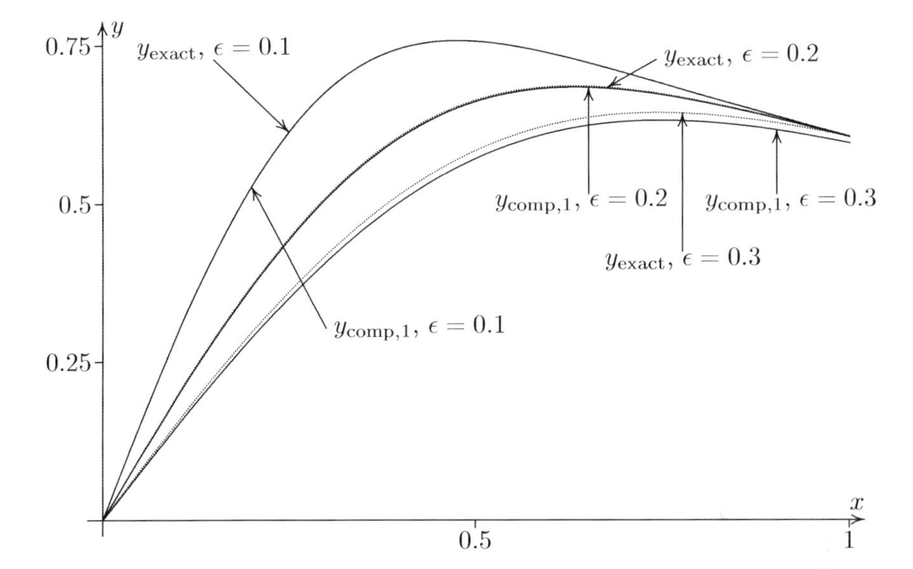

FIGURE 7.20: This compares the uniform composite approximations $y_{\text{comp},1}$ to the equation $\epsilon y'' + 2xy' + xy = 0$, with $y(0) = 0$ and $y(1) = e^{-1/2}$. The exact solutions are shown with a dotted line. One needs a magnifying glass to discern the difference when $\epsilon = 0.1$. This is a much better approximation than the zeroth order shown in figure 7.9.

must be a boundary layer at $x = 0$. To find the singular distinguished limit, we substitute $x = \delta t$ to obtain the equation

$$\frac{\epsilon Y''(t)}{\delta^2} - \delta t^2 Y'(t) - 4Y(t) = 0.$$

Since $\delta t^2 Y'(t) \ll 4Y(t)$, we must have $\delta = O(\sqrt{\epsilon})$, and the equation $Y'' - 4Y = 0$ has the general solution $Y = C_2 e^{-2t} + C_3 e^{2t}$.

To match the inner and outer solutions to first order, we compute

$$y_{\text{in},0}(x) = C_2 e^{-2x/\sqrt{\epsilon}} + C_3 e^{2x/\sqrt{\epsilon}} \sim C_3 e^{2x/\sqrt{\epsilon}}.$$

On the other hand,

$$Y_{\text{out},0} t = C_1 e^{4/(t\sqrt{\epsilon})} \sim C_1 e^{4/x}.$$

The only way for these to match is if both C_1 and C_3 are zero. So the outer solution is 0, and the inner solution is $Y = C_2 e^{-2t}$, which we can use the initial condition $y(0) = 1$ to determine that $C_2 = 1$. However, the outer solution no longer can satisfy the boundary condition at $x = 1$, so there must be *another* boundary layer at $x = 1$.

To find this inner solution, we substitute $x = 1 - \delta s$, to produce the equation

$$\frac{\epsilon \widetilde{Y}''(s)}{\delta^2} + (1 - \delta s)^2 \frac{\widetilde{Y}'(t)}{\delta} - 4\delta Y(s) = 0.$$

Here, we used $\widetilde{Y}(s)$ to distinguish between the two inner solutions. Since $\widetilde{Y}'(s)/\delta \gg 4\widetilde{Y}(s)$, we must have $\widetilde{Y}'(s)/\delta \sim \epsilon \widetilde{Y}''(s)/\delta^2$, so $\delta = O(\epsilon)$. The inner equation becomes $\widetilde{Y}_0''(s) + \widetilde{Y}_0'(s) = 0$, which has the solution $\widetilde{Y}_0 = C_4 e^{-s} + C_5$. To satisfy the condition $Y_0(0) = y(1) = 1$, we find$C_5 = 1 - C_4$. To match the outer solution, we convert to the variable x to form

$$\widetilde{Y}_0(x) = C_4 e^{(x-1)/\epsilon} + 1 - C_4.$$

To leading order in ϵ, this becomes $1 - C_4$, which must match to the outer solution of 0. Thus, $C_4 = 1$.

Since we have two inner solutions, we find the uniform composite approximation by

$$y_{\text{comp}} = y_{\text{out}} + y_{\text{in,left}} + y_{\text{in,right}} - y_{\text{leftmatch}} - y_{\text{rightmatch}}.$$

So $y_{\text{comp},0} = e^{-2x/\sqrt{\epsilon}} + e^{(x-1)/\epsilon}$. Figure 7.21 compares this approximation to the exact solution for $\epsilon = 0.05$ and 0.1. ▯

It should be noted that a boundary layer does not always have to appear at one of the endpoints, but could also appear in the *interior* of the interval. Such interior boundary layers are marked by the values of x for which the coefficient functions of the differential equation become 0.

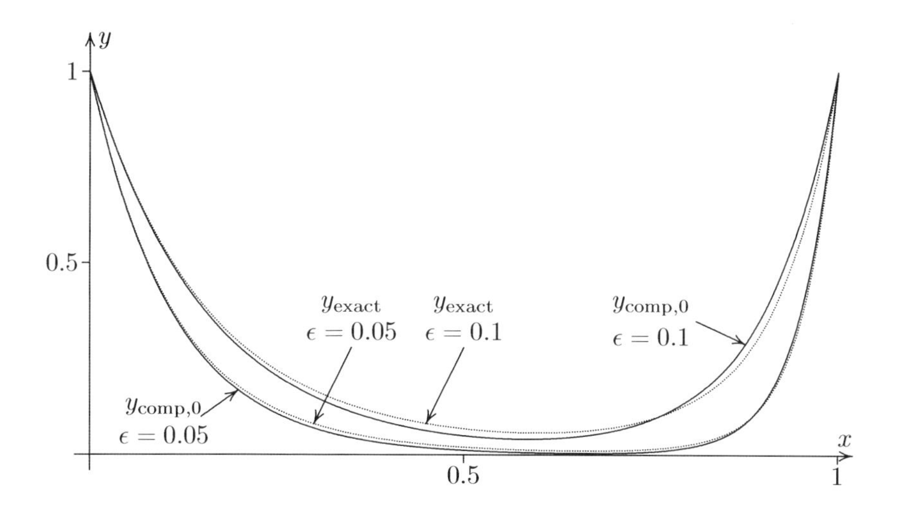

FIGURE 7.21: This compares the leading uniform composite approximations for the equation $\epsilon y'' - x^2 y' - 4y = 0$ with $y(0) = y(1) = 1$. The exact solutions, shown with a dotted line, are very close to the composite approximations for $\epsilon = 0.05$ and $\epsilon = 0.1$.

Example 7.24

Find the leading order perturbation solution for the problem

$$\epsilon y'' + xy' + xy = 0, \qquad y(-1) = e/2, \quad y(1) = 2/e.$$

SOLUTION: The unperturbed problem, $xy' + xy = 0$, has the general solution $y = C_1 e^{-x}$. This cannot satisfy both initial conditions, so there is a boundary layer somewhere in the problem. Although we cannot apply proposition 7.1 directly to this problem, we can argue that if the interval had been from $1/2 \le x \le 1$, the proposition precludes the possibility of a boundary layer near 1. Likewise, if the interval had been $-1 \le x \le -1/2$, proposition 7.1 would indicate that a boundary layer cannot be near $x = -1$. In fact, the only intervals for which the proposition cannot be applied are those containing the origin, so there must be a boundary layer near $x = 0$.

Since we have an internal boundary layer, there will be *two* outer solutions, the left solution being $y = C_1 e^{-x}$, and the right solution being $y = C_2 e^{-x}$. Each will satisfy one of the boundary conditions. So for the left outer solution, we must have $y_{\text{out,left}}(-1) = e/2$, so $C_1 = 1/2$. Likewise, for the right outer solution, we have $y_{\text{out,right}}(1) = 2/e$, so $C_2 = 2$.

If we let $x = \delta t$, then the equation becomes

$$\frac{\epsilon Y''(t)}{\delta^2} + tY'(t) + \delta t Y(t) = 0.$$

Since $\delta tY \ll tY'$, the distinguished limit is $\delta = O(\sqrt{\epsilon})$. The inner equation becomes $Y''(t) + tY'(t) = 0$, so

$$Y' = Ce^{-t^2/2}.$$

This can be integrated by letting $u = t/\sqrt{2}$, resulting in $Y(t) = C_3 + C_4\mathrm{erf}(t/\sqrt{2})$, where we incorporated the $\sqrt{\pi/2}$ into the constant C_4.

When we convert the inner solution back into the variable x, we get

$$Y_{\mathrm{in},0}(x) = C_3 + C_4\mathrm{erf}(x/\sqrt{2\epsilon}).$$

When we compute this to leading order in ϵ, the answer depends on whether x is positive or negative. In fact,

$$(Y_{\mathrm{in},0})_{\mathrm{out},0} \sim \begin{cases} C_3 + C_4 & \text{if } x > 0 \\ C_3 - C_4 & \text{if } x < 0 \end{cases}.$$

On the other hand, the left outer solution becomes $e^{-t\sqrt{\epsilon}}/2 \sim 1/2$, and the right outer solution becomes $2e^{-t\sqrt{\epsilon}} \sim 2$. So we solve the system of equations $C_3 + C_4 = 2$, $C_3 - C_4 = 1/2$. This produces $C_3 = 5/4$ and $C_4 = 3/4$. Hence, the inner solution is given by

$$Y_{\mathrm{in},0}(x) = \frac{5 + 3\mathrm{erf}(x/\sqrt{2\epsilon})}{4}.$$

Combining the inner and outer solutions to form a composite solution is tricker in this case, since there are two outer solutions. We cannot simply add all of the solutions, and subtract the matching functions, since the outer solutions only approach the match in the inner region. The goal is to find a function that is close to $e^{-x}/2$ for $x < 0$, and close to $2e^{-x}$ for $x > 0$. But the two outer solutions only differ by a constant. (This will always be true, since the two solutions stem from the same differential equation.) Hence, we can replace the constant with the inner solution, to produce

$$y_{\mathrm{comp},0} = e^{-x}\frac{5 + 3\mathrm{erf}(x/\sqrt{2\epsilon})}{4}.$$

It is easy to check that for $x > 0$ and ϵ small, this approaches $2e^{-x}$, but for $x < 0$, it approaches $e^{-x}/2$. Also, when x is small, $e^{-x} \sim 1$, so we reproduce the inner solution. Figure 7.22 shows that the composite approximation comes very close to the exact solution for $\epsilon = 0.02$ and $\epsilon = 0.01$. ▯

Although we have covered many boundary layer problems in this section, we have only scratched the surface to the variety of boundary layers possible. It is even possible for there to be a boundary layer within a boundary layer. Although the computational aspects can become unwieldy, the techniques for dealing with these different situations will be the same. Some of these unusual cases will be covered in the exercises.

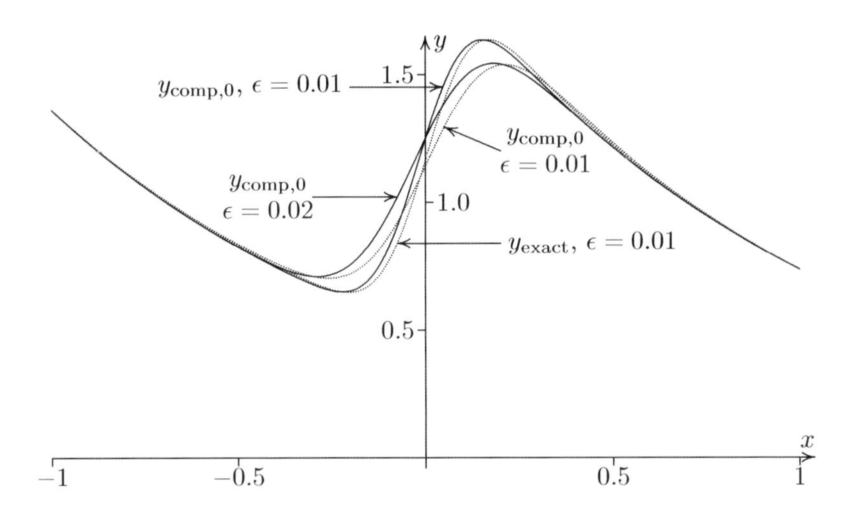

FIGURE 7.22: The solution to the equation $\epsilon y'' + xy' + xy = 0$, with $y(-1) = e/2$, $y(1) = 2/e$, has an internal boundary layer near $x = 0$. The leading order composite approximation closely follows the exact solution, shown with a dotted line. Note that as ϵ gets smaller, the slope at 0 goes to ∞.

Problems for §7.4

1 Show that if we substitute $x = s\sqrt{\epsilon}$ into the inner and outer solutions to order ϵ of example 7.16, and re-expand to order ϵ assuming that $s = O(1)$, then the inner and outer solutions do not match in the ϵ order term.

2 Show that if we substitute $x = s\epsilon^{2/3}$ into the inner and outer solutions to order ϵ^2 of example 7.16, and re-expand to order ϵ^2 assuming that $s = O(1)$, then the inner and outer solutions do not match in the ϵ^2 order term.

3 Suppose that the outer solution to a function is

$$y_{\text{out}}(x) = e^{-x} + \epsilon(1-x)e^{-x} + \epsilon^2(x^2 - 6x + 5)e^{-x}/2 + O(\epsilon^3),$$

whereas the inner solution, where $x = \epsilon t$, is given by

$$y_{\text{in}}(t) = K_\epsilon\left((1 - e^{-t}) - \epsilon t(1 + e^{-t}) + \epsilon^2 \frac{t^2 - t^2 e^{-t} - 2t - 2te^{-t}}{2} + O(\epsilon^3)\right).$$

Use asymptotic matching to determine K_ϵ to order ϵ^2.

4 Find $y_{\text{comp},0}$, $y_{\text{comp},1}$, and $y_{\text{comp},2}$ for the function described in problem 3.

5 Suppose that the outer solution to a function is

$$y_{\text{out}}(x) = \frac{2}{1+x} + \epsilon\frac{3 - 2x - x^2}{2(1+x)^3} + \epsilon^2\frac{21 - 8x - 8x^2 - 4x^3 - x^4}{4(1+x)^5} + O(\epsilon^3),$$

whereas the inner solution, where $x = \epsilon t$, is given by

$$y_{\text{in}}(t) = 1 - t\epsilon + \epsilon^2(t^2 - 2t) + K_\epsilon\left((1 - e^{-t}) + \epsilon\frac{t^2e^{-t} - 2t}{2} + \epsilon^2\frac{8t^2 - 16t - t^4e^{-t}}{8} + O(\epsilon^3)\right).$$

Use asymptotic matching to determine K_ϵ to order ϵ^2.

6 Find $y_{\text{comp},0}$ and $y_{\text{comp},1}$ for the function described in problem 5.

For problems **7** through **12**: Use Van Dyke's method to match the inner and outer solutions up to order ϵ^2 for the following problems. Then find the composite approximations $y_{\text{comp},0}$, $y_{\text{comp},1}$, and $y_{\text{comp},2}$ to the solution to the equation. Note that the inner and outer solutions were found in problems 3 through 14 of section 7.3.

7	$\epsilon y'' + y' + y = 0,$	$y(0) = 1,\ y(1) = 2$
8	$\epsilon y'' + 2y' - 2y = 0,$	$y(0) = 2,\ y(1) = 1$
9	$\epsilon y'' + (x + 1)y' - 2y = 0,$	$y(0) = 0,\ y(1) = 4$
10	$\epsilon y'' + (x + 1)y' + y = 0,$	$y(0) = 2,\ y(1) = 1/2$
11	$\epsilon y'' + y' + 2xy = 0,$	$y(0) = 1,\ y(1) = 1$
12	$\epsilon y'' + y' - 2xy = 0,$	$y(0) = 2,\ y(1) = 1$

13 Show that the series in equation 7.24 can be summed, and that the sum is in fact an exact solution to equation 7.23, except that $y(1) \neq 0$.

14 Use the result of problem 13 to find $y_{\text{comp},\infty,2}$ for the problem in example 7.18. Here, we define $y_{\text{comp},\infty,2}$ as $y_{\text{out},\infty} + y_{\text{in},2} - (y_{\text{out},\infty})_{\text{in},2}$, where $y_{\text{out},\infty}$ is the sum of all of the terms in the geometric series. In what sense is $(y_{\text{out},\infty})_{\text{in},2} = (y_{\text{in},2})_{\text{out},\infty}$?

15 Consider the perturbation problem

$$\epsilon y'' + 2xy' - xy = 0, \qquad y(0) = 1, \quad y(1) = \sqrt{e}.$$

Find the outer solution to order ϵ for this equation.

16 Consider the perturbation problem

$$\epsilon y'' + 2xy' - xy = 0, \qquad y(0) = 1, \quad y(1) = \sqrt{e}.$$

Find the inner solution to order $\sqrt{\epsilon}$ for this equation, in terms of the error function $\text{erf}(x)$, and an arbitrary constant K_ϵ.

17 Use the results of problems 15 and 16 to find the non-uniform composite approximation $y_{\text{comp},1,1/2}$. Note that this function becomes undefined as $x \to 0$.

18 Because the values for K_0 and K_1 were 0 in problem 17, it is particularly easy to find the next term in the inner solution for problem 16. Show that we can express the next term y_2 as

$$K_2 \text{erf}(t) + \int_0^t e^{-s^2} \int_0^s \frac{r^2}{2} e^{r^2}\, dr\, ds.$$

19 Use integration by parts to show that the integral in problem 18 is asymptotic to

$$\frac{t^2}{8} - \frac{\ln(t)}{8} + C + \frac{1}{32t^2} + \cdots.$$

The value of C turns out to be $\phi(1/2)/16 = -(2\ln 2 + \gamma)/16$.

20 Use the results of problems 15 through 19 to find the uniform composite approximation $y_{\text{comp},1}$ for the problem

$$\epsilon y'' + 2xy' - xy = 0, \qquad y(0) = 1, \quad y(1) = \sqrt{e}.$$

Do we have to add an $\epsilon \ln(\epsilon)$ term to one of the series?

21 The problem of the heat flow outside of a heated unit sphere can be modeled by the equation

$$f''(r) + 2f'(r)/r + \epsilon f'(r) f(r) = 0,$$

where r is the distance to the center of the sphere. We assume that $f = 0$ on the surface of the sphere, that is, when $r = 1$, and that f approaches 1 as $r \to \infty$ [11]. Show that this model can be converted to the problem in example 7.21.

22 Consider the non-linear problem

$$\epsilon y''(x) + y'(x)y(x) = 0, \qquad y(0) = -1, \quad y(1) = 1.$$

This equation could be used to describe the motion of a mass moving in a medium with damping proportional to the displacement, with either the mass small or the damping large [18]. Show that the boundary layer of width ϵ occurs near $x = 1/2$, and find the leading order for the outer layers and the inner layer, and put this information together to form $y_{\text{comp},0}$.

Hint: The inner layer equation is as complicated as the original, but it can be integrated to produce a separable equation.

23 A more complicated variation to problem 22 is the Lagerstrom-Cole equation [15]

$$\epsilon y''(x) + y'(x)y(x) - y(x) = 0, \qquad y(0) = -1, \quad y(1) = 1.$$

Symmetry still dictates that the boundary layer is at $x = 1/2$, but now the equation cannot be solved in closed form. Find the leading order for the two outer layers and the inner layer, and put this information together to form $y_{\text{comp},0}$. See the hint for problem 22.

24 If we change the boundary conditions for problem 23:

$$\epsilon y''(x) + y'(x)y(x) - y(x) = 0, \qquad y(0) = -1, \quad y(1) = 3/2,$$

then the boundary layer will no longer be at 1/2. Where will the boundary layer be? Find $y_{\mathrm{comp},0}$.

Hint: Assume the boundary layer occurs at $x = \alpha$. The value of α will be revealed as we match the inner and outer solutions.

Chapter 8

WKBJ Theory

WKBJ theory is a method of finding the global behavior of the solution to a linear differential equation for which the highest derivative is multiplied by a small parameter ϵ. In a sense, boundary layer theory can be considered as a special case of WKBJ theory, although boundary layer theory can apply to non-linear equations, whereas WKBJ theory cannot. The acronym stands for the main mathematicians that developed the theory, Gregor Wentzel, Henrik Kramers, Marcel Louis Brillouin, and Sir Harold Jeffreys. The latter actually developed the theory three years earlier, in 1923, as a general method for approximating the solutions to linear second order equations, including the Schrödinger equation. In spite of his precedence, Jeffreys contribution is often ignored by many sources, so the theory is sometimes referred to as WKB theory. Actually, there are many mathematicians that preceded all four of these mathematicians, including Francesco Carlini (1817), Joseph Liouville (1837), George Green (1837), Lord Rayleigh (1912) and Richard Gans (1915).

To introduce the basic concepts of WKBJ theory, let us consider two similar perturbation problems with two very different behaviors

$$\epsilon y'' + y = 0, \qquad y(0) = 0, \quad y(1) = 1,$$

and

$$\epsilon y'' - y = 0, \qquad y(0) = 0, \quad y(1) = 1.$$

In both cases, letting $x = \delta t$ shows that the singular distinguished limit occurs when $\delta = O(\sqrt{\epsilon})$, and the equations become

$$Y''(t) \pm Y(t) = 0.$$

Since $Y(0) = 0$, we have the solution to the equation $Y'' + Y = 0$ is $Y(t) = C_1 \sin(t)$, whereas the solution to $Y'' - Y = 0$ is $Y = C_2 \sinh(t)$. The exact solutions (provided $\epsilon \neq 1/(\pi^2 n^2)$ for some integer n) are therefore

$$y(x) = \frac{\sin(x/\sqrt{\epsilon})}{\sin(1/\sqrt{\epsilon})} \quad \text{and} \quad \frac{\sinh(x/\sqrt{\epsilon})}{\sinh(1/\sqrt{\epsilon})}.$$

In spite of the similarity of the equations, the graphs are vastly different, as we can see in figures 8.1 and 8.2.

Figure 8.2 shows a typical boundary layer. As $\epsilon \to 0$, the function breaks down and becomes discontinuous at a single point. This type of behavior is

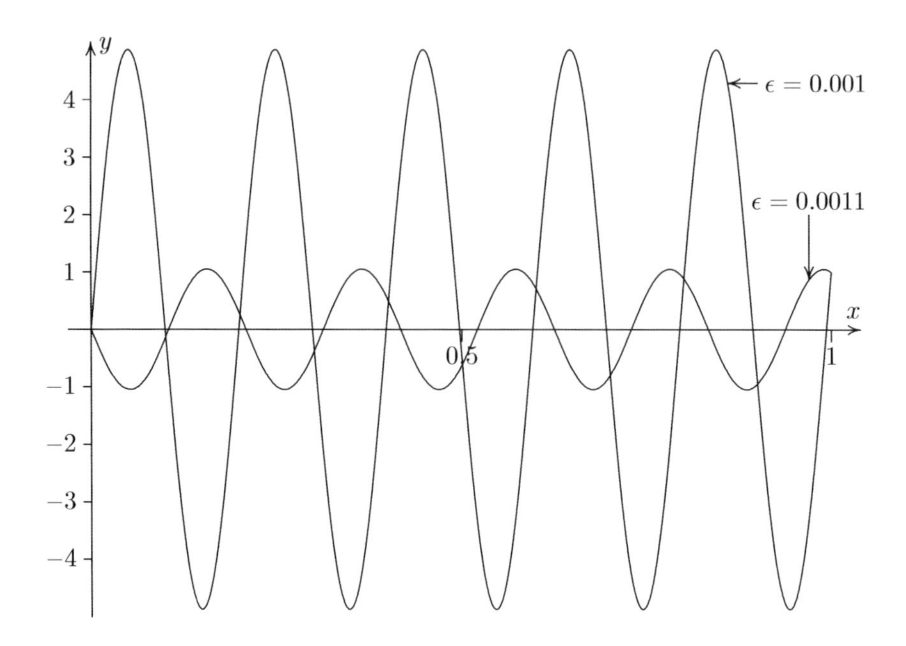

FIGURE 8.1: Solutions to $\epsilon y'' + y = 0$, $y(0) = 0$ and $y(1) = 1$. Note that a slight change in ϵ can create a huge change in the solution.

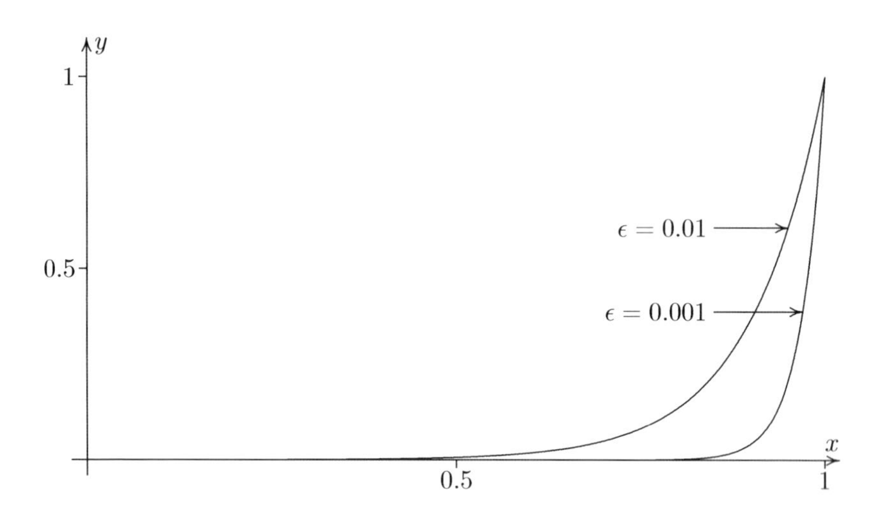

FIGURE 8.2: Solutions to $\epsilon y'' - y = 0$, $y(0) = 0$ and $y(1) = 1$. This time, there is a boundary layer near $x = 1$. Notice how a sign difference radically changed the behavior of the solutions.

called *dissipative*, since the solution dissipates, or decays exponentially, away from the point of rapid change.

On the other hand, in figure 8.1, we see that the solution becomes rapidly oscillatory as ϵ gets small. In fact, the function becomes discontinuous on the entire interval in the limit as $\epsilon \to 0$. This behavior is called a *dispersive* behavior. This type of behavior is characterized by rapid oscillations, in which the frequency and amplitude are slowly changing. We encountered dispersive behavior before in section 2.5 on the method of stationary phase.

8.1 The Exponential Approximation

It may seem ironic that we can obtain two radically different behaviors from a simple sign change in the differential equation. However, both of these behaviors can be expressed in terms of exponential functions.

$$\frac{\sin(x/\sqrt{\epsilon})}{\sin(1/\sqrt{\epsilon})} = C_1 e^{ix/\sqrt{\epsilon}} + C_2 e^{-ix/\sqrt{\epsilon}}, \qquad \frac{\sinh(x/\sqrt{\epsilon})}{\sinh(1/\sqrt{\epsilon})} = C_3 e^{x/\sqrt{\epsilon}} + C_4 e^{-x/\sqrt{\epsilon}}$$

for some constants C_1, C_2, C_3, and C_4. We see that the dissipative behavior comes from the exponents being real, whereas the dispersive behavior is caused from the exponents being purely complex. This suggests that we should consider solutions of the form

$$y(x) = e^{S(x)/\delta}.$$

When $S(x)$ is real, then δ becomes the thickness of the boundary layer. When $S(x)$ is purely complex, then y will experience oscillatory behavior with wavelengths of order δ. Finally, when $S(x)$ is nearly constant, then $y(x)$ is well behaved, as in the outer solution. Thus, we can model all types of behaviors with this exponential model. However, $S(x)$ still implicitly depends on δ. Hence, the strategy is to expand $S(x)$ in a perturbation series with powers of δ.

$$S(x) = S_0(x) + \delta S_1(x) + \delta^2 S_2(x) + \cdots.$$

Thus, we make the educated guess, or *ansatz*, that $S(x)$ is in this form, and solve for S_0, S_1, etc.

We will proceed in two steps. First, we will substitute

$$\boxed{y \to e^{S(x)/\delta},\ y' \to \frac{S'(x)}{\delta} e^{S(x)/\delta},\ y'' \to \left(\frac{S''(x)}{\delta} + \frac{(S'(x)}{\delta^2}\right) e^{S(x)/\delta},\ \text{etc.}} \tag{8.1}$$

If this seems vaguely familiar, it is because this is extremely similar to the substitution we used in section 5.1 to convert a linear equation to a non-linear equation, but whose leading behavior was easy to find. In equation 8.1, there are extra factors of δ, whose size is to be determined via dominant balance. As long as we start with a *linear homogeneous* equation, the $e^{S(x)/\delta}$ will cancel out of the equation.

Example 8.1
Consider the perturbation problem

$$\epsilon y''(x) + \cos^2(x)y(x) = 0, \qquad y(0) = 0, \quad y(1) = 1.$$

Use the WKBJ substitution to determine the singular distinguished limit, and the new non-linear equation for $S(x)$.
SOLUTION: Letting $y = e^{S(x)/\delta}$ produces

$$\epsilon\left(\frac{S''(x)}{\delta} + \frac{(S'(x))}{\delta^2}\right)e^{S(x)/\delta} + \cos^2(x)e^{S(x)/\delta} = 0.$$

Canceling out the exponential functions, we have

$$\epsilon S''(x)/\delta + \epsilon(S'(x))^2/\delta^2 + \cos^2(x) = 0.$$

We can assume that S' and S'' are of order 1, so $\epsilon S''(x)/\delta \ll \epsilon(S'(x))^2/\delta^2$. Thus, the singular distinguished limit is $\delta = O(\sqrt{\epsilon})$, and if we replace δ with $\sqrt{\epsilon}$, we get the new regular perturbation problem for $S(x)$:

$$(S'(x))^2 + \sqrt{\epsilon}S''(x) + \cos^2(x) = 0.$$ ∎

The next stage of the WKBJ process is to assume that $S(x)$ has a perturbation series of the form

$$S(x) \sim S_0(x) + \delta S_1(x) + \delta^2 S_2(x) + \delta^3 S_3(x) + \cdots.$$

However, the new equation for $S(x)$ will be a *non-linear* equation, so we will need to compute expressions such as

$$(S'(x))^2 \sim (S_0')^2 + 2\delta S_0'S_1' + \delta^2(2S_0'S_2' + (S_1')^2) + \delta^3(2S_0'S_3' + 2S_1'S_2') + \cdots.$$

Thus, we will have a sequence of non-linear equations to solve to find S_0, S_1, S_2, etc. However, these equations can typically be solved *algebraically* for the S_n', which then can be integrated to find $S_n(x)$.

Example 8.2
Find the first four terms for the series expansion of the $S(x)$ function for the equation

$$(S'(x))^2 + \delta S''(x) + \cos^2(x) = 0.$$

SOLUTION: Equating powers of δ, we get

$$\begin{aligned}(S_0') + \cos^2(x) &= 0,\\ 2S_0'S_1' + S_0'' &= 0,\\ 2S_0'S_2' + (S_1')^2 + S_1'' &= 0,\\ 2S_0'S_3' + 2S_1'S_2' + S_2'' &= 0.\end{aligned}$$

These equations can be solved S_0', S_1', etc., to produce

$$\begin{aligned}S_0' &= \pm i\cos x\\ S_1' &= \tan(x)/2\\ S_2' &= \pm i(3\sec^3 x - \sec x)/8\\ S_3' &= \tan x(\sec^2 x - 6\sec^4 x)/8.\end{aligned}$$

Integrating these, we get

$$\begin{aligned}S_0 &= \pm i\sin x + C_0\\ S_1 &= \ln(\sec x)/2 + C_1\\ S_2 &= \pm i(3\sec x\tan x + \ln(\sec x + \tan x))/16 + C_2\\ S_3 &= (\sec^2 x - 3\sec^4 x)/16 + C_3.\end{aligned}$$

All of the constants can combine to form a single constant, that will depend on δ. Thus, we have

$$\begin{aligned}S(x) \sim C_\delta \pm i\sin x + \delta\ln(\sec x)/2 \pm \delta^2 i(3\sec x\tan x + \ln(\sec x + \tan x))/16\\ + \delta^3(\sec^2 x - 3\sec^4 x)/16 + O(\delta^4).\end{aligned}$$ □

We still have to exponentiate

$$S(x)/\delta \sim S_0(x)/\delta + S_1 + \delta S_2 + \delta^2 S_3 + \cdots \qquad (8.2)$$

to produce the approximation for $y(x)$. In this case, some of the terms are purely imaginary, while others are real. The first term, with δ in the denominator, is the rapidly varying component of the solution. If S_0 is real, this term generates a boundary layer. On the other hand, if S_0 is purely complex, this term creates the rapid oscillating that we have seen in the dispersive behavior.

The second term in equation 8.2, S_1, does not involve δ, and often this term will involve a logarithm, so the exponential will be a slowly moving function. This term produces the outer solution in the case of dissipative behavior, but when the behavior is dispersive, this term gives the *enveloping function* of the highly oscillating curve. In this case, $S_1 = \ln(\sec x)/2$, so the exponential is

$$e^{\ln(\sec(x))/2} = \sqrt{\sec x}.$$

Terms beyond the second term will be small, so these will be corrections to the leading order.

Also, there will be two solutions added together, which in this case are complex conjugates of each other. We can utilize the fact that

$$C_1 e^{if(x)} + C_2 e^{-if(x)} = K_1 \sin(f(x)) + K_2 \cos(f(x))$$

for different constants K_1 and K_2. Thus, we have

$$y \sim \sqrt{\sec x}e^{\epsilon(\sec^2 x - 3\sec^4 x)/16} \times \left(K_1 \sin\left(\frac{\sin x}{\sqrt{\epsilon}} + \sqrt{\epsilon}\frac{3\sec(x)\tan(x) + \ln(\sec(x)+\tan(x))}{16}\right) + K_2 \cos\left(\frac{\sin x}{\sqrt{\epsilon}} + \sqrt{\epsilon}\frac{3\sec(x)\tan(x) + \ln(\sec(x)+\tan(x))}{16}\right)\right).$$

Notice that we replaced δ with $\sqrt{\epsilon}$, since that is the distinguished limit for this example. The initial condition $y(0) = 0$ indicates that $K_2 = 0$. We can then use the condition $y(1) = 1$ to determine the constant K_1. This produces the third order WKBJ approximation

$$y_{\text{WKBJ},3} = \frac{\sqrt{\sec x}}{\sqrt{\sec 1}} e^{\epsilon(\sec^2 x - 3\sec^4 x - \sec^2 1 + 3\sec^4 1)/16} \times \frac{\sin\left(\frac{\sin x}{\sqrt{\epsilon}} + \sqrt{\epsilon}\frac{3\sec(x)\tan(x)+\ln(\sec(x)+\tan(x))}{16}\right)}{\sin\left(\frac{\sin 1}{\sqrt{\epsilon}} + \sqrt{\epsilon}\frac{3\sec(1)\tan(1)+\ln(\sec(1)+\tan(1))}{16}\right)}. \tag{8.3}$$

Figures 8.3 and 8.4 compare the WKBJ approximations to the exact solutions. For $\epsilon = 0.005$, the results are fairly impressive. However, we cannot say that the approximations are truly asymptotic to the exact solutions due to the issue brought up in subsection 2.4.4: the zeros are non-coincident. The way to fix this is to express the exact solution by

$$e^{A(x,\epsilon) - A(1,\epsilon)} \frac{\sin(\omega(x,\epsilon))}{\sin(\omega(1,\epsilon))}.$$

Then we can accurately say that

$$A(x,\epsilon) \sim \ln(\sec x)/2 + \epsilon\frac{\sec^2 x - 3\sec^4 x}{16} + \cdots,$$

and

$$\omega(x,\epsilon) \sim \frac{\sin x}{\sqrt{\epsilon}} + \sqrt{\epsilon}\frac{3\sec(x)\tan(x) + \ln(\sec(x)+\tan(x))}{16} + \cdots.$$

Since we know that $y(0) = 0$, there will not be a phase shift δ, nor will there be a cosine component that we saw in subsection 2.4.4. We divided by $\sin(\omega(1,\epsilon))$ so that the solution is guaranteed to satisfy $y(1) = 1$.

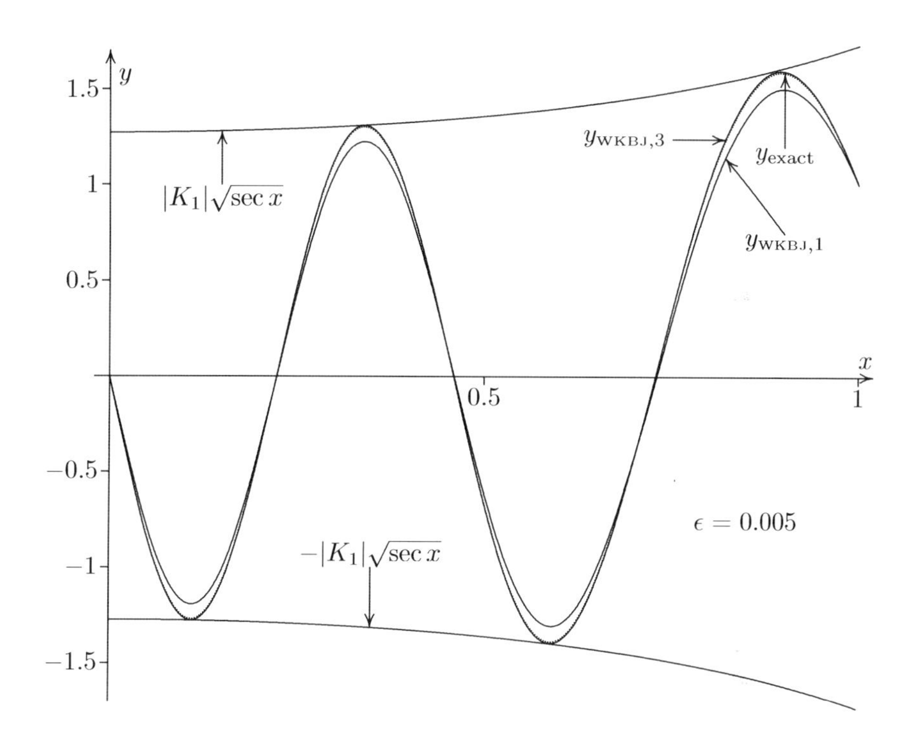

FIGURE 8.3: This shows the WKBJ approximations to $\epsilon y'' + \cos^2(x) y = 0$, $y(0) = 0$ and $y(1) = 1$, with $\epsilon = 0.005$. The exact solution, shown with a dotted line, is almost indistinguishable from the third order WKBJ approximation. Even the first order approximation is a good representation of the exact solution. The envelope of the solution $\pm K_1 \sqrt{\sec x}$ is obtained from the S_1 term of the WKBJ approximation.

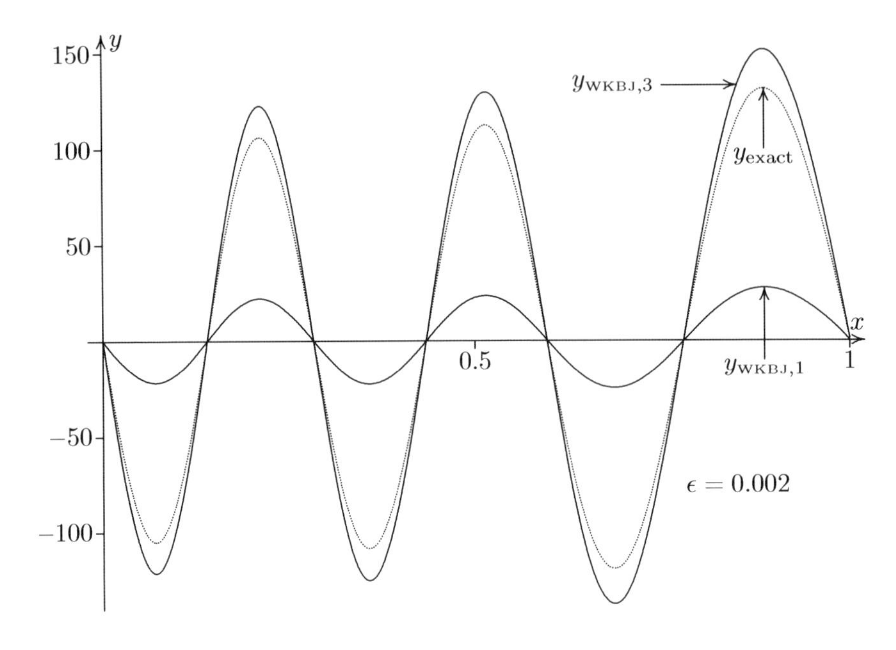

FIGURE 8.4: Same as figure 8.3, but with $\epsilon = 0.002$. For this value of ϵ, $\sin(1)/\sqrt{\epsilon}$ is very close to a multiple of π, so K_1 becomes huge. Although it appears that the WKBJ approximation is much worse than in figure 8.3, bear in mind that a very small change in the argument of the sin function results in a huge difference in the estimation of K_1. In fact, the WKBJ approximations are excellent estimates for a scale multiple of the exact solution.

TABLE 8.1: Although the approximations in figure 8.4 seem way off, in fact the WKBJ approximations are very close to a multiple of the true solution. To show evidence of this, we can compare the locations of the zeros and extrema of the exact solutions to the WKBJ approximations.

Zeros of y_{exact}	Zeros of $y_{\text{WKBJ},1}$	Zeros of $y_{\text{WKBJ},3}$	Extrema of y_{exact}	Extrema of $y_{\text{WKBJ},1}$	Extrema of $y_{\text{WKBJ},3}$
0.140891	0.140963	0.140891	0.0703414	0.0703769	0.0703413
0.284672	0.284828	0.284671	0.21245	0.212562	0.212449
0.434814	0.435087	0.434813	0.359114	0.359325	0.359113
0.596319	0.596784	0.596316	0.51447	0.51483	0.514468
0.778007	0.778878	0.777997	0.685157	0.685798	0.685151
1.00058	1.00279	1.0005	0.884792	0.886186	0.884761

The differential equation has no solution when $\omega(1,\epsilon)$ is a multiple of π. Figure 8.4 demonstrates the effect when $\omega(1,\epsilon)$ is close to, but not equal to, a multiple of π. Although the WKBJ approximation is fairly far away from the curve, the estimation of $\omega(x,\epsilon)$ is very accurate. This can be detected by comparing the locations of the zeros and the local extrema of the exact solution verses the WKBJ approximation, shown in table 8.1.

Example 8.3
Find the first order WKBJ approximation to the problem studied in example 7.18,

$$\epsilon y'' - (x+1)y' + 2y = 0, \qquad y(0) = 1, \quad y(1) = 0.$$

SOLUTION: If we let $y = e^{S(x)/\delta}$, we find that

$$\epsilon\left(\frac{S''(x)}{\delta} + \frac{(S'(x))^2}{\delta^2}\right) - (x+1)\frac{S'(x)}{\delta} + 2 = 0.$$

Since $S''/\delta \ll (S')^2/\delta^2$, we have the distinguished limit being $\delta = O(\epsilon)$. Setting $\delta = \epsilon$, the equation simplifies to

$$\epsilon S'' + (S')^2 - (x+1)S' + 2\epsilon = 0. \tag{8.4}$$

The zeroth order equation, $(S_0')^2 = (x+1)S_0'$, has two solutions, $S_0' = 0$ or $S_0' = x+1$. Thus, $S_0 = C_0$ or $S_0 = x^2/2 + x + C_0$.

To find the next order of perturbation, we consider the terms of order ϵ in equation 8.4. This produces

$$S_0'' + 2S_0'S_1' - (x+1)S_1' + 2 = 0.$$

When $S_0' = 0$, this simplifies to $S_1' = 2/(x+1)$, so $S_1 = 2\ln(x+1) + C_1$. However, if $S_0' = x+1$, then $S_1' = -3/(x+1)$, so $S_1 = -3\ln(x+1) + C_1$.

All arbitrary constants can be combined together, and will be exponentiated to form a multiplicative constant, so we have two solutions

$$y = K_1 e^{2\ln(x+1)} + K_2 e^{(x^2+2x)/(2\epsilon) - 3\ln(x+1)},$$

where K_1 and K_2 can depend on epsilon. This can simplify to

$$y = K_1(x+1)^2 + \frac{K_2}{(x+1)^3} e^{(x^2+2x)/(2\epsilon)}.$$

Now we can impose the initial conditions $y(0) = 1$ and $y(1) = 0$. The first condition states that $K_1 + K_2 = 1$, but the second is a bit more complicated:

$$4K_1 + K_2 e^{3/(2\epsilon)}/8 = 0.$$

In order to have these terms be of the same order, we clearly need $K_1 \gg K_2$. In fact K_2 would be of subdominant order compared to K_1. Thus, $K_1 \sim 1$, and $K_2 \sim -32e^{-3/(2\epsilon)}$. So we have

$$y \sim (x+1)^2 - \frac{32}{(x+1)^3} e^{(x^2+2x-3)/(2\epsilon)}. \tag{8.5}$$

Notice that as we find the constants using the initial conditions, we can drop terms that are of subdominant order. Even though this result is different from the solution of example 7.18, we can construct one solution from the other. The second term of equation 8.5 is subdominant unless x is close to 1. But when x is close to 1,

$$-\frac{32}{(x+1)^3} e^{-(1-x)(x+3)/(2\epsilon)} \sim -\frac{32}{8} e^{-(1-x)(4)/(2\epsilon)} = -4e^{-2(1-x)/\epsilon}.$$

Thus, equations 8.5 and 7.26 differ by a subdominant amount. So we see that the WKBJ approximation is equivalent to the first order boundary layer solution to leading order. This property can be seen in the graphs of the two, shown in figure 8.5. Notice that when we used WKBJ theory, we did not have to locate the boundary layer, since the theory took care of the boundary layer for us. Neither did we have to perform an asymptotic matching between the outer and inner solutions. ☐

The graph of figure 8.4 almost looks as though the solutions pass through both (0,0) and (1,0). This is very suggestive, since it means that $\epsilon = 0.002$ is very close to a solution to the eigenvalue problem

$$\epsilon y'' + \cos^2(x) y = 0, \qquad y(0) = y(1) = 0.$$

In fact, WKBJ theory is a powerful method of approximating the solutions to Sturm-Liouville eigenvalue problems, as in equation 4.13. If we consider the case where $\mu(x) = 1$ and $q(x) = 0$, the problem can be expressed as

$$y''(x) + Er(x)y(x) = 0, \qquad y(a) = y(b) = 0.$$

We know that there will only be solutions for discrete values of E, namely the eigenvalues E_1, E_2, $E_3 \ldots$. As it stands, this is not a perturbation problem,

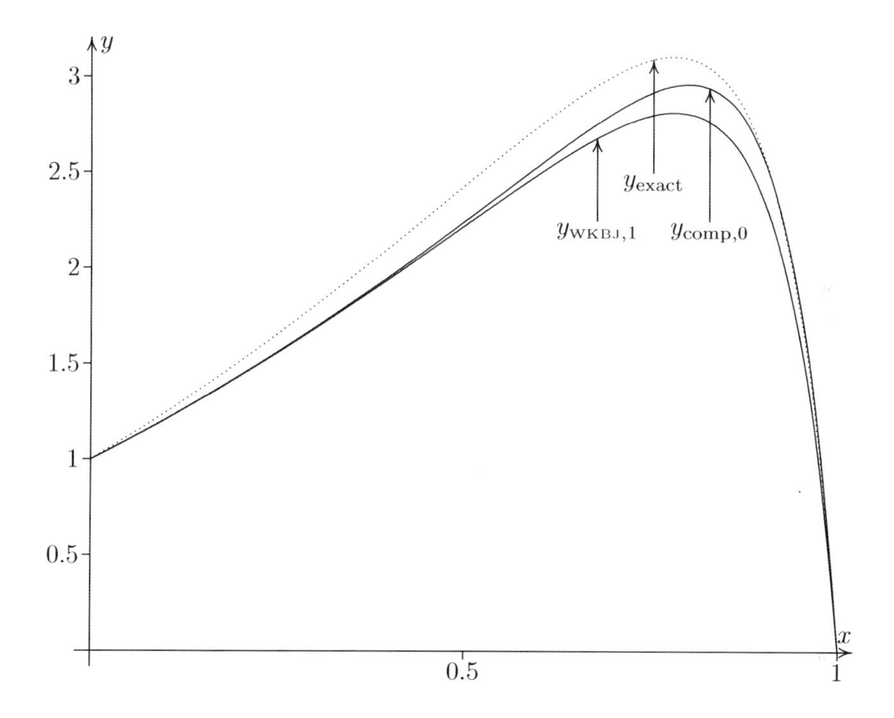

FIGURE 8.5: This compares the first order WKBJ approximations to $\epsilon y'' - (x+1)y' + 2y = 0$, $y(0) = 1$ and $y(1) = 0$, with $\epsilon = 0.15$. For comparison purposes, the leading order composite approximation is also shown, obtained through boundary layer theory. Although the WKBJ solution is not identical to the boundary layer solution, they have the same outer and inner expansions to first order.

but we are mainly interested in finding the eigenfunctions for the larger eigenvalues. We can assume that $E_n \to \infty$ (which will be verified by the WKBJ results), so if we let $\epsilon = 1/E$, we can rewrite the equation as

$$\epsilon y'' + r(x)y(x) = 0, \qquad y(a) = y(b) = 0.$$

Example 8.4

Use first order WKBJ theory to find the approximate eigenvalues and eigenfunctions for the Sturm-Liouville problem

$$y'' + E(x+1)^2 y = 0, \qquad y(0) = y(1) = 0.$$

SOLUTION: Letting $\epsilon = 1/E$, we have $\epsilon y'' + (x+1)^2 y = 0$. If we let $y = e^{S(x)/\delta}$, we obtain

$$\frac{\epsilon S''}{\delta} + \frac{\epsilon (S')^2}{\delta^2} + (x+1)^2 = 0.$$

The first term is clearly smaller than the second, so in order to get a balance, the distinguished limit must be $\delta = O(\sqrt{\epsilon})$. Letting $\delta = \sqrt{\epsilon}$ reduces the differential equation to

$$\sqrt{\epsilon} S'' + (S')^2 + (x+1)^2 = 0.$$

If we assume that $S(x) \sim S_0 + \sqrt{\epsilon} S_1 + \epsilon S_2 + \cdots$, we obtain the equations

$$\begin{aligned} (S_0')^2 + (x+1)^2 &= 0, \\ S_0'' + 2S_0'S_1' &= 0, \\ S_1'' + 2S_0'S_2' + (S_1')^2 &= 0, \quad \text{etc.} \end{aligned}$$

Solving these equations algebraically for S_0' and S_1', we obtain $S_0' = \pm(x+1)i$, and $S_1' = -1/(2x+2)$. These terms can be easily integrated to determine S_0 and S_1, so we find that the first order WKBJ approximation is given by

$$y_{\text{WKBJ},1} = \frac{C_1}{\sqrt{x+1}} e^{i(x^2/2+x)/\sqrt{\epsilon}} + \frac{C_1}{\sqrt{x+1}} e^{-i(x^2/2+x)/\sqrt{\epsilon}}.$$

We can rewrite the complex exponential functions as sines and cosines, changing the arbitrary constants.

$$y_{\text{WKBJ},1} = \frac{K_1}{\sqrt{x+1}} \sin\left(\frac{x^2+2x}{2\sqrt{\epsilon}}\right) + \frac{K_2}{\sqrt{x+1}} \cos\left(\frac{x^2+2x}{2\sqrt{\epsilon}}\right).$$

Since $y(0) = 0$, we have $K_2 = 0$. In order to have a non-zero solution with $y(1) = 0$, we must have $3/(2\sqrt{\epsilon})$ being a multiple of π. Setting $3/(2\sqrt{\epsilon}) = n\pi$, and solving for ϵ, we get $\epsilon = 9/(4n^2\pi^2)$. Since $E = 1/\epsilon$, the approximate eigenvalues are $E_n \sim 4n^2\pi^2/9$. The eigenfunctions become

$$y_n \sim \frac{K_1}{\sqrt{x+1}} \sin\left(\frac{n\pi(x^2+2x)}{3}\right). \tag{8.6}$$

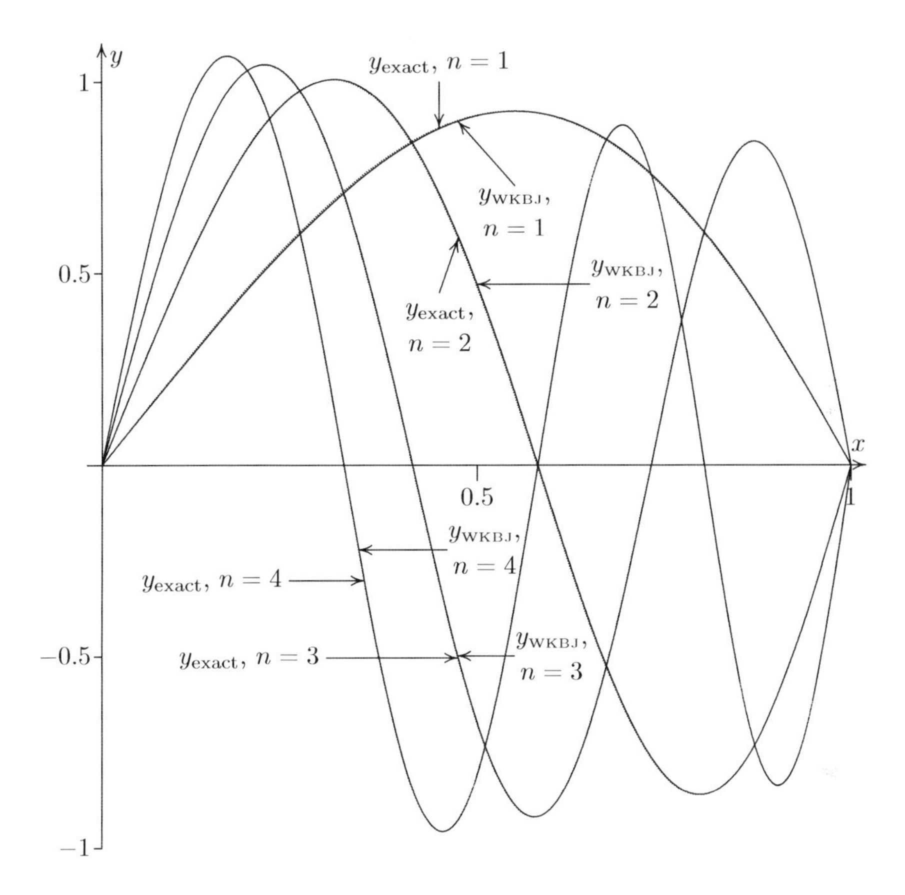

FIGURE 8.6: This compares the WKBJ approximations to the eigenfunctions to $y'' + E(x+1)^2y = 0$, $y(0) = y(1) = 0$ with the exact solutions. The first four modes are shown, with the exact solution shown with a dotted line. The WKBJ approximation is so close, it is almost impossible to distinguish between the two without a magnifying glass, especially for the $n = 4$ mode. (The lines look jagged on one side.) Note that all solutions have been normalized as to satisfy equation 4.15.

TABLE 8.2: This compares the exact eigenvalues to the WKBJ approximation, $4n^2\pi^2/9$. The second order WKBJ approximations are also included, which subtracts 3/16 (see problem 19).

n	$4n^2\pi^2/9$	$4n^2\pi^2/9 - 3/16$	Exact eigenvalue	Absolute error
1	4.38649	4.19899	4.24506	0.046071
2	17.54596	17.35846	17.37747	0.019005
3	39.47842	39.29092	39.30100	0.010078
4	70.18385	69.99635	70.00251	0.0061575
5	109.66227	109.47477	109.47889	0.0041222
6	157.91367	157.72617	157.72911	0.0029410
7	214.93805	214.75055	214.75275	0.0021986
8	280.73541	280.54791	280.54962	0.0017033
9	355.30576	355.11826	355.11962	0.0013571
10	438.64908	438.46158	438.46269	0.0011060

The graphs of the eigenfunctions are shown in figure 8.6. Note the remarkable closeness between the WKBJ approximations to the exact solutions. Since ϵ decreases as n increases, we know that further eigenfunctions will have even better accuracy.

We can also compare the exact eigenvalues with the WKBJ approximations. Table 8.2 compares the exact eigenvalues to the first and second order WKBJ approximations. ▯

We see from these examples that WKBJ theory can cover a variety of different perturbation problems. Ironically, the procedure is the same regardless of whether the solutions have a dissipative or dispersive behavior. There still is the issue of when this procedure will correctly give the asymptotic approximation to the solution, which we will cover in the next section.

Problems for §8.1

For problems **1** through **10**: Determine the first order WKBJ approximation for the following perturbation problems.

1 $\epsilon y'' + y = 0,\quad y(0) = 0,\ \ y(1) = 1$

2 $\epsilon y'' - y = 0,\quad y(0) = 1,\ \ y(1) = 1$

3 $\epsilon y'' + y' + y = 0,\quad y(0) = 0,\ \ y(1) = 1$

4 $\epsilon y'' + (x+1)y' + y = 0,\quad y(0) = 1,\ \ y(1) = 1$

5 $\epsilon y'' - (1+x)^2 y = 0,\quad y(0) = 1,\ \ y(1) = 1$

6 $\epsilon y'' + \cos(x)y' + \sin(x)y = 0,\quad y(0) = 0,\ \ y(\pi/4) = 1$

7 $\epsilon y'' + xy = 0,\quad y(1) = 0,\ \ y(4) = 1$

8 $\epsilon y'' + (x+1)y = 0,\quad y(0) = 0,\ \ y(15) = 1$

9 $\epsilon y'' + (1+\cos x)y = 0,\quad y(0) = 0,\ \ y(\pi/2) = 1$

10 $\epsilon y'' + \sec^2(x)y = 0,\quad y(0) = 0,\ \ y(\pi/6) = \sqrt[4]{3}$

11 Even though the functions in equation 8.6 are only the approximate eigenfunctions, show that they still satisfy the orthonormal condition given in equation 4.14. Note that in example 8.4, $r(x) = (x+1)^2$.

12 Find the value of K_1 in equation 8.6 so that the approximate eigenvalues satisfy the orthonormal property given in equation 4.15.

For problems **13** through **18**: Use the first order WKBJ approximation to estimate the eigenvalues and eigenfunctions for the following Sturm-Liouville problems. You do not have to normalize the eigenfunctions.

13 $y'' + E\cos^2(x)y = 0,$ $\quad y(0) = y(\pi/3) = 0$
14 $y'' + E(x+1)^4 y = 0,$ $\quad y(0) = y(1) = 0$
15 $y'' + Ey/(x+1)^2 = 0,$ $\quad y(0) = y(1) = 0$
16 $y'' + Ey/(x+1) = 0,$ $\quad y(0) = y(1) = 0$
17 $y'' + (E\cos^2 x - \sin^2 x)y = 0,$ $\quad y(0) = y(\pi/3) = 0$
18 $y'' + (E(x+1)^2 - x)y = 0,$ $\quad y(0) = y(1) = 0$

19 For example 8.4, find the next WKBJ approximation term S_2. By folding this term into the WKBJ approximation, how does this affect the approximate eigenvalues?

For problems **20** through **25**: Use the first second order WKBJ approximation find the correction for the approximate eigenvalues E_n in the following Sturm-Liouville problems. Note that only S_0 and S_2 will contribute to the eigenvalues.

20 $y'' + E\cos^2(x)y = 0,$ $\quad y(0) = y(\pi/3) = 0$
21 $y'' + E(x+1)^4 y = 0,$ $\quad y(0) = y(1) = 0$
22 $y'' + Ey/(x+1)^2 = 0,$ $\quad y(0) = y(1) = 0$
23 $y'' + Ey/(x+1) = 0,$ $\quad y(0) = y(1) = 0$
24 $y'' + (E(x+1)^2 - x)y = 0,$ $\quad y(0) = y(1) = 0$
25 $y'' + (E(x+1)^4 - x^2)y = 0,$ $\quad y(0) = y(1) = 0$

8.2 Region of Validity

A natural question that arises is to determine the conditions for the WKBJ approximation to be an accurate asymptotic approximation to the solution. We begin the analysis by finding the first order WKBJ approximation to the general second order linear equation.

Example 8.5
Determine the first order WKBJ solution, up to the arbitrary constants, for

the generalized second order perturbation problem

$$\epsilon y'' + p(x)y' + q(x)y = 0,$$

where we assume that $p(x) \neq 0$.

SOLUTION: If we substitute $y = e^{S(x)/\delta}$, we obtain the equation

$$\epsilon\left(\frac{S''}{\delta} + \frac{(S')^2}{\delta^2}\right) + p(x)\frac{S'}{\delta} + q(x) = 0.$$

Since we are assuming that $p(x) \neq 0$, we have $p(x)S'/\delta \gg q(x)$. Also, $S''/\delta \ll (S')^2/\delta^2$, so the equation becomes

$$\frac{\epsilon(S')^2}{\delta^2} \sim -\frac{p(x)S'}{\delta}.$$

For these to be the same size, we need to have $\delta = O(\epsilon)$, and if we set $\delta = \epsilon$, the differential equation becomes

$$\epsilon S'' + (S')^2 + p(x)S' + \epsilon q(x) = 0.$$

Next, we assume that $S(x) \sim S_0 + \epsilon S_1 + \epsilon^2 S_2 + \cdots$. The only tricky term is $(S')^2 \sim (S_0')^2 + 2\epsilon S_0' S_1' + \epsilon^2(2S_0'S_2' + (S_1')^2) + \cdots$. Thus, the set of equations to solve are as follows:

$$\begin{aligned} (S_0')^2 + p(x)S_0' &= 0, \\ S_0'' + 2S_0'S_1' + p(x)S_1' + q(x) &= 0, \\ S_1'' + 2S_0'S_2' + (S_1')^2 + p(x)S_2' &= 0, \text{ etc.} \end{aligned}$$

These equations can be solved algebraically for S_0', S_1', etc, which we can integrate to obtain the original series. The first equation has two solutions, $S_0' = 0$ or $S_0' = -p(x)$. The two cases have to be handled separately for the second equation. When $S_0' = 0$, then $S_1' = -q(x)/p(x)$. On the other hand, if $S_0' = -p(x)$, then $S_1' = (q(x) - p'(x))/p(x)$. By integrating these, we get two possibilities for the first two terms of the series

$$S(x) \sim -\epsilon \int \frac{q(x)}{p(x)}\,dx + C_0 \qquad \text{or}$$

$$S(x) \sim -\int p(x)\,dx + \epsilon \int \frac{q(x)}{p(x)} - \epsilon \ln(p(x)) + C_1.$$

So the first order WKBJ approximation is found by computing $e^{S(x)/\epsilon}$ for the two possibilities.

$$\boxed{y_{\text{WKBJ},1} = K_1 e^{-\int q(x)/p(x)\,dx} + \frac{K_2}{p(x)} e^{\int q(x)/p(x)\,dx} e^{-\int p(x)/\epsilon\,dx}.} \tag{8.7}$$

□

In computing the general WKBJ approximation for the second order equation over an interval, we find that the solution has a singularity if $p(x) = 0$, so the approximation would not be uniformly valid. In fact, if $p(x) = 0$ at one of the endpoints, then we will not be able to match the WKBJ solution with the initial conditions. This is a rather surprising result, since many of the boundary layer problems such as example 7.17 could be readily solved in spite of $p(x) = 0$ at one of the endpoints. This indicates that there are perturbation problems that cannot be solved directly with a single WKBJ approximation.

In the case where $p(x)$ is exactly zero, the analysis changes dramatically. In this case, the equation is a Schrödinger equation, which encompasses many eigenvalue problems. These equations warrant separate analysis.

Example 8.6

Determine the first order WKBJ solution, up to the arbitrary constants, for the generalized Schrödinger perturbation problem

$$\epsilon y'' + q(x)y = 0.$$

SOLUTION: When we substitute $y = e^{S(x)/\delta}$, we obtain

$$\epsilon\left(\frac{S''}{\delta} + \frac{(S')^2}{\delta^2}\right) + q(x) = 0.$$

Since $\epsilon S''/\delta \ll \epsilon(S')^2/\delta^2$, it is clear that the distinguished limit is $\delta = O(\sqrt{\epsilon})$, as opposed to the distinguished limit of example 8.5. The new equation for $S(x)$ becomes

$$\delta S'' + (S')^2 + q(x) = 0.$$

Letting $S(x) = S_0(x) + \delta S_1(x) + \delta^2 S_2(x) + \cdots$, we obtain the equations

$$\begin{aligned} (S_0')^2 + q(x) &= 0 \\ S_0'' + 2S_0'S_1' &= 0 \\ S_1'' + 2S_0'S_2' + (S_1')^2 &= 0. \end{aligned}$$

Hence, we have $S_0' = \pm\sqrt{-q(x)}$, and $S_1' = -q'(x)/(4q(x))$. Integration gives us the series for $S(x)$:

$$S(x) \sim \int \sqrt{-q(x)}\,dx - \frac{\delta}{4}\ln(q(x)) + \cdots.$$

If $q(x) > 0$, there will be complex solutions, resulting in trigonometric functions produced by calculating $e^{S(x)/\delta}$. In fact, we would have

$$y_{\text{WKBJ},1} \sim \frac{K_1}{\sqrt[4]{q(x)}}\cos(\omega(x)) + \frac{K_2}{\sqrt[4]{q(x)}}\sin(\omega(x)), \tag{8.8}$$

where

$$\omega(x) \sim \int \sqrt{q(x)}\, dt/\sqrt{\epsilon}.$$

On the other hand, if $q(x) < 0$, there will be two exponential solutions

$$y_{\text{WKBJ},1} \sim \frac{K_1}{\sqrt[4]{-q(x)}} e^{\int \sqrt{-q(x)}\, dx/\sqrt{\epsilon}} + \frac{K_2}{\sqrt[4]{-q(x)}} e^{-\int \sqrt{-q(x)}\, dx/\sqrt{\epsilon}}. \tag{8.9}$$

□

In this example, we see that if $q(x) = 0$ within the interval of definition, there will be a singularity in both equations 8.8 and 8.9. So once again, there are restrictions on whether the WKBJ approximation is valid.

Even when $p(x)$ is never 0 for example 8.5 or $q(x) \neq 0$ for example 8.6, there still is an issue over whether the WKBJ solution is a *uniform* approximation to the exact solution. Otherwise, we could have the situation that occurred in figure 7.18. To prevent this, we must have

$$\frac{S_0(x)}{\delta} \gg S_1(x) \gg \delta S_2(x) \gg \delta^2 S_3(x) \gg \cdots \quad \text{as } \delta \to 0 \tag{8.10}$$

where each asymptotic comparison is independent of x on the interval. If this is the case, we say that the perturbation series for $S(x)/\delta$ is *uniformly asymptotic in x as $\delta \to 0$*. This requirement is equivalent to saying the ratios of two consecutive non-zero terms are *bounded*. That for each n, there is an M_n such that

$$\frac{|S_{n+1}(x)|}{|S_n(x)|} \leq M_n \qquad \text{for all } x \text{ in the interval.} \tag{8.11}$$

For example, extending the series solution for example 8.4, we find that the two possible series for $S(x)$ are

$$S(x) \sim 2\epsilon \ln(1+x) - \frac{\epsilon^2}{(1+x)^2} - \frac{\epsilon^3}{2(x+1)^4} + \cdots \tag{8.12}$$

and

$$S(x) \sim (2x + x^2)/2 - 3\epsilon \ln(x+1) + \frac{6\epsilon^2}{(1+x)^2} + \frac{27\epsilon^3}{(x+1)^4} + \cdots. \tag{8.13}$$

As it stands, these series are not uniformly asymptotic, since the first ratio

$$\frac{-1/(1+x)^2}{2\ln(1+x)}$$

approaches ∞ as $x \to 0$. However, we can add an arbitrary constant to any of the S_n, thereby avoiding the situation where S_n has a root at a point where

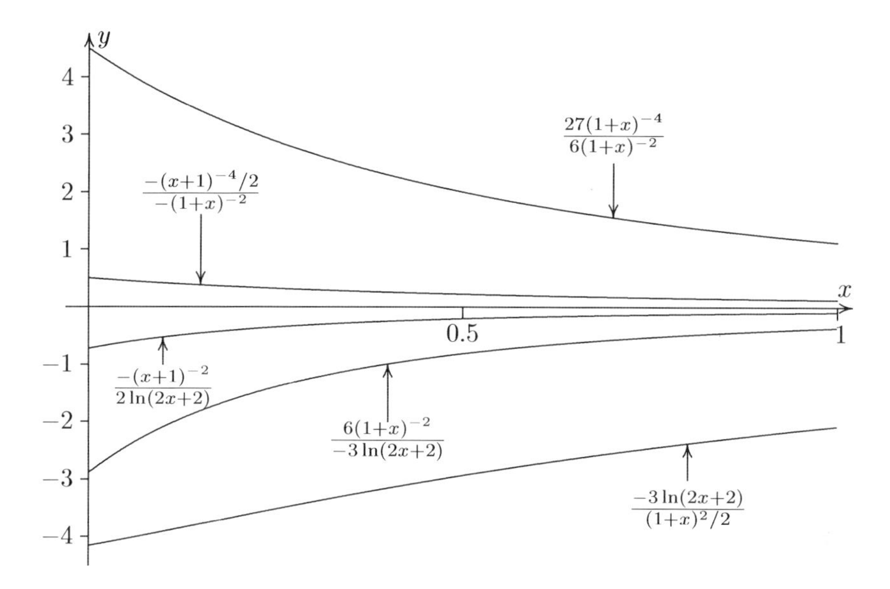

FIGURE 8.7: A perturbation series is *uniformly asymptotic in x* if the ratios of the coefficient functions are all bounded. This shows that indeed, by adding appropriate constants to the series, the ratios of consecutive terms are all bounded. Note that the bounds may change for each ratio.

S_{n+1} is non-zero. For example, we can modify the series in equations 8.12 and 8.13 to

$$S(x) \sim 2\epsilon \ln(2x+2) - \frac{\epsilon^2}{(x+1)^2} - \frac{\epsilon^3}{2(x+1)^4} + \cdots, \quad \text{and} \tag{8.14}$$

$$S(x) \sim (x+1)^2/2 - 3\epsilon \ln(2x+2) + \frac{6\epsilon^2}{(x+1)^2} + \frac{27\epsilon^3}{(x+1)^4} + \cdots. \tag{8.15}$$

By altering the series by only a constant, we have made the two series uniformly asymptotic in x. To see this, we look at the ratios of two consecutive coefficient functions.

$$\frac{-(1+x)^{-2}}{2\ln(2+2x)}, \quad \frac{-(1+x)^{-4}/2}{-(1+x)^{-2}},$$

$$\frac{-3\ln(2x+2)}{(x+1)^2/2}, \quad \frac{6(1+x)^{-2}}{-3\ln(2x+2)}, \quad \text{and} \quad \frac{27(1+x)^{-4}}{6(1+x)^{-2}}.$$

Figure 8.7 shows that all of these ratios are indeed bounded on the interval from 0 to 1.

But there is another issue that could cause a WKBJ approximation to be invalid. Even if we have a uniformly valid perturbation series for $S(x)$, this must be exponentiated to produce the WKBJ approximation. So unless the first term that we discarded approaches 0 as $\delta \to 0$, the exponential will not be asymptotic to the desired function. This is essentially the same principle as equation 5.3. For example, if we just keep the first term S_0, and form the approximation $e^{S_0(x)/\delta}$, we will not have an asymptotic expression of $y(x)$, because the second term $S_1(x)$ will not depend on δ, and will almost never be a constant. However, this approximation $e^{S(x)/\delta}$ will usually reflect some of the structure of $y(x)$, so this expression is given a name: the *geometrical optics*.

If we only keep the S_0 and S_1 terms, we have what is referred to as the *physical optics*: $e^{S_0(x)/\delta + S_1(x)}$. Usually, the physical optics will be a valid asymptotic approximation because the next term in the series, $\delta S_2(x)$ appears to be going to 0 as $\delta \to 0$. But actually, this will only be the case if $S_2(x)$ is bounded. So now only must we check to see if the ratios of two consecutive terms are bounded, but the first neglected term must also be bounded for the WKBJ approximation to be asymptotically valid.

Example 8.7
Test the validity of the physical optics WKBJ approximation for the perturbation problem

$$\epsilon x^2 y'' - (\ln(2x))^2 y = 0, \qquad y(1) = 1, \quad y(\infty) = 0.$$

SOLUTION: We can apply the calculations of example 8.6, with $q(x) = (\ln(2x))^2/x^2$. We find that

$$S_0 = \pm \int \sqrt{-q(x)}\, dx = \pm \int \frac{\ln(2x)}{x} = \pm \frac{\ln(2x)^2}{2} + C_0.$$

$$S_1 = -\ln(q(x))/4 = \ln(x)/2 - \ln(\ln(2x))/2.$$

But is the physical optics approximation with these two terms a valid asymptotic approximation? The equation for S_2' is

$$S_1'' + 2S_0' S_2' + (S_1')^2 = 0,$$

So we find that

$$S_2' = \frac{\pm(\ln(2x)^2 - 3}{8x \ln(2x)^3}, \qquad \text{so } S_2 = \frac{\pm 3}{16 \ln(2x)^2} \pm \frac{\ln(\ln(2x))}{8}.$$

We discover that $S_2(x)$ is *not* bounded in the interval from 1 to ∞. Hence, the physical optics approximation will not be asymptotic to the solution.

If we consider the second order WKBJ approximation, which includes the S_2 term, will this be an asymptotic approximation to the solution? To check, we must look at the next term, S_3. The equation for S_3 will be

$$S_2'' + 2S_0' S_3' + 2S_1' S_2' = 0,$$

so

$$S_3' = \frac{\ln(2x)^2 - 6}{8x\ln(x)^5}, \qquad \text{so } S_3 = \frac{3}{16\ln(2x)^4} - \frac{1}{16\ln(2x)^2}.$$

Hence, $S_3(x)$ is indeed bounded on the interval $[1, \infty)$. We also find that the ratios $S_2(x)/S_1(x)$ and $S_3(x)/S_2(x)$ are bounded on this interval, so equation 8.10 is satisfied. So we can find an asymptotic approximation by considering $e^{S_0/\delta + S_1 + \delta S_2}$. This produces

$$y \sim C_1 e^{\ln(2x)^2/(2\sqrt{\epsilon})}\sqrt{\frac{x}{\ln(2x)}}\ln(2x)^{\sqrt{\epsilon}/8}e^{3\sqrt{\epsilon}/(16\ln(2x)^2)}$$
$$+ C_2 e^{-\ln(2x)^2/(2\sqrt{\epsilon})}\sqrt{\frac{x}{\ln(2x)}}\ln(2x)^{-\sqrt{\epsilon}/8}e^{-3\sqrt{\epsilon}/(16\ln(2x)^2)}.$$

The condition $y(\infty) = 0$ forces $C_1 = 0$, and the condition $y(1) = 1$ allows us to solve for C_2. Thus, we have

$$y_{\text{WKBJ},2} = \frac{e^{-\ln(2x)^2/(2\sqrt{\epsilon})}\sqrt{x\ln 2}\,\ln(2x)^{-\sqrt{\epsilon}/8}e^{-3\sqrt{\epsilon}/(16\ln(2x)^2)}}{e^{-(\ln 2)^2/(2\sqrt{\epsilon})}\sqrt{\ln(2x)}(\ln 2)^{-\sqrt{\epsilon}/8}e^{-3\sqrt{\epsilon}/(16(\ln 2)^2)}}.$$

The solution is shown in figure 8.8. In spite of the fact that the first order WKBJ approximation appears to be very close to the true solution, the relative error

$$\frac{y_{\text{exact}} - y_{\text{WKBJ},1}}{y_{\text{exact}}}$$

will actually approach ∞ as $x \to \infty$. ∎

Notice that in figure 8.8, the third order WKBJ approximation is actually worse than the second order. We have observed this phenomenon before, and gave the explanation that the WKBJ series was in fact divergent. When an asymptotic series is uniformly asymptotic in x, we can use the bounds in the ratios S_i/S_{i-1} to determine the *optimal asymptotic approximation* for the WKBJ approximation, that is, for a given ϵ, we can determine the WKBJ order that best approximates the exact solution. Let us define

$$M_n = \max_x\{|S_{n+1}(x)|/|S_n(x)|\}.$$

This notation means that we are to find the maximum value of the expression for all x that is in the interval of the problem. Thus, M_n is the smallest value for which equation 8.11 is satisfied.

Looking back at example 8.4, we can examine figure 8.7 to see that the maximum ratio always occurs at the endpoint $x = 0$. Thus, we can compute the maximum values for the series in equation 8.14 to be $M_1 = 1/(2\ln(2)) \approx 0.721348$ and $M_2 = 1/2$. For equation 8.15, we find the maximum values are much larger: $M_0 = 6\ln(2) \approx 4.15888$, $M_1 = 2/\ln(2) \approx 2.88539$, and

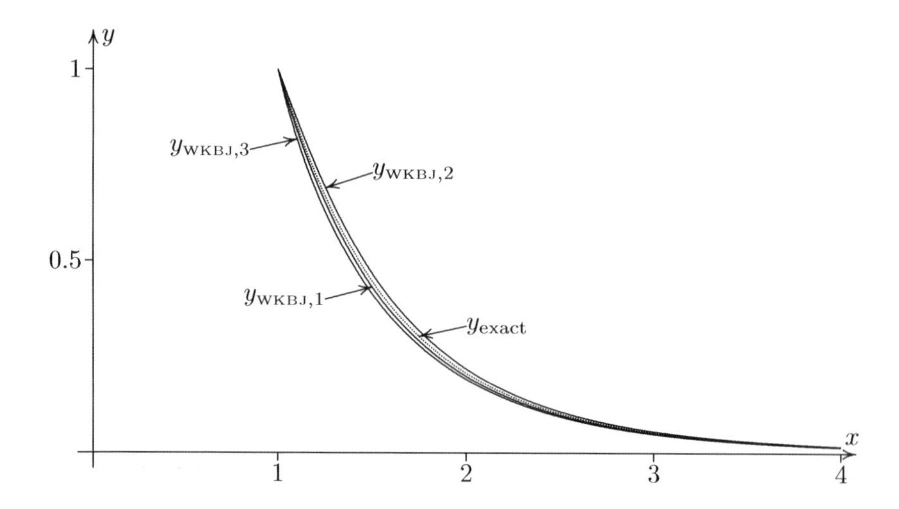

FIGURE 8.8: This shows the WKBJ approximations to the equation $\epsilon y'' - (\ln(2x)/x)^2 y = 0$, $y(1) = 1$, and $y(\infty) = 0$, with $\epsilon = 0.2$. The exact solution is shown with a dotted line. Even though the first order WKBJ approximate appears to be an excellent approximation to the true solution, the *relative* error approaches ∞ as $x \to 0$. Higher WKBJ approximates do not have this problem.

$M_2 = 4.5$. Part of the reason for the differences is that equation 8.14 is a convergent series for $\delta \le 1$, whereas equation 8.15 diverges for all δ.

Recall from subsection 2.2.1 that the optimal asymptotic approximation was found by truncating the series at the point where the terms stopped decreasing. For perturbation series, this occurs at the point where $\delta M_n \ge 1$. That is, if $\delta M_n > 1$ for some n, then the $(n+1)^{\text{st}}$ term will be larger than the nth term for some value of x. Hence, we would keep the S_n term, but the S_{n+1} term would actually make the approximation worse. We can use this in two different ways: for a given WKBJ order, we can determine how small δ must be for this order to be a valid approximation. Likewise, for a given δ, we can determine the optimal WKBJ asymptotic approximation for this δ.

Example 8.8

Consider the WKBJ approximation for the problem in example 8.4,

$$\epsilon y'' - (x+1)y' + 2y = 0, \qquad y(0) = 1, \quad y(1) = 0,$$

with $\epsilon = 0.25$. How many terms should be kept for the optimal WKBJ approximation?

SOLUTION: The asymptotic series for the two possible $S(x)$ has already been given in equations 8.14 and 8.15. Since the first of these converges, we only

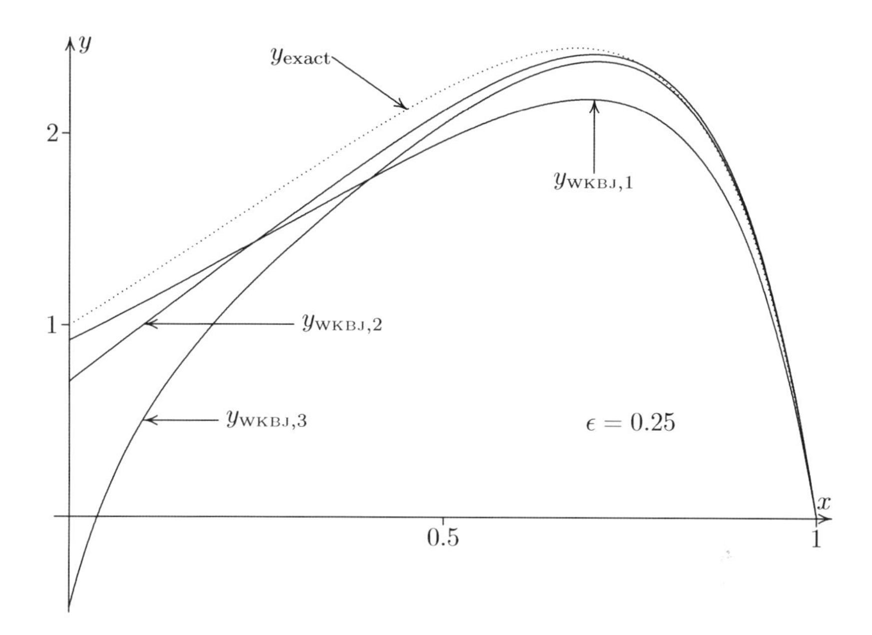

FIGURE 8.9: This demonstrates the optimal truncation rule for WKBJ approximations. For the perturbation problem $\epsilon y'' - (x+1)y' + 2y = 0$, $y(0) = 1, y(1) = 0$, the optimal truncation rule indicates that for $\epsilon = 0.25$, $y_{\text{WKBJ},2}$ is a better approximation than both $y_{\text{WKBJ},1}$ and $y_{\text{WKBJ},3}$.

have to consider the second. In this example, $\delta = \epsilon$, and we have already computed the bounds $M_0 \approx 4.15888$, $M_1 \approx 2.88539$, and $M_2 = 4.5$. Since we must include the first two terms, S_0 and S_1, we don't have to consider M_0. Since $M_1\epsilon \approx 0.721348 < 1$, we see that the S_2 term will be smaller than the S_1 term, so it should be included in the optimal WKBJ approximation. However, $M_2\epsilon = 1.125 > 1$, so the S_3 term will be larger for some values of x. Thus, for this value of ϵ, the optimal WKBJ approximation would be $y_{\text{WKBJ},2}$. Figure 8.9 shows that indeed, $y_{\text{WKBJ},3}$ is a worse approximation than $y_{\text{WKBJ},2}$.

$\square$

Figure 8.9 is a bit disconcerting. Why is there so much inaccuracy near the endpoint at $x = 0$, especially with $y_{\text{WKBJ},3}$? The constants for the WKBJ approximates were chosen so that $y(1)$ would be exactly 0, and that $y(0)$ would be subdominantly small. The WKBJ approximates that satisfy these initial conditions were given by

$$y_{\text{WKBJ},1} = (x+1)^2 - \frac{32e^{(x^2+2x-3)/(2\epsilon)}}{(x+1)^3},$$

$$y_{\text{WKBJ},2} = (x+1)^2 \exp\left(\epsilon - \frac{\epsilon}{(x+1)^2}\right) - \frac{32}{(x+1)^3} \exp\left(\frac{x^2+2x-3}{2\epsilon} + \frac{6\epsilon}{(x+1)^2} - \frac{3\epsilon}{4}\right),$$

and

$$y_{\text{WKBJ},3} = (x+1)^2 \exp\left(\epsilon - \frac{\epsilon}{(1+x)^2} + \frac{\epsilon^2}{2} - \frac{\epsilon^2}{2(1+x)^4}\right) - \frac{32}{(x+1)^3} \exp\left(\frac{x^2+2x-3}{2\epsilon} + \frac{6\epsilon}{(1+x)^2} - \frac{3\epsilon}{4} + \frac{27\epsilon^2}{(1+x)^4} - \frac{39\epsilon^2}{32}\right).$$

We can verify that

$$y_{\text{WKBJ},1}(0) = 1 - 32e^{-3/(2\epsilon)}, \qquad y_{\text{WKBJ},2}(0) = 1 - 32e^{-3/(2\epsilon)+21\epsilon/4}, \quad \text{and}$$

$$y_{\text{WKBJ},3}(0) = 1 - 32e^{-3/(2\epsilon)+21\epsilon/4+825\epsilon^2)/32}.$$

Indeed, the addition term is subdominant as $\epsilon \to 0$, but with ϵ as large as 0.25, these terms are no longer small. In fact, as $\epsilon \to 1$, these terms become huge!

One way to minimize the effect of the dramatically increasing error as δ increases is to expand the series in the exponential function. Because the S_1 term is not going to 0 as $\delta \to 0$, this must be left in the exponent of the exponential. Besides, this term will often involve a logarithm, and exponentiating S_1 will give us the outer solution to the perturbation problem. But we can expand the exponential of the terms that follow:

$$e^{\delta S_2 + \delta^2 S_3 + \delta^3 S_4 + \cdots} \sim 1 + \delta S_2 + \delta^2\left(\frac{S_2^2}{2} + S_3\right) + \delta^3\left(\frac{S_2^3}{6} + S_2 S_3 + S_4\right) + \cdots. \tag{8.16}$$

The plan then is to adjust the arbitrary constants K_ϵ so that the error is at most the same order of epsilon as the first term left out of the series. But in fact, the error at the endpoints will be much smaller than this, once we re-expand the series to incorporate the terms from the constants. See problem 22 for an explanation. This new series will be referred to as the *WKBJ series solution.*

Example 8.9

Find the WKBJ series solution to the problem introduced in example 8.4,

$$\epsilon y'' - (x+1)y' + 2y = 0, \qquad y(0) = 1, \quad y(1) = 0,$$

whose WKBJ expansions are given by

$$S(x) \sim 2\epsilon \ln(1+x) - \frac{\epsilon^2}{(1+x)^2} - \frac{\epsilon^3}{2(x+1)^4} + \cdots$$

and

$$S(x) \sim (2x + x^2)/2 - 3\epsilon \ln(x+1) + \frac{6\epsilon^2}{(1+x)^2} + \frac{27\epsilon^3}{(x+1)^4} + \cdots.$$

SOLUTION: We exponentiate $e^{S(x)/\epsilon}$, using equation 8.16 to simplify from the S_2 term onward. Thus, we have

$$\begin{aligned} y &\sim K_1(1+x)^2 \left(1 - \frac{\epsilon}{(1+x)^2} + \cdots\right) \\ &+ K_2 \frac{e^{(2x+x^2)/(2\epsilon)}}{(1+x)^3} \left(1 + \frac{6\epsilon}{(1+x)^2} + \frac{45\epsilon^2}{(1+x)^4} + \cdots\right). \end{aligned}$$

We still find that K_2 must have order $O(e^{-3/(2\epsilon)})$ in order for $y(1) = 0$. Hence, the K_2 will be subdominant at the endpoint $x = 0$. We calculate that $y(0) \sim K_1(1-\epsilon)$, so we must have $K_1 \sim 1 + \epsilon + \epsilon^2 + \cdots$. Then in order for $y(1) = O(\epsilon^3)$, we must have

$$K_2 \sim -e^{-3/(2\epsilon)}(32 - 24\epsilon - 30\epsilon^2).$$

Thus, we have

$$\begin{aligned} y &\sim (1+\epsilon+\epsilon^2+\cdots)(1+x)^2 \left(1 - \frac{\epsilon}{(1+x)^2} + \cdots\right) \\ &- (32 - 24\epsilon - 30\epsilon^2 + \cdots)\frac{e^{(2x+x^2-3)/(2\epsilon)}}{(1+x)^3} \left(1 + \frac{6\epsilon}{(1+x)^2} + \frac{45\epsilon^2}{(1+x)^4} + \cdots\right). \end{aligned} \tag{8.17}$$

Each term now involves two different series. We can now multiply the two series together for each term. This last step actually increases the accuracy, as indicated by problems 22 and 23.

$$\begin{aligned} y &\sim (1+x)^2 \left(1 + \frac{x^2+2x}{(1+x)^2}\epsilon + \frac{x^2+2x}{(1+x)^2}\epsilon^2 + \cdots\right) + \frac{e^{(2x+x^2-3)/(2\epsilon)}}{(1+x)^3} \\ &\times \left(-32 + \left(24 - \frac{192}{(1+x)^2}\right)\epsilon + \left(30 - \frac{1440}{(1+x)^4} + \frac{144}{(1+x)^2}\right)\epsilon^2 + \cdots\right). \end{aligned} \tag{8.18}$$

To find the nth order WKBJ series approximation, we now can take the first n terms of the series. The results are shown in figure 8.10. Note that the solutions are much closer to the exact answer than in figure 8.9, because the error is not being exponentiated. We still see that $y_{\text{WKBJ},2}$ is closer to the exact answer than $y_{\text{WKBJ},1}$ or $y_{\text{WKBJ},3}$. ∎

Some notes need to be mentioned about this technique of creating a Taylor series beyond the physical optics solution. By truncating the series after we multiply the series for the K_1 and K_2 with the series from the exponentials,

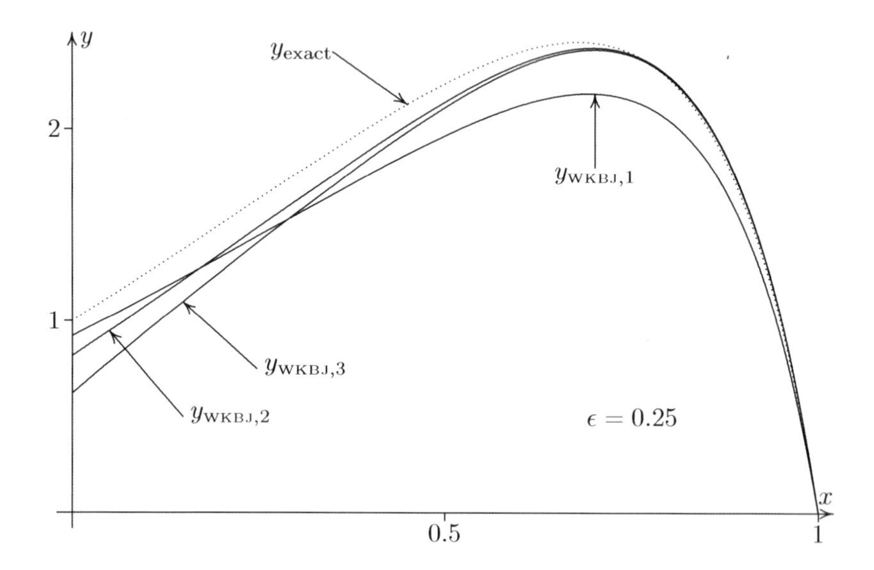

FIGURE 8.10: By expanding the small portions of the exponential functions, and then truncating the series to a certain order, we eliminate the massive error that took place in figure 8.9. There is still a small error at $x = 0$, but the error is of order $e^{-3/(2\epsilon)}$, so this will diminish quickly as ϵ gets smaller.

we are assured that the error at the endpoints will still be subdominant in ϵ. For example, all of the approximations in figure 8.10 go through the point $(1, 0)$. See problem 22 for the explanation.

In the event that some of the S_n are imaginary, as in example 8.1, the imaginary terms will contribute to the phase shift. Thus, only the *real* terms in the exponential would be expanded in the series, and the imaginary terms would end up inside the trig functions.

Most of the time, the series in the WKBJ approximation will diverge, hence the reason for finding the optimal asymptotic approximation to the series. One suggestion would be to rewrite the series into a continued fraction, as was done in section 3.5. Unfortunately, such expansions tend to have spontaneous poles, which would restrict the values of ϵ for which the expansion would be valid. If there are no poles on the entire interval, the results are fairly impressive, but this rarely happens unless ϵ is small.

Problems for §8.2

For problems **1** through **10**: Use one of the equations 8.7, 8.8, or 8.9 to find the first order WKBJ approximation for the following perturbation problems.

1	$\epsilon y'' + (x+1)y' + xy = 0,$	$y(0) = 1,$	$y(1) = 0$
2	$\epsilon y'' + e^x y' + y = 0,$	$y(0) = 0,$	$y(\infty) = 1$
3	$\epsilon y'' + x^2 y' + y = 0,$	$y(1) = 1,$	$y(\infty) = 1$
4	$\epsilon y'' + y'/(2x+1) + y = 0,$	$y(0) = 1,$	$y(1) = 0$
5	$\epsilon y'' + e^x y = 0,$	$y(0) = 0,$	$y(\ln 4) = 1$
6	$\epsilon y'' + (1+x)^2 y = 0,$	$y(0) = 1,$	$y(1) = 0$
7	$\epsilon y'' + (\sec^4 x) y = 0,$	$y(0) = 1,$	$y(\pi/4) = 1$
8	$\epsilon y'' - (1+x)^2 y = 0,$	$y(0) = 1,$	$y(1) = 0$
9	$\epsilon y'' - e^{-x} y = 0,$	$y(0) = 0,$	$y(4) = e$
10	$\epsilon y'' - y/(1+x)^2 = 0,$	$y(0) = 1,$	$y(1) = 0$

11 Find the second order WKBJ correction, S_2, for the case where $S_0' = 0$ in the generalized problem of example 8.5.

12 Find the second order WKBJ correction, S_2, for the case where $S_0' = -p(x)$ in the generalized problem of example 8.5.

13 Find the second order WKBJ correction, S_2, for the generalized Schrödinger problem of example 8.6 in the case where $S_0' = \sqrt{-q(x)}$.

14 Consider the perturbation problem $\epsilon y'' + \epsilon p(x) y' + q(x) y = 0$. Show that the first order WKBJ solution is indistinguishable from example 8.6, in spite of the $p(x)$ term.

15 Find the second order WKBJ correction, S_2, for the equation in problem 14 in the case where $S_0' = \sqrt{-q(x)}$.

16 Show that the series in example 8.2 is *not* uniformly asymptotic in x.

17 Show that constants can be added to the S_n terms in example 8.2 that would make the series uniformly asymptotic in x.

18 Using the results of problem 17, determine how many terms would be included using the optimal approximation if $\epsilon = 0.25$.

19 Show that the problem in example 8.7 can in fact be solved exactly in terms of a parabolic cylinder function.

Hint: First substitute $s = \ln(2x)$ to eliminate the logarithms from the equation. Then apply the substitution suggested by problem 1 of section 7.2 to convert the equation to a Schrödinger equation. Finally, make a scale transformation $s = ct$ for some constant c (which can depend on ϵ) to convert the equation to the form of equation 5.7.

20 In example 8.7, we observed that the ratios S_1/S_0, S_2/S_1, and S_3/S_2 are all bounded on the interval $1 < x < \infty$. However, we will have to make an adjustment by adding a constant to one of the S_n if we are to form a uniformly asymptotic series in x. Which one needs adjusting, and why?

21 Using the results of example 8.7, determine how many terms would be included using the optimal approximation if $\epsilon = 1$.

22 Show that by truncating the series for $y_{\text{WKBJ},n}$ after we fully expand the series for both the exponential and the constants, the error at the endpoints will be of order $e^{-k/\delta}$ for some positive constant k.

23 Show that if we used equation 8.17 for a WKBJ approximation, that is, before we fully expanded the series and truncated them, the error at the endpoints would be of order ϵ^3.

24 Convert the two series in equation 8.18 into continued fraction representations.

25 Consider the perturbation problem

$$\epsilon y'' - xy = 0, \qquad y(1) = 1, \quad y(\infty) = 0.$$

First find the second order WKBJ approximation, and then show that we can solve this equation in terms of the Airy function. Finally, show that if we apply the expansion of the Airy function from equation 5.4, we get precisely the WKBJ solution.

26 The third order WKBJ solution to example 8.1 is given in equation 8.3. Convert this to a WKBJ series solution, as was done in example 8.9.

Note: We only expand the exponential if the terms are real.

27 For the equation in problem 3,

$$\epsilon y'' + x^2 y' + y = 0, \qquad y(1) = 1, \quad y(\infty) = 1$$

find S_0, S_1 and S_2, and verify that the physical optics WKBJ approximation is uniform in x.

28 In problem 27, the WKBJ approximation clearly breaks down at $x = 0$. Determine how close to $x = 0$ the physical optics WKBJ approximation remains valid.

Hint: As long as $S_0/\delta \gg S_1 \gg \delta S_2$ and $\delta S_2 \ll 1$, the physical optics solution will be valid. Of course if $S_0 = 0$, the condition $S_0/\delta \gg S_1$ does not have to be satisfied.

29 In example 8.7, the WKBJ approximation breaks down at $x = 1/2$. Determine how close to $x = 1/2$ the second order WKBJ approximation remains valid.

Hint: This time, we need $S_0/\delta \gg S_1 \gg \delta S_2 \gg \delta^2 S_3$ and $\delta^2 S_3 \ll 1$.

8.3 Turning Points

By observing the solution to example 8.6, we find that the WKBJ solution is not valid in the neighborhood of a point where $q(x) = 0$. A similar problem occurs in example 8.5 in the vicinity of a point where $p(x) = 0$. Ironically, the exact solution will usually not have any singularities at this point, even though the physical optics approximation has a discontinuity. The points where $q(x) = 0$ in the Schrödinger equation of example 8.6, or the points where $p(x) = 0$ in the general second order equation of example 8.5, are called the *turning points* of the equation. The terminology comes from Schrödinger's quantum mechanical equation to describe a particle with potential energy $V(x)$:

$$\epsilon y'' + (E - V(x))y = 0.$$

Here, E is the total energy of the particle. Under the classic model, the particle will move until it reaches the "turning point," that is, the point where $V(x) = E$, and then move in the opposite direction. The quantum mechanical model actually predicts that the particle has a small chance of getting past the point where the potential energy is higher than the total energy, a process known as *tunneling*.

Even though the terminology is rooted in the quantum mechanics model, often the behavior of the solution will undergo a transition in the vicinity of the turning point. For example, in the general Schrödinger equation, if $q(x)$ changes from positive to negative, the solutions will change from a sinusoidal

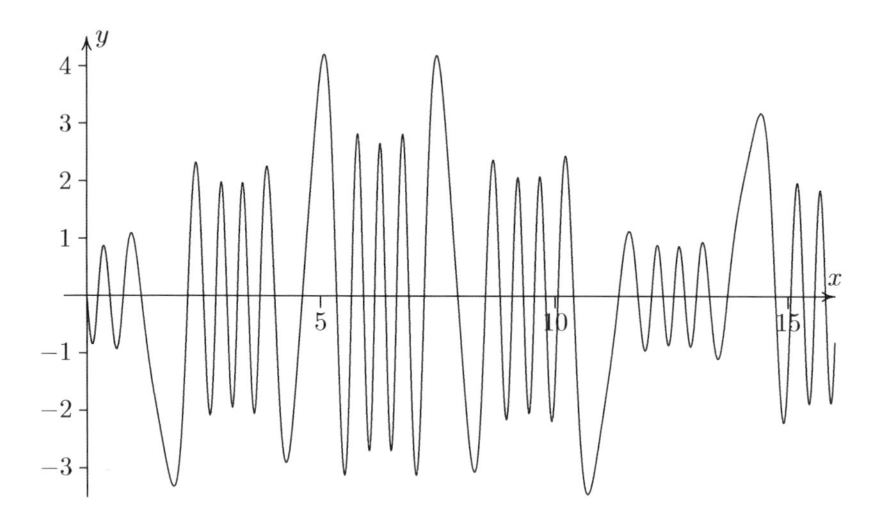

FIGURE 8.11: This shows the exact solution to $\epsilon y'' + \cos^2 xy = 0$, with $y(0) = 0$ and $y(1) = 1$. The value of ϵ is 0.0054. Whenever $\cos^2 x = 0$, there is a turning point in the equation, causing the behavior of the solution to change.

behavior to an exponential behavior. So we can think of the turning points as the places where the overall behavior of the solutions can turn.

The turning points are particularly evident if we plot the continuation of the exact solution to example 8.1, $\epsilon y'' + \cos^2(x)y = 0$ with $\epsilon = 0.0054$, we find that near each turning point $x = (2k+1)\pi/2$, the solution changes the amplitude. This is shown in figure 8.11. Both the phase shift and amplitude change across each of the turning points, yet the solution remains continuous. In fact, the solution is changing less rapidly near the turning points than elsewhere on the graph.

If there is a single turning point in the problem, we can approach the problem as if it were an interior boundary layer problem. For simplicity, we can assume that the turning point occurs at $x = 0$. We can then form a WKBJ approximation that is valid for $x > \delta_1$, where the relationship of δ_1 and ϵ will be determined later. There will be another WKBJ approximation valid in the region $x < -\delta_1$. Finally, for $x = O(\delta_2)$, we can solve the equation as we would for the inner solution of a boundary layer problem. These three pieces then have to be matched to form the complete solution.

Example 8.10

Consider the perturbation problem

$$\epsilon y'' + x(1+x^2)^2 y = 0, \qquad y(-1) = y(1) = 1.$$

There is a turning point at $x = 0$. Find the first order WKBJ approximations for the two regions $x < 0$ and $x > 0$, and then find the inner solution near $x = 0$.

SOLUTION: For $x > 0$, we can apply the result of equation 8.8 with $q(x) = x(x^2+1)^2$. In this case we can compute $\int \sqrt{q(x)}\,dx$ to be $2\sqrt{x}(x/3 + x^3/7)$. Thus, we find that in this region,

$$y_{\text{right}} \sim \frac{K_1 \sin(\omega(x))}{x^{1/4}\sqrt{1+x^2}} + \frac{K_2 \cos(\omega(x))}{x^{1/4}\sqrt{1+x^2}},$$

where

$$\omega(x) \sim \frac{2\sqrt{x}(7x+3x^3)}{21\sqrt{\epsilon}}.$$

For the region $x < 0$, $q(x)$ will be negative, so we can apply equation 8.9. The result is

$$y_{\text{left}} \sim \frac{K_3 e^{2\sqrt{-x}(7x+3x^3)/(21\sqrt{\epsilon})}}{(-x)^{1/4}\sqrt{1+x^2}} + \frac{K_4 e^{-2\sqrt{-x}(7x+3x^3)/(21\sqrt{\epsilon})}}{(-x)^{1/4}\sqrt{1+x^2}}.$$

Note that because of the exponential nature of this function, we can utilize the initial condition $y(-1) = 1$. Plugging in this value, we get

$$y_{\text{left}}(-1) = \frac{K_3}{\sqrt{2}} e^{-20/(21\sqrt{\epsilon})} + \frac{K_4}{\sqrt{2}} e^{20/(21\sqrt{\epsilon})}.$$

Clearly, the K_3 term will be subdominant, so we can get $y_{\text{left}}(-1) \sim 1$ by letting $K_4 = \sqrt{2}e^{-20/(21\sqrt{\epsilon})}$.

It is trickier to utilize an initial condition on a sinusoidal solution such as y_{right}, but it can be done with a phase shift. We can express the right WKBJ solution as

$$y_{\text{right}} \sim \frac{C_1 \sin(\omega(x) - \omega(1)) + C_2 \cos(\omega(x) - \omega(1))}{x^{1/4}\sqrt{1+x^2}}.$$

The advantage of expressing the solution in this form is that we can evaluate at $x = 1$ to obtain $y_{\text{right}}(1) = C_2/\sqrt{2}$. So the initial condition $y(1) = 1$ gives us that $C_2 \sim \sqrt{2}$.

Finally, we have the region where x is small, that is, $x \ll 1$. In this region, we can approximate the solution by

$$\epsilon y'' + xy = 0.$$

This equation can be solved via the Airy functions. Letting $x = -t\epsilon^{1/3}$, we find that $y''(x) = \epsilon^{-2/3} y''(t)$. So the new equation in t is merely $y'' = ty$, whose solution is $K_5\text{Ai}(t) + K_6\text{Bi}(t)$. Thus, we have that for small x,

$$y_{\text{in}} \sim K_5\text{Ai}(-\epsilon^{-1/3}x) + K_6\text{Bi}(-\epsilon^{-1/3}x).$$

□

At this point, the three solutions look radically different. The right solution is sinusoidal in nature, and the left solution is exponential. However, the inner solution involves the Airy function, which has both an exponential behavior and an oscillating behavior. If we consider the behavior of the inner solution as $x \gg \sqrt[3]{\epsilon}$, we can get the Airy function to match to two WKBJ solutions. The inner solution becomes the glue that allows us to attach the other two WKBJ solutions together.

For example, if $x < 0$ and $|x| \gg \sqrt[3]{\epsilon}$, then $t = -\epsilon^{-1/3}x \to \infty$. We can use the behavior for the Airy functions for large t, obtained through equation 5.4,

$$\mathrm{Ai}(t) \sim \frac{e^{-2t^{3/2}/3}}{2\sqrt{\pi}t^{1/4}}\left(1 - \frac{5}{48t^{3/2}} + \cdots\right) \qquad \text{as } t \to \infty,$$

and

$$\mathrm{Bi}(t) \sim \frac{e^{2t^{3/2}/3}}{\sqrt{\pi}t^{1/4}}\left(1 + \frac{5}{48t^{3/2}} + \cdots\right) \qquad \text{as } t \to \infty.$$

Thus, we see that for $x < 0$,

$$y_{\text{in}} \sim K_5 \frac{\epsilon^{1/12}e^{2x\sqrt{-x}/(3\sqrt{\epsilon})}}{2\sqrt{\pi}(-x)^{1/4}} + K_6 \frac{\epsilon^{1/12}e^{-2x\sqrt{-x}/(3\sqrt{\epsilon})}}{\sqrt{\pi}(-x)^{1/4}}. \tag{8.19}$$

This is starting to look like the asymptotic expansion of y_{left}. But there is an important issue. Will the WKBJ approximation be valid in a region where $x \ll 1$?

Recall from the last section that the WKBJ approximation was valid provided that $S_0/\delta \gg S_1 \gg \delta S_2$, and that $\delta S_2 \ll 1$. Since

$$S_0 = \pm 2\sqrt{-x}(7x + 3x^3)/21, \qquad S_1 = -\ln(x(1 + x^2)^2)/4,$$

and we can compute that $S_2 \sim \mp 10ix^{-3/2}/96$, these conditions are satisfied as long as $(-x) \gg \epsilon^{1/3}(\ln \epsilon)^{2/3}$. Actually, the exact point of validity is not important for the overlap hypothesis, only that there is *some* region of overlap, namely $\epsilon^{1/3}(\ln \epsilon)^{2/3} \ll |x| \ll 1$, for which both the WKBJ approximation and the inner solution are valid. This will allow us to use Van Dyke's matching rule, where the inner variable is $t = \pm x/\epsilon^{1/3}$. We have two expansions of the inner solution, one for $x < 0$, which we can naturally call the left expansion, and the other right expansion valid for $x > 0$. We can then match these to the inner approximations to the left and right WKBJ solutions.

Example 8.11

Use Van Dyke's matching rule to find the constants C_1, K_3, K_5, and K_6 from example 8.10.

SOLUTION: Equation 8.19 already has the inner solution expanded in the outer variable for $x < 0$, which we can refer to as $(y_{\text{in}})_{\text{left}}$. Thus, we need to consider the outer left solution expanded in the inner variable, $x = -t\epsilon^{1/3}$. Then

$$e^{2\sqrt{-x}(7x+3x^3)/(21\sqrt{\epsilon})} = e^{-2t^{3/2}/3-2t^{7/2}\epsilon^{2/3}/7} \sim e^{-2t^{3/2}/3} = e^{2x\sqrt{-x}/(3\sqrt{\epsilon})},$$

since we can clearly discard the $\epsilon^{2/3}$ term from the exponent, since it is approaching 0. Likewise, we can approximate

$$\sqrt{1+x^2} = \sqrt{1+t^2\epsilon^{2/3}} \sim 1.$$

So we have

$$(y_{\text{left}})_{\text{in}} = K_3(-x)^{-1/4}e^{2x\sqrt{-x}/(3\sqrt{\epsilon})} + K_4(-x)^{-1/4}e^{-2x\sqrt{-x}/(3\sqrt{\epsilon})}.$$

Comparing this equation with equation 8.19, we can determine that $K_3 = K_5\epsilon^{1/12}/(2\sqrt{\pi})$ and $K_4 = K_6\epsilon^{1/12}/\sqrt{\pi}$. Since we already know that $K_4 = e^{-20/(21\sqrt{\epsilon})}$, we find that $K_6 = \sqrt{2\pi}\epsilon^{-1/12}e^{-20/(21\sqrt{\epsilon})}$.

We can now find the behavior of the inner solution for $x > 0$, and match this to the right solution. In example 5.8, we found the behavior of the Airy functions for negative arguments:

$$\text{Ai}(x) \approx \frac{(-x)^{-1/4}}{\sqrt{\pi}} \sin\left(\frac{2}{3}(-x)^{3/2} + \frac{\pi}{4}\right),$$

and

$$\text{Bi}(x) \approx \frac{(-x)^{-1/4}}{\sqrt{\pi}} \cos\left(\frac{2}{3}(-x)^{3/2} + \frac{\pi}{4}\right).$$

Thus, we find that

$$(y_{\text{in}})_{\text{right}} \sim \frac{x^{-1/4}\epsilon^{1/12}}{\sqrt{\pi}}\left(K_5 \sin\left(\frac{2x^{3/2}}{3\sqrt{\epsilon}} + \frac{\pi}{4}\right) + K_6 \cos\left(\frac{2x^{3/2}}{3\sqrt{\epsilon}} + \frac{\pi}{4}\right)\right).$$

If we consider the right WKBJ solution for small values of x, we obtain

$$(y_{\text{right}})_{\text{in}} =\sim x^{-1/4}\left(C_1 \sin\left(\frac{2x^{3/2}}{3\sqrt{\epsilon}} - \omega(1)\right) + C_2 \cos\left(\frac{2x^{3/2}}{3\sqrt{\epsilon}} - \omega(1)\right)\right).$$

The phase angles are different between $(y_{\text{in}})_{\text{right}}$ and $(y_{\text{right}})_{\text{in}}$, but we can correct this using the formulas

$$\sin(\alpha - \beta) = \sin(\alpha)\cos(\beta) - \cos(\alpha)\sin(\beta),$$

and

$$\cos(\alpha - \beta) = \sin(\alpha)\sin(\beta) + \cos(\alpha)\cos(\beta),$$

where we will use $\alpha = 2x^{3/2}/(3\sqrt{\epsilon}) + \pi/4$, and $\beta = \pi/4 + \omega(1)$. Then

$$\begin{aligned}(y_{\text{right}})_{\text{in}} = &\sim C_1 x^{-1/4}\cos(\pi/4+\omega(1))\sin\left(\frac{2x^{3/2}}{3\sqrt{\epsilon}} - \frac{\pi}{4}\right)\\ &- C_1 x^{-1/4}\sin(\pi/4+\omega(1))\cos\left(\frac{2x^{3/2}}{3\sqrt{\epsilon}} - \frac{\pi}{4}\right)\\ &+ C_2 x^{-1/4}\sin(\pi/4+\omega(1))\sin\left(\frac{2x^{3/2}}{3\sqrt{\epsilon}} - \frac{\pi}{4}\right)\\ &+ C_2 x^{-1/4}\cos(\pi/4+\omega(1))\cos\left(\frac{2x^{3/2}}{3\sqrt{\epsilon}} - \frac{\pi}{4}\right).\end{aligned}$$

Thus,

$$K_5\epsilon^{1/12}/\sqrt{\pi} = C_1\cos(\pi/4+\omega(1)) + C_2\sin(\pi/4+\omega(1)) \text{ and}$$
$$K_6\epsilon^{1/12}/\sqrt{\pi} = C_2\cos(\pi/4+\omega(1)) - C_1\sin(\pi/4+\omega(1)).$$

We have already seen that K_4 is subdominantly small. As long as $\omega(1)+\pi/4$ doesn't happen to be a multiple of π, we can solve for $C_1 \sim \sqrt{2}\cot(\pi/4+\omega(1))$. Finally, we can solve for $K_5 \sim \sqrt{2\pi}\epsilon^{-1/12}\csc(\pi/4+\omega(1))$, which also can be used to find $K_3 \sim \csc(\pi/4+\omega(1))/\sqrt{2}$. □

Having found the constants for the different pieces of the WKBJ approximation, it is time to fit the pieces together to form a single function. At this point, we have

$$\begin{aligned}y_{\text{left}} \sim\; &\frac{\csc(\pi/4+\omega(1))}{\sqrt{2}(-x)^{1/4}\sqrt{1+x^2}}e^{2\sqrt{-x}(7x+3x^3)/(21\sqrt{\epsilon})}\\ &+ \frac{\sqrt{2}e^{-20/(21\sqrt{\epsilon})}}{(-x)^{1/4}\sqrt{1+x^2}}e^{-2\sqrt{-x}(7x+3x^3)/(21\sqrt{\epsilon})}.\end{aligned}$$

$$\begin{aligned}y_{\text{in}} \sim\; &\sqrt{2\pi}\epsilon^{-1/12}\csc(\pi/4+\omega(1))\mathrm{Ai}(-\epsilon^{-1/3}x)\\ &+ \sqrt{2\pi}\epsilon^{-1/12}e^{-20/(21\sqrt{\epsilon})}\mathrm{Bi}(-\epsilon^{-1/3}x).\end{aligned}$$

$$y_{\text{right}} \sim \frac{\sqrt{2}\cot(\pi/4+\omega(1))\sin(\omega(x)-\omega(1)) + \sqrt{2}\cos(\omega(x)-\omega(1))}{x^{1/4}\sqrt{1+x^2}}.$$

These three functions are shown in figure 8.12. Each function is close to the exact solution in the associated region.

Because the left and right WKBJ approximations are so different from each other, we have to be a bit more clever in combining these functions into a single uniform approximation. In 1935, Rudolph Langer came up with the brilliant idea of putting the expression $(3S_0/(2\delta))^{2/3}$ inside the Airy functions, where $S_0 = \int\sqrt{-q(x)}\,dx$. Then when the Airy functions are expanded as

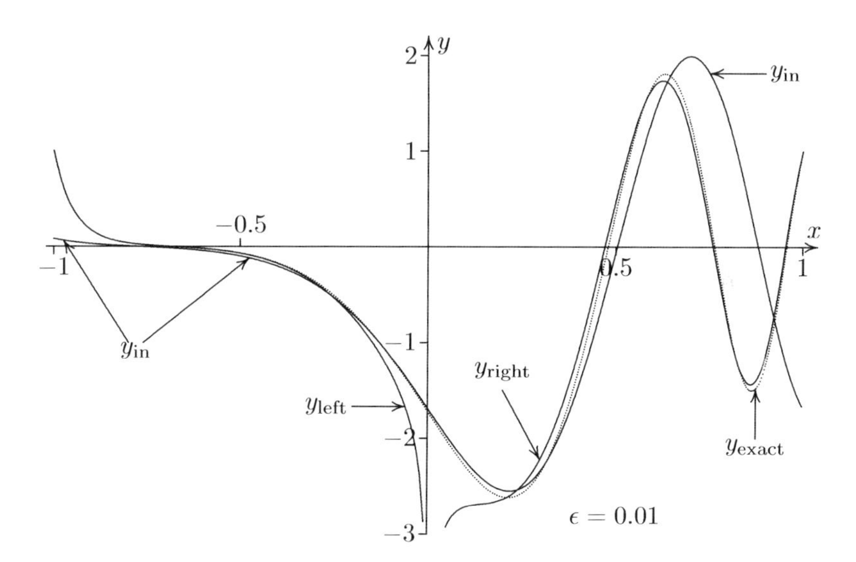

FIGURE 8.12: This shows the three regional WKBJ approximations, y_{left}, y_{in}, and y_{right}, for the equation $\epsilon y'' + x(1+x^2)^2 y = 0$, with $y(-1) = y(1) = 1$. Here, we used $\epsilon = 0.01$. The exact solution is shown with a dotted line. The left WKBJ approximation practically covers the exact solution for $x < -0.5$. The right WKBJ solution is also extremely close to the exact solution for $x > 0.5$. The inner solution is close to the exact solution between these, and acts as a bridge between the exponential behavior on the left, and the sinusoidal behavior on the right.

exponentials, the dominant behavior would become $e^{\pm S_0/\delta}$, which we see in the WKBJ solutions. In this example, $S_0 = 2\sqrt{-x}(7x+3x^3)/21$, which we can rewrite as $-2(-x)^{3/2}(1+3x^2/7)/3$. Hence, $(3S_0/(2\delta))^{2/3} = -\epsilon^{-1/3}x(1+3x^2/7)^{2/3}$. So we can replace the $-\epsilon^{-1/3}x$ inside the Airy functions for the inner solution, to produce

$$y \sim \sqrt{2\pi}\epsilon^{-1/12}\csc(\pi/4+\omega(1))\mathrm{Ai}(-\epsilon^{-1/3}x(1+3x^2/7)^{2/3}) \\ + \sqrt{2\pi}\epsilon^{-1/12}e^{-20/(21\sqrt{\epsilon})}\mathrm{Bi}(-\epsilon^{-1/3}x(1+3x^2/7)^{2/3}).$$

Clearly, when $x \ll 1$, this will be asymptotic to the inner solution. But when $x < 0$, then we can replace the Airy functions with their asymptotic behavior to get

$$\csc(\pi/4+\omega(1))\frac{(-x)^{-1/4}(1+3x^2/7)^{-1/6}}{\sqrt{2}}e^{2\sqrt{-x}(7x+3x^3)/(21\sqrt{\epsilon})} \\ + \sqrt{2}e^{-20/(21\sqrt{\epsilon})}(-x)^{-1/4}(1+3x^2/7)^{-1/6}e^{-2\sqrt{-x}(7x+3x^3)/(21\sqrt{\epsilon})}.$$

This is close to y_{left}, but we are off by a factor of $(1+3x^2/7)^{1/6}(1+x^2)^{-1/2}$. But this factor approaches 1 in the region $x \ll 1$, so we can multiply the inner solution by this factor, and it would still be valid. This gives us a composite uniform approximation

$$y_{\text{comp}} \sim \sqrt{2\pi}\epsilon^{-1/12}(1+3x^2/7)^{1/6}(1+x^2)^{-1/2} \\ \times \left(\csc(\pi/4+\omega(1))\mathrm{Ai}(-\epsilon^{-1/3}x(1+3x^2/7)^{2/3}) \right. \\ \left. + e^{-20/(21\sqrt{\epsilon})}\mathrm{Bi}(-\epsilon^{-1/3}x(1+3x^2/7)^{2/3})\right).$$

We still have to check that this matches the right WKBJ approximation. Replacing the Airy functions with their respective behaviors for $x > 0$, we obtain

$$y_{\text{comp}} \sim \sqrt{2}(1+x^2)^{-1/2}\csc(\pi/4+\omega(1))x^{-1/4}\sin\left(\frac{2\sqrt{x}(7x+3x^3)}{21\sqrt{\epsilon}}+\frac{\pi}{4}\right) \\ + \sqrt{2}(1+x^2)^{-1/2}e^{-20/(21\sqrt{\epsilon})}x^{-1/4}\cos\left(\frac{2\sqrt{x}(7x+3x^3)}{21\sqrt{\epsilon}}+\frac{\pi}{4}\right).$$

At first, this doesn't look like the right hand WKBJ solution, but notice that the cosine term is subdominant, and

$$\sin\left(\frac{2\sqrt{x}(7x+3x^3)}{21\sqrt{\epsilon}}+\frac{\pi}{4}\right) = \sin\left(\frac{2\sqrt{x}(7x+3x^3)}{21\sqrt{\epsilon}}-w(1)\right)\cos(w(1)+\pi/4) \\ + \cos\left(\frac{2\sqrt{x}(7x+3x^3)}{21\sqrt{\epsilon}}-w(1)\right)\sin(w(1)+\pi/4).$$

So indeed, the right WKBJ solution is reproduced by the composite solution. Figures 8.13 and 8.14 show that the composite WKBJ approximation can give very impressive accuracy.

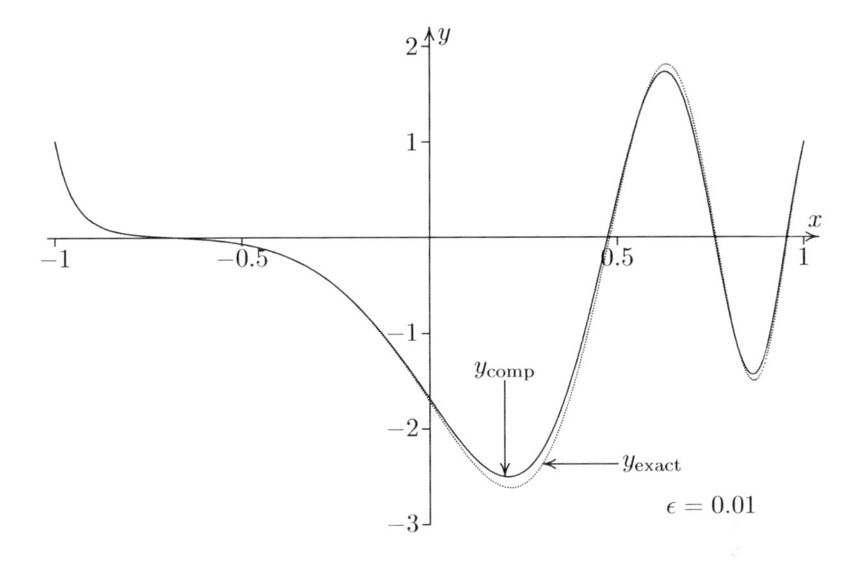

FIGURE 8.13: This compares the composite uniform WKBJ approximation for the equation $\epsilon y'' + x(1 + x^2)^2 y = 0$, $y(-1) = y(1) = 1$ with $\epsilon = 0.01$, with the exact solution, shown with a dotted line. Notice that the composite approximation mimics all of the features of the exact solution, such as the exponential decay, exponential growth, and sinusoidal behavior.

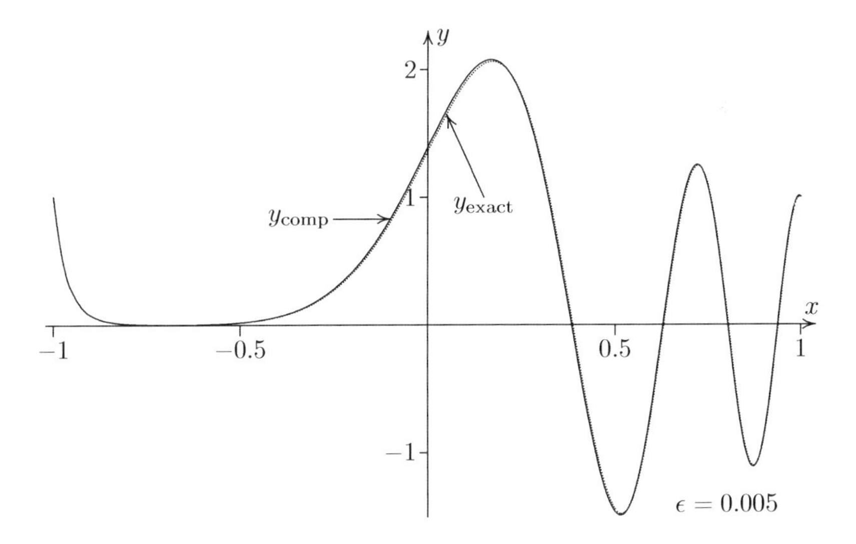

FIGURE 8.14: Same as figure 8.13, but with $\epsilon = 0.005$. This time, the two curves are nearly indistinguishable.

8.3.1 One Simple Root Turning Point Problem

We can generalize example 8.10 to cover a more general Schrödinger equation. For convenience, we will assume that $q(x)$ has only one simple root at $x = 0$, so that $q(x) \sim mx$ as $x \to 0$ for some slope $m \neq 0$. If $m > 0$ there will be oscillatory behavior on the right hand WKBJ approximation, as in example 8.10, but if $m < 0$ the oscillatory behavior will be on the left side. One last assumption we will make is that $q(x)$ does not approach 0 too quickly as $x \to \pm\infty$, say $qx \gg x^{-2}$ as $x \to \infty$. This is to insure the approximation is uniformly valid from $-\infty < x < \infty$. We will compute the case for $m < 0$, since the other case can be obtained through a simple substitution $x = -t$. (See problem 2.)

Example 8.12

Find the leading order WKBJ approximation to the turning point problem

$$\epsilon y'' + q(x)y = 0, \qquad y(a) = 0$$

where $a > 0$ (possibly ∞), and $q(x)$ is real, continuous, and has only one root at $x = 0$, at which the slope is negative.

SOLUTION: We can assume that $q'(0) = -b$, with $b > 0$. From the information given, we can infer that $q(x) < 0$ for $x > 0$, and $q(x) > 0$ for $x < 0$. Because only one initial condition is given, the solution will be determined up to a multiplicative constant. A second condition could easily be imposed at a point where $x < 0$.

In the region where $x > 0$, $q(x) < 0$, so we can apply the results of equation 8.9:

$$y_{\text{right}} \sim \frac{K_1}{\sqrt[4]{-q(x)}} e^{\int_0^x \sqrt{-q(t)}\,dt/\sqrt{\epsilon}} + \frac{K_2}{\sqrt[4]{-q(x)}} e^{-\int_0^x \sqrt{-q(t)}\,dt/\sqrt{\epsilon}}.$$

If we let $M = \int_0^a \sqrt{-q(t)}\,dt$, then the initial condition $y(a) = 0$ produces $K_1 e^{M/\sqrt{\epsilon}} + K_2 e^{-M/\sqrt{\epsilon}} = 0$. Since $M > 0$ (possibly even ∞), we see that K_1 must be subdominant to K_2. We know from experience of example 8.10 that K_2 will not be exponentially large. Thus, we can assume that $K_1 = 0$, which would only produce subdominant errors at $y(a)$.

For the region where $x \ll 1$, we can approximate the equation with

$$\epsilon y'' - bxy = 0,$$

which will simplify to the Airy equation $y'' = ty$ with the substitution $t = (b/\epsilon)^{1/3}x$. Thus, the inner solution is

$$y_{\text{in}} = K_3 \text{Ai}(b^{1/3}\epsilon^{-1/3}x) + K_4 \text{Bi}(b^{1/3}\epsilon^{-1/3}x).$$

This is fairly easy to match to the right hand solution. We know that when $x\epsilon^{-1/3}$ is large,

$$(y_{\text{in}})_{\text{right}} = K_3 \frac{1}{2\sqrt{\pi}} b^{-1/12} \epsilon^{1/12} x^{-1/4} e^{-2\sqrt{b}x^{3/2}/(3\sqrt{\epsilon})} + K_4 \frac{1}{\sqrt{\pi}} b^{-1/12} \epsilon^{1/12} x^{-1/4} e^{2x^{3/2}\sqrt{b}/(3\sqrt{\epsilon})}.$$

On the other hand, when x is close to 0, we can replace $q(x)$ with $-bx$. Thus,

$$(y_{\text{right}})_{\text{in}} = K_1 x^{-1/4} b^{-1/4} e^{2\sqrt{b}x^{3/2}/(3\sqrt{\epsilon})} + K_2 x^{-1/4} b^{-1/4} e^{-2\sqrt{b}x^{3/2}/(3\sqrt{\epsilon})}.$$

Hence, $K_4 = \sqrt{\pi} K_1 b^{-1/6} \epsilon^{-1/12}$ and $K_3 = 2\sqrt{\pi} K_2 b^{-1/6} \epsilon^{-1/12}$. But we have already established that $K_1 = 0$ from the initial conditions, so $K_4 = 0$.

In preparing to match the inner solution to the left hand solution, we consider the behavior as $x\epsilon^{-1/3} \to -\infty$.

$$(y_{\text{in}})_{\text{left}} = (K_3 w_1(x) + K_4 w_2(x)) \sin\left(\frac{2\sqrt{b}(-x)^{3/2}}{3\sqrt{\epsilon}} + \frac{\pi}{4}\right) + (K_4 w_1(x) - K_3 w_2(x)) \cos\left(\frac{2\sqrt{b}(-x)^{3/2}}{3\sqrt{\epsilon}} + \frac{\pi}{4}\right),$$

where

$$w_1(x) \sim (-x)^{-1/4} \epsilon^{1/12} b^{-1/12} / \sqrt{\pi}, \qquad \text{and}$$
$$w_2(x) \sim 5(-x)^{-7/4} \epsilon^{7/12} b^{-7/12} / (48\sqrt{\pi}).$$

Finally, we consider the region where $x < 0$. Then $q(x) > 0$, so we can use equation 8.8:

$$y_{\text{left}} \sim \frac{K_5}{\sqrt[4]{q(x)}} \cos(\omega(x)) + \frac{K_6}{\sqrt[4]{q(x)}} \sin(\omega(x)), \tag{8.20}$$

where

$$\omega(x) \sim \int_x^0 \sqrt{q(t)}\, dt / \sqrt{\epsilon}.$$

But we can add an arbitrary constant to $\omega(x)$, so we can insist that $\omega(0) = \pi/4$. Then for small x, we can replace $q(x)$ with $-bx$, so that

$$\omega(x) \sim 2\sqrt{b}(-x)^{3/2}/(3\sqrt{\epsilon}) + \pi/4.$$

Thus,

$$(y_{\text{left}})_{\text{in}} \approx K_5 b^{-1/4} (-x)^{-1/4} \cos\left(\frac{2\sqrt{b}(-x)^{3/2}}{3\sqrt{\epsilon}} + \frac{\pi}{4}\right) + K_6 b^{-1/4} (-x)^{-1/4} \sin\left(\frac{2\sqrt{b}(-x)^{3/2}}{3\sqrt{\epsilon}} + \frac{\pi}{4}\right).$$

Hence, we can match the left WKBJ solution to the inner solution to find that $K_5 = K_4 b^{1/6}\epsilon^{1/12}/\sqrt{\pi}$ and $K_6 = K_3 b^{1/6}\epsilon^{1/12}/\sqrt{\pi}$. But since $K_4 = 0$ we also have $K_5 = 0$.

In combining the three solutions into a composite approximation, we replace the argument of the Airy function in the inner solution with $(3S_0/(2\delta))^{2/3}$, where $S_0 = \int \sqrt{-q(x)}\, dx$. So the first estimate for the composite approximation is

$$y = K_3 \mathrm{Ai}\left(\left(\frac{3}{2\sqrt{\epsilon}}\int_0^x \sqrt{-q(t)}\, dt\right)^{2/3}\right).$$

Only if we take the asymptotic approximation of this as $x \to \infty$, we get

$$y \sim \frac{K_3}{2\sqrt{\pi}}\left(\frac{3}{2\sqrt{\epsilon}}\int \sqrt{-q(x)}\, dx\right)^{-1/6} e^{-\int_0^t \sqrt{-q(t)}\, dt/\sqrt{\epsilon}}.$$

Since we want this to match

$$y_{\text{right}} \sim \frac{K_3}{2\sqrt{\pi}} b^{1/6}\epsilon^{1/12}(-q(x))^{-1/4} e^{-\int_0^x \sqrt{-q(t)}\, dt/\sqrt{\epsilon}},$$

we apparently have to multiply y by

$$b^{1/6}\epsilon^{1/12}(-q(x))^{-1/4}\left(\frac{3}{2\sqrt{\epsilon}}\int_0^x \sqrt{-q(t)}\, dt\right)^{1/6}.$$

But can we multiply by this without affecting the inner approximation? If $x \ll 1$, we can replace $q(x)$ with $-bx$, and so the above expression is asymptotic to 1 for small x. Thus, we have

$$\boxed{y_{\text{comp}} \sim \frac{K_3 b^{1/6}}{\sqrt[4]{-q(x)}}\left(\frac{3}{2}\int_0^x \sqrt{-q(t)}\, dt\right)^{1/6} \mathrm{Ai}\left(\left(\frac{3}{2\sqrt{\epsilon}}\int_0^x \sqrt{-q(t)}\, dt\right)^{2/3}\right).} \tag{8.21}$$

This will give a uniform approximation provided that $q(x) \gg x^{-2}$ as $x \to \pm\infty$. Verifying that y_{comp} agrees with y_{left} for $x < 0$ is left as an exercise. (See problem 1.) □

8.3.2 Parabolic Turning Point Problems

The key to finding a WKBJ approximation through a turning point is that we can find the exact solution to the inner equation, so that we have global asymptotic information about the solutions to the equation. When there is a simple zero at the turning point for a Schrödinger equation, the WKBJ approximation will be expressed in terms of the Airy function. If $q(x)$ behaves like bx^2 near the turning point at 0, finding the inner solution requires solving an equation of the form

$$y'' \pm \frac{x^2}{4} y = 0.$$

The equation $y'' - x^2y/4 = 0$ is the parabolic cylinder equation for $v = -1/2$. All solutions have an exponential behavior both as $x \to \infty$ and $x \to -\infty$, so there will be no sinusoidal behavior. The solutions will be very similar to that of an internal boundary layer problem using boundary layer theory.

The more interesting case is when $q(x) = 0$, $q'(x) = 0$, and $q''(x) > 0$ at the turning point. Such a turning point is called a *parabolic turning point*. Analyzing such a point would require solving the equation

$$y'' + \frac{x^2}{4}y = 0. \tag{8.22}$$

This is a special case of the equation $y'' + (x^2/4 - a)y = 0$, which many references in literature refer to as the *modified parabolic cylindrical equation*. The solutions can be expressed in terms of the parabolic cylindrical functions using a complex argument.

$$y = C_1 D_{-1/2-ia}(e^{-i\pi/4}x) + C_2 D_{-1/2+ia}(e^{i\pi/4}x).$$

However, we consider the particular equation with $a = 0$, along with the particular solution

$$G_0(x) = \frac{e^{i\pi/8}D_{-1/2}(e^{-i\pi/4}x) + e^{-i\pi/8}D_{-1/2}(e^{i\pi/4}x)}{2}. \tag{8.23}$$

This solution has some extra properties. Both $G_{-1/2}(x)$ and $G_{-1/2}(-x)$ will be real solutions to equation 8.22, but these two solutions will be $\pi/2$ out of phase with each other as $x \to \infty$, although the amplitudes will be different. We can refer to this function as the *modified parabolic cylindrical function* of order 0.

We can determine (see problem 13) the full asymptotic expansion of the modified parabolic cylindrical function as $x \to \infty$ to be

$$G_0(x) = w_1(x)\frac{1}{\sqrt{x}}\cos\left(\frac{x^2+\pi}{4}\right) + w_2(x)\frac{1}{\sqrt{x}}\sin\left(\frac{x^2+\pi}{4}\right), \tag{8.24}$$

which is very similar to the asymptotic expansion as $x \to -\infty$ (see problem 14):

$$G_0(x) = w_1(x)\frac{1+\sqrt{2}}{\sqrt{-x}}\cos\left(\frac{x^2-\pi}{4}\right) + w_2(x)\frac{1+\sqrt{2}}{\sqrt{-x}}\sin\left(\frac{x^2-\pi}{4}\right), \tag{8.25}$$

where

$$w_1(x) \sim \sum_{n=0}^{\infty} \frac{(-1)^n(8n)!}{2^{10n}(2n)!(4n)!x^{4n}} \sim 1 - \frac{105}{128x^4} + \frac{675675}{32768x^8} - \cdots$$

and

$$w_2(x) \sim \sum_{n=0}^{\infty} \frac{(-1)^n(8n+4)!}{2^{10n+5}(2n+1)!(4n+2)!x^{4n+2}}$$
$$\sim \frac{3}{8x^2} - \frac{3465}{1024x^6} + \frac{43648605}{262144x^{10}} - \cdots.$$

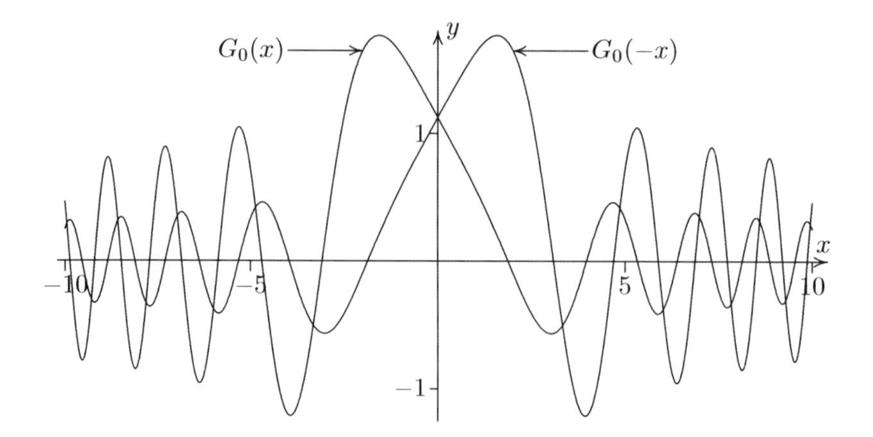

FIGURE 8.15: This shows the graph of the modified parabolic cylinder function, $G_0(x)$, along with its mirror image, $G_0(x)$. These two functions are out of phase by $\pi/2$ near $\pm\infty$, and the amplitudes differ by a factor of $1+\sqrt{2}$.

Figure 8.15 shows the graph of $G_0(x)$, along with $G_0(-x)$. We can use this function to find the WKBJ approximation about a turning point with a double root.

Example 8.13
Find the WKBJ approximation for the perturbation problem

$$\epsilon y'' + (x^2 + x^4)y = 0, \qquad y(-1) = 1, \quad y(1) = 0.$$

SOLUTION: We begin with the right WKBJ approximation. Since $q(x) > 0$ when $x \neq 0$, we can use equation 8.8, so we first compute

$$\int \sqrt{q(x)}\, dx = \int x\sqrt{1+x^2}\, dx = (1+x^2)^{3/2}/3 + C.$$

The solution can then be written as

$$y_{\text{right}} = \frac{K_1}{\sqrt[4]{x^2+x^4}} \cos(\omega(x)) + \frac{K_2}{\sqrt[4]{x^2+x^4}} \sin(\omega(x))$$

where $\omega(x) \sim (1+x^2)^{3/2}/(3\sqrt{\epsilon})$.

We can apply the initial condition $y(1) = 0$ by a judicious choice of the integration constant C. If we alter $\omega(x)$ by a constant, and have

$$\omega(x) \sim \frac{(1+x^2)^{3/2} - \sqrt{8}}{3\sqrt{\epsilon}},$$

then $\omega(1) = 0$, so that the condition $y(1) = 0$ forces $K_1 = 0$. Thus, we have

$$y_{\text{right}} = \frac{K_2}{\sqrt[4]{x^2 + x^4}} \sin\left(\frac{(1 + x^2)^{3/2} - \sqrt{8}}{3\sqrt{\epsilon}}\right).$$

Let us now consider the inner solution. When $x \ll 1$, the equation simplifies to

$$\epsilon y'' + x^2 y = 0.$$

This can be converted to the modified parabolic cylinder equation with the substitution $x = t\sqrt[4]{\epsilon/4}$. Thus, the inner solution is given by

$$y_{\text{in}} = K_3 G_0(x\sqrt[4]{4/\epsilon}) + K_4 G_0(-x\sqrt[4]{4/\epsilon}).$$

To match the inner solution to the right WKBJ approximation, we consider the asymptotic approximation of the inner solution as $x\epsilon^{-1/4} \to \infty$. Using equations 8.24 and 8.25, we have

$$\begin{aligned}(y_{\text{in}})_{\text{right}} &= \frac{K_3\sqrt[8]{\epsilon/4}}{\sqrt{x}} \cos\left(\frac{x^2}{2\sqrt{\epsilon}} + \frac{\pi}{4}\right) \\ &+ \frac{K_4(1 + \sqrt{2})\sqrt[8]{\epsilon/4}}{\sqrt{x}} \sin\left(\frac{x^2}{2\sqrt{\epsilon}} + \frac{\pi}{4}\right).\end{aligned}$$

To see that this matches the WKBJ solution for small x, we note that

$$(1 + x^2)^{3/2} \sim 1 + 3x^2/2 \text{ for } x \ll 1,$$

so

$$(y_{\text{right}})_{\text{in}} = \frac{K_2}{\sqrt{x}} \sin\left(\frac{3x^3 + 2 - 2\sqrt{8}}{6\sqrt{\epsilon}}\right).$$

The argument in the sine functions differ by a constant, but we can use a trig identity to show that

$$\begin{aligned}\sin\left(\frac{3x^3 + 2 - 2\sqrt{8}}{6\sqrt{\epsilon}}\right) &= \sin\left(\frac{x^2}{2\sqrt{\epsilon}} + \frac{\pi}{4}\right) \cos\left(\frac{\sqrt{8} - 1}{3\sqrt{\epsilon}} + \frac{\pi}{4}\right) \\ &- \cos\left(\frac{x^2}{2\sqrt{\epsilon}} + \frac{\pi}{4}\right) \sin\left(\frac{\sqrt{8} - 1}{3\sqrt{\epsilon}} + \frac{\pi}{4}\right).\end{aligned}$$

Thus,

$$K_3\sqrt[8]{\epsilon/4} = -K_2 \sin\left(\frac{\sqrt{8} - 1}{3\sqrt{\epsilon}} + \frac{\pi}{4}\right),$$

and

$$K_4(1 + \sqrt{2})\sqrt[8]{\epsilon/4} = K_2 \cos\left(\frac{\sqrt{8} - 1}{3\sqrt{\epsilon}} + \frac{\pi}{4}\right).$$

Now we will compute the left WKBJ approximation. The equation will be the same as for the right side, only replacing x with $-x$. Thus, we have

$$y_{\text{left}} = \frac{K_5}{\sqrt[4]{x^2+x^4}} \cos\left(\frac{(1+x^2)^{3/2}}{3\sqrt{\epsilon}} + C\right) + \frac{K_6}{\sqrt[4]{x^2+x^4}} \sin\left(\frac{(1+x^2)^{3/2}}{3\sqrt{\epsilon}} + C\right).$$

This time, we can choose the constant to be $\pi/4 - 1/(3\sqrt{\epsilon})$, so that the argument will match with the inner solution. That is, if we now compute

$$(y_{\text{left}})_{\text{in}} = \frac{K_5}{\sqrt{-x}} \cos\left(\frac{x^2}{2\sqrt{\epsilon}} + \frac{\pi}{4}\right) + \frac{K_6}{\sqrt{-x}} \sin\left(\frac{x^2}{2\sqrt{\epsilon}} + \frac{\pi}{4}\right),$$

we find that the trig functions are the same as

$$(y_{\text{in}})_{\text{left}} \approx \frac{K_3(1+\sqrt{2})\sqrt[8]{\epsilon/4}}{\sqrt{-x}} \sin\left(\frac{x^2}{2\sqrt{\epsilon}} + \frac{\pi}{4}\right) + \frac{K_4\sqrt[8]{\epsilon/4}}{\sqrt{-x}} \cos\left(\frac{x^2}{2\sqrt{\epsilon}} + \frac{\pi}{4}\right).$$

Thus, we see that $K_6 = K_3(1+\sqrt{2})\sqrt[8]{\epsilon/4}$, and $K_5 = K_4\sqrt[8]{\epsilon/4}$.

Finally, we can use the initial condition $y(-1) = 1$. This produces the equation

$$\frac{K_5}{\sqrt[4]{2}} \cos\left(\frac{\sqrt{8}-1}{3\sqrt{\epsilon}} + \frac{\pi}{4}\right) + \frac{K_6}{\sqrt[4]{2}} \sin\left(\frac{\sqrt{8}-1}{3\sqrt{\epsilon}} + \frac{\pi}{4}\right) = 1.$$

We can get all of the terms expressed using the constant K_2.

$$\frac{K_2}{1+\sqrt{2}} \cos^2\left(\frac{\sqrt{8}-1}{3\sqrt{\epsilon}} + \frac{\pi}{4}\right) - K_2(1+\sqrt{2}) \sin^2\left(\frac{\sqrt{8}-1}{3\sqrt{\epsilon}} + \frac{\pi}{4}\right) = \sqrt[4]{2}.$$

This can be solved numerically for K_2, which will determine all of the other constants. □

We can form the composite uniform approximation using Langer's method. Let us understand how this would be done in general. We want to replace the argument in the modified parabolic cylindrical functions with a function $f(x)$ so that, when expanded as $x\epsilon^{-1/4} \to \infty$, we see the same expression inside the trig functions as the $\omega(x)$ in the outer WKBJ solutions. That is, we need

$$\frac{f(x)^2}{4} = \omega(x) = \int_0^x \sqrt{q(t)}\,dt/\sqrt{\epsilon}.$$

This means that $f(x) = \sqrt{4\omega(x)}$, so we have our first attempt for a composite uniform approximation:

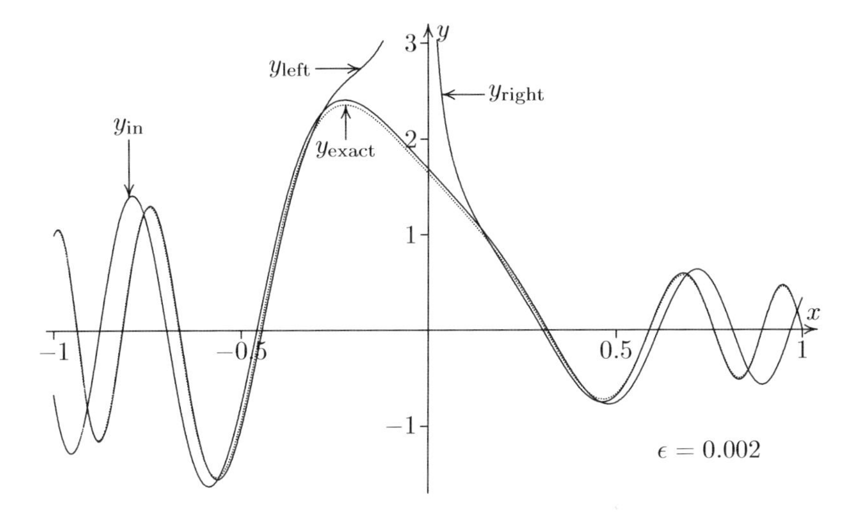

FIGURE 8.16: This graph compares the exact solution (shown with a dotted line) to the solution of $\epsilon y'' + (x^2 + x^4)y = 0$ with $y(-1) = 1$ and $y(1) = 0$, to the regional WKBJ approximations y_{left}, y_{in}, and y_{right}. Each of the regional approximations does an excellent job approaching the exact solution in its respective region, but diverges from the solution outside of this region. This graph uses $\epsilon = 0.002$.

$$y = K_3 G_0 \left(2\epsilon^{-1/4} \left(\int_0^x \sqrt{q(t)}\, dt \right)^{1/2} \right)$$
$$+ K_4 G_0 \left(-2\epsilon^{-1/4} \left(\int_0^x \sqrt{q(t)}\, dt \right)^{1/2} \right).$$

If we consider the approximation as $x > 0$ and $x \ll 1$, this will reproduce the inner solution. However, if we consider the limit as $x\epsilon^{-1/4} \to \infty$, we get

$$\frac{K_3}{\sqrt{2}} \epsilon^{1/8} \left(\int_0^x \sqrt{q(t)}\, dt \right)^{-1/4} \cos\left(\int_0^x \sqrt{q(t)}\, dt + \frac{\pi}{4} \right)$$
$$+ \frac{K_4}{\sqrt{2}} \epsilon^{1/8} \left(\int_0^x \sqrt{q(t)}\, dt \right)^{-1/4} \sin\left(\int_0^x \sqrt{q(t)}\, dt + \frac{\pi}{4} \right).$$

This has to match the known right WKBJ approximation

$$y_{\text{right}} = \frac{K_1}{\sqrt[4]{q(x)}} \cos\left(\int_0^x \sqrt{q(t)}\, dt + C \right) + \frac{K_2}{\sqrt[4]{q(x)}} \sin\left(\int_0^x \sqrt{q(t)}\, dt + C \right),$$

where C is chosen for convenience for the outer solution. In order to obtain this, we must multiply our first attempt by

$$\epsilon^{-1/8} q(x)^{-1/4} \left(\int_0^x \sqrt{q(t)}\, dt \right)^{1/4}.$$

But will this expression approach 1 as $x \to 0$? If we assume that $q(x) \sim bx^2$ where $b > 0$, then a simple computation shows that this expression approaches $\epsilon^{-1/8} b^{-1/8} / \sqrt[4]{2}$, so we will also have to multiply by the constant $\epsilon^{1/8} b^{1/8} \sqrt[4]{2}$. So our second attempt at a composite uniform approximation is

$$y = \frac{K_3 b^{1/8} \sqrt[4]{2}}{\sqrt[4]{q(x)}} \left(\int_0^x \sqrt{q(t)}\, dt \right)^{1/4} G_0 \left(2\epsilon^{-1/4} \left(\int_0^x \sqrt{q(t)}\, dt \right)^{1/2} \right)$$
$$+ \frac{K_4 b^{1/8} \sqrt[4]{2}}{\sqrt[4]{q(x)}} \left(\int_0^x \sqrt{q(t)}\, dt \right)^{1/4} G_0 \left(-2\epsilon^{-1/4} \left(\int_0^x \sqrt{q(t)}\, dt \right)^{1/2} \right).$$

There is one more complication. When we computed the function $f(x)$, we took a square root, which by default gives a positive value. However, if x is negative, then we need to have $f(x)$ be negative too, so that this will match the inner solution for $x < 0$. Technically, $f(x) = \pm\sqrt{4\omega(x)}$, where we take the negative root if $x < 0$. One way to express this in a definitive way is to multiply by the sign function,

$$\operatorname{sgn}(x) = \begin{cases} 1 & \text{if } x > 0, \\ 0 & \text{if } x = 0, \\ -1 & \text{if } x < 0. \end{cases}$$

Thus we have

$$y_{\text{comp}} = \frac{K_3 b^{1/8}\sqrt[4]{2}}{\sqrt[4]{q(x)}}\left|\int_0^x \sqrt{q(t)}\,dt\right|^{1/4} G_0\left(2\,\text{sgn}(x)\epsilon^{-1/4}\left|\int_0^x \sqrt{q(t)}\,dt\right|^{1/2}\right)$$
$$+ \frac{K_4 b^{1/8}\sqrt[4]{2}}{\sqrt[4]{q(x)}}\left|\int_0^x \sqrt{q(t)}\,dt\right|^{1/4} G_0\left(-2\,\text{sgn}(x)\epsilon^{-1/4}\left|\int_0^x \sqrt{q(t)}\,dt\right|^{1/2}\right). \quad (8.26)$$

This issue did not arise in equation 8.21, since we took a cube root to find $f(x)$ instead of a square root.

In order to match the composite uniform approximation with the left and right WKBJ approximations, we can expand this expression both as $x \to \infty$ and $x \to -\infty$. The results are

$$(y_{\text{comp}})_{\text{right}} \sim \frac{K_3 b^{1/8}\epsilon^{1/8}}{\sqrt[4]{2q(x)}}\cos\left(\frac{1}{\sqrt{\epsilon}}\int_0^x \sqrt{q(t)}\,dt + \frac{\pi}{4}\right) \quad (8.27)$$
$$+ \frac{K_4 b^{1/8}\epsilon^{1/12}}{\sqrt[4]{2q(x)}}(1+\sqrt{2})\sin\left(\frac{1}{\sqrt{\epsilon}}\int_0^x \sqrt{q(t)}\,dt + \frac{\pi}{4}\right),$$

and

$$(y_{\text{comp}})_{\text{left}} \sim \frac{K_3 b^{1/8}\epsilon^{1/8}}{\sqrt[4]{2q(x)}}(1+\sqrt{2})\sin\left(\frac{1}{\sqrt{\epsilon}}\int_x^0 \sqrt{q(t)}\,dt + \frac{\pi}{4}\right)$$
$$+ \frac{K_4 b^{1/8}\epsilon^{1/12}}{\sqrt[4]{2q(x)}}\cos\left(\frac{1}{\sqrt{\epsilon}}\int_x^0 \sqrt{q(t)}\,dt + \frac{\pi}{4}\right). \quad (8.28)$$

Example 8.14

Find the composite uniform WKBJ approximation for the problem in example 8.13.

SOLUTION: Since $q(x) = x^2 + x^4$, we have

$$\int_0^x \sqrt{q(t)}\,dt = \frac{(1+x^2)^{3/2}-1}{3}.$$

Also note that $q(x) \sim x^2$ as $x \to 0$, so $b = 1$. Thus, we have

$$y_{\text{comp}} = K_3\left(\frac{2((1+x^2)^{3/2}-1)}{3(x^2+x^4)}\right)^{1/4} G_0\left(\frac{2\,\text{sgn}(x)\sqrt{(1+x^2)^{3/2}-1}}{\epsilon^{1/4}\sqrt{3}}\right)$$
$$+ K_4\left(\frac{2((1+x^2)^{3/2}-1)}{3(x^2+x^4)}\right)^{1/4} G_0\left(\frac{-2\,\text{sgn}(x)\sqrt{(1+x^2)^{3/2}-1}}{\epsilon^{1/4}\sqrt{3}}\right),$$

where K_3 and K_4 are defined in example 8.13. The graph of the composite uniform approximation is given in figure 8.17. ☐

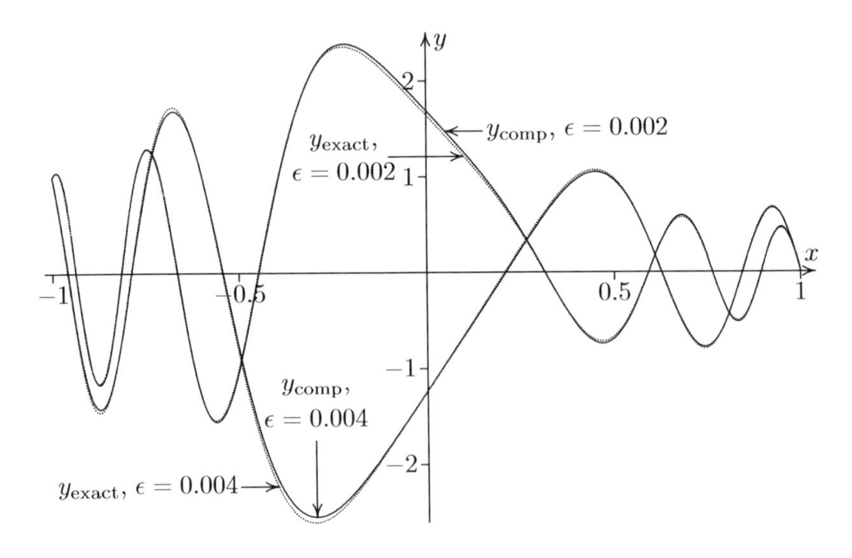

FIGURE 8.17: This compares the composite uniform WKBJ approximation for the equation $\epsilon y'' + (x^2 + x^4)y = 0$, with $y(-1) = 1$ and $y(1) = 0$. Even though the solutions can vastly differ for different ϵ, the composite approximations continue to closely follow the exact solutions.

8.3.3 The Two-turning Point Schrödinger Equation

Finally, let us consider an application that has *two* turning points in the problem. The time-dependent Schrödinger equation

$$i\hbar \frac{\partial}{\partial t}\psi(x,t) = \frac{-\hbar^2}{2m}\frac{\partial^2}{\partial x^2}\psi(x,t) + V(x)\psi(x,t)$$

models a single (non-relativistic) particle of mass m moving in an electric field with potential energy $V(x)$, but no magnetic field. The symbol $\hbar$ is the reduced Planck constant, an extremely small physical constant

$$\hbar \approx 1.054572 \times 10^{-34}\text{m}^2\text{kg/s}.$$

Although this is a partial differential equation, we can eliminate the time variable by assuming that the solution is of the form

$$\psi(x,t) = y(x)e^{-iEt/\hbar}.$$

Here, E is the total energy of the state $\phi(x,t)$. The resulting equation is called the *time-independent Schrödinger equation*

$$\frac{\hbar^2}{2m}y''(x) + (E - V(x))y(x) = 0.$$

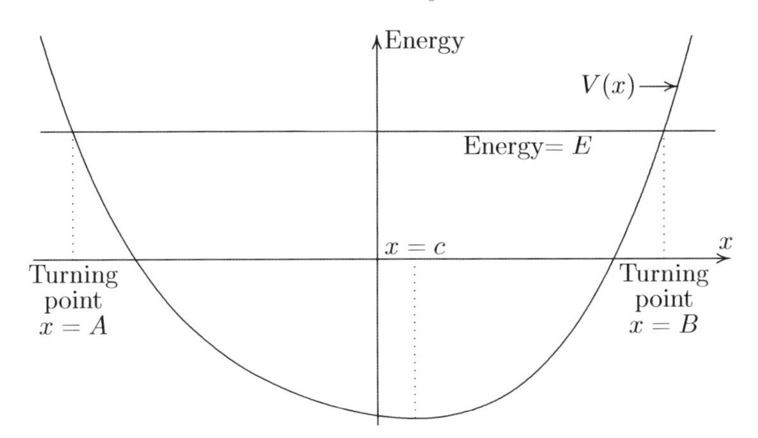

FIGURE 8.18: This shows the schematic diagram for a particle in potential energy well $V(x)$. Classical mechanics would state that the energy can be any value $\geq V(c)$, and that the particle would be confined to the region between the turning points A and B. But quantum mechanics shows that E can only have certain values, namely the eigenvalues of equation 8.29. Also, quantum mechanics allows for a slight possibility of the particle going *beyond* the turning points.

Even if m is the mass of an electron, $\hbar^2/(2m)$ will be a small quantity, which we can call ϵ. If we assume that $V(x)$ increases to ∞ as $x \to \pm\infty$, then the particle's presence will be less and less likely as $x \to \infty$, giving us the initial conditions $y(-\infty) = y(\infty) = 0$. This sets up an eigenvalue problem for the energy E.

$$\epsilon y'' + (E - V(x))y = 0 \qquad y(-\infty) = y(\infty) = 0. \tag{8.29}$$

Only certain values of E will allow for a non-zero solution to this equation. This will represent the possible energy levels of the particle in the electric field.

Let us make some simplifying assumptions. We will assume that $V(x)$ is continuous, and has a single local minimum at $x = c$, and no local maximum. Thus, for any energy E above the local minimum, $E = V(x)$ at precisely two points, which we will call A and B. These will be the two turning points in the equation. From figure 8.18, it is clear that $A \leq c \leq B$.

The plan is to consider each turning point separately. We will use the turning point at $x = A$ to find a composite uniform approximation valid from $x = -\infty$ up to a point slightly less than $x = B$, and another from a point slightly greater than $x = A$ to $x = \infty$. These two WKBJ approximations must match on the interval between A and B. Since $q(x) = E - V(x)$, the slope of $q(x)$ will be negative at the turning point $x = B$. Thus, we can apply the results of example 8.12, only shifting the result so that the turning point is at B.

$$y_{B,\text{comp}} \sim \frac{C_1}{\sqrt[4]{V(x)-E}}\left(\frac{3}{2}\int_B^x \sqrt{V(x)-E}\,dt\right)^{1/6} \times \text{Ai}\left(\left(\frac{3}{2\sqrt{\epsilon}}\int_B^x \sqrt{V(x)-E}\,dt\right)^{2/3}\right).$$

In the region between A and B, we can use the left hand WKBJ approximation from example 8.12:

$$y_{B,\text{left}} \approx \frac{C_1\epsilon^{1/12}}{\sqrt{\pi}\sqrt[4]{E-V(x)}}\sin\left(\frac{1}{\sqrt{\epsilon}}\int_x^B \sqrt{E-V(t)}\,dt+\frac{\pi}{4}\right).$$

For the turning point at $x=A$, we can use the results of problem 2, shifted so that the turning point occurs at $x=A$.

$$y_{A,\text{comp}} \sim \frac{C_2}{\sqrt[4]{E-V(x)}}\left(\frac{3}{2}\int_A^x \sqrt{E-V(x)}\,dt\right)^{1/6} \times \text{Ai}\left(-\left(\frac{3}{2\sqrt{\epsilon}}\int_A^x \sqrt{E-V(x)}\,dt\right)^{2/3}\right).$$

Again, we can consider the region between A and B, which will be the region to the right of A. Using the asymptotic expansion of the Airy function, we have

$$y_{A,\text{right}} \approx \frac{C_2\epsilon^{1/12}}{\sqrt{\pi}\sqrt[4]{E-V(x)}}\sin\left(\frac{1}{\sqrt{\epsilon}}\int_A^x \sqrt{E-V(t)}\,dt+\frac{\pi}{4}\right).$$

Notice that $y_{A,\text{right}}$ and $y_{B,\text{left}}$ are almost the same except the arguments in the sine functions are different. One of the arguments is increasing with x, while the other is decreasing. We can fix this problem by rewriting $y_{B,\text{left}}$ as

$$y_{B,\text{left}} \approx \frac{-C_1\epsilon^{1/12}}{\sqrt{\pi}\sqrt[4]{E-V(x)}}\sin\left(\frac{1}{\sqrt{\epsilon}}\int_B^x \sqrt{E-V(t)}\,dt-\frac{\pi}{4}\right),$$

where we added a negative sign both inside and outside of the sine function. Now the two functions can agree if the arguments differ by a multiple of π.

$$\left(\frac{1}{\sqrt{\epsilon}}\int_A^x \sqrt{E-V(t)}\,dt+\frac{\pi}{4}\right)-\left(\frac{1}{\sqrt{\epsilon}}\int_B^x \sqrt{E-V(t)}\,dt-\frac{\pi}{4}\right)=n\pi.$$

This simplifies to

$$\int_A^B \sqrt{E-V(t)}\,dt=\sqrt{\epsilon}\left(n+\frac{1}{2}\right)\pi.$$

Solving this equation for E gives us the possible energy levels of the particle.

Example 8.15

Find the possible approximate energy levels for the potential function $V(x) = x^2$.

SOLUTION: We first need to compute the turning points A and B in terms of E. This is accomplished by solving the equation $V(x) = x^2 = E$, so $A = -\sqrt{E}$ and $B = \sqrt{B}$.

The next step is to compute the integral

$$\int_{-\sqrt{E}}^{\sqrt{E}} \sqrt{E - t^2}\, dt = \frac{t\sqrt{E - t^2}}{2} + \frac{E}{2}\sin^{-1}\left(\frac{t}{\sqrt{E}}\right)\Bigg|_{-\sqrt{E}}^{\sqrt{E}} = \frac{E\pi}{2}.$$

Thus, we need to have

$$\frac{E\pi}{2} = \sqrt{\epsilon}\left(n + \frac{1}{2}\right)\pi,$$

making $E = (2n+1)\sqrt{\epsilon}$. This is positive for $n = 0, 1, 2, \ldots$. ▯

We can compute $y_{A,\mathrm{comp}}$ and $y_{B,\mathrm{comp}}$ for the case $V(x) = x^2$. The only complication is that the integral for $\sqrt{x^2 - E}$ is handled differently in the case of $x < \sqrt{E}$ and $x > \sqrt{E}$. Thus, we have to describe $y_{B,\mathrm{comp}}$ in two pieces. If $x < \sqrt{E}$, we have

$$\begin{aligned} y_{B,\mathrm{comp}} &= \frac{C_1}{\sqrt[4]{E - x^2}}(3(E\cos^{-1}(x/\sqrt{E}) - x\sqrt{E - x^2})/4)^{1/6} \\ &\quad \times \mathrm{Ai}\left(-\left(\frac{3}{4\sqrt{\epsilon}}(E\cos^{-1}(x/\sqrt{E}) - x\sqrt{E - x^2})\right)^{2/3}\right). \end{aligned}$$

If $x > \sqrt{E}$, we have a slightly different formula

$$\begin{aligned} y_{B,\mathrm{comp}} &= \frac{C_1}{\sqrt[4]{x^2 - E}}(3(x\sqrt{x^2 - E} - E\cosh^{-1}(x/\sqrt{E}))/4)^{1/6} \\ &\quad \times \mathrm{Ai}\left(-\left(\frac{3}{4\sqrt{\epsilon}}(x\sqrt{x^2 - E} - E\cosh^{-1}(x/\sqrt{E}))\right)^{2/3}\right). \end{aligned}$$

The graphs of $y_{B,\mathrm{comp}}$ are shown in figure 8.19, which compares this approximation to the exact solution. Notice that the composite approximation breaks down near the turning point at $-\sqrt{E}$.

We have seen from this section the power of the WKBJ approximations for solving a variety of perturbation problems. There is one more technique to consider, which we will explore in the final chapter.

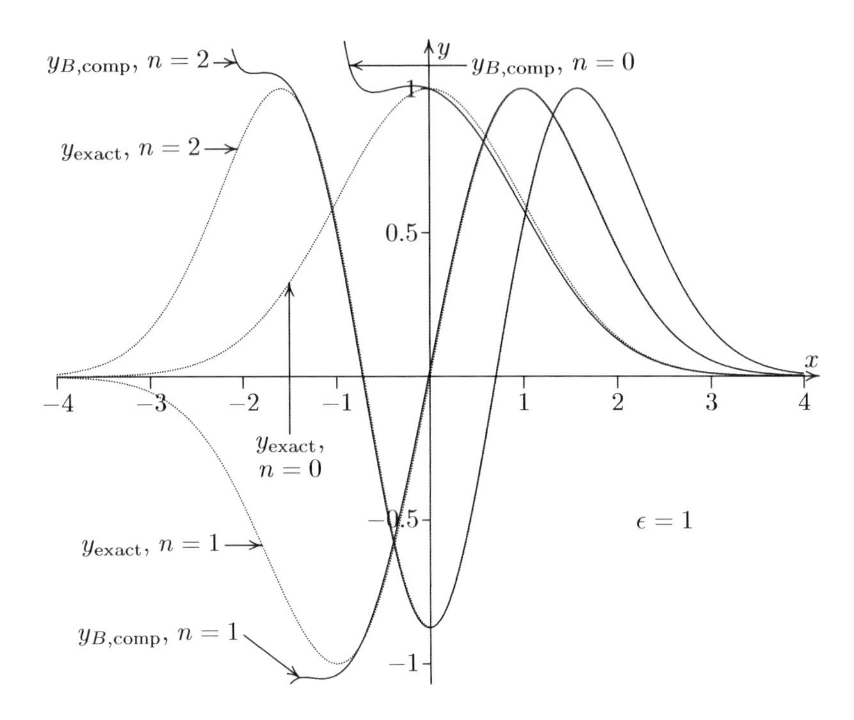

FIGURE 8.19: This shows the WKBJ approximations to the first three eigenfunctions from the time-independent Schrödinger equation with potential $V(x) = x^2$, using $\epsilon = 1$. The approximations $y_{A,\text{comp}}$ are not shown, since these would be mirror images of $y_{B,\text{comp}}$. The exact eigenfunctions are shown with dotted lines. Note that the approximations become more accurate as n increases.

Problems for §8.3

1 Verify the y_{comp} in equation 8.21 agrees with y_{left} of equation 8.20 for $x < 0$.

Hint: Be careful with the negative signs. $\int_0^x \sqrt{-q(t)}\,dt = e^{3\pi i/2} \int_x^0 \sqrt{q(t)}\,dt$.

2 Use the result of example 8.12 to find the composite uniform WKBJ approximation to the problem

$$\epsilon y'' + q(x)y = 0 \qquad q(a) = 0$$

where $a < 0$, and $q(x)$ is real, continuous, and has one root at $x = 0$, whose slope is positive.

Hint: Substitute $x = -t$ in example 8.12. Also see the hint for problem 1.

For problems **3** through **8**: Use either equation 8.12 or the result of problem 2 to find the composite uniform WKBJ approximations to the following one-turning-point problems.

3	$\epsilon y'' - x(1+x^2)^2 y = 0,$	$y(-1) = 1,$	$y(1) = 0$
4	$\epsilon y'' - x(3+7x^2)^2 y = 0,$	$y(-1) = 2,$	$y(4) = 0$
5	$\epsilon y'' - x(1+x+x^2)^2 y = 0,$	$y(-1) = 1,$	$y(\pi) = 0$
6	$\epsilon y'' + x(1+x^4)^2 y = 0,$	$y(-1) = 0,$	$y(1) = 1$
7	$\epsilon y'' + x(3+7x^2)^2 y = 0,$	$y(-\infty) = 0,$	$y(2) = 1$
8	$\epsilon y'' + x\cos^2(x^{3/2}) y = 0,$	$y(-3) = 0,$	$y(1) = 1$

9 Find the approximate eigenvalues to the eigenvalue problem

$$y'' + Ex(1+x^2)^2 y = 0 \qquad y(-\infty) = 0, \quad y(1) = 0.$$

Hint: Let $\epsilon = 1/E$.

10 Find the exact value for $G_0(x)G_0'(-x) - G_0'(x)G_0(-x)$, where $G_0(x)$ is defined in equation 8.23.

Hint: Use theorem 5.1.

11 Use the identity $D_{-1/2}(ix) + D_{-1/2}(-ix) = D_{-1/2}(x) + D_{-1/2}(-x)$ to show that $G_0(ix) + G_0(-ix) = G_0(x) + G_0(-x)$.

12 Use the identity $iD_{-1/2}(ix) - iD_{-1/2}(-ix) = D_{-1/2}(-x) - D_{-1/2}(x)$ to show that $iG_0(ix) - iG_0(-ix) = G_0(-x) - G_0(x)$.

13 Use equation 5.14, the full asymptotic expansion of the parabolic cylinder equation, to prove equation 8.24.

14 Equation 5.14 is not valid if $\arg(x) = 3\pi/4$, so instead we can use the full asymptotic series

$$D_v(x) \sim e^{-x^2/4} x^v \sum_{n=0}^{\infty} \frac{(-1)^n \Gamma(2n-v)}{\Gamma(-v) 2^n n! x^{2n}}.$$

$$- \frac{e^{i\pi v}\sqrt{2\pi}}{\Gamma(-v)} e^{x^2/4} x^{-v-1} \sum_{n=0}^{\infty} \frac{\Gamma(2n+v+1)}{\Gamma(v+1) 2^n n! x^{2n}}. \qquad (8.30)$$

This is valid for $\pi/4 < \arg(x) < 5\pi/4$. Use this to prove equation 8.25.

Hint: The complex conjugate of the equation 8.30 is valid for $-5\pi/4 < \arg(x) < -\pi/4$.

For problems **15** through **22**: Use equation 8.26 to find the composite uniform WKBJ approximations to the following one-turning-point problems.

15 $\epsilon y'' + \sin^2(x) y = 0, \quad y(-\pi/2) = 1, \quad y(\pi/2) = 0$
16 $\epsilon y'' + \sinh^2(x) y = 0, \quad y(-\ln 2) = 1, \quad y(\ln 2) = 0$
17 $\epsilon y'' + (2x + x^2)^2 y = 0, \quad y(-1) = 1, \quad y(1) = 0$
18 $\epsilon y'' + \tan^2(x) y = 0, \quad y(-\pi/4) = 1, \quad y(\pi/4) = 0$
19 $\epsilon y'' + \sin^2(x) y = 0, \quad y(-\pi/2) = 0, \quad y(\pi/2) = \sqrt{2}$
20 $\epsilon y'' + (x^2 + x^4) y = 0, \quad y(-\sqrt{3}) = 0, \quad y(\sqrt{3}) = 1$
21 $\epsilon y'' + \sinh^2(x) y = 0, \quad y(-\ln(2)) = 0, \quad y(\ln(3)) = \sqrt{3}$
22 $\epsilon y'' + x^2 e^x y = 0, \quad y(-2) = 0, \quad y(2) = 1$

23 Find the approximate eigenvalues to the eigenvalue problem

$$y'' + E(\sin^2 x) y = 0 \qquad y(-\pi/2) = 0, \quad y(\pi/2) = 0.$$

Hint: $\tan^{-1}\left(\frac{1}{1+\sqrt{2}}\right) = \pi/8$.

24 Show that the equation 8.29 using $V(x) = x^2$ can in fact be solved *exactly*, and that the energy levels found in example 8.15 are in fact exact.

For problems **25** through **28**: Find the possible approximate energy levels for Schrödinger's time independent equation using the following potential functions.

25 $V(x) = |x|$
26 $V(x) = x + 3|x|$
27 $V(x) = -x^{-2/3}$
28 $V(x) = \text{sgn}(x-1) + \text{sgn}(-x-1)$

29 Find the eigenfunction of the lowest energy level of Schrödinger's time independent equation for the potential $V(x) = |x|$. Use the approximate energy level from problem 25.

Chapter 9

Multiple-Scale Analysis

Multiple-scale analysis is really a collection of techniques for finding approximate solutions to perturbation problems. These incorporate the ideas of boundary layer theory and WKBJ theory, plus introduce some new concepts. The goal for multiple-scale analysis is to construct a uniformly valid approximation even for an infinite interval.

9.1 Strained Coordinates Method (Poincaré-Lindstedt)

To introduce the concepts of multiple-scale analysis, we will use a familiar problem from physics—the simple pendulum. We assume that the cord supporting the bob is massless, and always remains taut. The bob is a point mass, that there is no friction or air resistance. Even with these simplifying assumptions, the equation is difficult to solve.

The force diagram is given in figure 9.1. The direction of motion will always be tangent to the circle created by the pendulum, and the tangential force of gravity will be pointing toward the rest state. By Newton's second law of motion, the force which acts on the pendulum, $-mg\sin\theta$, will equal $ms''(t)$, where s is the position measured in arc length. Since arc length is $\ell\theta$, where θ is measured in radians, we have $s''(t) = \ell\theta''(t)$. This gives us the equation

$$\theta''(t) + \frac{g}{\ell}\sin(\theta) = 0. \tag{9.1}$$

By introducing a scaling factor $x = t\sqrt{g/\ell}$, we can simplify the equation further to $\theta''(x) + \sin(\theta) = 0$. This equation can be converted to a separable equation, but the solution would involve elliptic functions (see problem 1). Rather than finding an exact solution, let us turn this into a perturbation problem.

We can introduce the ϵ in the initial conditions: $\theta(0) = \epsilon$, $\theta'(0) = 0$, so that the pendulum will begin at rest with a small angle. Since θ will always be small, we can expand $\sin(\theta)$ in a Taylor series

$$\theta'' + \theta - \frac{\theta^3}{6} + \frac{\theta^5}{120} + \cdots = 0.$$

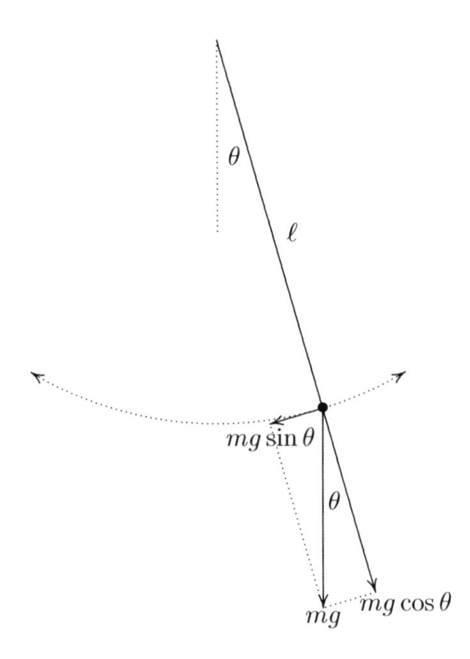

FIGURE 9.1: This shows the forces acting on a simple pendulum of length ℓ. The path of the pendulum sweeps out an arc of a circle. The angle θ, measured in radians, gives the displacement of the pendulum. The vertical arrow is the gravitational force acting on the bob, which is resolved into components parallel and perpendicular to the bob's instantaneous motion.

Let us try to find a regular perturbation solution for θ. We assume that

$$\theta(x) = \theta_0(x) + \epsilon\theta_1(x) + \epsilon^2\theta_2(x) + \epsilon^3\theta_3(x) + \cdots.$$

But the initial conditions force $\theta = O(\epsilon)$, so $\theta_0 = 0$. The equations for the other functions are

$$\begin{aligned} &\theta_1'' + \theta_1 = 0 && \theta(0) = 1,\ \theta'(0) = 0; \\ &\theta_2'' + \theta_2 = 0 && \theta(0) = 0,\ \theta'(0) = 0; \\ &\theta_3'' + \theta_3 = \theta_1^3/6 && \theta(0) = 0,\ \theta'(0) = 0; \cdots. \end{aligned}$$

The first equation easily produces $\theta_1 = \cos(x)$, which agrees with our experience of seeing near-harmonic motion. The second equation shows that $\theta_2 = 0$, which is to be expected, since by symmetry θ should be an odd function of ϵ. The third equation requires us to solve

$$\theta_3'' + \theta_3 = \cos^3(x)/6. \tag{9.2}$$

This can be solved using variation of parameters, equation 5.17, to produce

$$\theta_3 = C_1\cos(x) + C_2\sin(x) + \frac{-\cos(x)}{6}\int \sin(x)\cos^3(x)\,dx + \frac{\sin(x)}{6}\int \cos^4(x)\,dx.$$

The solution that satisfies $\theta(0) = \theta'(0) = 0$ is

$$\theta_3 = \frac{\cos(x)}{192} - \frac{\cos^3(x)}{192} + \frac{\cos(x)\sin(x)^2}{64} + \frac{x\sin(x)}{16}.$$

The last term is very disconcerting. It predicts that the amplitude of the pendulum will start to grow for large x, which we know is not the case. This term is the result of *resonance*. Resonance occurs when the forcing function, that is, the inhomogeneous term of the equation, has a frequency that matches the natural frequency of the associated homogeneous solution. As a result, the system will keep absorbing more and more energy.

At first it may be difficult to see why there is resonance in equation 9.2. However, a product of trigonometric functions can be converted to a sum of functions with different frequencies, using the identities

$$\begin{aligned} \sin(\alpha)\sin(\beta) &= \frac{\cos(\alpha-\beta) - \cos(\alpha+\beta)}{2}, \\ \cos(\alpha)\cos(\beta) &= \frac{\cos(\alpha-\beta) + \cos(\alpha+\beta)}{2}, \\ \sin(\alpha)\cos(\beta) &= \frac{\sin(\alpha-\beta) + \sin(\alpha+\beta)}{2}. \end{aligned} \tag{9.3}$$

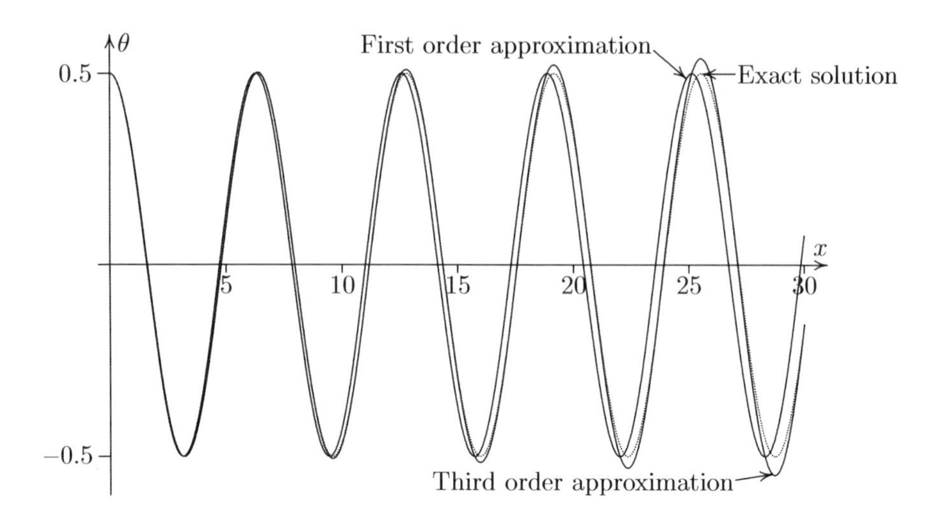

FIGURE 9.2: This compares the exact solution to the pendulum equation, $\theta'' + \sin(\theta) = 0$, with $\theta(0) = \epsilon = 0.5$ radians, $\theta'(0) = 0$. The first order approximation $\epsilon\cos(x)$ has a slightly different period. The third order approximation will grow indefinitely as $x \to \infty$, but the period is much closer to the exact solution.

For example,

$$\cos^3 x = \cos x \frac{1+\cos(2x)}{2} = \frac{\cos x}{2} + \frac{\cos x}{4} + \frac{\cos(3x)}{4}$$
$$= \frac{3\cos x + \cos(3x)}{4}.$$

Thus, we see that the inhomogeneous term has a component that matches the frequency of the homogeneous equation solution.

When resonance occurs, there will be terms in the inhomogeneous solution that grow faster than the associated homogeneous solution by at least a factor of x. Such terms are called *secular terms*. (The term is derived from the French word *siecle* meaning century, since in astronomical applications, it would take a century for the effect to be noticed [14].) Secular terms prevent the solution from being uniformly valid for all t, since the amplitude in the approximation will slowly grow, whereas the true solution has a constant amplitude.

So why do the secular terms appear? They are attempting to slightly change the frequency of the solution. This can be illustrated in figure 9.2. The ϵ^3 correction has the wrong amplitude, but a much more accurate period than the first order perturbation solution.

As this point we have a paradox. Even though the exact solution is bounded, the corrections to the first order behavior are all unbounded. In fact, the secular terms of $\theta_5, \theta_7, \ldots$ grow even more rapidly than the θ_3 term as $x \to \infty$. Yet each term in the perturbation series is suppose to give a more accurate approximation to the exact solution. The explanation is that even though each of the secular terms are unbounded as $x \to \infty$, the *sum* of all of these terms may indeed be bounded. For example, if we considered just the highest order secular terms for θ_n, we would have the series

$$\epsilon \cos(x) + \epsilon^3 \frac{x \sin(x)}{16} - \epsilon^5 \frac{x^2 \cos(x)}{512} - \epsilon^7 \frac{x^3 \sin(x)}{24576} + \epsilon^9 \frac{x^4 \cos x}{1572864} + \cdots.$$

We can separate the $\cos(x)$ terms from the $\sin(x)$ terms, allowing this can be written as

$$\epsilon \cos(x) \left(1 - \left(\frac{x\epsilon^2}{16} \right)^2 \frac{1}{2!} + \left(\frac{x\epsilon^2}{16} \right)^2 \frac{1}{4!} + \cdots \right)$$
$$+ \epsilon \sin(x) \left(\left(\frac{x\epsilon^2}{16} \right) - \left(\frac{x\epsilon^2}{16} \right)^3 \frac{1}{3!} + \cdots \right).$$

There are enough terms computed to see the pattern, the series for $\sin(x\epsilon^2/16)$ and $\cos(x\epsilon^2/16)$. In fact one can prove by induction that this pattern will continue. Thus, we can sum all of these secular terms to produce

$$\epsilon \cos(x) \cos\left(\frac{x\epsilon^2}{16} \right) + \epsilon \sin(x) \sin\left(\frac{x\epsilon^2}{16} \right) = \epsilon \cos\left(x - \frac{x\epsilon^2}{16} \right).$$

So we see that even though the secular terms become unbounded as $x \to \infty$, the sum of all of these terms can remain bounded, and in fact change the period of the solution slightly to produce a uniform approximation to the exact solution. Had this approximation been included in figure 9.2, it would have been indistinguishable from the exact solution.

Although we have been able to sum many of the secular terms to produce a bounded approximation, this is a very inefficient method. We need a strategy that will totally avoid the secular terms from the beginning. Such a method is the method of *strained coordinates*. Poincaré discussed this method in detail in a 1892 treatise on celestial mechanics, giving credit to an obscure 1882 work by Lindstedt. Actually, the idea originated in 1847 when Stokes used this method to solve a wave propagation problem [23]. Credit has also been given to Lighthill, who generalized the method in 1949, and to Kuo, who applied the method to viscous flow problems. Some references refer to this technique as the PLK method.

The strategy is to allow the frequency to be a function of ϵ by making a scale transformation $\tau = \omega(\epsilon)x$, so that $\theta'(x) = \omega\theta'(\tau)$, and $\theta''(x) = \omega^2\theta''(\tau)$. We now have two perturbation series to find, one for θ and one for $\omega = \omega_0 + \omega_1\epsilon + \omega_2\epsilon^2 + \cdots$, with only one equation, but we add the stipulation that *no secular terms should appear*. We can assume that $\omega_0 = 1$, so that the original equation is produced as $\epsilon \to 0$.

Example 9.1
Apply the strained coordinate method for the pendulum problem

$$\theta'' + \sin(\theta) = 0, \qquad \theta(0) = \epsilon, \quad \theta'(0) = 0.$$

SOLUTION: We apply the substitution $\tau = \omega(\epsilon)x$ to the equation to produce

$$\omega^2\theta''(\tau) + \sin(\theta(\tau)) = 0.$$

Thus, we have

$$(1 + \omega_1\epsilon + \omega_2\epsilon^2)^2(\theta_1''(\tau)\epsilon + \theta_2''(\tau)\epsilon^2 + \theta_3''(\tau)\epsilon^3) \\ + \theta_1(\tau)\epsilon + \theta_2(\tau)\epsilon^2 + \epsilon^3\left(\theta_3(\tau) - \frac{\theta_1(\tau)^3}{6}\right) + \cdots = 0.$$

The terms of order ϵ produce the equation $\theta_1'' + \theta_1 = 0$, with $\theta_1(0) = 1$, $\theta_1'(0) = 0$, so we have $\theta_1 = \cos(\tau)$, as before. The terms of order ϵ^2 give us

$$2\omega_1\theta_1'' + \theta_2'' + \theta_2 = 0, \qquad \theta_2(0) = \theta_2'(0) = 0.$$

Since $\theta_1'' = -\cos(\tau)$, we have to solve the inhomogeneous equation $\theta_2'' + \theta_2 = \omega_1\cos(\tau)$. This would produce a secular term unless $\omega_1 = 0$, which in turn makes $\theta_2 = 0$. In fact, it is not hard to see that the even θ's and odd ω's will go to 0.

The ϵ^3 terms produce the equation

$$2\omega_2\theta_1'' + \theta_3'' + \theta_3 - \theta_1^3/6 = 0,$$

giving us the inhomogeneous equation

$$\theta_3'' + \theta_3 = 2\omega_2\cos(\tau) + \cos^3(\tau)/6.$$

Using trig identities in equation 9.3, we find that $\cos^3(\tau)/6 = \cos(\tau)/8 + \cos(3\tau)/24$. In order to avoid secular terms, we need to have $\omega_2 = -1/16$. Then the equation for θ_3 becomes $\theta_3'' + \theta_3 = \cos(3\tau)/24$, which has the solution $\theta_3 = (\cos(\tau) - \cos(3\tau))/192$. So we have the third order correction

$$\theta(x) \sim \epsilon\cos\left(x - \frac{\epsilon^2 x}{16}\right) + \frac{\epsilon^3}{192}\left(\cos\left(x - \frac{\epsilon^2 x}{16}\right) - \cos\left(3x - \frac{3\epsilon^2 x}{16}\right)\right).$$

We can denote the third order strained coordinate approximation by $\theta_{\text{strain},3}$.

Notice that the period perturbation is the same as predicted by summing the most secular terms of the original perturbation series. The graph of the strained coordinate approximation is shown in figure 9.3. □

The strained coordinate method can be used for a variety of different perturbation problems for which at least one solution is known to be periodic.

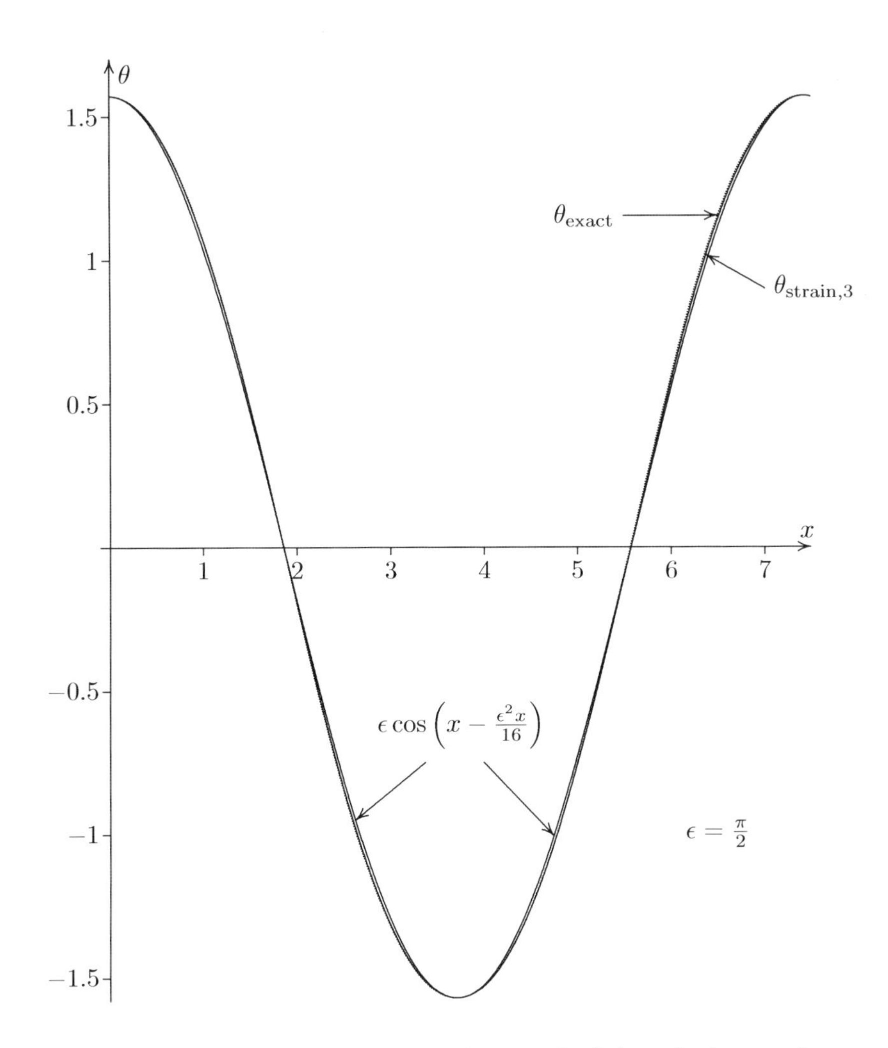

FIGURE 9.3: This compares a single period of the solution to the pendulum problem, $\theta'' + \sin(\theta) = 0$, to the third order strained coordinate approximation $\theta_{\text{strain},3}$. Even with a 180° swing, $\epsilon = \pi/2$, the approximate solution is difficult to discern from the exact solution, shown with a dotted line. The cosine curve with the same frequency is also shown, which is a poorer approximation than $\theta_{\text{strain},3}$.

The strategy of making the substitution $\tau = \omega(\epsilon)t$ reflects how the period changes slightly with ϵ. This substitutions creates the following replacements

$$\boxed{t \to \frac{\tau}{\omega(\epsilon)}, \qquad y'(t) \to \omega(\epsilon)y'(\tau), \qquad y''(t) \to \omega(\epsilon)^2 y''(\tau).} \tag{9.4}$$

Example 9.2

Find the first and second order strained coordinate approximation for the undamped Duffing equation

$$y''(t) + y(t) + \epsilon y(t)^3 = 0, \qquad y(0) = 0, \quad y'(0) = 1.$$

This equation describes a mass-spring system in which the spring's stiffness does not quite obey Hooke's law.

SOLUTION: It is possible to prove that the solutions to this equation are periodic, by converting the equation to a separable equation. (See problem 2.) Substituting $\tau = \omega(\epsilon)t$ gives us the equation

$$\omega^2 y''(\tau) + y(\tau) + \epsilon y(\tau)^3 = 0.$$

Next, we expand $\omega = 1+\omega_1\epsilon+\omega_2\epsilon^2+\cdots$ and $y = y_0(\tau)+y_1(\tau)\epsilon+y_2(\tau)\epsilon^2+\cdots$. However, the initial condition $y'(0) = 1$ when $t = 0$ also needs to be converted to the new variable. Since

$$\frac{dy}{dt} = \frac{dy}{d\tau} \cdot \frac{d\tau}{dt} = \omega(\epsilon)\frac{dy}{d\tau},$$

we find that

$$y'(\tau) = \frac{y'(t)}{\omega(\epsilon)} \sim (1 - \omega_1\epsilon + (\omega_1^2 - \omega_2)\epsilon^2 + \cdots)y'(t).$$

So the initial condition $y'(t) = 1$ when $t = 0$ converts to the equations

$$y_0'(\tau) = 1, \qquad y_1'(\tau) = -\omega_1, \qquad y_2'(\tau) = \omega_1^2 - \omega_2,$$

at the point where $\tau = 0$. So we need to solve the following equations:

$$\begin{aligned} y_0'' + y_0 &= 0, \quad y_0(0) = 0, \quad y_0'(0) = 1; \\ y_1'' + 2\omega_1 y_0'' + y_1 + y_0^3 &= 0, \quad y_1(0) = 0, \quad y_1'(0) = -\omega_1; \\ y_2'' + 2\omega_1 y_1'' + (\omega_1^2 + 2\omega_2)y_0'' + y_2 + 3y_0^2 y_1 &= 0, \quad y_2(0) = 0, \quad y_2'(0) = \omega_1^2 - \omega_2. \end{aligned}$$

The equation for y_0 is easy to solve, producing $y_0 = \sin(\tau)$. The second equation becomes

$$y_1'' + y_1 = 2\omega_1 \sin(\tau) - \sin^3(\tau).$$

Using equation 9.3 we find that $\sin^3(\tau) = 3\sin(\tau)/4 - \sin(3\tau)/4$, we can prevent the secular terms if we let $\omega_1 = 3/8$. Then we have to solve $y_1'' + y_1 =$

$\sin(3\tau)/4$ with the initial conditions $y_1(0) = 0$, $y_1'(0) = -3/8$, which has the solution $y_1 = -(9\sin(\tau) + \sin(3\tau))/32$. This gives us

$$y_{\text{strain},1} = \sin\left(t + \frac{3\epsilon t}{8}\right) - \frac{\epsilon}{32}\left(9\sin\left(t + \frac{3\epsilon t}{8}\right) + \sin\left(3t + \frac{9\epsilon t}{8}\right)\right).$$

If we plug in the known functions, and apply trig identities in equation 9.3, we get the following equation for y_2:

$$y_2'' + y_2 = 2\omega_2 \sin(\tau) + \frac{69\sin(\tau)}{128} - \frac{3\sin(3\tau)}{8} - \frac{3\sin(5\tau)}{128}.$$

We can avoid secular terms by letting $\omega_2 = -69/256$. Then the initial conditions for the equation become $y_2(0) = 0$, $y_2'(0) = 105/256$, so the solution is

$$y_2 = (271\sin(\tau) + 48\sin(3\tau) + \sin(5\tau))/1024.$$

This gives us the second order approximation

$$\begin{aligned} y_{\text{strain},2} &= \sin\left(t + \frac{3\epsilon t}{8} - \frac{69\epsilon^2 t}{256}\right) \\ &- \frac{\epsilon}{32}\left(9\sin\left(t + \frac{3\epsilon t}{8} - \frac{69\epsilon^2 t}{256}\right) + \sin\left(3t + \frac{9\epsilon t}{8} - \frac{207\epsilon^2 t}{256}\right)\right) \\ &+ \frac{\epsilon^2}{1024}\left(271\sin\left(t + \frac{3\epsilon t}{8} - \frac{69\epsilon^2 t}{256}\right) + 48\sin\left(3t + \frac{9\epsilon t}{8} - \frac{207\epsilon^2 t}{256}\right)\right. \\ &\qquad \left. + \sin\left(5t + \frac{15\epsilon t}{8} - \frac{345\epsilon^2 t}{256}\right)\right). \end{aligned}$$

The graphs of the first and second order approximations for $\epsilon = 0.3$ are shown in figure 9.4. Although the accuracy is not as impressive as the pendulum example, it is clear that the second order approximation is giving a better approximation of both the period and amplitude than the first order. □

The method of strained coordinates will only work if the solution is periodic. However, if a differential equation has just *one* special solution that is periodic, we can apply the method of strained coordinates to approximate this one solution. An example of such an equation is the Rayleigh oscillator.

Example 9.3

The Rayleigh oscillator is governed by the equation

$$y'' + y + \epsilon\left(\frac{(y')^3}{3} - y'\right). \tag{9.5}$$

For any $\epsilon > 0$, the solutions will converge to a single oscillatory solution. For example, figure 9.5 shows a phase-plane diagram of two different solutions for

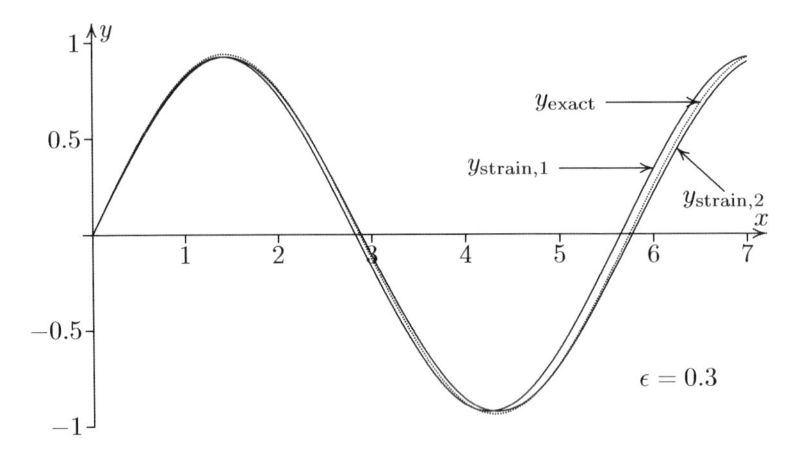

FIGURE 9.4: This shows the strained coordinate approximation to Duffing's equation $y'' + y + \epsilon y^3 = 0$, with $y(0) = 0$ and $y'(0) = 1$. The first and second order approximations are shown, with $\epsilon = 0.3$.

$\epsilon = 0.25$. This gives strong evidence that a periodic solution exists. Find the second order strained coordinate approximation for the periodic solution.

SOLUTION: The difficulty in this example is that we are not given initial conditions. We will always be able to shift the starting position so that $y'(0) = 0$, but then $y(0)$ will be unknown. Hence, we will have to consider a third perturbation series for $y(0) = a_0 + a_1\epsilon + a_2\epsilon^2 + \cdots$.

As before, we substitute $\tau = \omega(\epsilon)t$ into the equation to obtain

$$\omega(\epsilon)^2 y''(\tau) + y(\tau) + \epsilon\omega(\epsilon)^3(y'(\tau))^3 - \epsilon\omega(\epsilon)y'(\tau) = 0.$$

Expanding in orders of ϵ produces

$$\begin{aligned} y_0'' + y_0 = 0, \quad & y_0(0) = a_0,\ y_0'(0) = 0; \\ y_1'' + y_1 + 2\omega_1 y_0'' + (y_0')^3/3 - y_0' = 0, \quad & y_1(0) = a_1,\ y_1'(0) = 0; \\ y_2'' + y_2 + 2\omega_1 y_1'' + (2\omega_2 + \omega_1^2)y_0'' + \omega_1(y_0')^3 + (y_0')^2 y_1' - \omega_1 y_0' - y_1' = 0; & \\ & y_2(0) = a_2,\ y_2'(0) = 0. \end{aligned}$$

The first equation is easy to solve, producing $y_0 = a_0\cos(\tau)$. Plugging this into the second equation produces

$$y_1'' + y_1 = 2a_0\omega_1\cos(\tau) + \left(\frac{a_0^3}{4} - a_0\right)\sin(\tau) - \frac{a_0^3}{12}\sin(3\tau).$$

To avoid secular terms, we need to have both $2a_0\omega_1 = 0$ and $a_0^3/4 - a_0 = 0$. The latter is solved if $a_0 = 0$ or 2. If $a_0 = 0$, we obtain the trivial solution $y = 0$, so we throw out that possibility and determine that $a_0 = 2$. Also, we must have $\omega_1 = 0$ to avoid cosine terms. Hence, we are solving the simpler

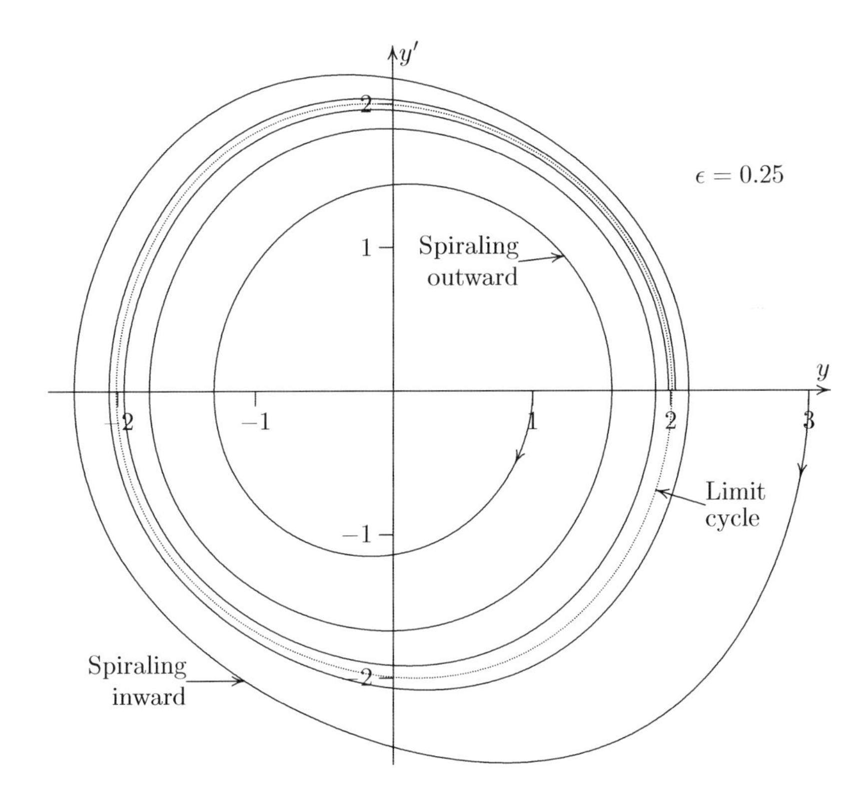

FIGURE 9.5: This shows a phase-plane diagram (y' vs. y) for the Rayleigh oscillator $y'' + y + \epsilon((y')^3/3 - y') = 0$ with $\epsilon = 0.25$. With the initial conditions $y(0) = 1$, $y'(0) = 0$, the solution spirals counterclockwise outward towards a limit cycle. When $y(0) = 3$ and $y'(0) = 0$, the solution spirals inward toward the same limit cycle.

equation $y_1'' + y = -2\sin(3\tau)/3$. Using the initial conditions $y(0) = a_1$, $y'(0) = 0$ gives us

$$y_1 = a_1 \cos(\tau) + \frac{\sin(3\tau)}{12} - \frac{\sin \tau}{4}.$$

We now can move on to the last equation. Since $\omega_1 = 0$ and $a_0 = 2$, this reduces to

$$y_2'' + y_2 = \left(4\omega_2 + \frac{1}{4}\right)\cos(\tau) + 2a_1 \sin(\tau) - a_1 \sin(3\tau) - \frac{\cos 3\tau}{2} + \frac{\cos 5\tau}{4}.$$

To avoid secular terms, we require having $a_1 = 0$ and $w_2 = -1/16$. Then the equation simplifies to $y_2'' + y_2 = \cos(5\tau)/4 - \cos(3\tau)/2$, with the initial conditions $y(0) = a_2$, $y'(0) = 0$ which solves to

$$y_2 = \left(a_2 - \frac{5}{96}\right)\cos(\tau) + \frac{\cos(3\tau)}{16} - \frac{\cos(5\tau)}{96}.$$

Unfortunately, in order to determine $y_{\text{strain},2}$, we need to determine the value of a_2. Hence, we will need to at least begin the next order evaluation. Fortunately, $w_1 = 0$, which helps simplify the equation for y_3.

$$y_3'' + y_3 + 2\omega_2 y_1'' + 2\omega_3 y_0'' + \omega_2 (y_0')^3 + (y_0')^2 y_2' + (y_1')^2 y_0' - \omega_2 y_0' - y_2' = 0.$$

After replacing the known values, this becomes

$$\begin{aligned} y_3'' + y_3 = {} & \left(2a_2 - \frac{17}{48}\right)\sin(\tau) + 4\omega_3 \cos(\tau) \\ & + \left(\frac{17}{48} - a_2\right)\sin(3\tau) - \frac{\sin 5\tau}{3} + \frac{\sin(7\tau)}{12}. \end{aligned}$$

To prevent secular terms, we must have $a_2 = 17/96$ and $\omega_3 = 0$. This gives us the last bit of information we need to compute

$$\begin{aligned} y_{\text{strain},2} = {} & 2\cos\left(t - \frac{t\epsilon^2}{16}\right) + \frac{\epsilon}{12}\sin\left(3t - \frac{3t\epsilon^2}{16}\right) - \frac{\epsilon}{4}\sin\left(t - \frac{t\epsilon^2}{16}\right) \\ & + \frac{\epsilon^2}{8}\cos\left(t - \frac{t\epsilon^2}{16}\right) + \frac{\epsilon^2}{16}\cos\left(3t - \frac{3t\epsilon^2}{16}\right) - \frac{\epsilon^2}{96}\cos\left(5t - \frac{5t\epsilon^2}{16}\right). \end{aligned}$$

Figure 9.6 shows the closeness to the stained coordinate approximation and the exact solution in the phase-plane. Even with ϵ as large as 0.5, the accuracy is remarkable. ☐

The method of strained coordinates relies on the assumption that there is a solution that is periodic, so that the exact solution depends to all orders on only the single variable τ. If there is not a periodic solution, then the method fails. In fact, there are examples for which the method of strained coordinates

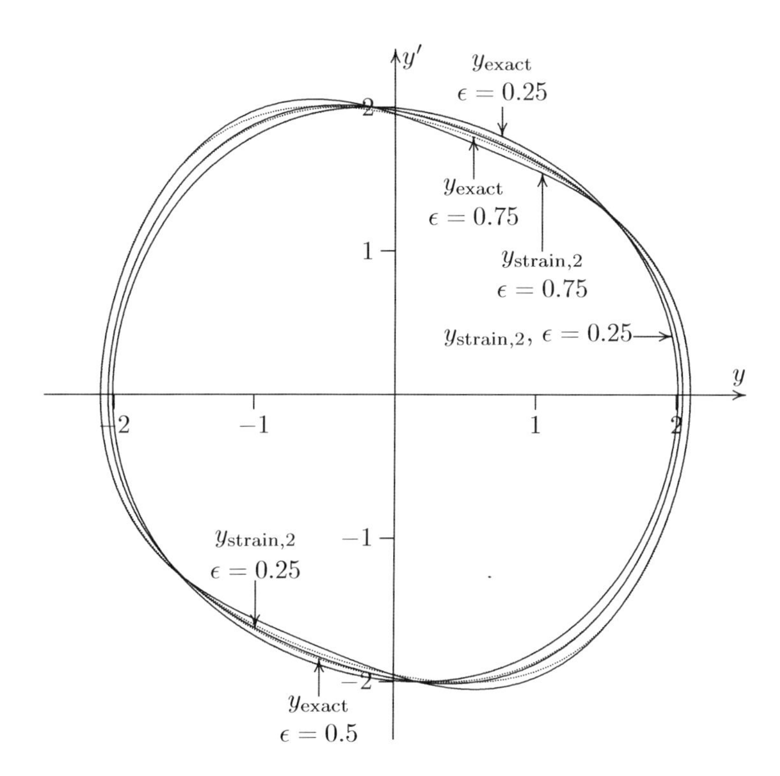

FIGURE 9.6: This compares the strained coordinate approximation to the exact Rayleigh oscillator cycles for various ϵ. The exact solutions are shown with a dotted line. When $\epsilon = 0.25$, the approximation is virtually indistinguishable from the exact solution. As ϵ increases, the cycle becomes more peanut-shaped, but the approximation follows the exact solution closely, especially near $y = \pm 2$.

seem to produce a periodic solution, when in fact no periodic solution exists [26].

For non-periodic solutions, there is little hope of expressing the answer in terms of a single strained variable, but it may be possible to express the answer in terms of *two* different variables. We will explore these methods in the next two subsections.

Problems for §9.1

1 Show that the equation $\theta''(x) + \sin(\theta) = 0$ can be converted to a separable equation by multiplying by θ', and integrating.

2 Show that the Duffing equation $y''(x) + y + \epsilon y^3 = 0$ can be converted to a separable equation by multiplying by y', and integrating. From this equation, show that the solutions are periodic.

For problems **3** through **12**: Use equation 9.3 to convert the following trigonometric expressions into a form where each term involves either $\cos(nx)$ or $\sin(nx)$ for some integer n.

3 $\cos^2(x)$
4 $\cos(x)\sin(x)$
5 $\cos(x)\sin^2(x)$
6 $\cos^2(x)\sin(x)$
7 $\cos^4(x)$
8 $\cos^2(x)\sin^2(x)$
9 $\cos^2(x)\sin^3(x)$
10 $\cos^5(x)$
11 $\sin^5(x)$
12 $\cos(x)\sin^4(x)$

13 The van der Pol equation is given by $y'' + y - \epsilon y'(1 - y^2) = 0$. Show that the Rayleigh oscillator in equation 9.5 can be converted to the van der Pol equation by taking the derivative, and setting $y'(t)$ to $w(t)$.

14 For the van der Pol equation, $y'' + y - \epsilon y'(1 - y^2) = 0$, with $y(0) = a$, $y'(0) = 0$, determine the approximate starting value of a that produces a periodic solution, and find the second order strained coordinate approximation to that solution.

15 Find the third order strained coordinate approximation to the equation

$$y'' + y\cos(y) = 0, \qquad y(0) = \epsilon, \quad y'(0) = 0.$$

For problems **16** through **23**: The following equations are known to produce periodic functions for all ϵ. Use the method of strained coordinates to find $y_{\text{strain},2}$, the second order strained coordinate approximation.

16 $\epsilon y'' + y + \epsilon y(y') = 0, \quad y(0) = 1, \quad y'(0) = 0$
17 $\epsilon y'' + y + \epsilon(y')^2 = 0, \quad y(0) = 1, \quad y'(0) = 0$
18 $\epsilon y'' + y + \epsilon y(y')^2 = 0, \quad y(0) = 1, \quad y'(0) = 0$
19 $\epsilon y'' + y + \epsilon y(y' + 1) = 0, \quad y(0) = 1, \quad y'(0) = 0$
20 $\epsilon y'' + y + \epsilon y(y')^2 = 0, \quad y(0) = 0, \quad y'(0) = 1$
21 $\epsilon y'' + y + \epsilon y(y' - 1) = 0, \quad y(0) = 0, \quad y'(0) = 1$
22 $\epsilon y'' + y + \epsilon y^4 = 0, \quad y(0) = 0, \quad y'(0) = 1$
23 $\epsilon y'' + y + \epsilon(y^2 + y) = 0, \quad y(0) = 0, \quad y'(0) = 1$

24 Find the second order strained coordinate approximation to find the periodic solution to the van der Pol equation

$$y'' + y + \epsilon y'(y^2 - 1) = 0.$$

Assume that $y(0) = a = a_0 + \epsilon a_1 + \epsilon^2 a_2 + \cdots$ and $y'(0) = 0$. Note that one must start the third order approximation in order to determine a_2.

9.2 The Multiple-Scale Procedure

In order to approximate solutions that are not periodic, yet have an oscillating behavior, we will have to consider *two* different time scales simultaneously. To illustrate the issues, consider the linear oscillator with just a tiny amount of damping.

$$y'' + 2\epsilon y' + y = 0, \qquad y(0) = 1, \quad y'(0) = 0. \tag{9.6}$$

The exact solution can be shown to be

$$e^{-\epsilon t}\cos(t\sqrt{1-\epsilon^2}) + \frac{\epsilon e^{-\epsilon t}}{\sqrt{1-\epsilon^2}}\sin(t\sqrt{1-\epsilon^2}).$$

The fluctuations in the solution occur as $t = O(1)$, but the decay does not occur until $t = O(1/\epsilon)$. Hence, the behavior of the solution occurs on two different time scales. If we try to solve this problem as a regular perturbation problem, we find that

$$y(t) \sim \cos(t) + \epsilon(\sin(t) - t\cos(t) + \frac{\epsilon^2}{2}(t^2\cos(t) - t\sin(t)) + \cdots,$$

showing that there are secular terms. This is not surprising, since the solution is not periodic. For the same reason, applying the method of strained coordinates is fruitless.

If we make the substitution $s = \epsilon t$, we get the solution to be

$$y = e^{-s}\cos(s\sqrt{1-\epsilon^2}/\epsilon) + \frac{\epsilon e^{-s}}{\sqrt{1-\epsilon^2}}\sin(s\sqrt{1-\epsilon^2}/\epsilon).$$

On this scale, there are no issues with the exponential functions, but the arguments of the trig functions now approach infinity as $\epsilon \to 0$. This means that the function is now varying rapidly for a small change in s. We have encountered such functions before with WKBJ approximations, such as figure 8.3. However, if we want to consider the limit as $\epsilon \to 0$ with s fixed, we find that this limit does not exist. In short, there is no outer expansion for this function.

In order to develop a general method for solving such equations, we need to consider both time frames simultaneously. That is, we can let t denote the "fast time," and $s = \epsilon t$ denote "slow time." We then will find an expansion of the solution in terms of *both* t and s, treating them as independent variables. Hence, the solution to leading order to equation 9.6 can be written

$$y(t, s) = e^{-s} \cos(t\sqrt{1 - \epsilon^2}).$$

In general, we assume that the solution is a perturbation series of the form

$$y(t) \sim Y_0(t, s) + \epsilon Y_1(t, s) + \epsilon^2 Y_2(t, s) + \cdots. \tag{9.7}$$

In order to take derivatives, we must use the multi-variable chain rule

$$\frac{d}{dt} Y_0(t, s) = \frac{\partial Y_0}{\partial t}\frac{dt}{dt} + \frac{\partial Y_0}{\partial s}\frac{ds}{dt} = \frac{\partial Y_0}{\partial t} + \epsilon \frac{\partial Y_0}{\partial s}.$$

Notice that there is a difference between the full derivative of Y_0 with respect to t, and the partial derivative with respect to t. Higher derivatives can also be calculated.

$$\frac{d^2}{dt^2} Y_0(t, s) = \frac{\partial^2 Y_0}{\partial t^2} + 2\epsilon \frac{\partial^2 Y_0}{\partial t \partial s} + \epsilon^2 \frac{\partial^2 Y_0}{\partial s^2}.$$

Hence, the equations produced for $Y_0, Y_1, \ldots$ will be *partial differential equations*. Normally, partial differential equations are much more complicated than ordinary equations, but in this case, each equation will only involve derivatives in one variable. For example, if we plug equation 9.7 into equation 9.6, and equate powers of ϵ, we obtain

$$\begin{aligned}
\frac{\partial^2 Y_0}{\partial t^2} + Y_0 &= 0,\\
\frac{\partial^2 Y_1}{\partial t^2} + Y_1 &= -2\frac{\partial^2 Y_0}{\partial t \partial s} - 2\frac{\partial Y_0}{\partial t},\\
\frac{\partial^2 Y_2}{\partial t^2} + Y_2 &= -2\frac{\partial^2 Y_1}{\partial t \partial s} - \frac{\partial^2 Y_0}{\partial s^2} - 2\frac{\partial Y_1}{\partial t} - 2\frac{\partial Y_0}{\partial s}.
\end{aligned}$$

The general solution to the first equation can be expressed by $A_0(s)\cos(t) + B_0(s)\sin(t)$, where now A_0 and B_0 are functions of s instead of constants. This can be plugged into the second equation, which now becomes

$$\frac{\partial^2 Y_1}{\partial t^2} + Y_1 = [-2B_0(s) - 2B_0'(s)]\cos(t) + [2A_0(s) + 2A_0'(s)]\sin(t).$$

Once again, we will apply the principle that we must avoid secular terms in the t variable. Hence, the terms each pair of brackets must add up to 0. So we have the following equations for A_0 and B_0:

$$A_0'(s) = -A_0(s), \qquad B_0'(s) = -B_0(s).$$

We also have the initial conditions $Y_0(0,0) = 1$ and $(\partial Y_0/\partial t)(0,0) = 0$. This gives us $A_0(s) = e^{-s}$ and $B_0(s) = 0$. Hence, we have $Y_0 = e^{-s}\cos(t)$.

The second equation now simplifies to

$$\frac{\partial^2 Y_1}{\partial t^2} + Y_1 = 0,$$

which has the general solution $Y_1(s,t) = A_1(s)\cos(t) + B_1(s)\sin(t)$. We can now plug this into the last of the partial differential equations to produce

$$\frac{\partial^2 Y_2}{\partial t^2} + Y_2 = [e^{-s} - 2B_1(s) - 2B_1'(s)]\cos(t) + [2A_1(s) + 2A_1'(s)]\sin(t).$$

To avoid secular terms, we need to have $A_1'(s) = -A_1(s)$, and $B_1'(s)+B_1(s) = e^{-s}/2$.

The initial condition $y(0) = 1$ gives us that $Y_1(0,0) = 0$ in the epsilon order term, but the initial condition $y'(0) = 0$ is a bit harder to convert because of the multi-variable chain rule. We need the epsilon order of $y'(t)$, namely $\partial Y_0/\partial s + \partial Y_1/\partial t$, to be 0 when $s = t = 0$. This gives us enough information to determine that $A_1(s) = 0$ and $B_1(s) = (1 + s/2)e^{-s}$. Thus, we have the series multiple-scale expansion as

$$y \sim e^{-s}\cos(t) + \epsilon\left(1 + \frac{s}{2}\right)e^{-s}\sin(t) + \cdots. \tag{9.8}$$

There are several observations that we can make from this example. First of all, we used the equation for $Y_2(s,t)$ in order to determine $Y_1(s,t)$. Hence, we must find the equations for one more order that we are asked to find. Also, all of the terms in the expansion will have the same frequency, so we will not be able to model changing frequencies as we did for the strained coordinate method. In the next section, we will show how to combine the strained coordinate method to the multiple-scale technique.

The multiple-scale method is most appropriate for equations of the form

$$y'' + y + \epsilon f(y, y') = 0, \tag{9.9}$$

for which the unperturbed solution will have a constant frequency. Such equations are called *weakly non-linear autonomous systems*. For such systems, even the leading order approximation can give impressive results.

Example 9.4

Find the leading order multiple-scale approximation for the perturbation problem

$$y'' + y + \epsilon y^2 y' = 0, \qquad y(0) = 1, \quad y'(0) = 0.$$

SOLUTION: Even though we need only the leading order, we must compute the differential equations to order ϵ. Thus, we will assume

$$y(t) = Y_0(t,s) + \epsilon Y_1(t,s) + O(\epsilon^2),$$

where $s = \epsilon t$. Using the multi-variable chain rule, we have

$$y'(t) = \frac{\partial Y_0}{\partial t} + \epsilon\left(\frac{\partial Y_0}{\partial s} + \frac{\partial Y_1}{\partial t}\right) + O(\epsilon^2),$$

and

$$y''(t) = \frac{\partial^2 Y_0}{\partial t^2} + \epsilon\left(2\frac{\partial^2 Y_0}{\partial s \partial t} + \frac{\partial^2 Y_1}{\partial t^2}\right).$$

Expanding the equation up to order ϵ, we get the two equations

$$\frac{\partial^2 Y_0}{\partial t^2} + Y_0 = 0,$$
$$\frac{\partial^2 Y_1}{\partial t^2} + Y_1 = -2\frac{\partial^2 Y_0}{\partial s \partial t} - Y_0^2\frac{\partial Y_0}{\partial t}.$$

The first of these is easily solved to produce $Y_0 = A_0(s)\cos(t) + B_0(s)\sin(t)$. Plugging this into the second equation produces

$$\begin{aligned}\frac{\partial^2 Y_1}{\partial t^2} + Y_1 = {} & A_0^3\cos^2(t)\sin(t) - A_0^2 B_0\cos^3(t) - 2A_0B_0^2\cos^2(t)\sin(t) \\ & + 2A_0^2B_0\cos(t)\sin^2(t) - B_0^3\cos(t)\sin^2(t) + A_0B_0^2\sin^3(t) \\ & + 2A_0'\sin(t) - 2B_0'\cos(t).\end{aligned}$$

Next, we use the identities

$$\begin{aligned}\cos^3(t) &= \frac{3\cos(t) + \cos(3t)}{4}, \qquad (9.10)\\ \cos^2(t)\sin(t) &= \frac{\sin(t) + \sin(3t)}{4},\\ \cos(t)\sin^2(t) &= \frac{\cos(t) - \cos(3t)}{4},\\ \sin^3(t) &= \frac{3\sin(t) - \sin(3t)}{4},\end{aligned}$$

to give us the equation

$$\begin{aligned}\frac{\partial^2 Y_1}{\partial t^2} + Y_1 = {} & \frac{1}{4}\left(\left[-8B_0' - A_0^2B_0 - B_0^3\right]\cos(t) + \left[8A_0' + A_0^3 + A_0B_0^2\right]\sin(t)\right. \\ & \left. - 3A_0^2B_0\cos(3t) + B_0^3\cos(3t) + A_0^3\sin(3t) - 3A_0B_0^2\sin(3t)\right).\end{aligned}$$

In order to avoid secular terms for Y_1, both expressions in brackets must be zero. Thus, we have a system of first order equations

$$\begin{aligned} A_0' &= -\frac{A_0}{8}(A_0^2 + B_0^2), \\ B_0' &= -\frac{B_0}{8}(A_0^2 + B_0^2). \end{aligned}$$

At first this looks like we have gained very little progress, since we still have to solve a system of non-linear equations. However, if we switch to polar coordinates,

$$\begin{aligned} A_0(s) &= r_0(s)\cos(\theta_0(s)), \\ B_0(s) &= r_0(s)\sin(\theta_0(s)), \end{aligned} \tag{9.11}$$

we find that

$$\begin{aligned} r_0'(s) &= \frac{A_0(s)A_0'(s) + B_0(s)B_0'(s)}{\sqrt{A(s)^2 + B(s)^2}}, \\ \theta_0'(s) &= \frac{A(s)B'(s) - B(s)A'(s)}{A(s)^2 + B(s)^2}. \end{aligned}$$

Hence, the equations convert to

$$r_0'(s) = \frac{-r_0^3}{8} \qquad \theta_0'(s) = 0.$$

This is easily solvable to produce

$$r_0(s) = \frac{2}{\sqrt{s + C_1}}, \qquad \theta_0(s) = C_2.$$

Finally, we can utilize the initial conditions $y(0) = 1$, and $y'(0) = 0$ to determine that $r_0(0) = 1$ and $\theta_0(0) = 0$. Hence, we have

$$A_0(s) = \frac{2}{\sqrt{s+4}}, \qquad \text{and} \qquad B_0(s) = 0.$$

Thus, the multiple-scale approximation is given by

$$y \sim \frac{2\cos(t)}{\sqrt{s+4}} = \frac{2\cos(t)}{\sqrt{\epsilon t + 4}}.$$

The approximation is compared to the exact solution in figure 9.7. We see that the decay is algebraic rather than exponential, mainly because the damping term $\epsilon y' y^2$ decreases as the amplitude decreases. Nonetheless, the solutions decay to 0 as $t \to \infty$. □

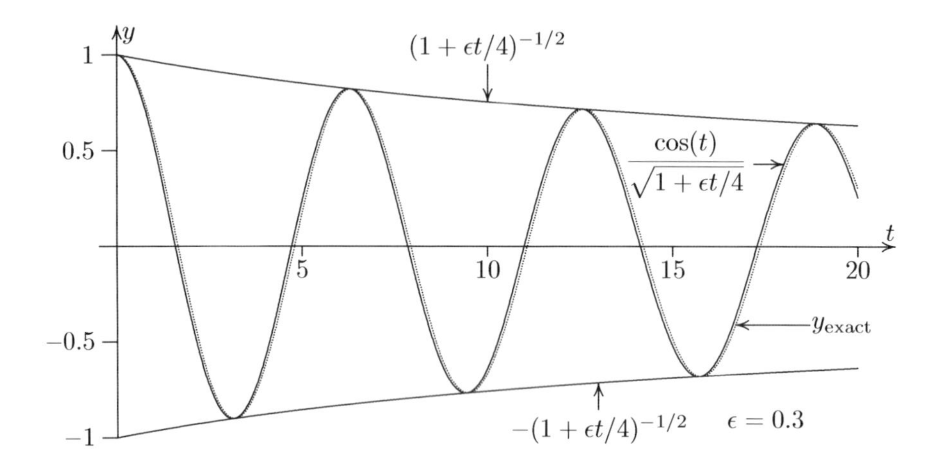

FIGURE 9.7: This compares the leading order multiple-scale approximation to the exact solution to the equation $y'' + y + \epsilon y^2 y' = 0$, with $y(0) = 1$ and $y'(0) = 0$. Here, we used $\epsilon = 0.3$. Note that the solution is enveloped between the functions $y = \pm 2/\sqrt{\epsilon t + 4}$.

Example 9.5
The van der Pol oscillator, given by the equation

$$y'' + \epsilon y'(y^2 - 1) + y = 0, \tag{9.12}$$

has been used to model the human heartbeat for large ϵ. This model has had a long history of applications, since an electrical circuit which exhibits the behavior of this equation can be constructed using an inductor and a capacitor, along with a tunnel diode or active resistor. Determine the leading order multiple-scale approximation for this equation with the initial conditions $y(0) = 1$, $y'(0) = 0$.
SOLUTION: We begin by assuming that

$$y(t) = Y_0(t, s) + \epsilon Y_1(t, s) + O(\epsilon^2),$$

where $s = \epsilon t$. Substituting this expression into equation 9.12 gives us

$$\frac{\partial^2 Y_0}{\partial t^2} + Y_0 = 0,$$
$$\frac{\partial^2 Y_1}{\partial t^2} + Y_1 = -2\frac{\partial^2 Y_0}{\partial s \partial t} - Y_0^2 \frac{\partial Y_0}{\partial t} + \frac{\partial Y_0}{\partial t}.$$

Plugging the solution to the first equation, $Y_0 = A_0(s)\cos(t) + B_0(s)\sin(t)$ into the second equation gives us

$$\frac{\partial^2 Y_1}{\partial t^2} + Y_1 = A_0^3 \cos^2(t)\sin(t) - A_0^2 B_0 \cos^3(t) - 2A_0 B_0^2 \cos^2(t)\sin(t)$$

$$+ 2A_0^2 B_0 \cos(t) \sin^2(t) - B_0^3 \cos(t) \sin^2(t) + A_0 B_0^2 \sin^3(t)$$
$$- A_0 \sin(t) + B_0 \cos(t) + 2A_0' \sin(t) - 2B_0' \cos(t).$$

After applying some trig identities, this becomes

$$\frac{\partial^2 Y_1}{\partial t^2} + Y_1 = \frac{1}{4}\Big(\left[-8B_0' - A_0^2 B_0 - B_0^3 + 4B_0\right] \cos(t)$$
$$+ \left[8A_0' + A_0^3 + A_0 B_0^2 - 4A_0\right] \sin(t)$$
$$- 3A_0^2 B_0 \cos(3t) + B_0^3 \cos(3t) + A_0^3 \sin(3t) - 3A_0 B_0^2 \sin(3t)\Big).$$

To avoid secular terms, the expression in each bracket must be 0. Thus, we have the system

$$A_0' = -\frac{A_0}{8}(A_0^2 + B_0^2) + \frac{A_0}{2},$$
$$B_0' = -\frac{B_0}{8}(A_0^2 + B_0^2) + \frac{B_0}{2}.$$

Converting to polar coordinates, we have

$$r_0'(s) = -\frac{r_0^3}{8} + \frac{r_0}{2}, \qquad \theta'(s) = 0.$$

This has the solution

$$r = \frac{2}{\sqrt{1 + C_1 e^{-s}}}, \qquad \theta_0(s) = C_2.$$

Hence,

$$Y_0(t, s) = \frac{2\cos(C_2)\cos(t) + 2\sin(C_2)\sin(t)}{\sqrt{1 + C_1 e^{-s}}}.$$

The initial conditions $y(0) = 1$, $y'(0) = 0$ translates to

$$Y_0(0, 0) = 1, \qquad (\partial Y_0/\partial t)(0, 0) = 0.$$

Thus, we see that $C_1 = 3$ and $C_2 = 0$. So we have the leading order estimate

$$y \approx \frac{2\cos t}{\sqrt{1 + 3e^{-\epsilon t}}}.$$

Figure 9.8 compares the multiple-scale approximation to the exact solution. Even with $\epsilon = 0.25$, the approximation follows fairly closely to the exact solution. □

Even though the multiple-scale method can be used for solutions that are not periodic, it does not account for the fact that the frequency will change as a function of ϵ, as the strained coordinate method does. In the next section, we will combine the two methods to produce a technique that will allow for both non-periodic solutions and fluctuating periods.

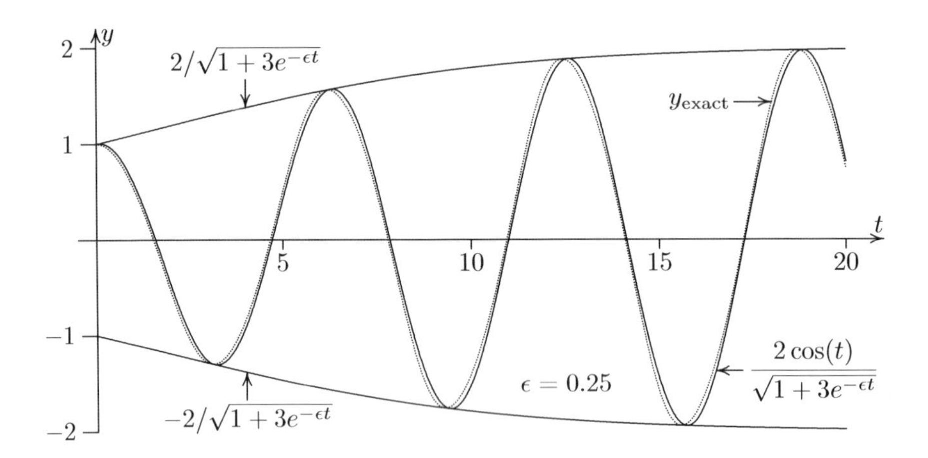

FIGURE 9.8: Graph comparing the leading order multiple-scale approximation to van der Pol's equation, $y'' + y + \epsilon y'(y^2 - 1) = 0$, with $y(0) = 1$, $y'(0) = 0$. The value of ϵ is 0.25. The solution is enveloped by the equations $y = \pm 2/\sqrt{1 + 3e^{-\epsilon t}}$.

Problems for §9.2

1 Find the next order approximation for the problem in example 9.4.

2 Show that the method of strained coordinates fails in finding an approximation to equation 9.6. What happens in attempting to find the first order approximation?

For problems **3** through **14**: For the following perturbation problems, find the leading order multiple-scale approximation, $y_{\text{multi},0}$, without the strained coordinates.

3 $y'' + y + \epsilon(y^2 + 2y) = 0, \quad y(0) = 1, \quad y'(0) = 0$

4 $y'' + y + \epsilon(2y + 2y') = 0, \quad y(0) = 1, \quad y'(0) = 0$

5 $y'' + y + \epsilon y^3 = 0, \quad y(0) = 1, \quad y'(0) = 0$

6 $y'' + y + \epsilon(y')^3 = 0, \quad y(0) = 1, \quad y'(0) = 0$

7 $y'' + y + \epsilon y^2 y' = 0, \quad y(0) = 1, \quad y'(0) = 0$

8 $y'' + y + \epsilon(y^3 + y) = 0, \quad y(0) = 1, \quad y'(0) = 0$

9 $y'' + y + \epsilon(y^3 - y) = 0, \quad y(0) = 1, \quad y'(0) = 0$

10 $y'' + y + \epsilon(y^2 y' + y) = 0, \quad y(0) = 0, \quad y'(0) = 1$

11 $y'' + y + \epsilon(y^2 y' - y') = 0, \quad y(0) = 0, \quad y'(0) = 1$

12 $y'' + y + \epsilon(y(y')^2 - y) = 0, \quad y(0) = 0, \quad y'(0) = 2$

13 $y'' + y + \epsilon(y(y')^2 + y') = 0, \quad y(0) = 1, \quad y'(0) = 0$

14 $y'' + y + \epsilon(y^3 + y') = 0, \quad y(0) = 1, \quad y'(0) = 0$

For problems **15** through **20**: For the following perturbation problems, find the first order multiple-scale approximation, $y_{\text{multi},1}$, without the strained coordinates.

15 $y'' + y + \epsilon y(y') = 0,\quad y(0) = 1,\quad y'(0) = 0$

16 $y'' + y + \epsilon(y^2 + y) = 0,\quad y(0) = 1,\quad y'(0) = 0$

17 $y'' + y + \epsilon(y')^3 = 0,\quad y(0) = 1,\quad y'(0) = 0$

18 $y'' + y + \epsilon y'(y^2 + (y')^2) = 0,\quad y(0) = 1,\quad y'(0) = 0$

19 $y'' + y + \epsilon y^2 y' = 0,\quad y(0) = 0,\quad y'(0) = 1$

20 $y'' + y + \epsilon(y')^3 = 0,\quad y(0) = 0,\quad y'(0) = 1$

21 Find the first order multiple-scale approximation for the Rayleigh equation, $y'' + y + \epsilon(y')^3/3 - \epsilon y' = 0$, with $y(0) = 1$ and $y'(0) = 0$.

9.3 Two-Variable Expansion Method

The difficulty with the multiple-scale method from the last section is that the frequency of the solutions will never be a function of ϵ, no matter how many orders we expand the approximation. We would like to modify the multiple-scale method so that it will also have all of the advantages of the strained coordinate method. The strategy is to alter the fast time variable to $\tau = \omega(\epsilon)t$, where

$$\omega(\epsilon) \sim 1 + \omega_2\epsilon^2 + \omega_3\epsilon^3 + \cdots.$$

The slow time variable does not have to be a function of ϵ, so we will let $s = \epsilon t$ as before. This methodology is called the *two-variable expansion method.*

Notice that the ω_1 term is missing. The reason for this is that any $\omega_1\epsilon$ terms can be absorbed by the slow variable. For example,

$$\begin{aligned}\cos(t + \omega_1 t\epsilon + \omega_2 t\epsilon^2 + \cdots) &= \cos(t + \omega_2 t\epsilon^2 + \cdots)\cos(\omega_1 s)\\ &\quad - \sin(t + \omega_2 t\epsilon^2 + \cdots)\sin(\omega_1 s).\end{aligned}$$

The change of variables $\tau = \omega(\epsilon)t$, $s = \epsilon t$ allows us to compute the derivatives with respect to t:

$$\begin{aligned}\frac{\partial y(\tau, s)}{\partial t} &= \frac{\partial y}{\partial \tau}\frac{d\tau}{dt} + \frac{\partial y}{\partial s}\frac{ds}{dt}.\\ &= \omega(\epsilon)\frac{\partial y}{\partial \tau} + \epsilon\frac{\partial y}{\partial s}.\end{aligned} \tag{9.13}$$

Likewise,

$$\frac{\partial^2 y(\tau, s)}{\partial t^2} = \omega(\epsilon)^2\frac{\partial^2 y}{\partial \tau^2} + 2\epsilon\omega(\epsilon)\frac{\partial^2 y}{\partial \tau \partial s} + \epsilon^2\frac{\partial^2 y}{\partial s^2}. \tag{9.14}$$

Since we are adding new unknown constants into the methodology, we also need more constraints that will allow us to solve for these unknowns. The constraint will be as follows:

No term in $Y_n(\tau, s)$ can have a different order when expressed in terms of t and ϵ.

So not only are secular terms of the form $t\sin(t)$ forbidden, but even the $se^{-s}/2$ term that appeared in equation 9.8 would be disallowed, since this term becomes $\epsilon te^{-\epsilon t}$ when expanded into the original variables, so the order of this term changes. This is best demonstrated with an example.

Example 9.6

Find the first order two-variable expansion to the perturbation problem

$$y'' + y + \epsilon((y')^2 + y') = 0, \qquad y(0) = 1, \quad y'(0) = 0.$$

SOLUTION: Using equations 9.13 and 9.14, the differential equation becomes

$$\omega(\epsilon)^2 \frac{\partial^2 y}{\partial \tau^2} + 2\epsilon\omega(\epsilon)\frac{\partial^2 y}{\partial \tau \partial s} + \epsilon^2 \frac{\partial^2 y}{\partial s^2} + y + \epsilon\left(\omega \frac{\partial y}{\partial \tau} + \epsilon \frac{\partial y}{\partial s}\right)^2 + \epsilon\omega\frac{\partial y}{\partial \tau} + \epsilon^2 \frac{\partial y}{\partial s} = 0.$$

Finally, we replace y with $Y_0(\tau, s) + \epsilon Y_1(\tau, s) + \epsilon^2 Y_2(\tau, s)$ and ω with $1 + \omega_2\epsilon^2$, and collect powers of ϵ to produce

$$\begin{aligned}
\frac{\partial^2 Y_0}{\partial \tau^2} + Y_0 &= 0, \\
\frac{\partial^2 Y_1}{\partial \tau^2} + Y_1 &= -2\frac{\partial^2 Y_0}{\partial s \partial \tau} - \frac{\partial Y_0}{\partial \tau} - \left(\frac{\partial Y_0}{\partial \tau}\right)^2, \quad \text{and} \\
\frac{\partial^2 Y_2}{\partial \tau^2} + Y_2 &= -2\frac{\partial^2 Y_1}{\partial s \partial \tau} - \frac{\partial Y_1}{\partial \tau} - 2\frac{\partial Y_0}{\partial \tau} \cdot \frac{\partial Y_1}{\partial \tau} \\
&\quad - 2\omega_2 \frac{\partial^2 Y_0}{\partial \tau^2} - \frac{\partial^2 Y_0}{\partial s^2} - 2\frac{\partial Y_0}{\partial s} \cdot \frac{\partial Y_0}{\partial t} - \frac{\partial Y_0}{\partial s}.
\end{aligned}$$

The first equation is easy to solve:

$$Y_0(\tau, s) = A_0(s)\cos(\tau) + B_0(s)\sin(\tau),$$

where $A(s)$ and $B(s)$ are arbitrary functions of s. Plugging this into the second equation produces

$$\begin{aligned}
\frac{\partial^2 Y_1}{\partial \tau^2} + Y_1 &= A_0\sin(\tau) - B_0\cos(\tau) - (A_0\sin(\tau) - B_0\cos(\tau))^2 \\
&\quad + 2A_0'\sin(\tau) - 2B_0'\cos(\tau).
\end{aligned}$$

We are only concerned with terms involving $\sin(\tau)$ or $\cos(\tau)$, since these terms would produce secular terms as we compute Y_1. Since

$$(A_0\sin(\tau) - B_0\cos(\tau))^2 = \frac{A_0^2 + B_0^2}{2} + \frac{B_0^2 - A_0^2}{2}\cos(2\tau) - 2A_0B_0\sin(2\tau),$$

these terms will not produce secular terms. Hence, to get the $\cos(\tau)$ and $\sin(\tau)$ terms to cancel, we need

$$2A_0'(s) = A_0, \qquad 2B_0'(s) = B_0.$$

For the initial conditions, we have $Y_0(0,0) = 1$, and $(\partial Y_0/\partial\tau)(0,0) = 0$. Thus, $A_0(0) = 1$ and $B_0(0) = 0$. The differential equations can easily be solved to give us $A_0(s) = e^{-s/2}$ and $B_0(s) = 0$. Hence, $Y_0(\tau, s) = e^{-s/2}\cos(\tau)$. This will greatly simplify the equation for Y_1:

$$\frac{\partial^2 Y_1}{\partial\tau^2} + Y_1 = \frac{e^{-s}}{2}\cos(2\tau) - \frac{e^{-s}}{2}.$$

We can now solve for Y_1, which by design will avoid secular terms.

$$Y_1 = -\frac{e^{-s}}{2} - \frac{e^{-s}}{6}\cos(2\tau) + A_1(s)\cos(\tau) + B_1(s)\sin(\tau).$$

The initial condition $y(0) = 1$ gives us $Y_1(0,0) = 0$, hence $A_1(0) = 2/3$, but the other initial condition $y'(0) = 0$ is trickier. We must have each order in epsilon of equation 9.13 be 0 when $\tau = s = 0$. So $(\partial Y_1/\partial\tau)(0,0) + (\partial Y_0/\partial s)(0,0) = 0$. This gives us $B_1(0) = 1/2$.

In order to determine the differential equations for which $A_1(s)$ and $B_1(s)$ satisfy, we evaluate the third equation using our knowledge of Y_0 and Y_1. After using trigonometric identities, we obtain

$$\begin{aligned}\frac{\partial^2 Y_2}{\partial\tau^2} + Y_2 &= [2A_1' + A_1]\sin(\tau) - e^{-s/2}A_1 + e^{-s/2}A_1\cos(2\tau) - \frac{e^{-3s/2}}{3}\cos(3\tau)\\ &+ \left[\frac{e^{-3s/2}}{3} + \frac{e^{-s/2}}{4} + 2\omega_2 e^{-s/2} - B_1 - 2B_1'\right]\cos(\tau)\\ &- \frac{e^{-s}}{6}\sin(2\tau) + e^{-s/2}B_1\sin(2\tau).\end{aligned}$$

In order to prevent secular terms, the expressions in the brackets must go to zero. In particular, A_1 satisfies $A_1'(s) = -A_1(s)/2$, and with the initial condition $A(0) = 2/3$, we get $A_1(s) = 2e^{-s/2}/3$.

The equation for $B_1(s)$ becomes

$$B_1'(s) + \frac{B_1 s}{2} = \frac{e^{-3s/2}}{6} + \frac{e^{-s/2}}{8} + \omega_2 e^{-s/2}.$$

Since $e^{-s/2}$ solves the homogeneous equation, we must avoid such terms in the inhomogeneous equation if we are to avoid secular terms. This can be done if $\omega_2 = -1/8$. Then the solution to the equation for B_1 which satisfies $B(0) = 1/2$ is

$$B_1(s) = \frac{2e^{-s/2}}{3} - \frac{e^{-3s/2}}{6}.$$

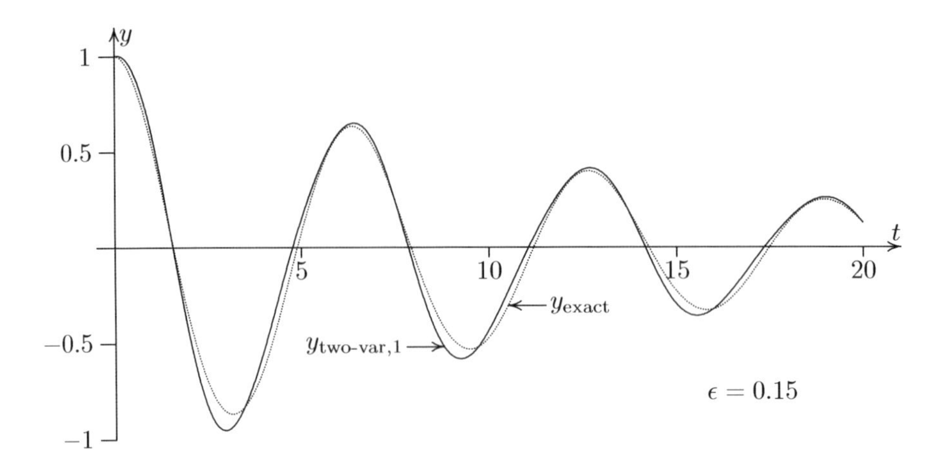

FIGURE 9.9: This graph compares the first order two-variable expansion to the equation $y'' + y + \epsilon((y')^2 + y') = 0$, with $y(0) = 1$, $y'(0) = 0$, using $\epsilon = 0.15$.

Thus, we have

$$Y_1 = -\frac{e^{-s}}{2} - \frac{e^{-s}}{6}\cos(2\tau) + \frac{2e^{-s/2}}{3}\cos(\tau) + \frac{2e^{-s/2}}{3}\sin(\tau) - \frac{e^{-3s/2}}{6}\sin(\tau).$$

This gives us

$$y_{\text{two-var},1} = e^{-\epsilon t/2}\cos\left(t - \frac{\epsilon^2 t}{8}\right) - \epsilon\frac{e^{-\epsilon t}}{2} - \epsilon\frac{e^{-\epsilon t}}{6}\cos\left(2t - \frac{\epsilon^2 t}{4}\right) + \epsilon\frac{2e^{-\epsilon t/2}}{3}\cos\left(t - \frac{\epsilon^2 t}{8}\right) + \epsilon\frac{4e^{-\epsilon t/2} - e^{-3\epsilon t/2}}{6}\sin\left(t - \frac{\epsilon^2 t}{8}\right).$$

The graph of this approximation is shown in figure 9.9. Notice that the approximation mimics both the decay rate and the phase change of the exact solution. □

At first it seems like a fortunate coincidence that the unknown constant ω_2 appeared with the secular terms, so we could simultaneously determine ω_2 and eliminate the secular terms. However, there is a simple argument that shows that this will always happen. Every term in the solution $y(t)$ will be of an order ϵ^n for some n, when expressed in terms of t and ϵ alone. For each term, there is a unique way of expressing it in terms of the fast and slow variables so that the partial derivatives with respect to both variables are of order 1. Thus, each of the $Y_n(\tau, s)$ will be uniquely determined, and the order will be consistent if in fact Y_n is expressed solely in terms of t and ϵ. This argument shows that indeed the secular terms can be totally avoided through a proper assignment of $\omega(\epsilon)$. Hence, each of the ω_n can be uniquely determined through the avoidance of secular terms [14].

Example 9.7
Find the second order two-variable expansion for the equation

$$y'' + y + \epsilon y^2 = 0, \qquad y(0) = 0, \quad y'(0) = 1.$$

SOLUTION: Using equation 9.14, we have

$$\omega(\epsilon)^2 \frac{\partial^2 y}{\partial \tau^2} + 2\epsilon\omega(\epsilon)\frac{\partial^2 y}{\partial \tau \partial s} + \epsilon^2 \frac{\partial^2 y}{\partial s^2} + y + \epsilon y^2 = 0.$$

Replacing $\omega(\epsilon)$ with $1+\omega_2\epsilon^2+\omega_3\epsilon^3$ and y with $Y_0(\tau,s)+\epsilon Y_1(\tau,s)+\epsilon^2 Y_2(\tau,s)+\epsilon^3 Y_3(\tau,s)$, and collecting like powers of ϵ, we obtain the equations

$$\begin{aligned}
&\frac{\partial^2 Y_0}{\partial \tau^2} + Y_0 = 0, \qquad (9.15)\\
&\frac{\partial^2 Y_1}{\partial \tau^2} + Y_1 = -2\frac{\partial^2 Y_0}{\partial s \partial \tau} - Y_0^2,\\
&\frac{\partial^2 Y_2}{\partial \tau^2} + Y_2 = -2\frac{\partial^2 Y_1}{\partial s \partial \tau} - 2Y_0Y_1 - 2\omega_2\frac{\partial^2 Y_0}{\partial \tau^2} - \frac{\partial^2 Y_0}{\partial s^2},\\
&\frac{\partial^2 Y_3}{\partial \tau^2} + Y_3 = -2\frac{\partial^2 Y_2}{\partial s \partial \tau} - 2Y_0Y_2 - 2\omega_2\frac{\partial^2 Y_1}{\partial \tau^2} - \frac{\partial^2 Y_1}{\partial s^2}\\
&\qquad\qquad - Y_1^2 - 2\omega_3\frac{\partial^2 Y_0}{\partial \tau^2} - 2\omega_2\frac{\partial^2 Y_0}{\partial s \partial \tau}.
\end{aligned}$$

The first equation gives us $Y_0 = A_0(s)\cos(\tau) + B_0(s)\sin(\tau)$, where A_0 and B_0 are arbitrary functions of s. Plugging this into the second equation produces

$$\begin{aligned}
\frac{\partial^2 Y_1}{\partial \tau^2} + Y_1 = &-A_0(s)^2\cos^2(\tau) - 2A_0(s)B_0(s)\cos(\tau)\sin(\tau) - B_0(s)^2\sin^2(\tau)\\
&+ 2\sin(\tau)A_0'(s) - 2\cos(\tau)B_0'(s),
\end{aligned}$$

which, after applying some standard trig identities, becomes

$$\begin{aligned}
\frac{\partial^2 Y_1}{\partial \tau^2} + Y_1 = &\frac{B_0(s)^2}{2}\cos(2\tau) - \frac{A_0^2+B_0^2}{2} - \frac{A_0(s)^2}{2}\cos(2\tau)\\
&- A_0(s)B_0(s)\sin(2\tau) + 2A_0'(s)\sin(\tau) - 2B_0'(s)\cos(\tau).
\end{aligned}$$

In order to determine $A_0(s)$ and $B_0(s)$, we examine the terms containing $\cos(\tau)$ or $\sin(\tau)$, since these will cause secular terms in computing Y_1. But in this case there are only two such terms, so we have $A_0'(s) = 0$ and $B_0'(s) = 0$. The initial condition tell us that $Y_0(0,0) = 0$, and $(\partial Y_0/\partial \tau)(0,0) = 1$, which translates to $A(0) = 0$ and $B(0) = 1$. Thus, $Y_0(\tau,s)$ is simply $\sin(\tau)$.

Solving for $A_0(s)$ and $B_0(s)$ simplifies the equation for Y_1:

$$\frac{\partial^2 Y_1}{\partial \tau^2} + Y_1 = \frac{1}{2}\cos(2\tau) - \frac{1}{2}.$$

The solution to this equation is

$$Y_1 = A_1(s)\cos(\tau) + B_1(s)\sin(\tau) - \frac{1}{6}\cos(2\tau) - \frac{1}{2}.$$

If we utilize our knowledge of Y_0 and Y_1, the third part of equation 9.15 becomes, after some trig identities, becomes

$$\begin{aligned}\frac{\partial^2 Y_2}{\partial \tau^2} + Y_2 = B_1(s)\cos(2\tau) - B_1(s) + \frac{5}{6}\sin(\tau) + 2\omega_2\sin(\tau)\\ - A_1(s)\sin(2\tau) + \frac{1}{6}\sin(3\tau) + 2A_1'(s)\sin\tau - 2B_1'(s)\cos(\tau).\end{aligned}$$

In order to avoid secular terms, we must have $2A_1'(s) + 5/6 + 2\omega_2 = 0$, and $2B_1'(s) = 0$. This gives us $A_1(s) = (5/12 + \omega_2)s + C$, which has a secular term unless $\omega_2 = -5/12$. The initial conditions are a bit trickier, since we need $Y_1(0,0) = 0$, but also the ϵ term of

$$\frac{dy}{dt} = \omega(\epsilon)\frac{\partial y}{\partial \tau} + \epsilon\frac{\partial y}{\partial s}$$

must be 0 at $\tau = 0$ and $s = 0$. Hence, $(\partial Y_1/\partial \tau)(0,0) + (\partial Y_0/\partial s)(0,0) = 0$. This gives us $A_1 = 2/3$ and $B_1 = 0$. So $Y_1 = 2/3\cos(t)$.

Having found ω_2, $A_1(s)$ and $B_1(s)$ simplifies the equation for Y_2:

$$\frac{\partial^2 Y_2}{\partial \tau^2} + Y_2 = \frac{1}{6}\sin(3\tau) - \frac{2}{3}\sin(2\tau).$$

The general solution to this equation is given by

$$Y_2 = A_2(s)\cos(s) + B_2(s)\sin(s) + \frac{2}{9}\sin(2\tau) - \frac{1}{48}\sin(3t).$$

To finish the second order approximation, we need to find $A_2(s)$ and $B_2(s)$, meaning that we have to use the last part of equation 9.15. This produces the equation

$$\begin{aligned}\frac{\partial^2 Y_2}{\partial \tau^2} + Y_2 = \frac{3}{16}\cos(2\tau) - \frac{35}{72} - B_2(s) + B_2(s)\cos(2\tau) + \frac{1}{3}\cos(3\tau)\\ - \frac{5}{144}\cos(4\tau) + 2\omega_3\sin(\tau) - A_2(s)\sin(2\tau)\\ + 2A_2'(s)\sin(\tau) - 2B_2'(s)\cos(\tau).\end{aligned}$$

To avoid secular terms, we must have $2\omega_3 + 2A_2'(s) = 0$ and $-2B_2'(s) = 0$. So A_2 and B_2 are constants, and $\omega_3 = 0$.

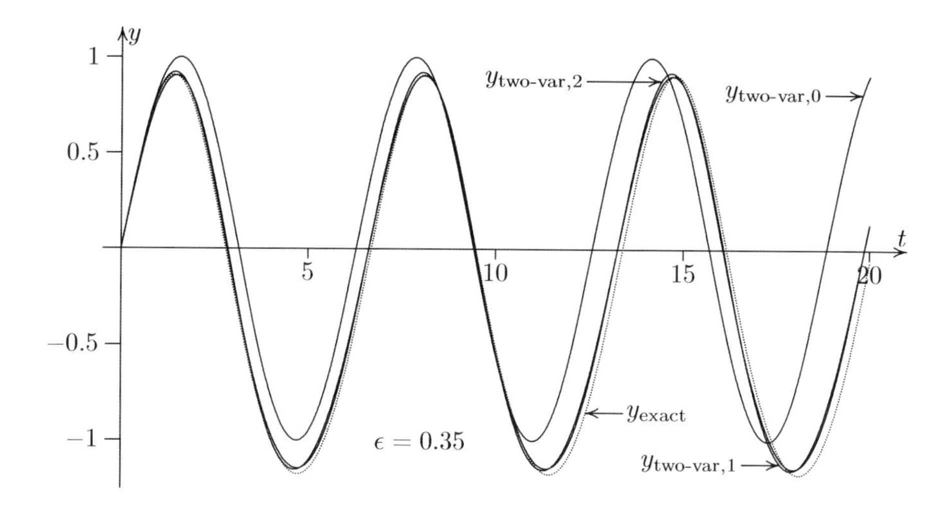

FIGURE 9.10: This graph compares the different two-variable expansions for the equation $y'' + y + \epsilon y^2 = 0$, with $y(0) = 0$, $y'(0) = 1$. Although the solution will always be periodic, the solution is lower than the unperturbed solution, which in this case coincides with $y_{\text{two-var},0}$. The higher order approximations closely approximate the exact solution, (shown with a dotted line) even with ϵ as large as 0.35.

Finally, for the initial conditions, we need $Y_2(0,0) = 0$ and

$$(\partial Y_2/\partial\tau)(0,0) + \omega_2(\partial Y_0/\partial\tau)(0,0) + (\partial Y_1/\partial s) = 0.$$

This gives us $A_2 = 0$ and $B_2 = 5/144$. So we have the final approximation

$$y_{\text{two-var},2} = \sin\left(t - \frac{5\epsilon^2 t}{12}\right) + \epsilon\left(\frac{2}{3}\cos\left(t - \frac{5\epsilon^2 t}{12}\right) - \frac{1}{2} - \frac{1}{6}\cos\left(2t - \frac{5\epsilon^2 t}{6}\right)\right)$$
$$+ \epsilon^2\left(\frac{5}{144}\sin\left(t - \frac{5\epsilon^2 t}{12}\right) + \frac{2}{9}\sin\left(2t - \frac{5\epsilon^2 t}{6}\right) - \frac{1}{48}\sin\left(3t - \frac{5\epsilon^2 t}{4}\right)\right).$$

Figure 9.10 compares the exact solution to the various two-variable expansions. It may seem odd that none of the $A(s)$'s or $B(s)$'s depended on s. But there is a simple explanation: the solutions are in fact periodic, so there is no dependence on the slow variable. This problem could in fact have been solved using the strained coordinate method, with exactly the same results. But without the prior knowledge of periodicity, the two-variable expansion method can be used, and this demonstrates that the solutions are indeed periodic. ☐

Example 9.8
Consider the Duffing equation with a damping term

$$y'' + y + \epsilon(y^3 + y') = 0, \qquad y(0) = 1, \quad y'(0) = 0.$$

Determine the leading order two-variable expansion to this equation.
SOLUTION: Using equations 9.13 and 9.14, we have

$$\omega(\epsilon)^2 \frac{\partial^2 y}{\partial \tau^2} + 2\epsilon\omega(\epsilon)\frac{\partial^2 y}{\partial \tau \partial s} + \epsilon^2 \frac{\partial^2 y}{\partial s^2} + y + \epsilon y^3 + \epsilon\omega(\epsilon)\frac{\partial y}{\partial \tau} + \epsilon^2 \frac{\partial y}{\partial s} = 0.$$

Replacing $\omega(\epsilon)$ with $1+\omega_2\epsilon^2+\omega_3\epsilon^3$ and y with $Y_0(\tau, s)+\epsilon Y_1(\tau, s)+\epsilon^2 Y_2(\tau, s)+\epsilon^3 Y_3(\tau, s)$, and collecting like powers of ϵ, we obtain the equations

$$\frac{\partial^2 Y_0}{\partial \tau^2} + Y_0 = 0,$$
$$\frac{\partial^2 Y_1}{\partial \tau^2} + Y_1 = -2\frac{\partial^2 Y_0}{\partial s \partial \tau} - \frac{\partial Y_0}{\partial \tau} - Y_0^3.$$

The first equation gives us $Y_0 = A_0(s)\cos(\tau) + B_0(s)\sin(\tau)$, where A_0 and B_0 are arbitrary functions of s. Plugging this into the second equation produces

$$\begin{aligned}\frac{\partial^2 Y_1}{\partial \tau^2} + Y_1 = &-A_0(s)^3\cos^3(\tau) - B_0(s)^3\sin^3(\tau) - 3A_0(s)^2 B_0(s)\cos^2(\tau)\sin(\tau)\\ &- 3A_0(s)B_0(s)^2\cos(\tau)\sin^2(\tau) + 2\sin(\tau)A_0'(s) - 2\cos(\tau)B_0'(s)\\ &+ A_0(s)\sin(\tau) - B_0(s)\cos(\tau)\end{aligned}$$

which, after applying the identities in equation 9.10, becomes

$$\begin{aligned}\frac{\partial^2 Y_1}{\partial \tau^2} + Y_1 = &\frac{1}{4}\left[8A_0' - 3B_0^3 - 3A_0^2 B_0 + 4A_0\right]\sin(\tau) - \frac{A_0^3}{4}\cos(3\tau)\\ &- \frac{1}{4}\left[8B_0' + 3A_0^3 + 3A_0 B_0^2 + 4B_0\right]\cos(\tau) + \frac{B_0^3}{4}\sin(3\tau)\\ &+ \frac{3A_0 B_0^2}{4}\cos(3\tau) - \frac{3A_0^2 B_0}{4}\sin(3\tau).\end{aligned}$$

To avoid secular terms, the expressions in the brackets must go to zero. Thus, we have the system of equations

$$A_0'(s) = \frac{3}{8}B_0(s)(A_0(s)^2 + B_0(s)^2) - \frac{A_0(s)}{2},$$

$$B_0'(s) = -\frac{3}{8}A_0(s)(A_0(s)^2 + B_0(s)^2) - \frac{B_0(s)}{2}.$$

The initial condition tell us that $Y_0(0, 0) = 1$, and $(\partial Y_0/\partial \tau)(0, 0) = 0$, which translates to $A_0(0) = 1$ and $B_0(0) = 0$. If we convert the equations to polar coordinates r_0 and θ_0 using equation 9.11, we get

$$r_0'(s) = -\frac{r_0(s)}{2} \qquad \theta_0'(s) = -\frac{3}{8}r_0(s)^2.$$

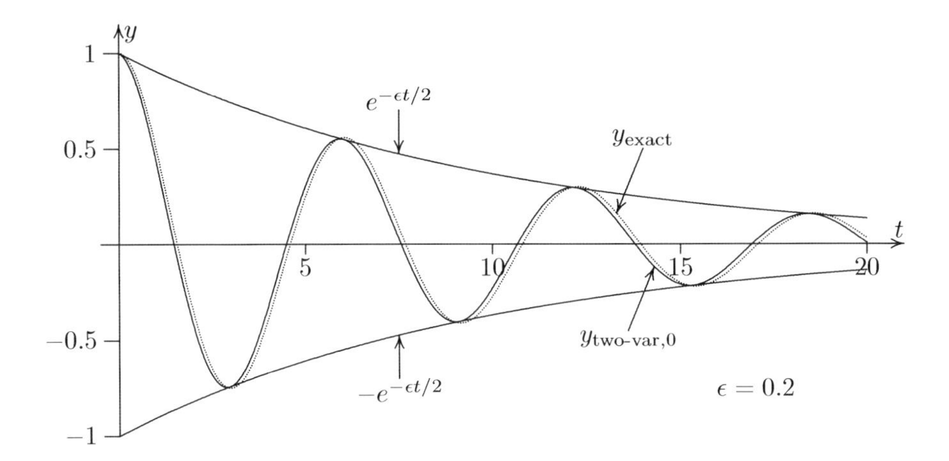

FIGURE 9.11: Here we compare the leading two-variable expansion for the equation $y'' + y + \epsilon(y^3 + y') = 0$, with $y(0) = 1$, $y'(0) = 0$, and $\epsilon = 0.2$. Notice that even the leading order accurately imitates the rate of decay and the changing of the period. The enveloping curves $y = \pm e^{-\epsilon t/2}$ are also shown for comparison.

The first equation tells us the $r_0 = C_1 e^{-s/2}$, and the initial conditions indicate that $C_1 = 1$. Then $\theta_0 = 3e^{-s}/8 + C_2$, and with $\theta_0(0) = 0$, we find that $C_2 = -3/8$. So

$$A_0(s) = e^{-s/2}\cos(3(e^{-s} - 1)/8), \qquad \text{and} \qquad B_0(s) = e^{-s/2}\sin(3(e^{-s} - 1)/8).$$

Thus, we have

$$Y_0 = e^{-s/2}\cos(3(e^{-s} - 1)/8)\cos(\tau) + e^{-s/2}\sin(3(e^{-s} - 1)/8)\sin(\tau).$$

This gives us the leading order two-variable expansion

$$y_{\text{two-var},0} \approx e^{-\epsilon t/2}\cos\left(t - \frac{3(e^{-\epsilon t} - 1)}{8}\right).$$

Figure 9.11 compares the leading order solution to the exact solution. In this case, even the leading order gives a very close approximation to the exact solution. □

Note that since no ω terms are involved in finding the leading order term, $y_{\text{two-var},0} = y_{\text{multi},0}$. However, higher order expansions will be different. Also, this example shows that the two-variable expansion method works even if the phase shift is of order ϵ, since the missing ω_1 term can be represented by trigonometric functions for $A_0(s)$ and $B_0(s)$. Unfortunately, if we continued to the first order approximation, the equations for A_1 and B_1 become unwieldy.

To end this chapter, we will do one more rather tricky example, but one that involves a very famous equation.

Example 9.9

Find the first order two-variable expansion to the van der Pol problem

$$y'' + \epsilon y'(y^2 - 1) + y = 0, \qquad y(0) = 1, \quad y'(0) = 0.$$

SOLUTION: Using equations 9.13 and 9.14, we find the differential equation becomes

$$\omega(\epsilon)^2 \frac{\partial^2 y}{\partial \tau^2} + 2\epsilon\omega(\epsilon)\frac{\partial^2 y}{\partial \tau \partial s} + \epsilon^2 \frac{\partial^2 y}{\partial s^2} + \epsilon\omega(\epsilon)\frac{\partial y}{\partial \tau}(y^2 - 1) + \epsilon^2 \frac{\partial y}{\partial s}(y^2 - 1) + y = 0.$$

If we replace y with $Y_0(\tau, s) + \epsilon Y_1(\tau, s) + \epsilon^2 Y_2(\tau, s)$ and collect powers of ϵ, we produce

$$\begin{aligned}
\frac{\partial^2 Y_0}{\partial \tau^2} + Y_0 &= 0, \\
\frac{\partial^2 Y_1}{\partial \tau^2} + Y_1 &= -2\frac{\partial^2 Y_0}{\partial s \partial \tau} - Y_0^2 \frac{\partial Y_0}{\partial \tau} + \frac{\partial Y_0}{\partial \tau}, \\
\frac{\partial^2 Y_2}{\partial \tau^2} + Y_2 &= -2\frac{\partial^2 Y_1}{\partial s \partial \tau} - Y_0^2 \frac{\partial Y_1}{\partial \tau} + \frac{\partial Y_1}{\partial \tau} - 2Y_0 Y_1 \frac{\partial Y_0}{\partial \tau} \\
&\quad - 2\omega_2 \frac{\partial^2 Y_0}{\partial \tau^2} - \frac{\partial^2 Y_0}{\partial s^2} - Y_0^2 \frac{\partial Y_0}{\partial s} + \frac{\partial Y_0}{\partial s}.
\end{aligned}$$

The first two equations are the same as example 9.5, so we know that

$$Y_0(\tau, s) = 2\cos(\tau)/\sqrt{1 + 3e^{-s}}.$$

Plugging this into the second equation gives us

$$\frac{\partial^2 Y_1}{\partial \tau^2} + Y_1 = 2\sin(3\tau)(1 + 3e^{-s})^{-3/2}.$$

Hence, we have

$$Y_1 = -\frac{\sin(3\tau)}{4(1 + 3e^(- s))^{3/2}} + A_1(s)\cos(\tau) + B_1(s)\sin(\tau).$$

For the initial conditions, we need

$$Y_1(0,0) = 0 \quad \text{and} \quad \frac{\partial Y_1}{\partial \tau}(0,0) + \frac{\partial Y_0}{\partial s}(0,0) = 0,$$

so $A_1(0) = 0$ and $B_1(0) = -9/32$.

In order to determine the differential equations for which $A_1(s)$ and $B_1(s)$ satisfy, we evaluate the third equation using our knowledge of Y_0 and Y_1. After using trigonometric identities, we obtain

$$\begin{aligned}\frac{\partial^2 Y_2}{\partial \tau^2} + Y_2 &= \left[\frac{2-3e^{-s}}{1+3e^{-s}}A_1(s) + 2A_1'(s)\right]\sin(\tau) + \frac{3A_1(s)}{1+3e^{-s}}\sin(3\tau) \\ &+ \left[\frac{3e^{-s}B_1(s)}{1+3e^{-s}} - 2B_1'(s) + \frac{4\omega_2}{\sqrt{1+3e^{-s}}} + \frac{1-12e^{-s}+18e^{-2s}}{4(1+3e^{-s})^{5/2}}\right]\cos(\tau) \\ &- \frac{3B_1(s)}{1+3e^{-s}}\cos(3\tau) + \frac{3+6e^{-s}}{4(1+e^{-s})^{5/2}}\cos(3\tau) + \frac{5}{4(1+e^{-s})^{5/2}}\cos(5\tau).\end{aligned}$$

In order to prevent secular terms, the expressions in the brackets must go to zero. In particular, A_1 must satisfy

$$\frac{2-3e^{-s}}{1+3e^{-s}}A_1(s) + 2A_1'(s) = 0 \qquad A_1(0) = 0,$$

which has the trivial solution $A_1(s) = 0$. The equation for $B_1(s)$ is more complicated.

$$\frac{3e^{-s}B_1(s)}{1+3e^{-s}} - 2B_1'(s) + \frac{4\omega_2}{\sqrt{1+3e^{-s}}} + \frac{1-12e^{-s}+18e^{-2s}}{4(1+3e^{-s})^{5/2}} = 0 \quad B_1(0) = -\frac{9}{32}. \tag{9.16}$$

The solution to this equation is (see problem 20)

$$B_1(s) = \frac{84/(3+e^s) - 39 + 8s + 64s\omega_2 - 4\ln((3+e^s)/4)}{32\sqrt{1+3e^{-s}}}. \tag{9.17}$$

We notice that there are some secular terms in this expression, such as the s terms in the numerator. But some secular terms are more difficult to spot. The term $\ln((3+e^s)/4)$ is actually asymptotic to s as $s \to \infty$, so we must rewrite this term as $s + \ln((1+3e^{-s})/4)$. Now the logarithm portion approaches a constant as $s \to \infty$, so this is not a secular term. Since the secular terms must cancel, we find that $4s + 64s\omega_2 = 0$, or $\omega_2 = -1/16$. Then B_1 simplifies to

$$B_1(s) = \frac{84/(3+e^s) - 39 - 4\ln((1+3e^{-s})/4)}{32\sqrt{1+3e^{-s}}}.$$

Thus, we have

$$Y_1 = -\frac{\sin(3\tau)}{4(1+3e^{-s})^{3/2}} + \frac{84/(3+e^s) - 39 - 4\ln((1+3e^{-s})/4)}{32\sqrt{1+3e^{-s}}}\sin(\tau).$$

This gives us

$$\begin{aligned}y_{\text{two-var},1} &= \frac{2}{\sqrt{1+3e^{-\epsilon t}}}\cos\left(t - \frac{\epsilon^2 t}{16}\right) - \frac{\epsilon}{4(1+3e^{-st})^{3/2}}\sin\left(3t - \frac{3\epsilon^2 t}{16}\right) \\ &+ \epsilon\frac{84/(3+e^s) - 39 - 4\ln((1+3e^{-s})/4)}{32\sqrt{1+3e^{-s}}}\sin\left(t - \frac{\epsilon^2 t}{16}\right).\end{aligned}$$

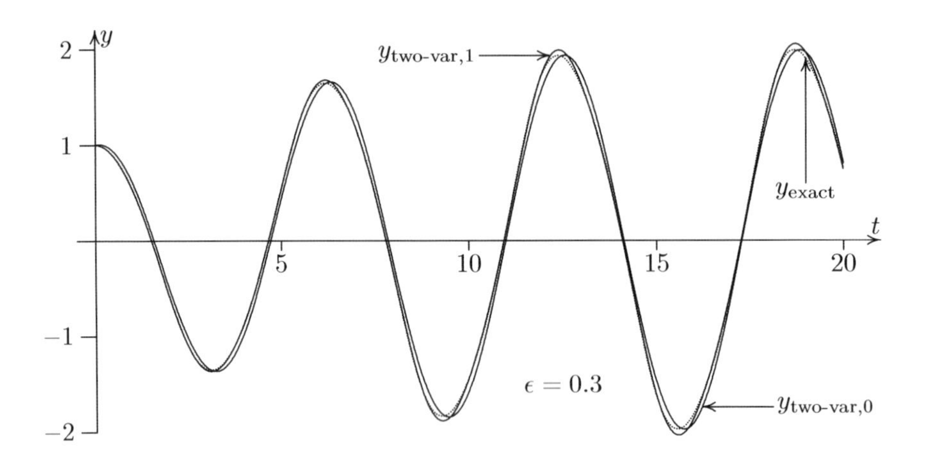

FIGURE 9.12: This graph compares the first order two-variable expansion for van der Pol's equation, $y'' + y + \epsilon y'(y^2 - 1) = 0$, with $y(0) = 1$, $y'(0) = 0$. This time, we used the value of ϵ is 0.3 to highlight the differences. The exact solution is shown with a dotted line. Although the leading order seems to have the phase angle slightly off, the first order approximation is only distinguishable from the exact solution near the peaks.

We can compare this result to the exact solution in figure 9.12. The accuracy is fairly impressive even for ϵ as large as 0.3. □

Problems for §9.3

1 Determine what the third derivative, $\partial^3 y(\tau, s)/\partial t^3$, would be under the substitutions $\tau = \omega(\epsilon)t$ and $s = \epsilon t$.

2 Find the first order two-variable expansion for the Rayleigh equation, $y'' + y + \epsilon(y')^3/3 - \epsilon y' = 0$, with $y(0) = 1$ and $y'(0) = 0$.

For problems **3** through **10**: Given the equation for $A_1(s)$, determine the value of ω_2 that will prevent secular terms in the solution, and solve for $A_1(s)$.

3 $A_1' + 2\omega_2 + 3/4 = 0,$ $A_1(0) = 3$

4 $A_1' - A_1 + e^s\omega_2 + 2 = 0,$ $A_1(0) = 4$

5 $A_1' - 2A_1 - 2e^{2s}\omega_2 + 4 = 0,$ $A_1(0) = 1$

6 $A_1' + sA_1 + 4s + \omega_2 e^{-s^2/2} = 0,$ $A_1(0) = 1$

7 $A_1' + A_1 + e^{-s}\omega_2 + 3/(1 + e^s) = 0,$ $A_1(0) = 2$

8 $(1 + e^s)A_1' - A_1 - e^s\omega_2 + 2e^{2s}/(1 + e^s) = 0,$ $A_1(0) = 2$

9 $2A_1' - A_1/(1 + e^s) + \omega_2(1 + e^{-s})^{-1/2} + 5(1 + e^{-s})^{-3/2} = 0,$ $A_1(0) = 1$

10 $A_1' - A_1 + \omega_2 e^s + \sqrt{1 + e^{2s}} = 0,$ $A_1(0) = 1$

For problems **11** through **18**: For the following perturbation problems, find the first order two-variable expansion, $y_{\text{two-var},1}$.

11 $y'' + y + \epsilon(y')y^2 = 0, \quad y(0) = 1, \quad y'(0) = 0$

12 $y'' + y + \epsilon y(y') = 0, \quad y(0) = 1, \quad y'(0) = 0$

13 $y'' + y + \epsilon(y')^2 = 0, \quad y(0) = 0, \quad y'(0) = 1$

14 $y'' + y + 2\epsilon y' = 0, \quad y(0) = 1, \quad y'(0) = 0$

15 $y'' + y + \epsilon y^2(y')^2 = 0, \quad y(0) = 1, \quad y'(0) = 0$

16 $y'' + y - \epsilon(y^2 + (y')^2) = 0, \quad y(0) = 1, \quad y'(0) = 0$

17 $y'' + y + \epsilon(y^2 + y') = 0, \quad y(0) = 1, \quad y'(0) = 0$

18 $y'' + y + \epsilon((y')^2 + y') = 0, \quad y(0) = 0, \quad y'(0) = 1$

19 Rework example 9.9 only with the initial conditions $y(0) = a$, $y'(0) = 0$ where a is an arbitrary positive constant. What happens when $a = 2$?

20 Show that the solution to the initial value problem given in equation 9.16 is given by equation 9.17.

Guide to the Special Functions

This appendix gives a guide to the special functions found throughout the book. This is designed to be used as a reference for these functions, and as such, will give further properties not mentioned in the text.

Lambert W Function

Definition:

$$W(x) \text{ solves the equation } W(x)e^{W(x)} = x, \qquad W(0) = 0.$$

Special values:

$$W(0) = 0, \quad W(e) = 1, \quad W'(0) = 1, \quad W(-\pi/2) = i\pi/2.$$

Taylor series:

$$W(z) = \sum_{n=1}^{\infty} \frac{(-n)^{n-1} z^n}{n!}, \qquad |z| < e^{-1}.$$

$$\left(\frac{W(z)}{z}\right)^r = \sum_{n=0}^{\infty} \frac{(-1)^n r(n+r)^{n-1} z^n}{n!}, \qquad r \neq 0 \text{ and } |z| < e^{-1}.$$

Gamma Function

Recursion formula:

$$\Gamma(z+1) = z\Gamma(z), \qquad \Gamma(1) = 1.$$

Integral representation, valid for $\mathrm{Re}(x) > 0$:

$$\Gamma(z) = \int_0^{\infty} t^{z-1} e^t \, dt.$$

Euler's product representation, valid everywhere except at the poles $z = 0, -1, -2, \ldots$:

$$\Gamma(z) = \frac{1}{z} \prod_{n=1}^{\infty} \left[\left(1 + \frac{1}{n}\right)^z \left(\frac{n}{n+z}\right) \right].$$

Weierstrass product representation, valid if $z \neq 0, -1, -2, \ldots$:

$$\Gamma(z) = \frac{e^{-\gamma z}}{z} \prod_{n=1}^{\infty} \frac{n e^{z/n}}{n+z}.$$

Duplication formula:

$$\Gamma(2n) = \frac{2^{2n-1}\Gamma(n)\Gamma(n+\frac{1}{2})}{\sqrt{\pi}}.$$

Reflection formula:

$$\Gamma(z)\Gamma(1-z) = \frac{\pi}{\sin(\pi z)}.$$

Beta integral, valid if $x > -1$ and $y > -1$:

$$\int_0^1 t^x(1-t)^y = \frac{\Gamma(x+1)\Gamma(y+1)}{\Gamma(x+y+2)}.$$

Stirling's asymptotic series, valid as $z \to \infty$ with $-\pi < \arg(z) < \pi$:

$$\Gamma(n) \sim \sqrt{2\pi} z^z e^{-z} z^{-1/2} \times \left(1 + \frac{1}{12z} + \frac{1}{288z^2} - \frac{139}{51840z^3} - \frac{571}{2488320z^4} + \frac{163879}{209018880z^5} + \cdots\right).$$

Continued fraction representation for Stirling's series, converging for $\mathrm{Re}(z) > 0$:

$$\Gamma(z) = \sqrt{2\pi} z^z e^{-z} z^{-1/2} \Big/ \left(1 + \frac{c_1}{z} \Big/ \left(1 + \frac{c_2}{z} \Big/ \left(1 + \frac{c_3}{z} \Big/ \left(1 + \frac{c_4}{z} \Big/ \left(1 + \cdots,\right.\right.\right.\right.\right.$$

with

$$c_1 = -\frac{1}{12}, \qquad c_2 = \frac{1}{24}, \qquad -c_3 = c_4 = \frac{293}{360},$$
$$-c_5 = c_6 = \frac{4406147}{14422632}, \qquad -c_7 = c_8 = \frac{805604978502319}{465939153254616},$$
$$-c_9 = c_{10} = \frac{522477249218864910554432538O7}{89810162621418253849277521320}.$$

Logarithmic asymptotic series, valid as $z \to \infty$ with $-\pi < \arg(z) < \pi$:

$$\ln(\Gamma(z)) \sim z\ln(z) - z + \frac{1}{2}\ln(2\pi) + \sum_{k=1}^{\infty} \frac{B_{2k}}{2k(2k-1)z^{2k-1}},$$

where B_n is the nth Bernoulli number. Continued fraction representation of logarithmic series, converging for $\mathrm{Re}(z) > 0$:

$$\ln(\Gamma(z)) = z\ln(z) - z + \frac{1}{2}\ln(2\pi) + \frac{1}{12z} \Big/ \left(1 + \frac{1}{30z^2} \Big/ \left(1 + \frac{53}{210z^2} \Big/ \left(1 + \frac{195}{371z^2} \Big/ \left(1 + \cdots.\right.\right.\right.\right.$$

Digamma Function

Definition:

$$\psi(z) = \frac{\Gamma'(z)}{\Gamma(z)}.$$

Recursion formula:

$$\psi(z+1) = \psi(z) + \frac{1}{z}, \qquad \psi(1) = -\gamma,$$

$$\text{where} \qquad \gamma = \lim_{n\to\infty}\left(1 + \frac{1}{2} + \frac{1}{3} + \cdots + \frac{1}{n} - \ln(n)\right) \approx 0.5772156649\ldots.$$

Frobenius series, converges for $|z| < 1$:

$$\psi(z) = -\frac{1}{z} - \gamma + \sum_{n=2}^{\infty}(-1)^n\zeta(n)z^{n-1}.$$

Asymptotic series, valid as $z \to \infty$ with $-\pi < \arg(z) < \pi$:

$$\psi(z) \sim \ln(z) + \sum_{n=1}^{\infty}\frac{(-1)^{n-1}B_n}{nz^n} \sim \ln(z) - \frac{1}{2z} - \frac{1}{12z^2} + \frac{1}{120z^4} - \frac{1}{252z^6} + \cdots.$$

Continued fraction representation, converging for $\mathrm{Re}(z) > 0$:

$$\ln(z) - \frac{1}{2z}\Big/\Big(1 - \frac{1}{6z}\Big/\Big(1 + \frac{1}{6z}\Big/\Big(1 - \frac{3}{5z}\Big/\Big(1 + \frac{3}{5z}\Big/\Big(1 - \frac{79}{126z}\Big/\Big(1 + \frac{79}{126z}\Big/\Big(\cdots.$$

Summation formulas, valid if $z \neq 0, -1, -2, \ldots$:

$$\psi(z) = -\frac{1}{z} - \gamma + \sum_{n=1}^{\infty}\frac{z}{n^2 + nz},$$

$$\psi'(z) = \sum_{n=0}^{\infty}\frac{1}{(n+z)^2}.$$

Integral formulas, valid if $\mathrm{Re}(z) > 0$:

$$\psi(z) = \int_0^{\infty}\left(\frac{e^{-t}}{t} - \frac{e^{-zt}}{1 - e^{-t}}\right) dt.$$

$$\psi(z) = -\gamma + \int_0^1 \frac{1 - t^{z-1}}{1 - t}\, dt.$$

Reflection formula:

$$\psi(1 - z) = \psi(z) + \pi\cot(\pi z).$$

Duplication formula:

$$\psi(2z) = \ln(2) + \frac{\psi(z) + \psi\left(z + \frac{1}{2}\right)}{2}.$$

Bernoulli Numbers and Polynomials

Generating functions:

$$\frac{x}{e^x - 1} = \sum_{n=0}^{\infty} \frac{B_n}{n!} x^n, \qquad \frac{te^{tx}}{e^t - 1} = \sum_{n=0}^{\infty} \frac{B_n(x)}{n!} t^n.$$

Initial terms:

$$B_n = \left\{1, -\frac{1}{2}, \frac{1}{6}, 0, -\frac{1}{30}, 0, \frac{1}{42}, 0, -\frac{1}{30}, 0, \frac{5}{66}, 0, -\frac{691}{2730}, 0, \frac{7}{6}, \ldots\right\}.$$

$$B_0(x) = 1, \quad B_1(x) = x - \frac{1}{2}, \quad B_2(x) = x^2 - x + \frac{1}{6}, \quad B_3(x) = x^3 - \frac{3x^2}{2} + \frac{x}{2}.$$

$$B_n = B_n(0) = (-1)^n B_n(1), \qquad B_{2n+1} = 0 \text{ for } n \geq 1.$$

Explicit formula (interpret 0^0 to be 1):

$$B_n = \sum_{k=0}^{n} \sum_{i=0}^{k} (-1)^i \binom{k}{i} \frac{i^n}{k+1}.$$

Difference equation:

$$B_n(x+1) - B_n(x) = nx^{n-1}.$$

Recursion formula, valid for $n \geq 1$:

$$B_n'(x) = nB_{n-1}(x), \qquad \text{with} \qquad \int_0^1 B_n(x)\, dx = 0.$$

Symmetry formulas:

$$B_n(1-x) = (-1)^n B_n(x), \qquad B_n(-x) = (-1)^n (B_n(x) + nx^{n-1}).$$

Error Function

Definitions:

$$\operatorname{erf}(z) = \frac{2}{\sqrt{\pi}} \int_0^z e^{-t^2}\, dt, \qquad \operatorname{erfc}(z) = 1 - \operatorname{erf}(z).$$

Maclaurin series, converges for all z:

$$\operatorname{erf}(z) = \frac{2}{\sqrt{\pi}} \sum_{n=0}^{\infty} \frac{(-1)^n z^{2n+1}}{n!(2n+1)}.$$

Asymptotic series, valid as $z \to \infty$ for $-3\pi/4 < \arg(z) < 3\pi/4$:

$$\operatorname{erf}(z) \sim 1 - \frac{e^{-x^2}}{z\sqrt{\pi}} \sum_{n=0}^{\infty} \frac{(-1)^n (2n)!}{4^n n! z^{2n}}.$$

Continued fraction representation, valid for $\mathrm{Re}(z) > 0$:

$$\mathrm{erf}(z) = 1 - \frac{e^{-x^2}}{z\sqrt{\pi}} \Big/ \Big(1 + \mathop{\mathbf{K}}_{n=1}^{\infty} \frac{n}{2z^2} \Big/ \Big(1.$$

Properties:

$$\mathrm{erf}(-z) = -\mathrm{erf}(z), \qquad \mathrm{erf}(0) = 0, \qquad \lim_{z \to +\infty} \mathrm{erf}(z) = 1.$$

$$\frac{d}{dz}\mathrm{erf}(z) = \frac{2}{\sqrt{\pi}} e^{-z^2}.$$

$$\int \mathrm{erf}(z)\, dz = z\,\mathrm{erf}(z) + \frac{e^{-z^2}}{\sqrt{\pi}} + C.$$

Integral Functions

Definitions:

$$\mathrm{Ein}(z) = \int_0^z \frac{1 - e^{-t}}{t}\, dt, \qquad E_1(z) = \int_z^\infty \frac{e^{-t}}{t}\, dt = \mathrm{Ein}(z) - \ln(z) - \gamma.$$

$$\mathrm{Si}(z) = \int_0^z \frac{\sin(t)}{t}\, dt.$$

$$\mathrm{Cin}(z) = \int_0^z \frac{1 - \cos(t)}{t}\, dt, \qquad \mathrm{Ci}(z) = -\int_z^\infty \frac{\cos(t)}{t}\, dt = -\mathrm{Cin}(z) + \ln(z) + \gamma.$$

$$\mathrm{Shi}(z) = \int_0^z \frac{\sinh(t)}{t}\, dt.$$

$$\mathrm{Chin}(z) = \int_0^z \frac{\cosh(t) - 1}{t}\, dt, \qquad \mathrm{Chi}(z) = \mathrm{Chin}(z) + \ln(z) + \gamma.$$

$$\mathrm{Thi}(z) = \int_0^z \frac{\tanh(t)}{t}\, dt.$$

$$\mathrm{Cthi}(z) = \int_0^z \frac{t\coth(t) - 1}{t^2}\, dt - \frac{1}{z}.$$

Stieltjes integral function:

$$S(x) = \int_0^\infty \frac{e^{-t}}{1 + xt}\, dt = \frac{e^{1/x}}{x} E_1\left(\frac{1}{x}\right).$$

Taylor and Frobenius series:

$$\mathrm{Ein}(z) = \sum_{n=1}^{\infty} \frac{(-1)^{n+1} z^n}{n\, n!}.$$

$$\mathrm{Si}(z) = \sum_{n=0}^{\infty} \frac{(-1)^n z^{2n+1}}{(2n+1)(2n+1)!}.$$

$$\mathrm{Cin}(z) = \sum_{n=1}^{\infty} \frac{(-1)^{n+1} z^{2n}}{(2n)(2n)!}.$$

$$\mathrm{Shi}(z) = \sum_{n=0}^{\infty} \frac{z^{2n+1}}{(2n+1)(2n+1)!}.$$

$$\mathrm{Chin}(z) = \sum_{n=1}^{\infty} \frac{z^{2n}}{(2n)(2n)!}.$$

$$\mathrm{Thi}(z) = \sum_{n=1}^{\infty} \frac{(16^n - 4^n) B_{2n} z^{2n-1}}{(2n-1)(2n)!}, \qquad |z| < \frac{\pi}{2}.$$

$$\mathrm{Cthi}(z) = \sum_{n=0}^{\infty} \frac{4^n B_{2n} z^{2n-1}}{(2n-1)(2n)!}, \qquad 0 < |z| < \pi.$$

Asymptotic series as $z \to \infty$:

$$E_1(z) \sim \frac{e^{-z}}{z} \sum_{n=0}^{\infty} \frac{(-1)^n n!}{z^n}, \qquad -3\pi/2 < \arg(z) < 3\pi/2.$$

$$\mathrm{Si}(z) = \pi/2 - w_1(z)\cos(z) - w_2(z)\sin(z), \quad \text{and}$$

$$\mathrm{Cin}(z) = \ln(z) + \gamma + w_2(x)\cos(z) - w_1 \sin(z), \quad \text{where}$$

$$w_1(x) \sim \sum_{n=0}^{\infty} \frac{(-1)^n (2n)!}{z^{2n+1}}, \qquad w_2 \sim \sum_{n=0}^{\infty} \frac{(-1)^n (2n+1)!}{z^{2n+2}}.$$

$$\mathrm{Shi}(z) \sim \mathrm{Chin}(z) \sim \frac{e^x}{2x} \sum_{n=0}^{\infty} \frac{n!}{z^n}, \qquad -\pi/2 < \arg(z) < \pi/2.$$

$$\mathrm{Thi}(z) \sim \ln(z) + (2\ln(2) + \gamma - \ln(\pi)) + \frac{e^{-2z}}{z} \sum_{n=0}^{\infty} \frac{(-1)^n n!}{2^n z^n}.$$

$$\mathrm{Cthi}(z) \sim \ln(z) + (\gamma - \ln(\pi)) - \frac{e^{-2z}}{z} \sum_{n=0}^{\infty} \frac{(-1)^n n!}{2^n z^n}.$$

Continued fractions representations, valid for $|\arg(z)| < \pi$:

$$E_1(z) = \frac{e^{-z}}{z} \Big/ \Big(1 + \mathop{\mathbf{K}}_{n=1}^{\infty} \frac{n}{z} \Big/ \Big(1 + \frac{n}{z} \Big/ \Big(1$$

$$= \frac{e^{-z}}{z} \Big/ \Big(1 + \frac{1}{z} \Big/ \Big(1 + \frac{1}{z} \Big/ \Big(1 + \frac{2}{z} \Big/ \Big(1 + \frac{2}{z} \Big/ \Big(1 + \frac{3}{z} \Big/ \Big(1 + \frac{3}{z} \Big/ \Big(1 + \cdots.$$

$$S(z) = 1\Big/\Big(1 + \mathop{\mathrm{K}}_{n=1}^{\infty} nz\Big/\Big(1 + nz\Big/\Big(1$$

$$= 1\Big/\Big(1 + z\Big/\Big(1 + z\Big/\Big(1 + 2z\Big/\Big(1 + 2z\Big/\Big(1 + 3z\Big/\Big(1 + 3z\Big/\Big(1 + \cdots.$$

Interrelationships:

$$\mathrm{Ein}(\pm iz) = \mathrm{Cin}(z) \pm i\mathrm{Si}(z).$$

$$\mathrm{Ein}(z) = \mathrm{Shi}(z) - \mathrm{Chin}(z).$$

$$\mathrm{Cin}(iz) = -\mathrm{Chin}(z), \qquad \mathrm{Si}(iz) = i\mathrm{Shi}(x)$$

$$\mathrm{Cthi}(2z) = \frac{\mathrm{Thi}(z) + \mathrm{Cthi}(z)}{2}.$$

Airy Functions

Differential equation:

$$y'' = xy.$$

Solutions are $\mathrm{Ai}(x)$ and $\mathrm{Bi}(x)$, the Airy functions of the first and second kind.

Maclaurin series (converges for all finite z):

$$\mathrm{Ai}(z) = \sum_{n=0}^{\infty} \frac{z^{3n}}{3^{2/3}9^n n!\Gamma\left(n+\frac{2}{3}\right)} - \sum_{n=0}^{\infty} \frac{z^{3n+1}}{3^{4/3}9^n n!\Gamma\left(n+\frac{4}{3}\right)},$$

$$\mathrm{Bi}(z) = \sum_{n=0}^{\infty} \frac{z^{3n}}{3^{1/6}9^n n!\Gamma\left(n+\frac{2}{3}\right)} + \sum_{n=0}^{\infty} \frac{z^{3n+1}}{3^{5/6}9^n n!\Gamma\left(n+\frac{4}{3}\right)}.$$

Asymptotic series as $z \to \infty$:

$$\mathrm{Ai}(x) \sim \frac{e^{-2z^{3/2}/3}z^{-1/4}}{2\sqrt{\pi}} \sum_{n=0}^{\infty} \frac{(-3)^n\Gamma\left(n+\frac{1}{6}\right)\Gamma\left(n+\frac{5}{6}\right)z^{-3n/2}}{2\pi 4^n n!}, \qquad |\arg(z)| < \pi,$$

$$\mathrm{Bi}(x) \sim \frac{e^{2z^{3/2}/3}z^{-1/4}}{\sqrt{\pi}} \sum_{n=0}^{\infty} \frac{3^n\Gamma\left(n+\frac{1}{6}\right)\Gamma\left(n+\frac{5}{6}\right)z^{-3n/2}}{2\pi 4^n n!}, \qquad |\arg(z)| < \frac{\pi}{3}.$$

Continued fraction representation, valid for $|\arg(z)| < \pi$:

$$\mathrm{Ai}(z) = \frac{e^{-2z^{3/2}/3}}{2\sqrt{\pi}x^{1/4}}\Big/\Big(1 + \frac{5}{48}z^{-3/2}\Big/\Big(1 + \frac{67}{96}z^{-3/2}\Big/\Big(1 + \frac{16247}{19296}z^{-3/2}\Big/\Big(1 + \cdots.$$

Asymptotic series valid for $|z| \to \infty$, $\pi/3 < \arg(z) < 5\pi/3$:

$$\mathrm{Ai}(z) = w_1(z)\cos\left(\frac{2(-z)^{3/2}}{3} - \frac{\pi}{4}\right) + w_2(z)\sin\left(\frac{2(-z)^{3/2}}{3} - \frac{\pi}{4}\right),$$

$$\mathrm{Bi}(z) = w_1(z)\cos\left(\frac{2(-z)^{3/2}}{3} + \frac{\pi}{4}\right) + w_2(z)\sin\left(\frac{2(-z)^{3/2}}{3} + \frac{\pi}{4}\right),$$

$$\text{where } w_1(z) \sim \frac{(-z)^{-1/4}}{\sqrt{\pi}} \sum_{n=0}^{\infty} \frac{3^{2n}\Gamma\left(2n+\frac{1}{6}\right)\Gamma\left(2n+\frac{5}{6}\right) z^{-3n}}{2\pi 4^{2n}(2n)!},$$

$$\text{and } w_2(z) \sim \frac{(-z)^{-7/4}}{\sqrt{\pi}} \sum_{n=0}^{\infty} \frac{3^{2n+1}\Gamma\left(2n+\frac{7}{6}\right)\Gamma\left(2n+\frac{11}{6}\right)}{z^{-3n}2\pi 4^{2n+1}(2n+1)!}.$$

Integral representation, valid for real x:

$$\mathrm{Ai}(x) = \frac{1}{\pi}\int_0^{\infty} \cos\left(\frac{t^3}{3} + xt\right) dt,$$

$$\mathrm{Bi}(x) = \frac{1}{\pi}\int_0^{\infty} e^{-t^3/3+xt} + \sin\left(\frac{t^3}{3} + xt\right) dt.$$

Contour integral representation, valid for all complex numbers:

$$\mathrm{Ai}(z) = \frac{1}{2\pi i}\int_{e^{-\pi i/3}\infty}^{e^{\pi i/3}\infty} e^{t^3/3-zt}\, dt,$$

$$\mathrm{Bi}(z) = \frac{1}{2\pi}\int_{-\infty}^{e^{\pi i/3}\infty} e^{t^3/3-zt}\, dt + \frac{1}{2\pi}\int_{-\infty}^{e^{-\pi i/3}\infty} e^{t^3/3-zt}\, dt.$$

The contour on the Airy function starts at the point at infinity with argument $-\pi/3$ and ends at the point of infinity with argument $\pi/3$.
Rotation identities, (Here, $\omega = e^{2\pi i/3}$):

$$\mathrm{Ai}(\omega z) = \frac{-\omega^2}{2}(\mathrm{Ai}(z) - i\mathrm{Bi}(z)), \qquad \mathrm{Ai}(\omega^2 z) = \frac{-\omega}{2}(\mathrm{Ai}(z) + i\mathrm{Bi}(z)),$$

$$\mathrm{Bi}(\omega z) = \frac{-\omega^2}{2}(\mathrm{Bi}(z) - 3i\mathrm{Ai}(z)), \qquad \mathrm{Bi}(\omega^2 z) = \frac{-\omega}{2}(\mathrm{Bi}(z) + 3i\mathrm{Ai}(z)).$$

Bessel Functions

Differential equation:

$$z^2y'' + zy' + (z^2 - v^2)y = 0.$$

Solutions are $J_v(z)$ and $Y_v(z)$, the Bessel functions of the first and second kind.

Frobenius series:

$$J_v(z) = \sum_{k=0}^{\infty} \frac{(-1)^k z^{2k+v}}{4^k 2^v k!\Gamma(k+v+1)},$$

$$Y_v(z) = \frac{J_v(z)\cos(v\pi) - J_{-v}(z)}{\sin(v\pi)} \quad \text{for } v \text{ non-integer.}$$

$$Y_n(z) = \frac{1}{\pi}\frac{\partial J_v(z)}{\partial v}\bigg|_{v=n} + \frac{(-1)^n}{\pi}\frac{\partial J_v(z)}{\partial v}\bigg|_{v=-n} \quad \text{for } n \text{ integer.}$$

$$Y_n(z) = \frac{2\ln\left(\frac{z}{2}\right)}{\pi}J_n(z) - \frac{2^n}{\pi}\sum_{k=0}^{n-1}\frac{(n-k-1)!z^{2k-n}}{4^k k!}$$
$$-\frac{1}{2^n\pi}\sum_{k=0}^{\infty}\frac{(-1)^k z^{2k+n}}{4^k k!(n+k)!}(\psi(k+1) + \psi(n+k+1)) \quad \text{for } n = 0, 1, 2, \ldots.$$

Asymptotic series, valid as $z \to \infty$ for $-\pi < \arg(z) < \pi$:

$$J_v(z) = \sqrt{\frac{2}{\pi z}}\left(w_1(z)\cos\left(z - \frac{\pi}{4} - \frac{v\pi}{2}\right) + w_2(z)\sin\left(z - \frac{\pi}{4} - \frac{v\pi}{2}\right)\right),$$

$$Y_v(z) = \sqrt{\frac{2}{\pi z}}\left(w_1(z)\cos\left(z - \frac{3\pi}{4} - \frac{v\pi}{2}\right) + w_2(z)\sin\left(z - \frac{3\pi}{4} - \frac{v\pi}{2}\right)\right),$$

$$\text{where } w_1(z) \sim \sum_{n=0}^{\infty}\frac{(-1)^n a_{2n}(v)}{z^{2n}}, \qquad w_2(z) \sim \sum_{n=0}^{\infty}\frac{(-1)^{n+1}a_{2n+1}(v)}{z^{2n+1}}$$

$$\text{with } a_n(v) = \frac{(4v^2 - 1^2)(4v^2 - 3^2)(4v^2 - 5^2)\cdots(4v^2 - (2n-1)^2)}{n!8^n}.$$

$$J_v(z) \pm iY_v(z) \sim \sqrt{\frac{2}{\pi z}}e^{\pm i(z - \pi/4 - \pi v/2)}\sum_{n=0}^{\infty}\frac{(\pm i)^n a_n(v)}{z^n}.$$

Analytical continuation:

Both $J_v(z)$ and $Y_v(z)$ have branch cuts along the negative real axis. To extend to other sheets of the Riemann surface, for integer m,

$$J_v(ze^{m\pi i}) = e^{mv\pi i}J_v(z).$$

$$Y_v(ze^{m\pi i}) = e^{-mv\pi i}Y_v(z) + \frac{2i\sin(mv\pi)}{\tan(v\pi)}J_v(z), \quad \text{for non-integer } v.$$

$$Y_n(ze^{m\pi i}) = (-1)^{mn}(Y_n(z) + 2imJ_n(z)), \quad \text{for integer } n.$$

Integral representations, valid for $\mathrm{Re}(z) > 0$:

$$J_v(z) = \frac{1}{\pi}\int_0^{\pi}\cos(z\sin t - vt)\,dt - \frac{\sin(v\pi)}{\pi}\int_0^{\infty}e^{-z\sinh t - vt}\,dt.$$

$$Y_v(z) = \frac{1}{\pi}\int_0^{\pi}\sin(z\sin t - vt)\,dt - \frac{1}{\pi}\int_0^{\infty}(e^{vt} + e^{-vt}\cos(v\pi))e^{-z\sinh t}\,dt.$$

Functional relations:

$$J_{-n}(z) = (-1)^n J_n(z) \qquad Y_{-n}(z) = (-1)^n Y_n(z) \qquad \text{for integer } n,$$

$$J_v(z) = \frac{x}{2v}(J_{v-1}(z) + J_{v+1}(z)) \qquad Y_v(z) = \frac{x}{2v}(Y_{v-1}(z) + Y_{v+1}(z)),$$

$$J_v'(z) = \frac{1}{2}(J_{v+1}(z) - J_{v-1}(z)) \qquad Y_v'(z) = \frac{1}{2}(Y_{v-1}(z) - Y_{v+1}(z)).$$

Half-integer values:

$$J_{1/2}(z) = Y_{-1/2}(z) = \sqrt{\frac{2}{\pi z}}\sin(z), \qquad J_{-1/2}(z) = -Y_{1/2}(z) = \sqrt{\frac{2}{\pi z}}\cos(x)$$

$$J_{3/2}(z) = -Y_{-3/2}(z) = \sqrt{\frac{2}{\pi z}}\left(-\cos(z) + \frac{\sin z}{z}\right),$$

$$J_{-3/2}(z) = Y_{3/2}(z) = \sqrt{\frac{2}{\pi z}}\left(-\sin(z) - \frac{\cos z}{z}\right).$$

Continued fraction representation, valid if $J_{v-1}(z) \neq 0$ and $v \neq 0, -1, -2, \ldots$:

$$\frac{J_v(z)}{J_{v-1}(z)} = \frac{z}{2v}\Big/\Big(1 + \mathop{\mathrm{K}}_{n=0}^{\infty} \frac{-z^2}{4(v+n)(v+n+1)}\Big/\Big(1$$

$$= \frac{z}{2v}\Big/\Big(1 - \frac{z^2}{4v(v+1)}\Big/\Big(1 - \frac{z^2}{4(v+1)(v+2)}\Big/\Big(1 - \cdots.$$

Modified Bessel Functions

Differential equation:

$$z^2y'' + zy' - (z^2 + v^2)y = 0.$$

Solutions are $I_v(z)$ and $K_v(z)$, the modified Bessel functions of the first and second kind.

Frobenius series:

$$I_v(z) = \sum_{k=0}^{\infty} \frac{z^{2k+v}}{4^k 2^v k!\Gamma(k+v+1)},$$

$$K_v(z) = \frac{\pi(I_{-v}(z) - I_v(z))}{2\sin(v\pi)} \qquad \text{for } v \text{ non-integer.}$$

$$K_n(z) = \frac{(-1)^{n-1}}{2}\left(\left.\frac{\partial I_v(z)}{\partial v}\right|_{v=n} + \left.\frac{\partial I_v(z)}{\partial v}\right|_{v=-n}\right) \qquad \text{for } n \text{ integer.}$$

$$K_n(z) = (-1)^{n+1} \ln\left(\frac{z}{2}\right) I_n(z) + 2^{n-1} \sum_{k=0}^{n-1} \frac{(-1)^k (n-k-1)! z^{2k-n}}{4^k k!}$$
$$+ \frac{(-1)^n}{2^{n+1}} \sum_{k=0}^{\infty} \frac{z^{2k+n}}{4^k k!(n+k)!} (\psi(k+1) + \psi(n+k+1)) \quad \text{for } n = 0, 1, 2, \ldots.$$

Asymptotic series, valid as $z \to \infty$:

$$I_v(z) \sim \frac{e^z}{\sqrt{2\pi z}} \sum_{k=0}^{\infty} \frac{(-1)^k a_k(v)}{z^k}, \qquad -\pi/2 < \arg(z) < \pi/2,$$

$$K_v(z) \sim \frac{\sqrt{\pi} e^{-z}}{\sqrt{2z}} \sum_{k=0}^{\infty} \frac{a_k(v)}{z^k}, \qquad -3\pi/2 < \arg(z) < 3\pi/2,$$

$$\text{with } a_n(v) = \frac{(4v^2 - 1^2)(4v^2 - 3^2)(4v^2 - 5^2) \cdots (4v^2 - (2n-1)^2)}{n! 8^n}.$$

Analytical continuation:

Both $I_v(z)$ and $K_v(z)$ have branch cuts along the negative real axis. To extend to other sheets of the Riemann surface, for integer m,

$$I_v(ze^{m\pi i}) = e^{mv\pi i} I_v(z).$$

$$K_v(ze^{m\pi i}) = e^{-mv\pi i} K_v(z) - \frac{\pi i \sin(mv\pi)}{\sin(v\pi)} I_v(z), \qquad \text{for non-integer } v.$$

$$K_n(ze^{m\pi i}) = (-1)^{mn} K_n(z) + (-1)^{n(m-1)-1} m\pi i I_n(z)), \qquad \text{for integer } n.$$

Integral representations, valid for $\mathrm{Re}(z) > 0$:

$$I_v(z) = \frac{1}{\pi} \int_0^{\pi} e^{z \cos t} \cos(vt)\, dt - \frac{\sin(v\pi)}{\pi} \int_0^{\infty} e^{-z \cosh t - vt}\, dt.$$

$$K_v(z) = \int_0^{\infty} e^{-z \cosh t} \cosh(vt)\, dt.$$

Functional relations:

$$I_{-n}(z) = I_n(z) \quad \text{for integer } n, \qquad K_{-v}(z) = K_v(z) \quad \text{for all } v,$$

$$I_v(z) = \frac{x}{2v}(I_{v-1}(z) - I_{v+1}(z)), \qquad K_v(z) = \frac{x}{2v}(K_{v-1}(z) - K_{v+1}(z)),$$

$$I'_v(z) = \frac{1}{2}(I_{v+1}(z) + I_{v-1}(z)), \qquad K'_v(z) = \frac{1}{2}(K_{v-1}(z) + K_{v+1}(z)).$$

Half-integer values:

$$I_{1/2}(z) = \sqrt{\frac{2}{\pi z}} \sinh(z), \quad I_{-1/2}(z) = \sqrt{\frac{2}{\pi z}} \cosh(x), \quad K_{\pm 1/2}(z) = \sqrt{\frac{\pi}{2z}} e^{-z},$$

$$I_{3/2}(z) = \sqrt{\frac{2}{\pi z}} \left(\cosh(z) - \frac{\sinh z}{z}\right),$$

$$I_{-3/2}(z) = \sqrt{\frac{2}{\pi z}} \left(\sinh(z) - \frac{\cosh z}{z}\right),$$

$$K_{\pm 3/2}(z) = \sqrt{\frac{\pi}{2z}} e^{-z} \left(1 + \frac{1}{x}\right).$$

Continued fraction representation, valid if $I_{v-1}(z) \neq 0$ and $v \neq 0, -1, -2, \ldots$:

$$\frac{I_v(z)}{I_{v-1}(z)} = \frac{z}{2v} \Big/ \Big(1 + \mathop{\mathbf{K}}_{n=0}^{\infty} \frac{z^2}{4(v+n)(v+n+1)} \Big/ \Big(1$$

$$= \frac{z}{2v} \Big/ \Big(1 + \frac{z^2}{4v(v+1)} \Big/ \Big(1 + \frac{z^2}{4(v+1)(v+2)} \Big/ \Big(1 + \cdots.$$

Connection to Bessel functions, valid for at least $\mathrm{Re}(z) > 0$:

$$I_v(z) = e^{\mp v\pi i/2} J_v(\pm iz), \qquad K_v(z) = \pm \frac{\pi i}{2} e^{\pm v\pi i/2} (J_v(\pm iz) \pm iY_v(\pm iz)).$$

Parabolic Cylinder Function

Differential equation:

$$y'' + \left(v + \frac{1}{2} - \frac{x^2}{4}\right) y = 0.$$

Solutions are $D_v(z)$ and $D_{-v-1}(iz)$, where $D_v(z)$ is the parabolic cylinder function.

Taylor series, converges for all z:

$$D_v(z) = 2^{v/2} \sqrt{\pi} \sum_{n=0}^{\infty} \frac{a_n z^n}{n!},$$

$$\text{where } a_0 = \frac{1}{\Gamma\left(\frac{1}{2} - \frac{v}{2}\right)}, \quad a_1 = \frac{-\sqrt{2}}{\Gamma\left(-\frac{v}{2}\right)},$$

and $a_{n+2} = -\left(v + \frac{1}{2}\right) a_n + \frac{1}{4} n(n-1) a_{n-2}$ for $n \geq 0$.

Asymptotic expansion valid as $z \to \infty$:

$$D_v(z) \sim z^v e^{-z^2/4} \sum_{n=0}^{\infty} \frac{(-1)^n c_n}{z^{2n}}, \qquad -3\pi/4 < \arg(z) < 3\pi/4,$$

$$\text{where } c_n = \frac{\Gamma(v+1)}{\Gamma(v-2n+1) 2^n n!} = \frac{v(v-1)(v-2)\cdots(v-2n+1)}{2^n n!}.$$

$$D_v(iz) \sim e^{i\pi v/2}\left(z^v e^{z^2/4}\sum_{n=0}^{\infty}\frac{c_n}{z^{2n}} + \frac{i\sqrt{2\pi}}{\Gamma(-v)}z^{-v-1}e^{-z^2/4}\sum_{n=0}^{\infty}\frac{(-1)^n d_n}{z^{2n}}\right),$$

$$\text{where } d_n = \frac{\Gamma(v+2n+1)}{\Gamma(v+1)2^n n!} = \frac{(v+1)(v+2)(v+3)\cdots(v+2n)}{2^n n!}.$$

Special values: (He(x) is the Hermite polynomial)

$$D_0(z) = e^{-z^2/4}, \quad D_1(z) = ze^{-z^2/4}, \quad D_2(z) = (z^2-1)e^{-z^2/4}, \ldots$$

$$D_n = \mathrm{He}_n(x)e^{-z^2/4}, \qquad D_{-1} = \sqrt{\frac{\pi}{2}}e^{z^2/4}\mathrm{erfc}\left(\frac{z}{\sqrt{2}}\right),$$

$$D_{-1/2}(z) = \sqrt{\frac{z}{2\pi}}K_{1/4}\left(\frac{z^2}{4}\right)$$

$$D_{1/2}(z) = \frac{z^{3/2}}{2\sqrt{2\pi}}\left(K_{1/4}\left(\frac{z^2}{4}\right) + K_{3/4}\left(\frac{z^2}{4}\right)\right).$$

Recursion formulas:

$$zD_v(z) = D_{v+1}(z) + vD_{v-1}(z).$$

$$D_v'(z) = vD_{v-1}(z) - \frac{z}{2}D_v(z).$$

The functional relation:

$$D_v(-x) = e^{-iv\pi}D_v(z) - \frac{i\sqrt{2\pi}}{\Gamma(-v)}e^{iv\pi/2}D_{-v-1}(-ix).$$

Modified Parabolic Cylinder Functions

Note: There is no standardized notation for the solution to the modified parabolic cylinder equation. This text introduces the following notation.
Differential equation:

$$y'' + \left(\frac{x^2}{4} - a\right)y = 0.$$

Solutions are $G_v(x)$ and $G_v(-x)$, where $G_v(x)$ is the modified parabolic cylinder function.

Relationship to parabolic cylinder functions:

$$G_v(x) = e^{\pi v/4}\frac{e^{\pi i/8}\omega^{1/2}D_{-1/2-iv}(e^{-\pi i/4}x) + e^{-\pi i/8}\omega^{-1/2}D_{-1/2+iv}(e^{\pi i/4}x)}{2},$$

where

$$\omega = \Gamma\left(\frac{1}{2} + iv\right)\sqrt{\frac{\cosh(\pi v)}{\pi}}. \qquad \text{(Note that } |\omega| = 1).$$

Asymptotic series, valid as $z \to \infty$ with $\mathrm{Re}(z) > 0$:

$$G_v(x) = \frac{w_1(z)}{\sqrt{z}} \cos\left(\frac{z^2+\pi}{4} - v\ln(z) - \frac{i}{2}\ln(\omega)\right) + \frac{w_2(z)}{\sqrt{z}} \sin\left(\frac{z^2+\pi}{4} - v\ln(z) - \frac{i}{2}\ln(\omega)\right),$$

$$\text{where } w_1(z) + iw_2(z) \sim \sum_{n=0}^{\infty} \frac{i^n\Gamma\left(2n+\frac{1}{2}-iv\right)}{2^n n!\Gamma\left(\frac{1}{2}-iv\right) z^{2n}}.$$

(Choose the branches of $\sqrt{\omega}$ and $\ln(\omega)$ so that they are continuous functions of v.)

Asymptotic series, valid as $z \to \infty$ with $\mathrm{Re}(z) < 0$:

$$G_v(x) = \frac{(e^{\pi v} + \sqrt{1+e^{2\pi v}})w_1(z)}{\sqrt{-z}} \cos\left(\frac{z^2-\pi}{4} - v\ln(-z) - \frac{i}{2}\ln(\omega)\right) + \frac{(e^{\pi v} + \sqrt{1+e^{2\pi v}})w_2(z)}{\sqrt{-z}} \sin\left(\frac{z^2-\pi}{4} - v\ln(-z) - \frac{i}{2}\ln(\omega)\right).$$

Riemann Zeta Function

Definition, valid for $\mathrm{Re}(z) > 1$:

$$\zeta(z) = \sum_{n=1}^{\infty} \frac{1}{n^z}.$$

Other summation formula, valid in $\mathrm{Re}(z) > 0$ and $z \neq 1 + 2\pi ki/(\ln 2)$ for integer k:

$$\zeta(z) = \frac{1}{1-2^{1-z}} \sum_{n=1}^{\infty} \frac{(-1)^{n+1}}{n^z}.$$

Reflection formulas, valid for $z \neq 0$ or 1:

$$\zeta(1-z) = 2(2\pi)^{-z}\cos(\pi z/2)\Gamma(z)\zeta(z),$$

$$\zeta(z) = 2(2\pi)^{z-1}\sin(\pi z/2)\Gamma(1-z)\zeta(1-z).$$

Special values:

$$\zeta(0) = -\frac{1}{2}, \quad \zeta(2) = \frac{\pi^2}{6}, \quad \zeta(4) = \frac{\pi^4}{90},$$

$$\zeta(-1) = -\frac{1}{12}, \quad \zeta(-2) = 0, \quad \zeta(-3) = \frac{1}{120}.$$

$$\zeta(2n) = \frac{(2\pi)^{2n}}{2(2n)!}|B_{2n}|, \qquad \zeta(-n) = -\frac{B_{n+1}}{n+1}, \qquad n = 1, 2, 3, \ldots.$$

Integral representations:

$$\zeta(z) = \frac{1}{\Gamma(z)} \int_0^\infty \frac{t^{z-1}}{e^t - 1}\, dt, \quad \text{valid if } \mathrm{Re}(z) > 1,$$

$$\zeta(z) = \frac{1}{(1 - 2^{1-z})\Gamma(z)} \int_0^\infty \frac{t^{z-1}}{e^t + 1}\, dt, \quad \text{valid if } \mathrm{Re}(z) > 0, z \neq 1.$$

Infinite product representation:

$$\zeta(z) = \prod_{p \text{ prime}} \frac{p^z}{p^z - 1} = \frac{2^z}{2^z - 1} \cdot \frac{3^z}{3^z - 1} \cdot \frac{5^z}{5^z - 1} \cdot \frac{7^z}{7^z - 1} \cdot \frac{11^z}{11^z - 1} \cdots.$$

Answers to Odd-Numbered Problems

Section 1.1

1) T 3) F 5) T 7) T 9) T 11) F

13) 4 15) $(\pi/2) - x$ 17) $3(x-1)^2$ 19) $(x-\pi)^2/2$

21) $1 + x + x^2/2$

23) $x - x^3/6 + x^5/120$

25) $1 - x^2/2 + x^4/24$

27) No, no matter how fast $f(x)$ grows, $[f(x)]^2$ will grow faster.

29) Yes, $f(x) = 0$.

31) $\lim_{x\to\infty} e^{bx}/e^{ax} = \lim_{x\to\infty} e^{(b-a)x} = 0$.

33) $\lim\limits_{x\to 0} \frac{\ln x}{x^{-a}} = \lim\limits_{x\to 0} \frac{1/x}{-ax^{-a-1}} = \lim\limits_{x\to 0} -\frac{x^a}{a} = 0$.

Section 1.2

1) 3/2 3) 1 5) $\sqrt{3}/2$ 7) 1

9) 2/3 11) $3^{20}/2^{30} \approx 3.247$ 13) 3/2 15) 2/3

17) $3/\ln 8$ 19) 1/2 21) $1/e$ 23) e^{-2}

25) $\lim\limits_{x\to a} \frac{f(x)h(x)}{g(x)h(x)} = \lim\limits_{x\to a} \frac{f(x)}{g(x)} = 1$.

27)

$$\lim_{x\to a} \frac{\ln(f(x))}{\ln(g(x))} = \lim_{x\to a} \frac{f'(x)/f(x)}{g'(x)/g(x)} = \lim_{x\to a} \frac{g(x)}{f(x)} \cdot \lim_{x\to a} \frac{f'(x)}{g'(x)}$$
$$= 1 \cdot \lim_{x\to a} \frac{f'(x)}{g'(x)} = \lim_{x\to a} \frac{f(x)}{g(x)} = 1.$$

Section 1.3

1) 1 3) 1/2 5) 1/2 7) 3/2

9) 1/4 11) 2/3 13) 4/3 15) 3/2

17) $3\ln(x) + 1/x - 1/(2x^2)$ 19) $x^2 + x + 3/2$

21) $x^2 + x^6/18 + x^8/30$ 23) $1 + x + 3x^2/2$

25) $x + 4/3 - 7/(9x)$ 27) $1/x + x/6 + 7x^3/360$

Section 1.4

1) $f^{-1} \sim \pm\sqrt{x} - 5/2pm13/8x^{-1/2} + \ldots$

3) $f^{-1} \sim \sqrt[3]{x} - 1/3x^{-1/3} - 1/3x^{-2/3} + \ldots$

5) $f^{-1} \sim \pm\sqrt[4]{x} - 1/(4\sqrt{x}) \mp 1/32x^{-5/4} + \ldots$

7) $f^{-1} \sim \ln(x) - 2\ln(\ln(x)) + 4\ln(\ln(x))/\ln(x) + \ldots$

9) $f^{-1} \sim \ln(x) - \ln(\ln(x))/x + \ln(\ln(x))^2/(2x^2) + \ldots$

11) $f^{-1} \sim \ln(x) - \ln(x)/x - \ln(x)^2/(2x^2) + \ldots$

13) $W(e^x)$ 15) $x - W(e^x)$ 17) $\sqrt{W(2e^{2x})/2}$

19)

$$A_6 = a_0^{-13}(132a_1^6 - 330a_0a_1^4a_2 + 180a_0^2a_1^2a_2^2 - 12a_0^3a_2^3 + 120a_0^2a_1^3a_3 - 72a_0^3a_1a_2a_3 + 4a_0^4a_3^2 - 36a_0^3a_1^2a_4 + 8a_0^4a_2a_4 + 8a_0^4a_1a_5 - a_0^5a_6)$$

21) $f^{-1} \sim x + x^3/2 + 17x^5/24 + 961x^7/720 + \ldots$
23) $f^{-1} \sim x/2 + x^2/16 - x^3/192 - x^4/3072 + \ldots$
25) $f^{-1} \sim x - x^4 + 3x^7 - 11x^{10} + \ldots$
27)

$$f^{-1}(x) = \sum_{n=0}^{\infty} \frac{(-1)^n(3n)!x^{2n+1}}{(2n+1)!n!}$$

Section 1.5

1) $y \sim x - 1/3$
3) $y \sim \pm x - 1/2$ or $y \sim 1 + 1/x^2$
5) $y \sim \sqrt[3]{x} - 1/(3x^{4/3})$
7) $y \sim \pm\sqrt{x} + 1/(2x)$ or $y \sim -1/x - 1/x^4$
9) $y \sim x \pm 1$ or $y \sim 1/x + 2/x^3$
11) $y \sim \pm x^{3/4} + \sqrt{x}/4$
13) $y \sim \sqrt{2x} \pm \sqrt[4]{x/8}$
15) $y \sim \sqrt{x} \pm \sqrt[4]{x}/2 + 1/4$
17) $y = x + 1/2$, $y = -x + 1/2$, and $y = -1$
19) $y = x^{3/2} - x + \sqrt{x}/2$, $y = -x^{3/2} - x - \sqrt{x}/2$, and $y = 0$
21) $y = -x + 1/2$, and $y = -x - 1/2$
23) $y = 2\sqrt{x} + \sqrt[4]{x}/\sqrt{2} + 1/4$, and $y = 2\sqrt{x} - \sqrt[4]{x}/\sqrt{2} + 1/4$
25) $y = x^{3/4} + \sqrt[4]{x}/4 + 1/4$, and $y = -x^{3/4} - \sqrt[4]{x}/4 + 1/4$

Section 2.1

1) $\int_0^1 (\cos t - 1)/t\, dt$.
3) $x - x^3/3 + x^5/10 - x^7/42 + \cdots$.
5) $x + x^5/10 + x^9/216 + x^{13}/9360 + \cdots$.
7) $1 + x/8 - x^2/56 + x^3/160 + \cdots$.
9) $-\ln x + x - 3x^2/4 + x^3/18 + \cdots$.
11) $-\ln x - x^2/4 + 25x^4/96 - x^6/4320 + \cdots$.
13) $\ln 2 + (1 - \ln 2)x + (2\ln 2 - 1)x^2/4 + (5 - 6\ln 2)x^3/36 + \cdots$.
15) $1 - 2x^2/3 + 7x^4/15 + \cdots$.
17) $1/x + \ln x + C - x/2 + x^2/12 - x^3/72 + x^4/480 - x^5/3600 + \cdots$. ($C$ turns out to be $\gamma - 1$.)
19) $-\ln x + C + x^2/12 - x^4/480 + \cdots$. ($C$ turns out to be $1 - \gamma$.)
21) $-\ln x + C + x^2/6 - x^4/20 + \cdots$. ($C$ turns out to be 1.)
23) $S(x) \sim \sum_{n=0}^{\infty}(-1)^n n! x^n = 1 - x + 2x^2 - 6x^3 + 24x^4 + \cdots$. The series diverges for all x.

Section 2.2

1) $\sin x/x - \cos x/x^2 - 2\sin x/x^3 + 6\cos x/x^4 + 24\sin x/x^5 + \cdots$

3) $-\sin x/x^2 + 2\cos x/x^3 + 6\sin x/x^4 - 24\cos x/x^5 - 120\sin x/x^6 + \cdots$
5) $e^x(1/x^2 + 2/x^3 + 6/x^4 + 24/x^5 + 120/x^6 + \cdots)$
7) $e^{x^2}(1/(2x) + 1/(4x^3) + 3/(8x^5) + 15/(16x^7) + 105/(32x^9) + \cdots)$.
9) $e^{x^3}(1/(3x^2) + 2/(9x^5) + 10/(27x^8) + 80/(81x^{11}) + 880/(243x^{14}) + \cdots)$.
11) $1/x - 2/x^3 + 24/x^5 - 720/x^7 + 40320/x^9 + \cdots$.
13) $1 + x/2 - x^2/4 + 3x^3/8 - 15x^4/16 + \cdots$.
15a) 0.0147517 15b) 0.000357022
17a) 0.666667 17b) 0.916667
19a) 1 19b) 0.666667
21a) 0.806699 21b) 0.907986
23) Substituting $u = t - 1/x$ into $\int_{1/x}^{\infty} e^{1/x}e^{-t}/(xt)\,dt$ gives the integral for $S(x)$.

Section 2.3

1) Letting $u = s^{n+1}$ and $dv = e^{-s}ds$,

$$f(n+1) = \int_0^\infty s^{n+1}e^{-s}\,ds = -s^{n+1}e^{-s}\Big|_0^\infty + \int_0^\infty (n+1)s^n e^{-s}\,ds = (n+1)f(n).$$

3) Substituting $u = s^2$, $du = 2s\,ds$ gives $\int_0^\infty (u^n/2)e^{-u}\,du = n!/2$.
5) $1/x - 2/x^3 + 24/x^5 + \cdots$.
7) $\sqrt{\pi/x}(1 + 1/(2x) + 3/(4x^2) + \cdots)$.
9) $-\sqrt{\pi}/(8x\sqrt{2x})(1 + 3/(8x) + 5/(16x^2) + \cdots)$.
11) $e^{-x}(1/x - 1/x^2 + 2/x^3 + \cdots)$.
13) $(1 + 1/n)^z(n/(n+z)) \sim 1 + (z^2 - z)/(2n^2)$ as $n \to \infty$. Since the sum $\sum_{n=1}^{\infty}(z^2 - z)/(2n^2)$ converges by the p-series test ($p = 2$), the infinite product converges.
15) $\Gamma(3n) = 3^{3n-(1/2)}\Gamma(n)\Gamma(n + (1/3))\Gamma(n + (2/3))/(2\pi)$.
17) $4\pi^{3/2}$.
19) $\frac{2^{1/4}}{3^{3/8}}\sqrt{\frac{\pi}{1+\sqrt{3}}}$.
21) $$\lim_{x\to\infty} \frac{2^{2x-1}x^x e^{-x}\sqrt{2\pi/x}(x+(1/2))^{x+1/2}e^{-x-1/2}\sqrt{2\pi/(x+1/2)}}{(2x)^{2x}e^{-2x}\sqrt{\pi/x}}$$

$$= \lim_{x\to\infty} \frac{(x+1/2)^x e^{-1/2}\sqrt{\pi}}{x^x} = \sqrt{\pi}.$$

23) Since $\Gamma'(n+1) = \Gamma(n) + n\Gamma'(n)$, $\Gamma'(n+1)/\Gamma(n+1) = \Gamma'(n)/\Gamma(n) + 1/n$. Induction establishes the first identity. Since $\ln(\Gamma(x)) = \ln(x!/x) \sim \ln(x^x e^{-x}\sqrt{2\pi/x}) = x\ln x - x - \ln x/2 + \ln(2\pi)/2$. Taking the derivate of both sides yields $\Gamma'(x)/\Gamma(x) \sim \ln x - 1/(2x)$.
25) $\int_x^\infty e^{-t}/t\,dt = -e^{-x}\ln x + \int_x^\infty e^{-t}\ln t\,dt$, so as $x \to 0$,

$$\int_x^\infty \frac{e^{-t}}{t}\,dt + \ln x = \int_x^\infty e^{-t}\ln t\,dt + \ln x(1 - e^{-x}) \sim \int_x^\infty e^{-t}\ln t\,dt$$

$$\sim \int_0^\infty e^{-t}\ln t\,dt = -\gamma.$$

27) $\Gamma(1/4)/(4x^{1/4}) + \Gamma(3/4)/(16x^{3/4}) + \Gamma(1/4)/(256x^{5/4}) + \cdots$.
29) $1/x^2 - 1/x^3 + 2/x^4 + \cdots$.
31) $2/x^2 - 12/x^4 + 240/x^6 + \cdots$.
33) $1/x - 2/x^3 + 24/x^5 + \cdots$.
35) $\Gamma(1/3)/(3x^{1/3}) - \Gamma(2/3)/(3x^{2/3}) + 1/(3x) + \cdots$.

Section 2.4

1) $-9/13 + 19i/13$ 3) $e^3\cos(2) - ie^3\sin(2) \approx -8.35853 - 18.2637i$
5) $2\cos(3\ln 2) + 2i\sin(3\ln 2) \approx -0.973989 + 1.74681i$ 7) $\ln 2 + 5i\pi/6 + 2k\pi i$
9) $\pm 2,\ 1 \pm i\sqrt{3},\ -1 \pm i\sqrt{3}$
11)

$$\begin{aligned}\sin(x+iy) &= \sin(x)\cos(iy) + \cos(x)\sin(iy)\\ &= \sin(x)\cosh(y) + i\cos(x)\sinh(y)\end{aligned}$$

13) If $\Delta y = 0$, $\Delta x \to 0$, the limit is $u_x + iv_x$. If $\Delta x = 0$, $\Delta y \to 0$, the limit is $v_y - iu_y$.
15) $u = x^3 - 3xy^2$, $v = 3x^2y - y^3$, $u_x = v_y = 3x^2 - 3y^2$, $u_y = -v_x = -6xy$.
17) Since $\zeta(z) = 2^z\eta(z)/(2^z - 2)$ for $\mathrm{Re}(z) > 1$, we can define this to be $\zeta(z)$ for $\mathrm{Re}(z) > 0$, $z \neq 1 + 2k\pi i/(\ln 2)$.
19) Clearly $\eta(1) = \ln 2$, and $\lim_{z\to 1}(z-1)2^z\ln 2/(2^z - 2) = 1$ by L'Hôpital's Rule. So the residue is 1.
21) Since the path does not matter in equation 2.20, we can pick $\Delta y = 0$, $\Delta x \to 0$, showing $F'(x) = \partial\phi_1/\partial x + i\partial\phi_2/\partial y = u(x,y) + iv(x,y) = f(x)$.
23) An anti-derivative of $f(x)$ is $-A/(2(z-z_0)^2) - B/(z-z_0) + R\ln(z-z_0) + \int g(z)\,dz$. Only the log term is affected by the moving to a different sheet of the Riemann surface, so the integral is $2\pi Ri$.
25) For the integer n, the residue is $1/(\pi n^2)$.
27) $-\pi/2 < \arg(x) < \pi/2$.
29) $-\pi/6 < \arg(x) < \pi/6$.
31) $-3\pi/2 < \arg(x) < 3\pi/2$.

Section 2.5

1) $-i\ln(2)e^{ix}/x + (e^{ix} - 2)/2x^2 + i(4 - e^{ix})/(4x^3) + \cdots$.
3) $2\sin(3x)/x + (\cos(3x) - 2)/4x^2 + \sin(3x)/32x^3 + \cdots$.
5)

$$\frac{-i\ln(2)e^{3ix}}{4x} + \frac{e^{3ix}(1-\ln(2)) - 8}{32x^2} + \frac{e^{3ix}(3i\ln(2) - 4i) + 128i}{256x^3} + \cdots.$$

7) $(1+i)\sqrt{\pi}/\sqrt{8y}$.
9) $(\sqrt{3} - i)\Gamma(1/3)\sqrt[3]{6/y}/12$.
11) $\ln(3)\Gamma(1/3)/\sqrt[6]{3y^2}$.
13) $\sqrt{\pi}/\sqrt{8y}$.
15) $(\cos(y) + \sin(y))\sqrt{\pi/y}$.

17) $J_n = A_n(x)\cos(x) + B_n(x)\sin(x), A_0 \sim B_0 \sim B_1 \sim 1/\sqrt{\pi x}, A_1 \sim A_2 \sim B_2 \sim -1/\sqrt{\pi x}$.
19) $(1+i\sqrt{3})\Gamma(2/3)/(6y^{2/3})$. 21) $(1+i)e^{2iy}\sqrt{3\pi}/\sqrt{8y}$.
23) $4\pi i e^{9iy/16}/\sqrt{3y} - 27i/(8y)$.

Section 2.6

1)

$$\frac{3i - 2ie^{ix}}{6x} + \frac{9 - 4e^{ix}}{36x^2} + \frac{8ie^{ix} - 27i}{108x^3} + \cdots.$$

3)

$$\frac{i + ie^{i\pi x}}{x} + \frac{i + ie^{i\pi x}}{x^3} + \cdots.$$

5)

$$\frac{-2ie^{4ix}}{x} + \frac{(i-1)\sqrt{\pi}}{\sqrt{8}x^{3/2}} + \frac{e^{4ix}}{4x^2} - \frac{ie^{4ix}}{32x^3} + \cdots.$$

7) $I(x) = -2\pi e^{-x} + A(x)\cos(x) + B(x)\sin(x)$, where $A(x) \sim 1/x + 1/x^2 + 1/x^3 + \cdots$ and $B(x) \sim -1/x + 1/x^3 + \cdots$.
9) $\int_{C_1} e^{ixt^3}\,dt \sim e^{\pi i/6}\Gamma(1/3)/(3x^{1/3})$ as before, but

$$\int_{C_2} e^{ixt^3}\,dt \sim \frac{-ie^{ix}}{3x}\int_0^{\epsilon} e^{-s}e^{-ix^2/3s + s^3/27x^2}\,ds.$$

Expanding $e^{-ix^2/3s+s^3/27x^2} \sim 1 - is^2/(3s) + (2s^3 - 3s^4)/(54x^2) + (is^6 - 2is^5)/(162x^3) + \cdots$ yields the same first five terms.
11) $e^{\pi i/8}\Gamma(5/4)/\sqrt[4]{x} - ie^{ix}/(4x) - 3e^{ix}/(16x^2) + 21ie^{ix}/(64x^3) + \cdots$.
13) $(1+i)\sqrt{\pi}/(4\sqrt{2x}) - ie^{ix}/(4x) - e^{ix}/(8x^2) + 3ie^{ix}/(16x^3) + \cdots$.
15) $I(x) = A(x)\cos(x - \pi/4) + B(x)\sin(x - \pi/4)$, where $A(x) \sim \sqrt{2\pi/x} - 9\sqrt{\pi}x^{-5/2}/(64\sqrt{2}) + \cdots$, and $B(x) \sim \sqrt{\pi}x^{-3/2}/(4\sqrt{2})$.
17) $(3i - ie^{2ix})/(3x) + (54 - 2e^{2ix})/(27x^2) + (4ie^{2ix} - 972i)/(81x^3) + \cdots$.
19) $(1+i)e^{-ix}\sqrt{\pi}/\sqrt{2x} - i/x - 1/(2x^2) + 3i/(4x^3) + \cdots$.
21) $e^{-2ix}(1+i)\sqrt{\pi}/\sqrt{6x} - i/(2x) + 5(i-1)\sqrt{\pi}x^{-3/2}/(144\sqrt{6}) - 1/(12\sqrt{3}x^2) - e^{-2ix}(1+i)385\sqrt{\pi}x^{-5/2}/(41472\sqrt{6}) + i/(9x^3) + \cdots$.
23)

$$(1 - i + (1+i)e^{-4ix})\sqrt{\pi/(6x)} + \frac{ie^{-4ix} - i}{9x}$$
$$+ ((5i - 5)e^{-4xi} - 5 - 5i)\sqrt{\pi/6}\frac{x^{-3/2}}{144} - \frac{4 + 4e^{-4ix}}{243x^2}$$
$$+ (385i - 385 - (385 + 385i)e^{-4ix})\sqrt{\pi/6}\frac{x^{-5/2}}{41472} + \frac{14i - 14ie^{-4ix}}{2187x^3} + \cdots.$$

25) $\mathrm{Ai}(x) \sim e^{-2x^{3/2}/3}/(2\pi)\sum_{n=0}^{\infty}(-1)^n\Gamma(3n+\frac{1}{2})x^{-(6n+1)/4}/(9^n(2n!))$ as $x \to \infty$

Section 3.1

1a) {0.69327731, 0.69310576, 0.69316334}; 3 places accuracy
1b) {0.69333333, 0.69308943, 0.69316940}; 3 places accuracy
3a) {1.0818717, 1.0818710, 1.0818716}; 7 places accuracy
3b) {1.0818713, 1.0818713, 1.0818713}; All numbers are exact.
5a) {0.37353103, 0.37355797, 0.37354760}; 5 places accuracy
5b) {0.37352309, 0.37356072, 0.37354646}; 5 places accuracy
7a) {0.55456399, 0.55446005, 0.55449929}; 4 places accuracy
7b) {0.55457376, 0.55445673, 0.55450070}; 4 places accuracy
9a) {Undefined, 20.171875, 20.0875}; 4 places accuracy
9b) {18.4, 19.75, 20.02}; 2 places accuracy
11a) {2.7184191, 2.7182850, 2.7182813}; 6 places accuracy
11b) {2.7160160, 2.7179950, 2.7182458}; 4 places accuracy
13a) {2.3847926, 2.3842838, 2.3842366}; 5 places accuracy
13b) {2.3823529, 2.3840028, 2.3842029}; 5 places accuracy
15a) {0.56875632, 0.56869227, 0.56869981}; 5 places accuracy
15b) {0.56859206, 0.56871283, 0.56869724}; 5 places accuracy

Section 3.2

1) $(1/(n^2-1)-1/n^2)/(1/(n+1)+1/(n-1)-2/n) = (1/2n)$.
3) $k = 3$, $A_8 = 1.08178416770$, $R(A_4) = 1.08221734466$; 20% of the error.
5) $k = 1/2$, $A_8 = 1.92667667971$, $R(A_4) = 2.54392638519$; 10% of the error.
7) $k = 1$, $A_8 = 0.556609240433$, $R(A_4) = 0.604488322135$; 16% of the error.
9) $k = 2$, $A_8 = 0.384567230155$, $R(A_4) = 0.386036730154$; 15% of the error.
11) $k = 3$, $A_8 = 0.354930889022$, $R(A_4) = 0.355040063645$; 19% of the error.
13) $k = 1$, $A_8 = 0.638415601177$, $R(A_4) = 0.579792230141$; 4% of the error.
15)

$$\begin{aligned}\int_a^b f(x)\,dx \approx \frac{2\Delta x}{45}&(7f(x_0)+32f(x_1)+12f(x_2)+32f(x_3)+14f(x_4)\\&+32f(x_5)+12f(x_6)+32f(x_7)+14f(x_8)+\cdots\\&+32f(x_{4n-3})+12f(x_{4n-2})+32f(x_{4n-1})+7f(x_{4n})).\end{aligned}$$

17) 7.3555556; 2 places accuracy
19) 1.2020661; 5 places accuracy
21) 2.3433607; no places of accuracy
23) 0.69314785; 5 places accuracy
25) 0.82699215; 5 places accuracy
27) 0.78541608; 4 places accuracy

Section 3.3

1) 1/2. 3) 0. 5) $k!/2^{k+1}$.
7) 1/2. 9) $\ln 2 - 3/4$.
11) Repeated Shanks transformation gets closer and closer to 1.
13a) $-1/2$ 13b) $-1/12$ 13c) 0 13d) 1/120
15) $f(x) = (xe^x - e^x + 1)/x^2$
17) $f(x) = (1/2 - x^2)e^{-x^2}$

Section 3.4

1) 1/3.

3) $\int_0^\infty (2+t)e^{-t}/(1+t)^2\,dt = -e^{-t}/(1+t)|_0^\infty = 1$.
5) If the original $\phi(x) = \sum_{n=0}^\infty a_n x^n/n!$, then inserting a leading 0 creates $\psi(x) = \sum_{n=1}^\infty a_{n-1}x^n/n! = \sum_{n=0}^\infty a_n x^{n+1}/(n+1)!$. Note that $\psi'(x) = \phi(x)$. So integrating by parts, $\int_0^\infty e^{-t}\psi(t)\,dt = -e^{-t}\psi(t)|_0^\infty + \int_0^\infty e^{-t}\psi'(t)\,dt = \int_0^\infty e^{-t}\phi(t)\,dt$, since $\psi(0) = 0$.
7) $\int_0^\infty e^{-t}/\sqrt{4xt+1}\,dt$, $f(1) \approx 0.545641360765$.
9) $\int_0^\infty e^{-t}(1-xt^2)/(1+xt^2)^2\,dt$, $f(1) \approx 0.343377961556$.
11) Letting $x = bt$, $dx = b\,dt$, we have the integral transforming to

$$\int_0^1 b^m t^m (b-bt)^n b\,dt = b^{m+n+1}\int_0^1 t^m(1-t)^n = \frac{b^{m+n+1}m!n!}{(m+n+1)!}.$$

13) If we let $\rho(t) = e^{-\sqrt{t}}/(2\sqrt{t})$, then $\int_0^\infty t^n\rho(t)\,dt = \int_0^\infty u^{2n}e^{-u}\,du = (2n)!$.
15) Since $\det\begin{vmatrix} 1 & 2 \\ 2 & 3 \end{vmatrix} = -1$, the matrix in equation 3.15 is not non-negative definite.
17) Since $f(x)$ and $g(x)$ satisfy the 4 properties of definition 3.1, it is clear that $f(x) + g(x)$ satisfy all 4 properties, too.
19) For any $\epsilon > 0$, $\int_{-\infty}^\infty f_\epsilon(x)\,dx = \int_{-\epsilon}^\epsilon 1/(2\epsilon) = 1$. If $x \neq 0$, then $f_\epsilon(x) = 0$ for all $\epsilon < |x|$.
21) $\int_{-\infty}^\infty \epsilon/(\pi(x^2+\epsilon^2))dx = (1/\pi)\tan^{-1}(x/\epsilon)\big|_{-\infty}^\infty = 1$. If $x \neq 0$, then $\epsilon/(\pi(x^2+\epsilon^2)) \sim \epsilon/(\pi x^2) \to 0$.

Section 3.5

1) $c_{2n} = (2n+1)/(8n-4)$, $c_{2n+1} = (2n-1)/(8n+4)$ for $n \geq 1$.

$$1\Big/\Big(1-\frac{x}{2}\Big/\Big(1+\frac{3x}{4}\Big/\Big(1+\frac{x}{12}\Big/\Big(1+\frac{5x}{12}\Big/\Big(1+\frac{3x}{20}\Big/\Big(1+\cdots.$$

3)

$$1\Big/\Big(1+\frac{x^2}{2}\Big/\Big(1-\frac{5x^2}{12}\Big/\Big(1+\frac{x^2}{100}\Big/\Big(1-\frac{313x^2}{6300}\Big/\Big(1+\frac{295x^2}{78876}\Big/\Big(1+\cdots.$$

5) $c_n = -1/(4n^2-1)$ for $n \geq 0$.

$$x\Big/\Big(1-\frac{x^2}{3}\Big/\Big(1-\frac{x^2}{15}\Big/\Big(1-\frac{x^2}{35}\Big/\Big(1-\frac{x^2}{63}\Big/\Big(1-\frac{x^2}{99}\Big/\Big(1+\cdots.$$

7) $c_n = 1/(4n^2-1)$ for $n \geq 1$.

$$x\Big/\Big(1+\frac{x^2}{3}\Big/\Big(1+\frac{x^2}{15}\Big/\Big(1+\frac{x^2}{35}\Big/\Big(1+\frac{x^2}{63}\Big/\Big(1+\frac{x^2}{99}\Big/\Big(1+\cdots.$$

9) $c_0 = 1$, $c_n = 2n$ for $n \geq 1$.

$$1\Big/\Big(1+2x\Big/\Big(1+4x\Big/\Big(1+6x\Big/\Big(1+8x\Big/\Big(1+\cdots.$$

$F_n \approx \{.33333, .71428, .44, .62963, .4864, \ldots\}$, Convergence is very slow, but it does converge to 0.545641360765.
11) $1 - x + x^2 - x^4 + x^5 - x^6 + \cdots$.
13) $c_0 = 1$, $c_{2n-1} = 2^{4n-3}$, $c_{2n} = 2^{4n-1} - 2^{2n-1}$ for $n \geq 1$.

$$1\Big/\Big(1+2x\Big/\Big(1+6x\Big/\Big(1+32x\Big/\Big(1+120x\Big/\Big(1+512x\Big/\Big(1+\cdots.$$

Since all of the c_n are positive, the series is Stieltjes.
15) Substituting $u = \log_4 t$, $du = dt/(t \ln 4)$, and completing the square, we get

$$2^{n^2} \ln 4 \int_{-\infty}^{\infty} \frac{2^{-(u-n)^2}}{2\sqrt{\pi} \ln 2}\, du.$$

Use the fact that $\int_{-\infty}^{\infty} e^{-v^2}\, dv = \sqrt{\pi}$.
17) $F_1 \approx .333333$, $F_2 \approx .777778$, $F_3 \approx .371429$, $F_4 \approx .741772$, $F_5 \approx .379416$. In fact, F_{2n} will converge to ≈ 0.73098 whereas F_{2n+1} converges to ≈ 0.38196. Since these are different, the sequence diverges.
19) $\sqrt{1 + 2x\sqrt{1 + 5x\sqrt{1 + 89x/10}}}$.
21) $\exp(x \exp(3x/2 \exp(77x/36)))$.

Section 3.6

1) $(1 + 5x/4 + 5x^2/16)/(1 + 3x/4 + x^2/16)$.
3) $(1 + x/2 + x^2/12)/(1 - x/2 + x^2/12)$.
5) $(1 + x^2/12)/(1 - 5x^2/12)$.
7) $(1 - 7x^2/60)/(1 + x^2/20)$.
9) $b_2 = 0, (1 - 7x/4 + 2x^2)/(1 - 3x/4 + x^2/4)$.
11) $b_2 = 0, (1 + x/4 + x^2/3)/(1 - 3x/4 + x^2/12)$.
13) $b_2 = 0, (2 + 17x/5 + x^2/6)/(1 + 11x/5 + 14x^2/15)$.
15) $P_1^0 = \frac{\sqrt{\pi}}{2}/(1 + \sqrt{\pi}x)$.
17) $P_4^3 = (x + 2x^3/3)/(1 + 4x^2/3 + 4x^4/3)$.
19) $P_2^3 = (1 + 3x/2 + 3x^2/2 + x^3)/(1 + 3x/2 + x^2)$.
21) $P_2^3 = (1 + 6x + 6x^2 + x^3)/(1 + 3x + x^2)$.

Section 4.1

1) $y' = 2/(x(c + \ln x)^2)$, $y'' = -4/(x^2(c + \ln(x))^3) - 2/(x^2(c + \ln(x))^2)$, so $y(x)y'(x)$ is the same as $xy''(x)$.
3) $y' = -a/(b+x)^2$, $y'' = 2a/(b+x)^3$, so $y''(x)y(x) = 2a^2/(x+b)^4 = 2[y'(x)]^2$.
5) $y = -1 + \sqrt{3}\tanh(\sinh^{-1}(1/\sqrt{2}) - \sqrt{3}\ln(x)/2)$.
7) $y = 9/(3 - 2x)$.
9) $F = g(x) - p_0(x)y - p_1(x)y' - \cdots p_{n-1}y^{(n-1)}$. Since all of the components of F are continuous near x_0, F is continuous. Also, $\partial F/\partial y = -p_0(x), \partial F/\partial y' =$

$-p_1(x), \ldots \partial F/\partial y^{(n-1)} = -p_{n-1}(x)$ are all continuous. So the conditions for theorem 4.1 are satisfied.
11) If $y = x^2$, $y'' = 2$ so $x^2 y'' = 2x^2$. If $y = 1/x$, $y'' = 2/x^3$ so $x^2 y'' = 2/x$. $W(x^2, 1/x) = -3 \neq 0$, so the general solution is $y = c_1 x^2 + c_2/x$.
13) $y = -x^2$.
15) $y_p = -x\cos(x)/2$.
17) $y_p = x^2 \ln(x)/3 - x^2/9$. Note that x^2 solves the homogeneous equation, so $y_p = x^2 \ln(x)/3$ also works.
19) Since $\mu = e^{\int p_1(x)\,dx}$, $(\mu(x)y')' = \mu(x)y''(x) + \mu(x)p_1(x)y'(x)$. So we need $q(x) = p_0(x)\mu(x)$ and $r(x) = -\mu(x)h(x)$.
21) If $y = \sqrt{x}\sin(\sqrt{E - 1/4}\ln x)$,

$$y' = (\sin(\sqrt{E-1/4}\ln x) + 2\sqrt{E-1/4}\cos(\sqrt{E-1/4}\ln x))/(2\sqrt{x}),$$

and $y'' = -E\sin(\sqrt{E-1/4}\ln x)/x^{3/2}$. If $y = \sqrt{x}\cos(\sqrt{E-1/4}\ln x)$,

$$y' = (\cos(\sqrt{E-1/4}\ln x) - 2\sqrt{E-1/4}\sin(\sqrt{E-1/4}\ln x))/(2\sqrt{x}),$$

and $y'' = -E\cos(\sqrt{E-1/4}\ln x)/x^{3/2}$.
23) Normalized eigenfunctions are $\sqrt{2x}\sin(n\pi\ln(x))$.
25) $E = 1 + n^2\pi^2/(\ln 2)^2$ for $n \geq 1$. Eigenfunctions are $\sin(n\pi\ln(x)/\ln 2)/x$.
27) 0 regular, ∞ irregular.
29) 0 irregular, ∞ regular.
31) ∞ irregular, $k\pi + \pi/2$ regular for integer k. (0 is ordinary).
33) $\pm i$ regular, ∞ regular.
35) 0 irregular, 1 regular. (∞ is ordinary).

Section 4.2

1) $y = 2/(K - x^2)$ or $y = 0$.
3) $y = Kx^x e^{-x}$.
5) $y = -\ln(K - e^x)$.
7) $y = (x^{3/2}/3 + K)^2$ or $y = 0$.
9) $y = 6/(5 - 2x^3)$.
11) $y = -1/\sqrt{5 - x^2}$.
13) $y = 1 - 2x$.
15) $y = 1 - e^{x^2/2 + x}$.
17) $y = 1 + Ce^{-x^2}$.
19) $y = x^2/4 + C/x^2$.
21) $y = x^2/2 - x/2 + 1/4 + Ce^{-2x}$.
23) $y = (e^x + C)/(x^2 + 1)$.
25) $y = 3x^3 - x^2$.
27) $y = (3x - 1 + 10e^{-3x})/9$.
29) $y = (2\sin^{-1}(x) + 1)/\sqrt{1 - x^2}$.
31) $y = (x^2 - 2x + 2 + 2e^{1-x})/x$.

Section 4.3

1) $y = a_0(1 + x^2/2 - x^4/24 + x^6/120 + \cdots) + a_1x$.
3) $y = a_0(1 + x^4/12 + \cdots) + a_1(x + x^5/20 + \cdots)$.
5) $y = a_0(1 + x^2/2 - x^3/6 + x^5/60 - x^6/360 + \cdots) + a_1(x - x^2/2 + x^3/6 - x^5/60 + x^6/360 + \cdots)$.
7) $y = a_0(1 - x^2/2 + x^4/24 - x^6/80 + \cdots) + a_1x$.
9) $y = a_0(1 - x^2) - a_1\sum_{n=0}^{\infty} x^{2n+1}/(4n^2 - 1)$.
11) $y = a_0\sum_{n=0}^{\infty} 3x^{2n}/((2n-1)(2n-3)) + a_1(x - x^3)$.
13) $y = a_0\sum_{n=0}^{\infty}(-1)^n(n+1)x^{2n} + a_1\sum_{n=0}^{\infty}(-1)^n(2n+3)x^{2n+1}/3$.
15) $\int x^3[y_2'/x]'\,dx = -x^2/(2-2x^2)$, so $y_2 = 3x/4 + (1-3x^2)\ln((1+x)/(1-x))/8$.
17) $y = a_0(1 - (x-1)^2/2 - (x-1)^4/24 + (x-1)^5/60 + \cdots) + a_1((x-1) + (x-1)^2/2 + (x-1)^4/24 - (x-1)^5/60 + \cdots)$.
19) $y = a_0(1 - (x-2)^2/2 - (x-2)^3/12 + 5(x-2)^4/96 + (x-2)^5/160 + \cdots) + a_1((x-2) + (x-2)^2/4 - (x-2)^3/6 - (x-2)^4/24 + (x-2)^5/96 + \cdots)$.
21) $y = a_0(1 + (x+1)^3/6 + (x+1)^4/6 + 17(x+1)^5/120 + \cdots) + a_1((x+1) + (x+1)^2 + 5(x+1)^3/6 + 3(x+1)^4/4 + 27(x+1)^5/40 + \cdots)$.
23) $y_p = x^2/2 - x^3/6 + x^5/120 + x^6/720 + \cdots$.
25) $y_p = x^3/6 - x^5/30 + \cdots$.
27) $y_p = x^3/6 - x^4/24 - x^5/30 + x^6/80 + \cdots$.
29) $\rho \geq (\sqrt{5} - 1)/2$.
31) $\rho \geq 5$.

Section 4.4

1) $y_1 = x^{1/2} - 2x^{3/2}/3 + 2x^{5/2}/15 - 4x^{7/2}/315 + 2x^{9/2}/2835 + \cdots$, $y_2 = 1 - 2x + 2x^2/3 - 4x^3/45 + 2x^4/315 + \cdots$.
3) $y_1 = x^{1/2} - x^{3/2}/5 + x^{5/2}/70 - x^{7/2}/1890 + x^{9/2}/83160 + \cdots$, $y_2 = 1/x + 1 - x/2 + x^2/18 - x^3/360 + \cdots$.
5) $y_1 = x^{2/3} - x^{8/3}/42 + x^{14/3}/6552 + \cdots$, $y_2 = x^{1/3} - x^{7/3}/30 + x^{13/3}/3960 + \cdots$.
7) $y_1 = x^{3/2} - x^{7/2}/18 + x^{11/2}/936 + \cdots$, $y_2 = 1/x + x/2 - x^3/24 + \cdots$.
9) $y_1 = x + 2x^2/7 + x^3/70$, $y_2 = x^{-1/3} - 10x^{2/3}/3 - 35x^{5/3}/18 - 14x^{8/3}/81 - 7x^{11/3}/3888 + \cdots$.
11) $y_1 = x^i + (i-1)x^{2+i}/8 + (1-3i)x^{4+i}/320 + \cdots$, $y_2 = x^{-i} - (1+i)x^{2-i}/8 + (1+3i)x^{4-i}/320 + \cdots$.
13) $y_1 = x + x^2/2 + x^3/12 + x^4/144 + \cdots$, $y_2 = \ln(x)y_1 + 1 - 3x^2/4 - 7x^3/36 + \cdots$.
15) $y_1 = x + x^2/2$, $y_2 = 2\ln(x)y_1 + 1 - 5x^2/2 - x^3/6 + \cdots$.
17) $y_1 = x^3 - x^5/10 + \cdots$, $y_2 = 1 + x^2/2 + \cdots$.
19) $y_1 = x^2 - x^3/3 + x^4/6 - x^5/30 + \cdots$, $y_2 = 1 + x + x^3/3 + \cdots$.
21) $y_1 = x - x^2/3 + x^3/24 - x^4/360 + \cdots$, $y_2 = -\ln(x)y_1/2 + 1/x + 1 - 2x^2/9 + \cdots$.
23) $y_1 = x^{3/2} - x^{5/2}/12 + x^{7/2}/384 - x^{9/2}/23040 + \cdots$, $y_2 = -\ln(x)y_1/32 + x^{-1/2} + x^{1/2}/4 - x^{5/2}/288 + \cdots$.

Section 5.1

1) $y \sim Kx^{\pm 1/8 - 1/2}e^{\pm x^2/2 + x/2}$.

3) $y \sim Kx^{-1}e^{-x}$ or $y = Kx$, which is an exact solution.
5) $y \sim Kx^{-1/4}e^{x^2/2 \pm 2x^{3/2}/3 \mp x^{1/2}}$.
7) $y \sim K(\ln(x))^{-1/2}e^{\pm x\ln(x) \mp x}$.
9) $y \sim Kx^{-2/3}e^{-\omega 3x^{5/3}/5}$, where $\omega = 1$ or $\omega = -1/2 \pm i\sqrt{3}/2$.
11) If $y = ue^x$, then $x^3u'' - x^2u' + xu + u = 0$. Letting $x = 1/t$ produces $t^2u''(t) + 3tu'(t) + (t+1)u(t) = 0$, with an indicial polynomial of $k^2 + 2k + 1$.
13) $y(x) = c_1\mathrm{Ai}((ax+b)/\sqrt[3]{a^2}) + c_2\mathrm{Ai}((ax+b)/\sqrt[3]{a^2})$.
15) $y \sim Kx^{3/4}e^{\pm 2x^{-1/2}}$.
17) $y \sim Kx^{2/3}e^{-1/x}$ or $y \sim Kx^{4/3}e^{2/x}$.
19) $y \sim Kx^{3/4}e^{-2/x \pm 4x^{-1/2}}$.
21) $y \sim Kx^4e^{1/x}$ or $y \sim Kx^{-2}$.
23) $y \sim Kx^{1/2}e^{1/x}$ or $y \sim Kx^{3/2}e^{-1/x}$.
25) $y \sim K/x^2$ with second order $-K(4 + 3\ln x)/(9x^3)$, or $y \sim K$, with second order $K\ln x/x$.
27) $y \sim K/x^3$ with second order $-Ke^{-x}/x^4$, or $y \sim K$, with second order $-Ke^{-x}/x$.
29) The eigenvalue problem is of Sturm-Liouville problem, with $q(x) = -x^2/4$ and $r(x) = \mu(x) = 1$, so all eigenvalues are real and positive. The general solution to the equation is $c_1D_{E-1/2}(x) + c_2D_{-E-1/2}(ix)$. The condition $y(\infty) = 0$ makes $c_2 = 0$, and $y(-\infty) = 0$ means that $D_{E-1/2}(-x) \to 0$ as $x \to \infty$. The functional relation shows this happens if and only if $E - 1/2$ is an integer. The eigenfunctions are $D_n(x)$ for $n = 0, 1, 2\ldots$. The integral comes from the orthogonal property of the eigenfunctions.

Section 5.2

1) $y \sim \sum_{n=0}^{\infty}(-1)^n n! x^n$ as $x \to 0$.
3) $y \sim \sum_{n=0}^{\infty}(2n)!/(2^n n!)x^{2n}$ or $y \sim \sum_{n=0}^{\infty} 2^n\Gamma(n+1/2)/\sqrt{\pi}x^{2n}$ as $x \to 0$.
5) $y \sim \sum_{n=0}^{\infty}(n+2)(n+1)(2n+1)!/(2(-6)^n n!)x^{2n/3}$ as $x \to 0$.
7) $y \sim \sum_{n=0}^{\infty} -(n+1)2^{n-1}\Gamma(n-1/2)/\sqrt{\pi}x^{-2n}$ as $x \to \infty$.
9) $y \sim \sum_{n=0}^{\infty}(n+1)(3/2)^n\Gamma(n+2/3)/\Gamma(2/3)x^{-3n/2}$ as $x \to \infty$.
11) The series truncates to give the exact solution $y_1 = x^{-1}e^{-1/x}(1 - 2x)$.
13) $y \sim x^3e^{-1/x}\sum_{n=0}^{\infty}(-1)^n(n+1)(n+2)!/2x^n$ as $x \to 0$.
15) $y \sim x^{5/4}e^{2/\sqrt{x}}\sum_{n=0}^{\infty}\Gamma(n-3/2)\Gamma(n+5/2)/(4^n\pi n!)x^{n/2}$ as $x \to 0$.
17) $y \sim x^{2/3}e^{-1/x}\sum_{n=0}^{\infty}\Gamma(n-1/3)\Gamma(n+2/3)/((-3)^n\Gamma(-1/3)\Gamma(2/3)n!)x^n$ as $x \to 0$.
19) $y \sim x^{-1}e^{-x}\sum_{n=0}^{\infty}(-1)^n(n+1)!x^{-n}$ as $x \to \infty$.
21) $y \sim x^{-3}e^{-x^3/3}\sum_{n=0}^{\infty}(-3)^n\Gamma(n+4/3)/\Gamma(4/3)x^{-3n}$ as $x \to \infty$.
23)

$$y \sim x^{-5/8}e^{(x-x^2)/2}\sum_{n=0}^{\infty}(-1)^n\frac{\Gamma(n+5/16)\Gamma(n+13/16)}{n!\Gamma(5/16)\Gamma(13/16)}x^{-2n}$$

as $x \to \infty$.

25)

$$y \sim c_1 x^{1/4} e^{2\sqrt{x}} \sum_{n=0}^{\infty} \frac{-\Gamma(n-1/2)\Gamma(n+3/2)}{4^n n! \pi} x^{-n/2}$$
$$+ c_2 x^{1/4} e^{-2\sqrt{x}} \sum_{n=0}^{\infty} \frac{-\Gamma(n-1/2)\Gamma(n+3/2)}{(-4)^n n! \pi} x^{-n/2}.$$

27)

$$y \sim c_1 x^{-1} e^{x^3/3} \sum_{n=0}^{\infty} \frac{3^n \Gamma(n+1/3)\Gamma(n+2/3)}{2^n n! \Gamma(1/3)\Gamma(2/3)} x^{-3n}$$
$$+ c_2 x^{-1} e^{-x^3/3} \sum_{n=0}^{\infty} \frac{(-3)^n \Gamma(n+1/3)\Gamma(n+2/3)}{2^n n! \Gamma(1/3)\Gamma(2/3)} x^{-3n}.$$

29) $y \sim c_1 x^{-5} \sum_{n=0}^{\infty} 2^{n+1}(n+1)(n+2)\Gamma(n+5/2)/(3\sqrt{\pi}) x^{-2n} + c_2 e^{-x^2/2}(x^4 - 6x^2 + 3)$.

Section 5.3

1) $y_p \sim 1/(2x) + x^2/4 + x^5/80 + \cdots$.
3) $y_p \sim 1/(2x) + (x - x\ln x)/2 + (6\ln x - 11)x^3/72 + \cdots$.
5) $y_p \sim 1/(2x) + (\ln x)/2 + (2\ln x - (\ln x)^2 - 2)x/4 + \cdots$.
7) $y_p \sim -1/x + \ln x + x\ln x + \cdots$.
9) $y_p \sim 1/x + (\ln x)^2/2 + x((\ln x)^2 - 4\ln x + 6)/2$.
11) $y_p \sim -1/x - 2/x^3 - 24/x^5 + \cdots$.
13) $y_p \sim -1/x - 2/x^2 - 12/x^3 + \cdots$.
15) $y_p \sim -1/x^2 + 2/x^4 - 6/x^5 + \cdots$.
17) $y_p \sim -1/(2x) - 1/(3x^2) - 1/(2x^3) + \cdots$.
19) $y = -x^{-1}\int e^x/(3x)\,dx + x^2 \int e^x/(3x^4)\,dx$.
21) $W(y_1, y_2) = Ce^{-x^2/2}$.
23) $W(y_1, y_2) = C/(1 + x^2)$.
25) $W(y_1, y_2, y_3) = C/x^2$.
27) $W(J_0(x), Y_0(x)) = 2/(\pi x)$.
29) $y_p = -e^x/2 \int_x^{\infty} e^{-t}/(1+t)\,dt - e^{-x}/2 \int_0^x e^t/(t+1)\,dt + e^{-x}/2 \int_0^{\infty} e^{-t}/(1+t)\,dt$. $y_p \sim -1/x + 1/x^2$ as $x \to \infty$, $y_p \sim -kx + x^2/2$ as $x \to 0$, where $k = \int_0^{\infty} e^{-t}/(1+t)\,dt$.

Section 5.4

1) $w(x) \sim x^2/2 + C + \cos(x^2 + 2C)/(4x^2) + \cdots$.
3) $y \sim ke^x - (2x^2 + 2 + 1)e^{-x}/(8k) + \cdots$, where $k > 0$.
5) $y \sim ke^x + (9x^3 + 9x^2 + 6x + 2)e^{-2x}/(81k^2) + \cdots$, where $k \neq 0$, or $y \sim x + 1/(3x^2) + \cdots$.
7) $y \sim k_1 e^x + (4x^3 + 6x^2 + 6x + k_2)e^{-x}/(48k_1)$ where $k_1 > 0$. Also, $y = \pm x$ is an exact solution.
9) $y \sim x^2/2 - 2/x + \cdots$, or $y \sim -1/x - 1/x^4$.

11) $y \sim ke^x + (2x+4)\sqrt{k}e^{x/2} + \cdots$ where $k > 0$, or $y \sim x^2 + 4x + \cdots$.
13) All solutions have the same asymptotic series: $y \sim x - 2 + \cdots$.
15) $y \sim -1/(x-c) + e^c(x-c)/3 + e^c(x-c)^2/4 + \cdots$.
17) $y \sim -1/(x-c) + 1/2 + (4c^2-1)(x-c)/12 + \cdots$.
19) $y \sim 6/(x-c)^2 - 1/2 + (x-c)^2/40 + \cdots$.
21) $y \sim (x-c)^{-2} - c(x-c)^2/10 - (x-c)^3/6 + k(x-c)^4 + \cdots$. Second order pole. This equation is the first Painlevé transcendent.
23) $y \sim (x-c)^{-2} - 1/12 + (1-24c)(x-c)^2/240 - (x-c)^3/6 + k(x-c)^4 + \cdots$. Second order pole.
25) $y \sim \pm[(x-c)^{-1} - c(x-c)/6 - (x-c)^2/4 + k(x-c)^3 + \cdots]$. First order pole. This equation is the second Painlevé transcendent.
27) $y \sim \pm[(x-c)^{-1} - c^2(x-c)/6 - c(x-c)^2/2 + \ln(c-x)(x-c)^3/5 + k(x-c)^3 + \cdots$. Not a true pole.
29) $y = \pm 4(x-c)^{3/4}$ special solutions, $y \sim ax^2 + bx + c - 1/(a^3x^3) + 3b/(2a^4x^4) + \cdots$.
31) $y = \pm 8(x-c)^{5/4}$ special solutions, $y \sim ax^2 + bx + c + 25/(a^3x) - 25b/(2a^4x^2) + \cdots$.
33) $y = 9(x-c)^{4/3}$ special solution, $y \sim ax^2 + bx - 72\ln(x)/a^2 + c - 36b/(a^3x) + \cdots$.

Section 6.1

1) Assuming true for the previous k,

$$\begin{aligned}\Delta^{k-1}(\Delta a_n) &= \sum_{i=0}^{k-1}(-1)^{k-i-1}\binom{k-1}{i}a_{n+i+1} - \sum_{i=0}^{k-1}(-1)^{k-i-1}\binom{k-1}{i}a_{n+i} \\ &= \sum_{i=1}^{k}(-1)^{k-i}\binom{k-1}{i-1}a_{n+i} - \sum_{i=0}^{k-1}(-1)^{k-i-1}\binom{k-1}{i}a_{n+i} \\ &= \sum_{i=0}^{k}(-1)^{k-i}\left[\binom{k-1}{i-1} + \binom{k-1}{i}\right]a_{n+i} \\ &= \sum_{i=0}^{k}(-1)^{k-i}\binom{k}{i}a_{n+i}.\end{aligned}$$

3) $y' = y/(x+1)^2$, ordinary point at infinity.
5) $x^2y'' + y = 0$, regular singular point at infinity.
7) $x^2y'' + x^2y' + y = 0$, irregular singular point at infinity.
9) $x^4y'' + 2x^3y' + y = 0$, ordinary point at infinity.
11) $n^2(n-1)^2(2n^2-2n-1)/12$.
13) $n(n+1)(2n+1)(3n^2+3n-1)/30$.
15) $((-1)^n(1-2n)-1)/4$.
17) $n/(1-2n)$.
19) $\ln(n+1)$.
21) $\tan(n)/\sin(1)$.

23) $a_n \sim Cn^2$.
25) $a_n \sim \pm\sqrt{2n}$.
27) $a_n \sim \pm 2n^{3/2}$.
29) $a_n \sim C_1 n^3 + C_2 n^{-2}$.
31) $a_n \sim C_1 n + C_2 n \ln n$.

Section 6.2

1) $a_n = a_0\sqrt{\pi}\Gamma(n+1)/(2^n\Gamma(n+(1/2)))$.
3) $a_n = a_0/(n^2+1)$.
5) $a_n = (6a_0 + 2n^3 + 3n^2 - 5n)/(6n+6)$.
7) $a_n = (n+1)(a_0 + n + 2 - 2\gamma - 2\psi(n+2))$.
9) $a_n = (4a_0 + 2n3^n + 3^n - 1)/(4n+4)$.
11) The IRA would be valued at \$1,413,772.70.
13) The monthly payment is \$444.89.
15) The present value is \$9007.35.
17) By comparison to a p-series, with $p = 2$, the series converges (absolutely). Thus, $H(z+1) - H(z) = \sum_{n=0}^{\infty} z/((n+1)(n+z+1)) - (z-1)/((n+1)(n+z)) = \sum_{n=0}^{\infty} 1/(n+z) - 1/(n+z+1)$, which telescopes to $1/z$.
19) If we let $f(z) = \psi(z) - H(z)$, by problem 17 we have $f(z+1) = f(z)$, so $f(z)$ is periodic. By problem 18, we see that $\lim_{z\to\infty}$ is a constant. Hence, $f(z)$ is a constant for all z, and $f(1) = -\gamma$. So $\psi(z) = -\gamma + H(z)$.
21) $\psi'(1) = \sum_{n=0}^{\infty} 1/(n+1)^2 = \zeta(2) = \pi^2/6$.
23) $2\psi(n+1) + 2\gamma - (9n^2 + 15n)/(2n^2 + 6n + 4)$.
25) $\psi(n - \frac{1}{2}) - 2\psi'(n+1) - \psi(1/2) - 2 + 2\psi'(1)$.
27) $\psi(n+i) + \psi(n-i) - \psi(i) - \psi(-i)$.
29) $3\psi'(1) - \psi(1/2) - 2 - \gamma$, or $2\ln 2 - 2 + \pi^2/2$.
31) $a_n = n!\left(1 + \sum_{k=0}^{n-1} k/(k+1)!\right) = 2\,n! - 1$.
33) $a_n = (n+1)\psi(n+1) + (n+1)\gamma + 2$.
35) $a_n = 1/n! \sum_{k=0}^{n-1} k!2^k$.

Section 6.3

1) $a_n \sim C2^n\sqrt{n}$.
3) $a_n \sim Cn^n(2/e)^n$.
5) $a_n \sim C_1 n + C_2 2^n/\sqrt{n}$.
7) $a_n \sim C_1 2^n n^2 + C_2 3^n/n^3$.
9) $a_n \sim C_1 n^n(3/e)^n/n^2 + C_2 n^n e^{-n}$.
11)

$$P_1^1 = \frac{1+24n}{-1+24n} \qquad P_2^2 = \frac{293+360n+8640n^2}{293-360n+8640n^2},$$

$$P_3^3 = \frac{4406147+120170160n+17720640n^2+425295360n^3}{-4406147+120170160n-17720640n^2+425295360n^3}.$$

13) Although the *relative* error gets better as n gets large, the *absolute* error will increase to the point that the error will be more than 1. For example, this algorithm will produce $8! = 40318$, when the true value is 40320.

15) $a_n \sim 1 + 2/n + 3/n^2 + 13/(3n^3) + \cdots$.
17) $a_n \sim 1 - 1/(2n^2) - 1/(2n^3) - 1/(8n^4) + \cdots$.
19) $a_n \sim 1 - 1/n + 1/(6n^3) + 1/(12n^4) + \cdots$.
21) $a_n \sim 1 + 1/n + 3/n^2 + 35/(3n^3) + \cdots$.
23) $a_n \sim 1 - 1/(2n) - 7/(8n^2) - 27/(16n^3) + \cdots$.
25) $a_n \sim C_1 2^n(1 - 3/n + 9/n^2 + \cdots) + C_2 3^n(1 + 2/n + 9/n^2 + \cdots)$.
27) $a_n \sim C_1 n^2(1-1/(3n)+1/(3n^2)+\cdots)+C_2 n(-2)^n(1-1/n+1/(2n^2)+\cdots)$.
29) $a_n \sim C_1 n\Gamma(n)(1 - 4/(3n) + 4/(3n^2) + \cdots) + C_2(-2)^n\Gamma(n)(1 - 1/n + 7/(9n^3) + \cdots)$.

Section 6.4

1) $f(-x) = -x/(e^{-x} - 1) - x/2 = xe^x/(e^x - 1) - x/2 = xe^x/(e^x - 1) - x(e^x - 1)/(e^x - 1) + x/2 = x/(e^x - 1) + x/2 = f(x)$. Hence, the Maclaurin series for $f(x)$ will have only even terms, and $x/(e^x - 1)$ only differs by $x/2$.
3) Replacing x with $2x$ in problem 2 results in

$$2x\coth(2x) = \sum_{n=0}^{\infty} B_{2n}4^{2n}x^{2n}/(2n)!.$$

Since $x\tanh(x) = 2x\coth(2x) - x\coth(x)$, $x\tanh(x) = \sum_{n=1}^{\infty} B_{2n}(4^{2n} - 2^{2n})x^{2n}/(2n)!$. The term for $n = 0$ is now 0, so it can be dropped. Replacing x with ix, and dividing by $-x$ gives the result.
5) $B_8 = -1/30$.
7) Since $te^t/(e^t - 1) = \sum_{n=0}^{\infty} B_n(1)t^n/n!$ and $t/(e^t - 1) = \sum_{n=0}^{\infty} B_n(0)t^n/n!$, the difference between the two series is just t. Thus, $B_n(1) = B_n(0)$ except for when $n = 1$.
9) Trivial if $n = 0$, so assume statement is true for previous n. $[B_n(1-x)]' = -B_n'(1 - x) = -nB_{n-1}(1 - x) = -n(-1)^{n-1}B_{n-1}(x) = (-1)^n B_n'(x)$. Thus, $B_n(1-x) = (-1)^n B_n(x)+C$ for some constant C. Since $B_n(1) = (-1)^n B_n(0)$, this constant must be 0.
11) $B_5(x) = x^5 - 5x^4/2 + 5x^3/3 - x/6$.
13) Let $\phi(x) = \sum_{k=1}^{\infty} B_k(-1)^{k-1}x^k/(k \cdot k!)$, so $\phi'(x) = \sum_{k=1}^{\infty} B_k(-x)^{k-1}/k!$, which by equation 6.25 is $1/(e^{-x} - 1) + 1/x$. Integrating, (and adjusting the constant of integration so that $\phi(0) = 0$), we find that $\phi(x) = \ln(x/(e^x - 1))$, so the Borel sum follows.
15) $n^6/6 - n^5/2 + 5n^4/12 - n^2/12$.
17) Sum ≈ 1.2020569013, true value $= \zeta(3) = 1.2020569032$
19) Sum ≈ 2.6123753420, true value $= \zeta(3/2) = 2.6123753487$
21) Sum ≈ 1.5819767029, true value $= e/(e - 1) = 1.5819767069$
23) $\mathrm{Cthi}(x) = \sum_{n=0}^{\infty} 2^{2n}B_{2n}x^{2n-1}/((2n - 1)(2n)!)$.
25) $\zeta(1/2) \approx -1.4603545$.
27) $C = \zeta(-1/2) \approx -0.207886225$.

Section 6.5

1) $\Gamma(n)/\Gamma(n+k-4)+(10-4k)\Gamma(n)/\Gamma(n+k-3)+(6k^2-25k+25)\Gamma(n)/\Gamma(n+k-2) + (-4k^3 + 18k^2 - 28k + 15)\Gamma(n)/\Gamma(n + k - 1) + (1 - k)^4\Gamma(n)/\Gamma(n + k)$.

3) $(k+1)B_{k+1} + (k - 2k^2 - 2)B_k + (k-1)^3 B_{k-1} = 0$, $a_n = B_0(1 + 2/n + 3/(n(n+1)) + 22/(3n(n+1)(n+2)) + \cdots$.
5) $-(k+2)B_{k+2} + 3k(k+1)B_{k+1} + (3k^2 - 3k^3 - k - 1)B_k + (k-1)^4 B_{k-1} = 0$, $a_n = B_0(1 - 1/(2n(n+1)) - 1/(n(n+1)(n+2)) - 21/(8n(n+1)(n+2)(n+3)) + \cdots$.
7) $-(k+2)B_{k+2} + (3k^2 + 3k - 1)B_{k+1} + (3k^2 - 3k^3)B_k + (k-1)^4 B_{k-1} = 0$, $a_n = B_0(1 - 1/n + 1/(2n(n+1)) + 5/(6n(n+1)(n+2)) + \cdots$.
9) $-(k+1)B_{k+1} + (4k^2 + k + 1)B_k + (1-k)(5k^2 - 4k + 1)B_{k-1} + 2(k-2)(k-1)^3 B_{k-2} = 0$, $a_n = B_0(1 + 1/n + 3/(n(n+1)) + 44/(3n(n+1)(n+2)) + \cdots$.
11) $-2(k+1)B_{k+1} + (7k^2 + k - 1)B_k + (1-k)(8k^2 - 7k + 2)B_{k-1} + 3(k-2)(k-1)^3 B_{k-2} = 0$, $a_n = B_0(1 - 1/(2n) - 7/(8n(n+1)) - 41/(16n(n+1)(n+2)) + \cdots$.
13) $\alpha = -1/2$, $4kB_k = (2k-1)(2k-3)B_{k-1}$, $a_n = B_0(n^{[-1/2]} - n^{[-3/2]}/4 - 3n^{[-5/2]}/32 - 15n^{[-7/2]}/128 + \cdots)$.
15) $\alpha = 1/3$, $9kB_k = (3k-5)(3k-4)B_{k-1}$, $a_n = B_0(n^{[1/3]} + 2n^{[-2/3]}/9 + 2n^{[-5/3]}/81 + 40n^{[-8/3]}/2187 + \cdots)$.
17) $\alpha = -2/3$, $9kB_k = (3k-1)(3k-4)B_{k-1}$, $a_n = B_0(n^{[-2/3]} - 2n^{[-5/3]}/9 - 10n^{[-8/3]}/81 - 400n^{[-11/3]}/2187 + \cdots)$.
19) $\alpha = -3/2$, $4kB_k = (2k-1)(2k+1)B_{k-2} + (2k-3)(2k+1)B_{k-1}$, $a_n = B_0(n^{[-3/2]} - 3n^{[-5/2]}/4 + 45n^{[-7/2]}/32 + 35n^{[-9/2]}/128 + \cdots)$.
21) $\alpha = -2$, $kB_k = k(k+1)B_{k-2} + (k+1)^2 B_{k-1}$, $a_n = B_0(n^{[-2]} + 4n^{[-3]} + 21n^{[-4]} + 128n^{[-5]} + \cdots)$.
23) $a_n = \sum_{k=0}^{\infty}(-1)^k n^{[-k-1]}$.
25)

$$(2k-3)(2k-1)^3 A_k - 8k(4k^2-1)A_{k+1} + 16k(k+2)A_{k+2} - (2k-2)(2k-1)^2 B_{k-2} + 16k^2 B_{k-1} - 8(k+1)B_k = 0,$$

$$a_n = A_0\left(n^{[3/2]} + \frac{3n^{[1/2]}}{2} + \frac{3\psi(n)n^{[-1/2]}}{8} + \frac{3\psi(n)n^{[-3/2]}}{16} - \frac{n^{[-3/2]}}{32} + \cdots\right).$$

27) $a_n = c_1(n^{[1]} + n^{[0]}) + c_2 2^n(n^{[-1/2]} + 7n^{[-3/2]}/4 + 237n^{[-5/2]}/32 + \cdots)$.
29) $a_n = c_1 2^n(n^{[2]} - 8n^{[1]} + 8n^{[0]}) + c_2 3^n(n^{[-3]} + 27n^{[-4]} + 648n^{[-5]} + \cdots)$.
31) $a_n = c_1\Gamma(n)(n^{[1/2]} - n^{[-1/2]}/8 + n^{[-3/2]}/128 + \cdots) + c_2\Gamma(n)3^n(n^{[-3/2]} + 27n^{[-5/2]}/8 + 2025n^{[-7/2]}/128 + \cdots)$.

Section 7.1

1) $x = \pm 1 - \epsilon/2 \pm \epsilon^2/8 + \cdots$.
3) $x = 1 + \epsilon/2 + 5\epsilon^2/8 + \cdots$, $x = 3 - 9\epsilon/2 + 27\epsilon^2/8 + \cdots$.
5) $x = -\epsilon/2 + 5\epsilon^2/8 + \cdots$, $x = 1 + 2\epsilon + 2\epsilon^2 + \cdots$, $x = 2 - 3\epsilon/2 - 21\epsilon^2/8 + \cdots$.
7) $x = 1 + \epsilon/2 + 7\epsilon^2/16 + \cdots$, $x = 2 - \epsilon - 2\epsilon^2/5 + \cdots$, $x = -3 - \epsilon/2 - 3\epsilon^2/80 + \cdots$.
9) $x = \pm 1 + \epsilon/6 \pm 7\epsilon^2/216 + \cdots$, $x = \pm 2 - \epsilon/6 \mp 13\epsilon^2/432 + \cdots$.
11) $x = -1 - \epsilon - 2\epsilon^2 + \cdots$, $x = \pm\sqrt{\epsilon} - \epsilon/2 \mp 3\epsilon^{3/2}/8 + \cdots$.
13) $x = -\epsilon - \epsilon^2 - 3\epsilon^3 + \cdots$, $x = -1 \pm \sqrt{\epsilon} \pm \epsilon^{3/2}/2 + \cdots$.
15) $x = \pm 2\sqrt{\epsilon} + 3\epsilon \pm 37\epsilon^{3/2}/4 + \cdots$, $x = 1 \pm \sqrt{\epsilon} - 3\epsilon + \cdots$.
17) $x = 1 + \omega\sqrt[3]{\epsilon} + \omega^2\epsilon^{2/3}/3 + \cdots$, where $\omega^3 = 1$.
19) $x = \pm 1 + \epsilon/2 \pm 5\epsilon^2/8 + \cdots$, $x = 1/\epsilon - \epsilon - 2\epsilon^3 + \cdots$.

21) $x = 2 + 9\epsilon + 108\epsilon^2 + \cdots$, $x = \pm\epsilon^{-1/2} - 1 \mp 3\sqrt{\epsilon}/2 + \cdots$.
23) $x = 1 + 3\epsilon/2 + 4\epsilon^2 + \cdots$, $x = 1/\epsilon - 2 - 8\epsilon^2 + \cdots$, $x = 2/\epsilon + 1 - 3\epsilon/2 + \cdots$.
25) $x = 1 + \epsilon/3 + 11\epsilon^2/27 + \cdots$, $x = -2 - 16\epsilon/3 - 1280\epsilon^2/27 + \cdots$, $x = \pm\epsilon^{-1/2} + 1/2 \mp 11\sqrt{\epsilon}/8 + \cdots$.
27) $f'(k) = (k-1)(k-2)(k-3)\cdots(1)(-1)(-2)\cdots(k-19)(k-20) = (k-1)!(-1)^k(20-k)!$. Equation 7.3 then shows $a_1 = k^{19}(-1)^{k+1}/((k-1)!(20-k)!)$.
29) $x = r - \epsilon g(r)/f'(r) + \epsilon^2(2f'(r)g(r)g'(r) - f''(r)[g(r)]^2)/(2[f'(r)]^3) + \cdots$.

Section 7.2

1) $u = \exp(-\int p(x)/2\,dx)$, the new equation becomes $f'' + (q(x) - p'(x)/2 - p(x)^2/4)f = 0$.
3) $y_0 = x$, $y_1 = x^4/(3{\cdot}4)$, $y_2 = x^7/(3{\cdot}4{\cdot}6{\cdot}7)$, $y_3 = x^{10}/(3{\cdot}4{\cdot}6{\cdot}7{\cdot}9{\cdot}10)$. Summing these gives exactly the same series as the second solution of example 4.18.
5) $y_0 = 1$, $y_1 = 1 - e^x$, $y_2 = e^{2x}/2 - e^x + 1/2$.
7) $y_0 = e^x$, $y_1 = (1-x)e^x - 1$, $y_2 = (x^2 - 3x + 3)e^x + x^2/2 - 3$.
9) $y_0 = \ln(x)$, $y_2 = x^2\ln(x)/2 + (1-x^2)/4$, $y_3 = x^4\ln(x)/8 + (4x^2 - 3x^4 - 1)/32$.
11) $y_0 = x$, $y_1 = (x-2)e^x + x + 2$, $y_2 = e^{2x}(x-3)/4 + xe^x + (x+3)/4$.
13) $y_0 = x^3/6$, $y_1 = -x^6/180$, $y_2 = x^9/12960$.
15) $y_0 = x$, $y_1 = (x^4 - x)/12$, $y_2 = (2x^7 - 7x^4 + 5x)/1008$.
17) $y_0 = x$, $y_1 = (5x - x^5)/20$, $y_2 = (x^9 - 18x^5 + 81x)/1440$.
19) $y_0 = (x^4 - x)/12$, $y_1 = (-2x^7 + 7x^4 - 5x)/1008$, $y_2 = (4x^{10} - 30x^7 + 75x^4 - 49x)/181440$.
21) $y = \sin(nx) + \epsilon(x(\pi - x)\cos(nx)/(4n) + (2nx - \pi)\sin(nx)/(8n^3)) + \cdots$.
23) $E_0 = 1$, $y_0 = \sin(x)$, $E_1 = 0$, $y_1 = (\pi - 2x)\sin(x)/4$, $E_2 = 1/4$, $y_2 = (\pi - 2x)^2\sin(x)/32$.
25) $E_0 = 1$, $y_0 = \sin(x)$, $E_1 = -1$, $y_1 = (4x^2 - \pi^2)\sin(x)/8$.
27)

$$\begin{aligned}\epsilon(y_1(v_0\sin(\alpha)/g)) &= -\epsilon v_0^4\cos^4(\alpha)\ln(\cos(\alpha)/(1+\sin(\alpha)))/(8g^2)\\ &\quad + \epsilon v_0^4\sin(\alpha)(\cos^2(\alpha) - 2)/(8g^2).\end{aligned}$$

29) $x(t) = v_0\cos(\alpha)(t - \epsilon t^2/2 + \epsilon^2 t^3/6 + \cdots)$, $y(t) = v_0\sin(\alpha)(t - \epsilon t^2/2 + \epsilon^2 t^3/6 + \cdots) + g(-t^2/2 + \epsilon t^3/6 - \epsilon^2 t^4/24)$.

Section 7.3

1) $y''(\delta) - y(\delta) = 0$, $y(0) = 1$, $y(1/\sqrt{\epsilon}) = 0$.
3) $y_{\text{out}} \sim 2e^{1-x} + \epsilon 2(1-x)e^{1-x} + \epsilon^2(x^2 - 6x + 5)e^{1-x} + \cdots$.
5) $y_{\text{out}} \sim (x+1)^2 + \epsilon(3 - 2x - x^2)/4 + \epsilon^2(x^2 + 2x - 3)/16 + \cdots$.
7) $y_{\text{out}} \sim e^{1-x^2} + 2\epsilon e^{1-x^2}(3x - 1 - 2x^3)/3 + 2\epsilon^2 e^{1-x^2}(x-1)^2(4x^4 + 8x^3 - 18x^2 - 40x - 17)/9 + \cdots$.
9) $y_{\text{in}}(t) \sim 1 + K_\epsilon(1 - e^{-t}) + \epsilon(-t - K_\epsilon t(1 + e^{-t})) + \epsilon^2(t^2/2 - t + K_\epsilon(t^2/2 - t - te^{-t} - t^2e^{-t}/2)) + \cdots$.
11) $y_{\text{in}}(t) \sim K_\epsilon((1 - e^{-t}) + \epsilon(2t + 3te^{-t} + t^2e^{-t}/2) + \epsilon^2(t^2 - 2t - 12te^{-t} - 6t^2e^{-t} - 3t^3e^{-t}/2 - t^4e^{-t}/8) + \cdots)$.

13) $y_{\text{in}}(t) \sim 1 + K_\epsilon(1 - e^{-t}) + \epsilon^2(2t - t^2 + K_\epsilon(2t - 2te^{-t} - t^2 - t^2e^{-t})) + \epsilon^4(10t^2 - 20t - 10t^3/3 + t^4/2 + K_\epsilon(-20t - 20te^{-t} + 10t^2 - 10t^2e^{-t} - 10t^3/3 - 10t^3e^{-t}/3 + t^4/2 - t^4e^{-t}/2)) + \cdots$.
15) Near $x = 1$, $\delta = O(\epsilon)$, and $y_{\text{in}} \sim C_1 + C_2e^{2(x-1)/\epsilon}$.
17) Near $x = 0$, $\delta = O(\sqrt[3]{\epsilon})$, and $y_{\text{in}} \sim C_1 + C_2\int_0^{x/\sqrt[3]{\epsilon}} e^{-t^3}\,dt$.
19) Near $x = 1$, $\delta = O(\sqrt{\epsilon})$, and $y_{\text{in}} \sim C_1 + C_2\int_0^{(1-x)/\sqrt{\epsilon}} e^{-t^2/2}\,dt$.
21) Near $x = 0$, $\delta = O(\sqrt{\epsilon})$, and $y_{\text{in}} \sim C_1 + C_2\int_0^{x/\sqrt{\epsilon}} e^{-t^2/2}\,dt$.
23)

$$y_{\text{unif}} \sim y_2 \exp\left(\int_x^{x_2} q(t)/p(t)\,dt\right) + \left(y_1 - y_2\exp\left(\int_{x_1}^{x_2} q(t)/p(t)\,dt\right)\right)e^{-p(x_1)(x-x_1)/\epsilon}.$$

25) $y_{\text{unif}} \sim 2e^{x/2} + (1 - 2e)e^{2(x-2)/\epsilon}$.
27) $y_{\text{unif}} \sim 2x^3 - 15e^{2(x-2)/\epsilon}$.
29) $y_{\text{unif}} \sim e^{x^2/2+x} - e^{3/2} + e^{3/2}\sqrt{2/\pi}\int_0^{(1-x)/\sqrt{\epsilon}} e^{-t^2/2}\,dt$.

Section 7.4

1) $y_{\text{out}}(s) \sim \sqrt{\epsilon}(1 - s\sqrt{\epsilon}/2 + (1+s^2)\epsilon/8) + \cdots$, whereas $y_{\text{in}}(s) \sim K_0 + K_0 s\sqrt{\epsilon}/2 + K_1\epsilon$. Only by including the ϵ^2 term of the inner solution do we get a match in the ϵ term for $x = O(\sqrt{\epsilon})$.
3) $K_\epsilon \sim 1 + \epsilon + 5\epsilon^2/2 + \cdots$.
5) $K_\epsilon \sim 1 + 3\epsilon/2 + 21\epsilon^2/4 + \cdots$.
7) $K_\epsilon = 2e - 1 + 2e\epsilon + 5e\epsilon^2$, $y_{\text{comp},0} = 2e^{1-x} - (2e - 1)e^{-x/\epsilon}$, $y_{\text{comp},1} = e^{1-x}(2 + 2\epsilon(1 - x)) + e^{-x/\epsilon}(1 - 2e - 2e\epsilon - x(2e - 1))$, $y_{\text{comp},2} = e^{1-x}(2 + 2\epsilon(1 - x) + \epsilon^2(x^2 - 6x + 5)) + e^{-x/\epsilon}(1 - 2e - 2e\epsilon - 5e\epsilon^2 + x(1 - 2e + \epsilon - 4e\epsilon) + x^2(1 - 2e)/2)$.
9) $K_\epsilon = 1 + 3\epsilon/4 - 3\epsilon^2/16$, $y_{\text{comp},0} = (1 + x)^2 - e^{-x/\epsilon}$, $y_{\text{comp},1} = (x + 1)^2 + \epsilon(3 - 2x - x^2)/4 + e^{-x/\epsilon}(3x - 1 - 3\epsilon/4 + x^2/(2\epsilon))$, $y_{\text{comp},2} = (x + 1)^2 + \epsilon(3 - 2x - x^2)/4 + \epsilon^2(x^2 + 2x - 3)/16 + e^{-x/\epsilon}(3x - 1 - 3\epsilon/4 + 3\epsilon^2/16 + x^2/(2\epsilon) - 39\epsilon x/4 - 45x^2/8 - 3x^3/(2\epsilon) - x^4/(8\epsilon^2))$.
11) $K_\epsilon = e - 1 - 2e\epsilon/3 - 34e\epsilon^2/9$, $y_{\text{comp},0} = e^{1-x^2} + (1 - e)e^{-x/\epsilon}$, $y_{\text{comp},1} = e^{1-x^2}(1 + 2\epsilon(3x - 1 - 2x^3)/3) + e^{-x/\epsilon}(1 - e - 2e\epsilon/3)$, $y_{\text{comp},2} = e^{1-x^2}(1 + 2\epsilon(3x - 1 - 2x^3)/3 + 2\epsilon^2(x - 1)^2(4x^4 + 8x^3 - 18x^2 - 40x - 17)/9) + e^{-x/\epsilon}(1 - e - 2e\epsilon/3 - 34e\epsilon^2/9 + (1 - e)(x^2 + 2\epsilon x))$.
13) The series sums to $(x^2 + 2x)/(1 - \epsilon) + 1$.
15) $y_{\text{out}} \sim e^{x/2} - \epsilon e^{x/2}\ln(x)/8$.
17) The matching shows that $K_0 = K_1 = 0$, so $y_{\text{in},1/2} = y_{\text{match}} = 1 + \sqrt{\epsilon}t/2$. Thus, $y_{\text{comp},1,1/2} = y_{\text{out},1} = e^{x/2} - \epsilon e^{x/2}\ln(x)/8$.
19) First, we see that $\int r^2e^{r^2}/2\,dr = re^{r^2}/4 - \int e^{r^2}/4\,dr = re^{r^2}/4 - e^{r^2}/(8r) - \int e^{r^2}/(8r^2) = \cdots$. Evaluating this from 0 to s, and multiplying by e^{-s^2} yields $s/4 - 1/(8s) - 1/(16s^3) +$ subdominant terms. Integrating this term by term gives the series shown.

21) Letting $r = 1/x$, $f'(r) \to -x^2 f'(x)$, and $f''(r) \to x^4 f''(x) + 2x^3 f'(x)$ converts the equation to $f''(x) = f'(x)f(x)/x^2$.
23) $y_{\text{out,left}} \sim x - 1$ and $y_{\text{out,right}} \sim x$. The inner equation, $Y''(t) + Y(t)Y'(t) + \epsilon Y(t) = 0$, has the same leading order as problem 22, so the solution is $y_{\text{in}} \sim k\tanh(tk/2)$ for some k. Since $\lim_{t\to+\infty} y_{\text{in}}$ must be $1/2$, $k = 1/2$. Since the outer solutions differ by a constant, we can form $y_{\text{comp},0} \sim x - 1/2 + \tanh((2x-1)/(8e))/2$.

Section 8.1

1) $y_{\text{WKBJ}} = \sin(x/\sqrt{\epsilon})/\sin(1/\sqrt{\epsilon})$ (Exact solution).
3) $y_{\text{WKBJ}} = e^{1-x} - e^{1+x-(x/\epsilon)}$.
5) $y_{\text{WKBJ}} = (\sqrt{2}e^{(x^2+2x-3)/(2\sqrt{\epsilon})} + e^{-(x^2+2x)/(2\sqrt{\epsilon})})/\sqrt{x+1}$.
7) $y_{\text{WKBJ}} = \sqrt{2}x^{-1/4}\sin(2(x^{3/2}-1)/(3\sqrt{\epsilon}))/\sin(14/(3\sqrt{\epsilon}))$.
9) $y_{\text{WKBJ}} = (1+\cos x)^{-1/4}\sin(2\sqrt{2}\sin(x/2)/\sqrt{\epsilon})/\sin(2/\sqrt{\epsilon})$.
11) By letting $u = (x^2+2x)/3$, $du = (2x+2)dx/3$, $\int_0^1 (x+1)\sin(n\pi(x^2+2x)/3)\sin(m\pi(x^2+2x)/3)\,dx = \int_0^1 3\sin(n\pi u)\sin(m\pi u)/2\,du = 0$.
13) $E_n \sim 4n^2\pi^2/3$, $y \sim K_1\sqrt{\sec x}\sin(2n\pi\sin(x)/\sqrt{3})$.
15) $E_n \sim n^2\pi^2/(\ln 2)^2$, $y \sim K_1\sqrt{x+1}\sin(n\pi\ln(x+1)/(\ln 2))$.
17) $E_n \sim 4n^2\pi^2/9$, $y \sim K_1/\sqrt{x+1}\sin(n\pi(x^2+2x)/3)$.
19) $S_2' = \pm 3i(x+1)^{-3}/8$, so $S_2 = \mp 3i(x+1)^{-2}/16$.

$$y_{\text{WKBJ},2} = K_1/\sqrt{x+1}\sin((x^2+2x)/(2\sqrt{\epsilon}) - 3\sqrt{\epsilon}(x+1)^{-2}/16 + 3\sqrt{\epsilon}/16).$$

$E \sim 4n^2\pi^2/9 - 3/16$.
21) $E_n \sim 9n^2\pi^2/49 - 1/4$.
23) $E_n \sim (3+2\sqrt{2})n^2\pi^2/4 + 3\sqrt{2}/32$.
25) $E_n \sim 9n^2\pi^2/49 + 11/28 - 6\ln(2)/7$.

Section 8.2

1) $y_{\text{WKBJ},1} \sim -e^{-3/(2\epsilon)}(x+1)e^{-x}/2 + e^{-x}e^{-(x^2+2x)/(2\epsilon)}$.
3) $y_{\text{WKBJ},1} \sim e^{1/x} + (e - e^2)e^{-1/x}e^{(1-x^3)/(3\epsilon)}/x^2$.
5) $y_{\text{WKBJ},1} \sim \sqrt{2}e^{-x/4}\sin((2e^{x/2}-2)/\sqrt{\epsilon})/\sin(2/\sqrt{\epsilon})$.
7) $y_{\text{WKBJ},1} \sim \cos(x)(\sqrt{2}\sin((\tan x)/\sqrt{\epsilon}) + \sin((1-\tan x)/\sqrt{\epsilon}))/\sin(1/\sqrt{\epsilon})$.
9) $y_{\text{WKBJ},1} \sim e^{x/4}(e^{2(e^{-2}-e^{-x/2})/\sqrt{\epsilon}} - e^{(2e^{-x/2}-4+2e^{-2})/\sqrt{\epsilon}})$.
11) $S_2 = \int (pq' - q^2 - qp')/p^3\,dx$.
13) $S_2 = \int (4qq'' - 5(q')^2)/(32q^2\sqrt{-q})\,dx$.
15) $S_2 = \int (4qq'' - 5(q')^2 - 16pq^2\sqrt{-q})/(32q^2\sqrt{-q})\,dx$.
17) One way: $S_0 = \pm i(\sin x + 1)$, $S_1 = \ln(2\sec x)/2$, $S_2 = \pm i(3\sec(x)\tan(x) + \ln(2\sec x + 2\tan x))/16$, $S_3 = (\sec^2 x - 3\sec^4 x)/16$.
19) $y = \sqrt{x}D_{-1/2-\sqrt{\epsilon}/8}(\sqrt{2}\ln(2x)/\sqrt[4]{\epsilon})/D_{-1/2-\sqrt{\epsilon}/8}(\sqrt{2}\ln(2)/\sqrt[4]{\epsilon})$.
21) $M_0 \approx .762849$, $M_1 \approx 1.87957$, and $M_2 \approx 1.98084$. Normally, because $M_1\epsilon > 1$, we would not include S_2, but in this example, S_2 is needed because this term doesn't approach 0. Thus, the optimal truncation is after the S_2 term.
23) For equation 8.17, $y(0) = 1 - \epsilon^3 + O(e^{-3/(2\epsilon)})$, so the error is $\sim -\epsilon^3$. Also, $y(1) = 209\epsilon^3/16 + 675\epsilon^4/64$, so this error is even larger.

25) $y_{\text{WKBJ},2} = x^{-1/4}e^{-2(x^{3/2}-1)/(3\sqrt{\epsilon})-5\sqrt{\epsilon}(x^{-3/2}-1)/48}$. The exact solution is $y = \text{Ai}(x\epsilon^{-1/3})/\text{Ai}(\epsilon^{-1/3})$. Since $\text{Ai}(x) \sim e^{-2x^{3/2}/3}/(2\sqrt{\pi}x^{1/4})(1-5x^{3/2}/48+\cdots)$, we can reproduce the WKBJ solution from the exact solution.
27) In one case, $S_0 = 0$, $S_1 = 1/x$, and $S_2 = (2+5x)/(10x^5)$. In the other case, $S_0 = -x^3/3$, $S_1 = -2\ln(x) - 1/x$ and $S_2 = (15x - 20x^2 - 2)/(10x^5)$. Clearly, $(2+5x)/(10x^4)$, $(6x\ln(x)+3)/x^4$, and $(15x-20x^2-2)/(10x^5)(-2\ln(x)-1/x)$ are bounded on $1 \le x < \infty$.
29) The WKBJ approximation remains valid as long as $x - 1/2 \gg \epsilon^{1/4}\sqrt{\ln \epsilon}$.

Section 8.3

1) Since $(e^{3\pi i/2})^{2/3} = e^{\pi i} = -1$, there will be a negative sign in the Airy function. Applying the results of example 5.8, we have

$$(y_{\text{comp}})_{\text{left}} \sim K_3 b^{1/6}\epsilon^{1/12}/\sqrt{\pi}(q(x))^{-1/4}\sin\left(\int_x^0 \sqrt{q(t)}\,dt/\sqrt{\epsilon} + \pi/4\right).$$

Since $K_6 = K_3 b^{1/6}\epsilon^{1/12}/\sqrt{\pi}$, this matches y_{left}.
3)

$$y_{\text{comp}} \sim \frac{(4/5)^{1/6}(7+3x^2)^{1/6}\text{Ai}(\epsilon^{-1/3}x(1+3x^2/7)^{2/3})}{\sqrt{1+x^2}\text{Ai}(-\epsilon^{-1/3}(10/7)^{2/3})}.$$

5)

$$y_{\text{comp}} \sim \frac{29^{-1/6}(35+21x+15x^2)^{1/6}\text{Ai}(\epsilon^{-1/3}x(1+3x/5+3x^2/7)^{2/3})}{\sqrt{1+x+x^2}\text{Ai}(-\epsilon^{-1/3}(29/35)^{2/3})}.$$

7)

$$y_{\text{comp}} \sim \frac{5^{-1/6}\sqrt{31}(1+x^2)^{1/6}\text{Ai}(\epsilon^{-1/3}x(3+3x^2)^{2/3})}{\sqrt{3+7x^2}\text{Ai}(-2\epsilon^{-1/3}15^{2/3})}.$$

9) $E \sim (20\pi(n+3/4)/21)^2$, where n is a positive integer.
11) Both sides simplify to $\cos(\pi/8)(D_{-1/2}(e^{i\pi/4}x) + D_{-1/2}(e^{-i\pi/4}x)$.
13) When $v = -1/2$, $\Gamma(2n-v)/\Gamma(-v) = (4n)!/(2^{4n}(2n)!)$. Thus,

$$G_0(x) \sim e^{i(x^2+\pi)/4}x^{-1/2}\sum_{n=0}^{\infty}\frac{(-i)^n(4n)!x^{-2n}}{2^{5n+1}(2n)!n!}$$
$$+ e^{-i(x^2+\pi)/4}x^{-1/2}\sum_{n=0}^{\infty}(i)^n\frac{(4n)!x^{-2n}}{2^{5n+1}(2n)!n!}.$$

Combining complex exponential into trig functions yields the result.
15)

$$y_{\text{comp}} \sim \frac{\sqrt{2}\epsilon^{-1/8}(1-\cos(x))^{1/4}}{d\sqrt{|\sin(x)|}}(G_0(-2\text{sgn}(x)\epsilon^{-1/4}\sqrt{1-\cos(x)})$$
$$-(1+\sqrt{2})\tan(1/\sqrt{\epsilon}+\pi/4)G_0(2\text{sgn}(x)\epsilon^{-1/4}\sqrt{1-\cos(x)}))$$

where $d = \cos(1/\sqrt{\epsilon}+\pi/4) - (1+\sqrt{2})^2\tan(1/\sqrt{\epsilon}+\pi/4)\sin(1/\sqrt{\epsilon}+\pi/4)$.
17)

$$y_{\text{comp}} \sim \frac{\sqrt{2}\epsilon^{-1/8}(x^2+x^3/3)^{1/4}}{d\sqrt{|2x+x^2|}}\left(G_0(-2\text{sgn}(x)\epsilon^{-1/4}\sqrt{x^2+x^3/3})\right.$$
$$\left. - (1+\sqrt{2})\tan(4/(3\sqrt{\epsilon})+\pi/4)G_0(2\text{sgn}(x)\epsilon^{-1/4}\sqrt{x^2+x^3/3})\right)$$

where $d = \cos(2/(3\sqrt{\epsilon})+\pi/4) - (1+\sqrt{2})^2\tan(4/(3\sqrt{\epsilon})+\pi/4)\sin(2/(3\sqrt{\epsilon})+\pi/4)$.
19)

$$y_{\text{comp}} \sim \frac{2\epsilon^{-1/8}(1-\cos(x))^{1/4}}{d\sqrt{|\sin(x)|}}\left(G_0(2\text{sgn}(x)\epsilon^{-1/4}\sqrt{1-\cos(x)})\right.$$
$$\left. - (1+\sqrt{2})\tan(1/\sqrt{\epsilon}+\pi/4)G_0(-2\text{sgn}(x)\epsilon^{-1/4}\sqrt{1-\cos(x)})\right)$$

where $d = \cos(1/\sqrt{\epsilon}+\pi/4) - (1+\sqrt{2})^2\tan(1/\sqrt{\epsilon}+\pi/4)\sin(1/\sqrt{\epsilon}+\pi/4)$.
21)

$$y_{\text{comp}} \sim \frac{2\sqrt{2}\epsilon^{-1/8}(\cosh(x)-1)^{1/4}}{d\sqrt{|\sinh(x)|}}\left(G_0(2\text{sgn}(x)\epsilon^{-1/4}\sqrt{\cosh(x)-1})\right.$$
$$\left. - (1+\sqrt{2})\tan(1/(4\sqrt{\epsilon})+\pi/4)G_0(-2\text{sgn}(x)\epsilon^{-1/4}\sqrt{\cosh(x)-1})\right)$$

where $d = \cos(2/(3\sqrt{\epsilon})+\pi/4) - (1+\sqrt{2})^2\tan(1/(4\sqrt{\epsilon})+\pi/4)\sin(2/(3\sqrt{\epsilon})+\pi/4)$.
23) $E \sim (n+3/4\pm 1/8)^2\pi^2$, where n is a positive integer.
25) $E \sim (3\sqrt{e}(2n+1)\pi/8)^{2/3}$, where n is a non-negative integer.
27) $E \sim -4/(\sqrt{\epsilon}(2n+1)\pi)$, where n is a non-negative integer.
29) $y = \text{Ai}(\epsilon^{-1/3}|x| - (3\pi/8)^{2/3})$.

Section 9.1

1) $\theta''\theta' + \theta'\sin(\theta) = 0$ can be written as $((\theta')^2/2 - \cos(\theta))' = 0$, so we have $\theta' = \pm\sqrt{C+2\cos\theta}$. This is now separable, but the integral can only be expressed via elliptic functions.
3) $1/2 + \cos(2x)/2$.
5) $\cos(x)/4 - \cos(3x)/4$.
7) $3/8 + \cos(2x)/2 + \cos(4x)/8$.
9) $\sin(x)/8 + \sin(3x)/16 - \sin(5x)/16$.
11) $5\sin(x)/8 + 5\sin(3x)/16 + \sin(5x)/16$.
13) Taking the derivative of the equation produces $y''' + y' + \epsilon((y')^2y'' - y'') = 0$.
15) $y_{\text{strain},3} = \epsilon\cos(\tau) + \epsilon^3(\cos(\tau) - \cos(3\tau))/64$, where $\tau \sim t - 3t\epsilon^2/16$.
17) $y_{\text{strain},2} = \cos(\tau) + \epsilon(4\cos(\tau) - 3 - \cos(2\tau))/6 + \epsilon^2(61\cos(\tau) - 48 - 16\cos(2\tau) + 3\cos(3\tau))/72$, where $\tau \sim t - t\epsilon^2/6$.
19) $y_{\text{strain},2} = \cos(\tau) + \epsilon(2\sin(\tau) - \sin(2\tau))/6 + \epsilon^2(32\cos(2\tau) - 23\cos(\tau) - 9\cos(3\tau) - 48\sin(\tau) + 24\sin(2\tau))/288$, where $\tau \sim t + t\epsilon/2 - t\epsilon^2/6$.

21) $y_{\text{strain},2} = \sin(\tau) + \epsilon(\sin(\tau) + \sin(2\tau))/6 + \epsilon^2(13\sin(\tau) + 40\sin(2\tau) + 9\sin(3\tau))/288$, where $\tau \sim t - t\epsilon/2 - t\epsilon^2/6$.
23) $y_{\text{strain},2} = \sin(\tau)+\epsilon(4\cos(\tau)-3-\cos(2\tau)-3\sin\tau)/6+\epsilon^2(144-192\cos(\tau)+48\cos(2\tau) + 59\sin(\tau) + 32\sin(2\tau) - 3\sin(3\tau))/144$, where $\tau \sim t + t\epsilon/2 - 13t\epsilon^2/24$.

Section 9.2

1) $y_{\text{multi},1} = 2\cos(t)/\sqrt{4+s} + \epsilon((56+21s)\sin(t) - 8\sin(3t))/(32(4+s)^{3/2})$.
3) $y_{\text{multi},0} = \cos(s)\cos(t) - \sin(s)\sin(t)$.
5) $y_{\text{multi},0} = \cos(3s/8)\cos(t) - \sin(3s/8)\sin(t)$.
7) $y_{\text{multi},0} = 2\cos(t)/\sqrt{s+4}$.
9) $y_{\text{multi},0} = \cos(s/8)\cos(t) + \sin(s/8)\sin(t)$.
11) $y_{\text{multi},0} = 2e^{s/2}\sin(t)/\sqrt{3+e^s}$.
13) $y_{\text{multi},0} = e^{-s/2}\cos((e^{-s}-1)/8)\cos(t) + e^{-s/2}\sin((e^{-s}-1)/8)\sin(t)$.
15) $y_{\text{multi},1} = \cos(t) + \epsilon((8+s)\sin(t) - 4\sin(3t))/24$.
17) $y_{\text{multi},1} = 2\cos(t)/\sqrt{4+3s}+\epsilon((72+45s)\sin(t)+8\sin(3t))/(32(4+3s)^{3/2})$.
19) $y_{\text{multi},1} = 2\sin(t)/\sqrt{4+s} + \epsilon((8-5s)\cos(t) - 8\cos(3t))/(32(4+s)^{3/2})$.
21) $y_{\text{multi},1} = 2\cos(t)/\sqrt{1+3e^{-s}}+\epsilon(1+3e^{-s})^{-3/2}(8\sin(3t)+(-243e^{-s}-69+24s(1+3e^{-s}) - 12(1+3e^{-s})\ln((3+e^s)/4))\sin(t))/96$.

Section 9.3

1) $\partial^3 y(\tau,s)/\partial t^3 = \omega(\epsilon)^3\partial^3 y/\partial\tau^3 + 3\epsilon\omega(\epsilon)^2\partial^3 y/\partial\tau^2\partial s + 3\epsilon^2\omega(\epsilon)\partial^3 y/\partial\tau\partial s^2 + \epsilon^3\partial^3 y/\partial s^3$.
3) $\omega_2 = -3/8$, $A_1(s) = 3$.
5) $\omega_2 = 0$, $A_1(s) = 2 - e^{2s}$.
7) $\omega_2 = -3$, $A_1(s) = 2e^{-s} - 3\ln((1+e^{-s})/2)$.
9) $\omega_2 = 5$, $A_1(s) = (\sqrt{2} - 5\ln((1+e^{-s})/2))/\sqrt{1+e^{-s}}$.
11) $y_{\text{two-var},1} =$

$$2\cos(\tau)/\sqrt{4+s} + \epsilon(4+s)^{-3/2}(-32\sin(3\tau) + (224 - 21s^2)\sin(\tau))/128,$$

where $\tau = t - 21t\epsilon^2/256$.
13) $y_{\text{two-var},1} = \sin(\tau) + \epsilon(2\cos(\tau) - 3 + \cos(2\tau))/6$, where $\tau = t - t\epsilon^2/6$.
15) $y_{\text{two-var},1} = \cos(\tau) + \epsilon(16\cos(\tau) - 15 - \cos(4\tau))/120$, where $\tau = t - 11t\epsilon^2/320$.
17) $y_{\text{two-var},1} =$

$$e^{-s/2}\sin(\tau) + \epsilon(5e^{-3s/2}\cos(\tau) - 6e^{-s} + 3e^{-s/2}\cos(\tau) - 2e^{-s}\cos(2\tau))/12,$$

where $\tau \sim t - t\epsilon^2/8$.
19)

$$y_{\text{two-var},1} = \frac{2a}{\sqrt{a^2+(4-a^2)e^{-\epsilon t}}}\cos(\tau) - \frac{a^3\epsilon}{4(a^2+(4-a^2)e^{-st})^{3/2}}\sin(3\tau)$$

$$+\,\epsilon\frac{21a^3-60a+(112a-28a^3)/(4-a^2+a^2e^s)}{32\sqrt{a^2+(4-a^2)e^{-s}}}\sin(\tau)$$
$$-\,\epsilon\frac{4\ln((a^2+(4-a^2)e^{-s})/4)}{32\sqrt{a^2+(4-a^2)e^{-s}}}\sin(\tau).$$

where $\tau \sim t-\epsilon^2 t/16$. When $a=2$, this simplifies to $2\cos(\tau)+3\epsilon\sin(\tau)/4-\epsilon\sin(3\tau)/4$.

Bibliography

The following list not only gives the books and articles mentioned in the text, but also additional references that may help students explore related topics.

1. N. I. Akhiezer, *The Classical Moment Problem and Some Related Questions in Analysis*, First English edition, Oliver and Boyd, Edinburgh, 1965.

2. G. D. Allen, C. K. Chui, W. R. Madych F. J. Narcowich, and P. W. Smith, Padé Approximation of Stieltjes Series, *Journal of Aproximation Theory,* Vol. 14, No. 4, 1975, pp. 302–316.

3. W. Balser, From Divergent Power Series to Analytic Functions, *Lecture Notes in Mathematics*, No. 1582, Springer-Verlag, Berlin, 1994.

4. C. Bender and S. Orszag, *Advanced Mathematical Methods for Scientists and Engineers,* Springer-Verlag, New York, 1999.

5. W. E. Boyce and R. C. DiPrima, *Elementary Differential Equations and Boundary Value Problems*, 9th ed., John Wiley and Sons, Hoboken, New Jersey, 2009.

6. J. W. Brown and R. V. Churchill, *Complex Variables and Applications*, 8th ed., McGraw Hill, New York, 2008.

7. W. Eckhaus, Fundamental Concepts of Matching, *SIAM Review*, Vol. 35, No. 3. (September 1994) pp. 431–439.

8. H. W. Gould, "Explicit formulas for Bernoulli numbers", *Amer. Math. Monthly*, Vol. 79, No. 1, 1972, pp. 44-51.

9. J. J. Gray, "Fuchs and the Theory of Differential Equations", *Bull. Amer. Math. Soc. (new series)*, Vol. 10, No. 1, 1984, pp. 1–26.

10. R. L. Herman, A Second Course in Ordinary Differential Equations: Dynamical Systems and Boundary Value Problems, 2008.

11. E. J. Hinch, *Perturbation Methods*, Cambridge University Press, 1991.

12. P. Holmes, "A nonlinear oscillator with a strange attractor", *Phil. Trans. Royal Soc., Series A*, Vol. 292, 1979, pp. 419–448.

13. S. Kaplun, *Fluid Mechanics and Singular Perturbations*, a collection of papers by S. Kaplun, edited by P. A. Lagerstrom, L. N. Howard, and C. S. Liu), Academic, New York, 1967.

14. J. Kevorkian and J. D. Cole, *Perturbation Methods in Applied Mathematics*, Springer-Verlag, New York, 1981.

15. P. A. Lagerstrom, *Matched Asymptotoic Expansions*, Springer-Verlag, New York, 1988.

16. D. H. Lehmer, "On the Maxima and Minima of Bernoulli Polynomials", *Amer. Math. Monthly*, Vol. 47, 1940, pp. 533-538.

17. R. K. Miller and A. N. Michel, Ordinary Differential Equations, Academic Press, New York, 1982.

18. R. E. O'Malley, Jr., *Singular Perturbation Methods for Ordinary Differential Equations*, Springer-Verlag, New York, 1991.

19. C. Mortici, "A new Stirling series as continued fraction", *Numerical Algorithms*, Vol. 56, No. 1, 2011, pp. 17–26.

20. W. Paulsen, "A new way to find the controlling factor of the solution to a difference equation," *Journal of the Korean Mathematical Society*, Vol. 36, No. 5, 1999, pp. 833–846.

21. D. J. Pengelley, "Dances between continuous and discrete: Euler's summation formula", in: Robert Bradley and Ed Sandifer (Eds), Proceedings, Euler $2K + 2$ Conference (Rumford, Maine, 2002), Euler Society, 2003.

22. H. L. Royden, *Real Analysis*, 3rd Ed., Prentice Hall, Upper Saddle River, New Jersey, 1988.

23. G. G. Stokes, "On the Theory of Oscillatory Waves", *Trans. Camb. Phil. Soc.*, Vol. 8, 1847, pp. 441-473.

24. E.T. Whittaker and G.N. Watson, *A course of Modern Analysis,* Cambridge University Press, 1963.

25. J. H. Wilkinson, "The evaluation of the zeros of ill-conditioned polynomials, Part I", *Numerische Mathematik*, Vol. 1, 1959, pp. 150–166.

26. M. Van Dyke, *Perturbation Methods in Fluid Mechanics*, Academic Press, New York, 1964.

Index

Page numbers are underlined in the index when they represent the definition or the main source of information about the subject indexed. Boldface page numbers refer to sections or subsections for which they entirely pertain to the topic. References to problems are in italics. Occasionally, both underlining and italics are appropriate, should a homework problem introduce a new concept.